Spin Observables
of Nuclear Probes

Spin Observables of Nuclear Probes

Edited by
Charles J. Horowitz,
Charles D. Goodman,
and George E. Walker

Indiana University
Bloomington, Indiana

Plenum Press • New York and London

Library of Congress Cataloging in Publication Data

Telluride International Conference on Spin Observables of Nuclear Probes (1988)
  Spin Observables of nuclear probes / edited by Charles J. Horowitz, Charles D.
Goodman, and George E. Walker.
      p.   cm.
  "Proceedings of the Telluride International Conference on Spin Observables of
Nuclear Probes, held March 14–17, 1988, in Telluride, Colorado" — T.p. verso.
  Bibliography: p.
  Includes index.
  ISBN-13: 978-1-4612-8073-6   e-ISBN-13: 978-1-4613-0769-3
  DOI: 10.1007/978-1-4613-0769-3
  1. Nuclear spin—Congresses. 2. Nuclear structure—Congresses. I. Horowitz,
Charles. II. Goodman, Charles. III. Walker, George E. IV. Title.
QC793.3.S8T45   1988   88-38971
539.7′23—dc19   CIP

Proceedings of the Telluride International Conference on
Spin Observables of Nuclear Probes, held March 14–17, 1988,
in Telluride, Colorado

© 1988 Plenum Press, New York

Softcover reprint of the hardcover 1st edition 1988

A Division of Plenum Publishing Corporation
233 Spring Street, New York, N.Y. 10013

PREFACE

The proceedings of the "International Conference on Spin Observables of Nuclear Probes" are presented in this volume. This conference was held in Telluride, Colorado, March 14–17, 1988, and was the fourth in the Telluride series of nuclear physics conferences.

A continuing theme in the Telluride conference series has been the complementarity of various intermediate–energy projectiles for elucidating the nucleon–nucleon interaction and nuclear structure. Earlier conferences have contributed significantly to an understanding of spin currents in nuclei, in particular the distribution of Gamow–Teller strength using charge–exchange reactions. The previous conference on "Antinucleon and Nucleon Nucleus Interactions" compared nuclear information from trational probes to recent results from antinucleon reactions. The 1988 conference on Spin Observables of Nuclear Probes, put special emphasis on spin observables and brought together experts using spin information to probe nuclear structure.

Spin observables have provided very detailed information about nuclear structure and reactions. Since the 1985 Telluride conference we have seen data from new focal plane polarimeters at LAMPF, TRIUMF, IUCF and elsewhere. In addition, spin observables provide an important common ground between electron and hadron scattering physics. In the future we look forward to new facilities such as NTOF for polarized neutron measurements at Los Alamos and a vigorous spin program at CEBAF.

The first day of the conference dealt with spin observables in both inclusive (morning) and coincidence reactions (afternoon). Charge exchange reactions were discussed on the second day and relativistic descriptions of nuclei on the third day. There has always been a close relationship between relativity and spin. Proton scattering spin obser- vables presented at the '85 conference have led to much theoretical work on relativistic models in the intervening years. Finally, spin observables in few body systems were presented on the fourth day.

We are very happy that five eastern bloc physicists from Dubna and Novosibirsk were able to attend the conference. We hope that this signals an increase in East/West physics collaborations in the future.

We wish to gratefully acknowledge the sponsorship of the National Science Foundation. We are also grateful for the support from Horshaw/ Filtrol, Inc. in sponsoring the reception for the conference participants.

CONTENTS

(* Asterisk indicates author who presented the talk)

NUCLEAR SPIN RESPONSE STUDIES IN INELASTIC
POLARIZED PROTON SCATTERING

Kevin W. Jones

Los Alamos National Laboratory
Los Alamos, N.M. 87544

ABSTRACT

Spin-flip probabilities S_{nn} have been measured for inelastic proton scattering at incident proton energies around 300 MeV from a number of nuclei. At low excitation energies S_{nn} is below the free value. For excitation energies above about 30 MeV for momentum transfers between about 0.35 fm^{-1} and 0.65 fm^{-1} S_{nn} exceeds free values significantly. These results suggest that the relative $\Delta S = 1/(\Delta S = 0 + \Delta S = 1)$ nuclear spin response approaches about 90% in the region of the enhancement. Comparison of the data with slab response calculations are presented. Decomposition of the measured cross sections into $\sigma(\Delta S = 0)$ and $\sigma(\Delta S = 1)$ permit extraction of nonspin-flip and spin-flip dipole and quadrupole strengths.

INTRODUCTION

Spin excitations in nuclei have been the subject of considerable theoretical and experimental work during the past six years. Indeed, much of the work presented here originated in the study of $M1$ transitions in nuclei, a subject addressed in detail in the 1982 Telluride Conference and in subsequent publications.[1-7] The initially strong focus on the distribution of $M1$ strength excited by proton inelastic scattering has evolved into a more general study of the distribution of spin excitations in nuclei excited with this probe. The program begun at the high resolution proton spectrometer facility (HRS) at LAMPF has been facilitated by the development of a polarimeter facility associated with the medium resolution spectrometer (MRS) at TRIUMF.[8,9] The combination of data available from the HRS facility at LAMPF and the MRS at TRIUMF forms a substantial contribution to the study of continuum excitations by inelastic proton scattering up to about 50 MeV in a variety of nuclei from ^{12}C to ^{90}Zr.

The proton is a complex but desirable probe of the nucleus, roughly equally scalar and vector in nature. As such it can be used to excite the full complexity of the nuclear continuum. By contrast, the charge-exchange reaction (p, n) is a probe of purely isovector character; consequently the excitation of the nucleus is selective. The complementarity of these and other probes should result in a detailed understanding of the isospin and spin response of the nucleus. Recent advances in charge-exchange studies addressing the distribution of Gamow-Teller and other isovector strength in nuclei are discussed by other contributors at this conference.

EXPERIMENTAL DETAILS

Experiments have been carried out at both the HRS at LAMPF and the MRS at TRI-UMF. Measurements at the HRS were made with a beam of 319-MeV protons polarized perpendicular to the reaction plane. Scattered particles were detected in the standard HRS focal-plane polarimeter[10] modified to minimize backgrounds inherent in small-angle inelastic scattering. The choice of 319 MeV was dictated by precession considerations in the spectrometer and the relative strength of spin-flip to nonspin-flip components of the effective nucleon-nucleon interaction. Measurements at the MRS were made with a beam of 290-MeV protons polarized perpendicular to the reaction plane. Scattered particles were detected in the standard MRS focal plane polarimeter. Beam polarizations at both facilities were monitored continuously with in-beam polarimeters, as well as with the quench-ratio technique at the HRS. Absolute normalization of measured cross sections was made by comparison of measured cross sections for proton scattering from hydrogen with extant data and phase shift analyses.[11] Data were taken for scattering from ^{12}C, ^{40}Ca, ^{48}Ca, ^{51}V, and ^{90}Zr at the HRS, and from ^{44}Ca and ^{54}Fe at the MRS.

Typical cross-section spectra for inelastic proton scattering around an incident energy of 300 MeV are shown in Fig. 1.

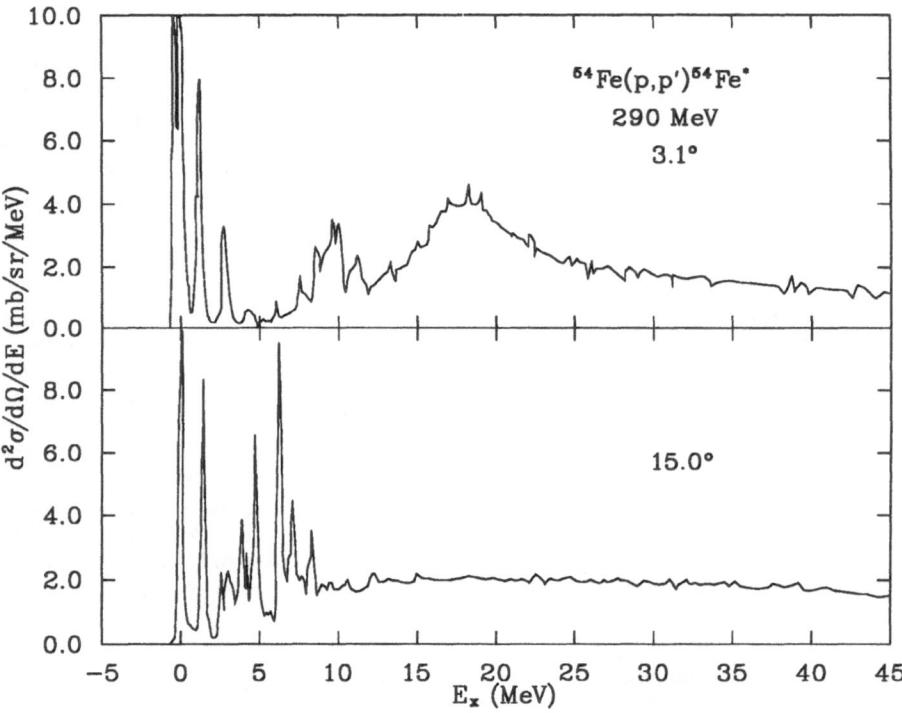

Fig. 1.   Differential cross section spectra for the ^{54}Fe$(p, p')^{54}$Fe* reaction at $T_p = 290$ MeV. The data were taken using the MRS facility at TRIUMF at laboratory scattering angles of 3.1° and 15°.

Preliminary spectra are shown for scattering from ^{54}Fe at laboratory scattering angles of 3.1° and 15°. The strongly quenched $M1$ resonances around 10 MeV are clearly seen at 3.1 degrees, together with discrete states at low excitation and the strong natural parity $\Delta S = 0$ giant resonances around 15-20 MeV, dominated by the Giant Dipole Resonance (GDR) at this angle. A continuum with little structure extends out to 45 MeV and beyond. The 15° spectrum shows more discrete states, no pronounced giant resonance structures, and the feature that the onset of the continuum is at a significantly lower excitation energy than at 3.1°. The continuum has generally been treated as background in the studies of giant resonances. Several authors have assumed the background to be due to single-step quasi-free scattering and have modeled it as such,[12,13] while other approaches emphasized the drawing of smooth, empirical lines to represent this background.[14,15]

The principal thrust of this work, then, is to determine the relative distribution of $\Delta S = 1$ and $\Delta S = 0$ excitations in the nuclear continuum up to about 50 MeV of excitation.

SPIN FLIP MEASUREMENTS

An experimental observable well suited for the study of the nuclear spin response is the transverse spin-flip probability, S_{nn}, of the inelastically scattered protons. This is simply the probability that an incoming proton with, for example, spin up (normal to the scattering plane), will interact with the target nucleus and leave with spin down. S_{nn} is related to the Wolfenstein parameter D by

$$S_{nn} = \frac{1}{2}(1 - D) \ . \tag{1}$$

This observable may be directly related to $\Delta S = 1$ strength and will be shown to be relatively insensitive to structure effects, reaction mechanisms, and the like.

Some comments about typical values for S_{nn} will be helpful in interpreting the data to be presented. Natural parity $\Delta S = 0$ excitations have values of $S_{nn} \simeq 0$ at these incident energies.[16] For discrete unnatural parity transitions such as 1^+ states, excited mostly by the $\sigma_1 \cdot \sigma_2$ term in the nucleon-nucleon (NN) potential, values of S_{nn} around 0.5 or higher are observed.[16,17] In general, $\Delta S = 1$ transitions exhibit values for S_{nn} which are large but essentially never unity. For free NN scattering around 300 MeV and over the momentum transfers of interest, the isospin-averaged value of S_{nn} for $p - p$ and $p - n$ scattering ranges from ~ 0.2 to ~ 0.35 (Ref. 11). From these values of S_{nn}, it can be determined that the (p, p') probe is roughly equally scalar and vector,

$$\frac{\sigma^f(\Delta S = 1)}{\sigma^f(\Delta S = 0) + \sigma^f(\Delta S = 1)} \simeq 0.5 \ , \tag{2}$$

where σ^f is the differential cross section for free scattering at a particular momentum transfer q.

Typical spectra of S_{nn} and σS_{nn} are shown in Fig. 2. The data are for scattering from ^{90}Zr at 3.0° in the laboratory; data up to 25 MeV of excitation with somewhat larger error bars have been published previously.[4]

The spectra show large amounts of spin excitation strength in the $M1$ region around 10 MeV and also up to at least 40 MeV of excitation. The giant resonance region is by no means exclusively $\Delta S = 0$, for example, and the spin-flip cross section falls off more slowly at high excitation energies than does the cross section itself.

3

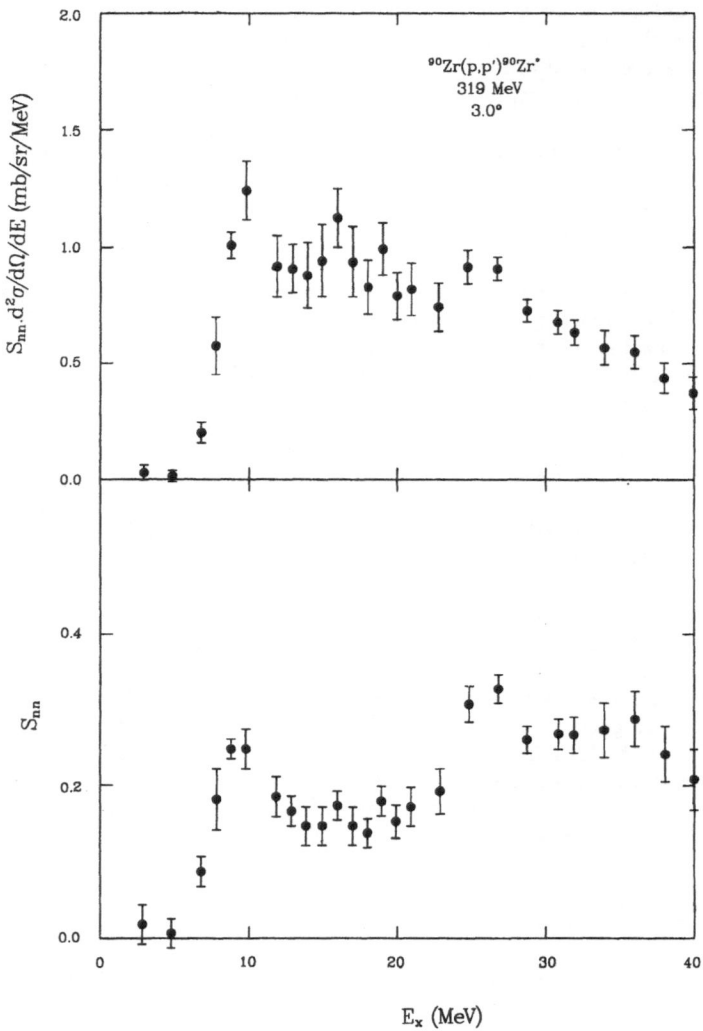

Fig. 2.   Spin-flip probability S_{nn} and spin-flip cross section σS_{nn} data for 319-MeV proton scattering from ^{90}Zr at a center-of-mass scattering angle of 3.0°.

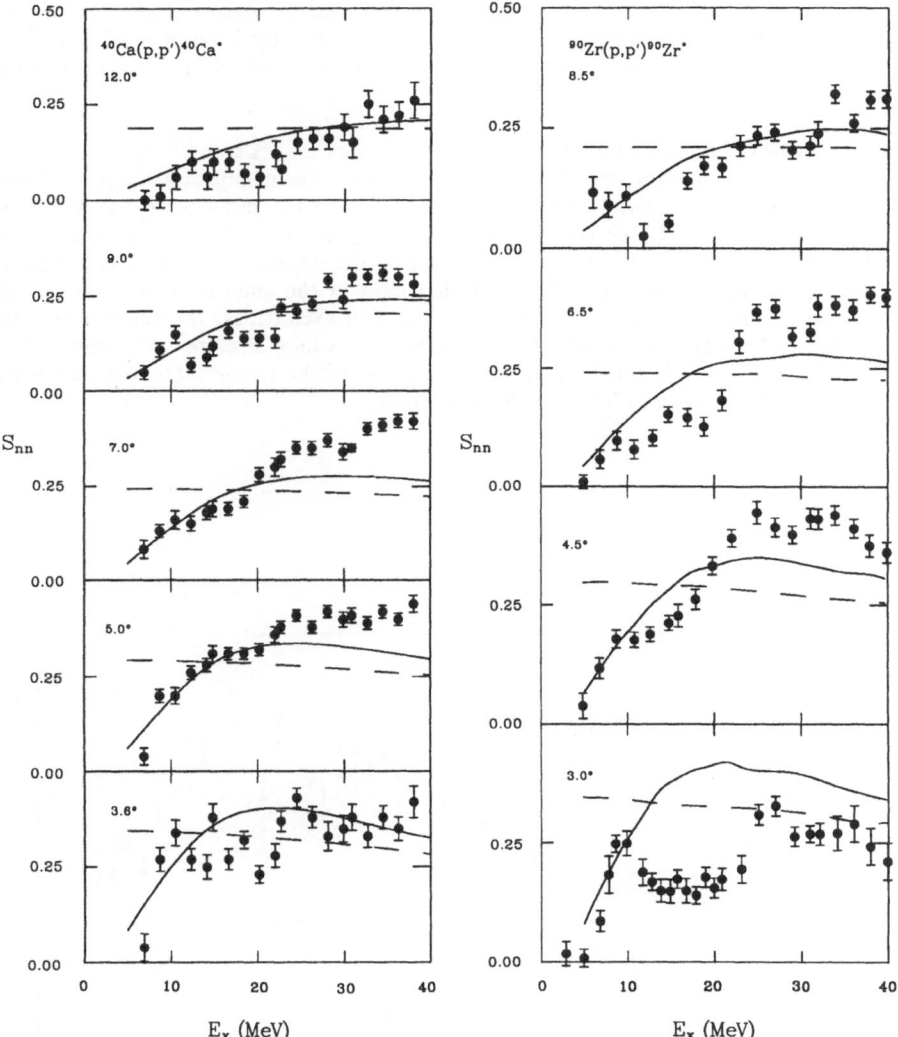

Fig. 3.   Spin-flip probability S_{nn} for the ${}^{40}Ca(p,p'){}^{40}Ca^*$ reaction at $T_p = 319$ MeV as a function of excitation energy at 3.6°, 5.0°, 7.0°, 9.0° and 12.0°. The solid line is the prediction of the Esbensen-Bertsch slab model described in the text. The dashed line corresponds to isospin-averaged free values for S_{nn}.

Fig. 4.   Spin-flip probability S_{nn} for the ${}^{90}Zr(p,p'){}^{90}Zr^*$ reaction at $T_p = 319$ MeV as a function of excitation energy at 3.0°, 4.5°, 6.5°, and 8.5°. The solid line is the prediction of the Esbensen-Bertsch slab model described in the text. The dashed line corresponds to isospin-averaged free values for S_{nn}.

Roughly similar behavior has been observed in all the nuclei we have studied, from ^{12}C to ^{90}Zr. Lower excitation regions (≤ 20 MeV) tend to be dominated by specific excitations — the $M1$ excitations in ^{48}Ca, ^{51}V, and ^{54}Fe, are particularly prominent, as are a variety of resonances in ^{12}C in the $20 - 30$ MeV range. Structures are observed at 7° laboratory in the 15 MeV range for all targets. In general, though, σS_{nn} tends to increase relative to σ as excitation energy increases, the ratio becoming roughly constant at an angle-dependent excitation energy.

The shapes of S_{nn} as a function of excitation energy for ^{40}Ca and ^{90}Zr shown in Figs. 3 and 4 are therefore not surprising. What is remarkable are the magnitudes of S_{nn}, particularly at laboratory angles of about 5° and 7° and for the higher excitation energies. The value of S_{nn} attained is about 0.4, and this is consistent for all nuclei studied thus far. Comparison with the free NN isospin-averaged values shown by the dashed lines indicates a significant enhancement. Also noteworthy is the fact that even in the giant-resonance region, where the only well-known structures correspond to $\Delta S = 0$ excitations, the value of S_{nn} is not much different from the free values. At low excitations, where discrete states dominate, S_{nn} is indeed much smaller than the free values, indicative of the strong attractive nature of the $\Delta S = 0$ components of the effective NN interaction.

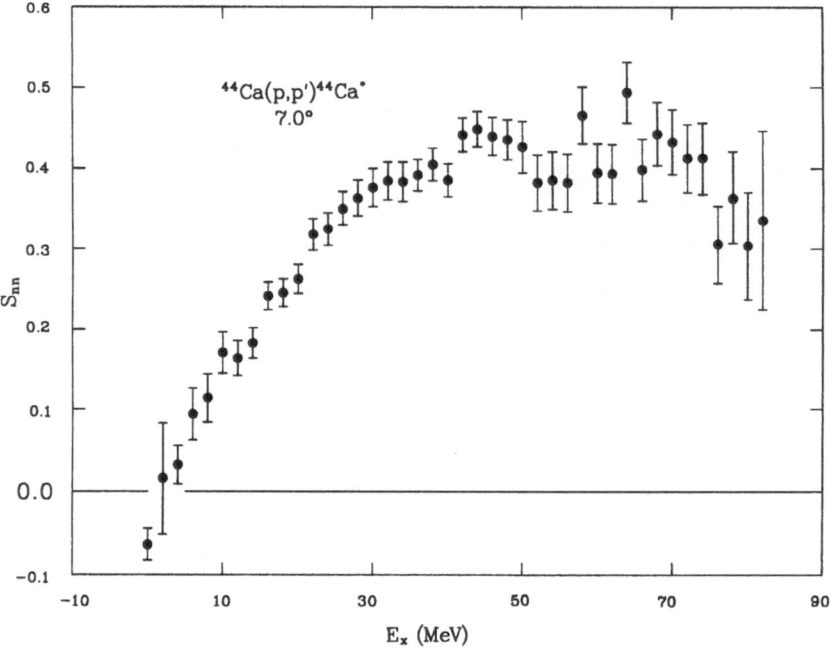

Fig. 5. Spin-flip probability S_{nn} for the ^{44}Ca$(p,p')^{44}$Ca* reaction at $T_p = 290$ MeV as a function of excitation energy at 7.0°.

Preliminary data for S_{nn} at 7° from ^{44}Ca, measured at TRIUMF, are shown in Fig. 5. The excitation energy here extends to about 80 MeV, and S_{nn} is seen to remain at a roughly constant value of about 0.4 over the entire range of excitation energy. At the largest angles measured (15° for ^{54}Fe and 18° for ^{12}C) values of S_{nn} are found to be similar to the free values at about 40 MeV; a similar observation can be made for the data at 3°.

THE RELATIVE NUCLEAR SPIN RESPONSE

Given the extensive S_{nn} data discussed above, what may we deduce about the nuclear spin response to the proton? As we shall see, S_{nn} is relatively insensitive to such effects as Fermi averaging, distortions, multi-step processes, and so on. We may therefore consider S_{nn} to be a robust variable. Given this understanding, we may write:

$$S_{nn} = \frac{\alpha \sigma_1}{(\sigma_0 + \sigma_1)} \tag{3}$$

where σ_i is the differential cross section for spin transfer $\Delta S = i$ scattering, and α is the spin-flip probability for a pure $\Delta S = 1$ scattering. Adopting a factorized approximation,[18-20] the partial cross sections may be written as

$$\sigma_i^A = N_{eff} \, f_i \, \sigma_i^f \quad . \tag{4}$$

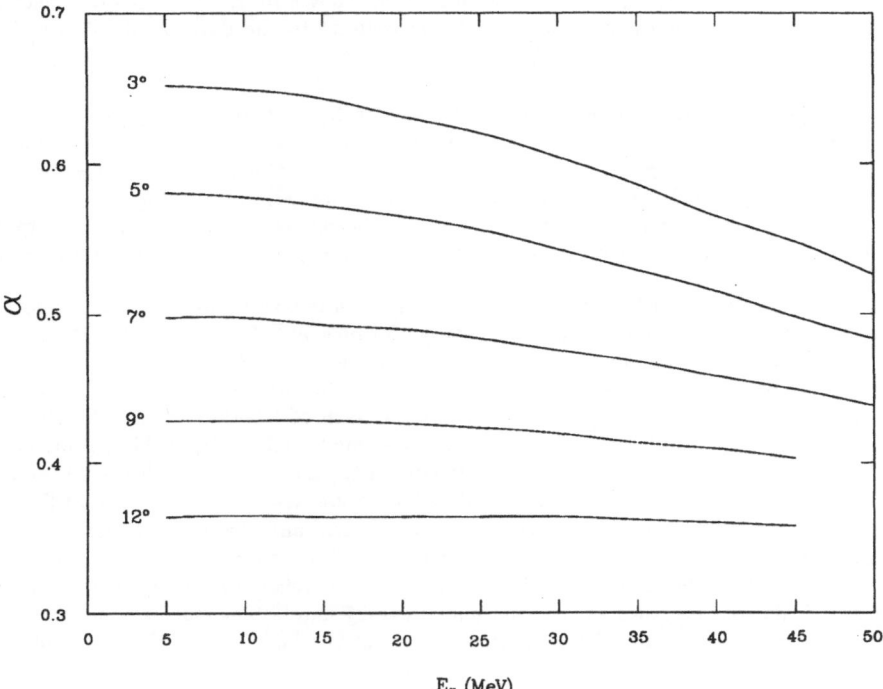

Fig. 6.   The value of α for quasifree scattering as a function of angle and excitation energy for an incident energy $T_p = 300$ MeV.

Here N_{eff} is the effective number of target nucleons participating in the reaction, and f_i is the modification of the free cross section σ_i^f in the channel i by the nucleus. f_i is then the nuclear response for the spin channel i. From Eqs. (3) and (4) we may deduce

$$\frac{f_1}{f_0} = \frac{\sigma_1^A}{\sigma_1^f} \bigg/ \frac{\sigma_0^A}{\sigma_0^f} = \frac{S_{nn}}{\alpha - S_{nn}} \bigg/ \frac{\sigma_1^f}{\sigma_0^f} \quad . \tag{5}$$

We may now define the relative nuclear spin response R_s as

$$R_s = \frac{f_1}{f_0 + f_1} \quad . \tag{6}$$

R_s is then the fraction of the nuclear response that corresponds to $\Delta S = 1$ excitations. If the nuclear medium were a non-interacting Fermi gas, then R_s would be 0.5. Since the free cross sections are known, measurement of S_{nn} enables calculation of R_s provided α is known. The quantity α is then a single model-dependent parameter required to determine R_s. Except in regions of the nuclear excitation dominated by resonances of particular structure, spin, and parity, α is determined mostly by the nature of the effective NN interaction.

The simplest determination of α may be made from free scattering using Eq. (3). Essentially identical values may be obtained from surface-response calculations described in the next section. The behaviour of α as a function of laboratory scattering angle and excitation energy is shown in Fig. 6. The trend of R_s as a function of excitation energy computed from the data for ^{40}Ca and ^{90}Zr is shown by the solid lines in Figs. 7 and 8. The solid lines have been drawn to guide the eye through the actual values shown by the data points. Because both theoretical and experimental uncertainties contribute to the data points shown, it is likely that only the general shape of the line is significant.

These figures dramatically illustrate our answer to the question posed previously: they show our measurement of the distribution of the relative spin-excitation strength in nuclei. These data present the first general picture of the nuclear spin response in this <u>nuclear</u> regime of q and E_x. Of particular interest is that R_s approaches 90% at the highest excitation energies for some angles, and that since $\sigma_1^f/\sigma_0^f \simeq 1$, the ratio of the cross sections actually excited in 300-MeV nuclear scattering also approaches about 90% at these excitation energies.

The conclusions drawn from the data thus far rest on two assumptions: S_{nn} is a robust variable, and $\alpha^A \simeq \alpha^f$. Calculations have been performed by Esbensen[21] in the framework of the surface-response model developed by Bertsch, Esbensen, and Scholten. In this model the nucleus is considered as a semi-infinite slab with RPA correlations and an absorbing surface; the absorption is treated via an approximate form of Glauber theory. Single hard scattering described by the free NN amplitudes is assumed, and Pauli blocking is included. Values of α determined from this model are very similar to those computed directly from free NN values. Predicted values of S_{nn} as calculated by Esbensen are shown as solid lines in Figs. 3 and 4. The general form of S_{nn} is described well, and an enhancement relative to the free values is predicted at the higher excitation energies. This enhancement is not sufficient, however, to account for the large values seen in the data. The relative response R_s determined from these calculations is shown as the solid line in Figs. 7 and 8. The calculations clearly do not predict the large fraction of $\Delta S = 1$ strength seen at the higher excitations around 7°.

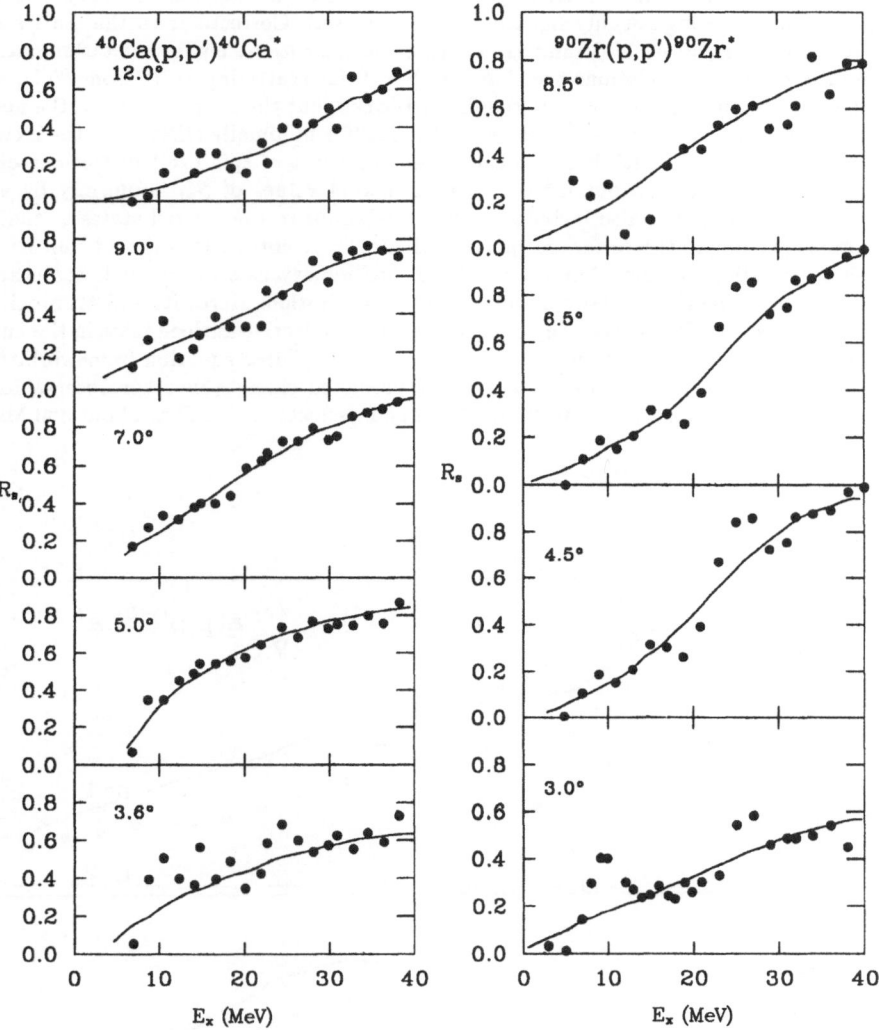

Fig. 7.   The nuclear spin response R_s determined from the S_{nn} data for $^{40}Ca(p,p')^{40}Ca^*$ as shown in Fig. 3. The lines have been drawn to guide the eye.

Fig. 8.   The nuclear spin response R_s determined from the S_{nn} data for $^{90}Zr(p,p')^{90}Zr^*$ as shown in Fig. 4. The lines have been drawn to guide the eye.

What then of corrections to this model? Several have been considered, and the effects of all appear to be small. Smith has considered the possibility of a second hard collision within the framework of the surface-response model.[22] The effect of this process is negligible for small E_x, increasing roughly linearly as E_x increases. Corrections to the cross section are about 5% at $E_x = 40$ MeV and are expected to increase at higher excitation energies. The results of these calculations are shown in Fig. 9 for scattering at 7° from ^{40}Ca. Note that the cross sections themselves are roughly described, but the simple model of the nucleus used precludes the description of resonances. S_{nn} itself is minimally affected by the inclusion of two-step processes. Smith has also calculated the effect of spin-orbit distortions on the values of S_{nn}; such distortions increase the calculated values of S_{nn} uniformly by about 0.03 (Ref. 22). Smith has also included 2p-2h correlations in the excited states to the RPA correlations existing in the surface response model.[22] These correlations spread spin strength to higher excitation energies, but the results of preliminary calculations indicate that this spreading is only significant above about 40 MeV of excitation. Horowitz and Murdock have studied the effects of Fermi averaging and changes in the effective nucleon mass in the nucleus that arise from a treatment of quasifree knockout within a Dirac equation framework.[23] The predicted changes in S_{nn} are very small, while predicted changes in other observables are quite large. Rees has also shown that Fermi averaging effects are small at about 300 MeV.[24]

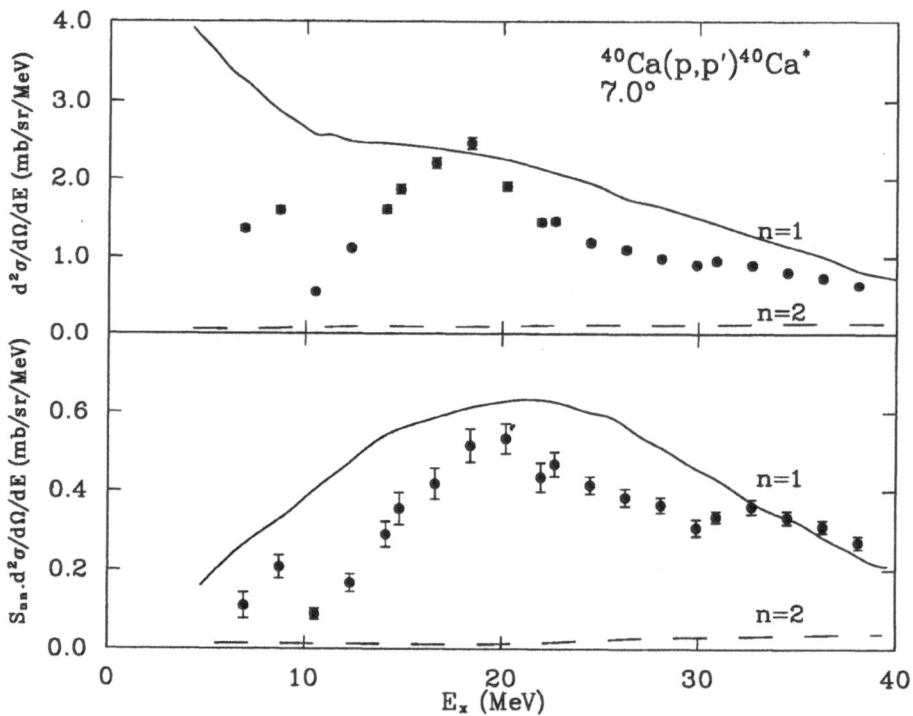

Fig. 9.   Differential cross sections σ and spin-flip cross sections σS_{nn} for inelastic scattering at 7° from ^{40}Ca at 319 MeV as a function of excitation energy. The solid and dashed lines are the results of one- and two-step calculations by Smith.

The net effect of all these possible corrections is small. The response R_s should then be a very good estimate of the actual relative nuclear spin response. Since all these calculations contain approximations, it is still conceivable that what is observed at high excitation energies is an artifact of the choice of the incident energy of the probe. The relative spin dependence of the free scattering is maximized at this energy. Given the assumption that the simple model presented here is valid, it is possible to predict a value for S_{nn} at a different incident energy. At 800 MeV, the free value of S_{nn} is about 0.05. If R_s is taken to be about 0.85, then a value of around 0.20 is predicted for S_{nn} at this energy, a much larger enhancement than at 300 MeV. Preliminary data taken at the HRS at 800 MeV and 4° from ^{40}Ca give an average value of $S_{nn} = 0.23 \pm 0.03$ over the excitation energy range 33 to 44 MeV in very good agreement with the prediction. The nuclear response appears energy independent, as it should.

GIANT RESONANCE STUDIES

As has been mentioned previously, analysis of backgrounds under giant resonances excited by inelastic proton scattering has been problematical.[12-15] The measurement of S_{nn} has revealed that much of the background underlying the giant dipole and quadrupole resonances at these small angles is from $\Delta S = 1$ transitions. The decomposition of the cross section into partial terms σ_1 ($\Delta S = 1$) and σ_0 ($\Delta S = 0$) holds the promise of allowing a much better estimate of this background than has previously been possible. As outlined above, only the one model-dependent parameter α is necessary.

Once the decomposition into σ_1 and σ_0 has been accomplished, contributions from various multipoles may be determined. Baker et al.[25] have performed such an analysis of the ^{40}Ca data shown above. Nonspin-flip dipole strength is found to be concentrated in the GDR at 20 MeV of excitation but persists to higher E_x, in agreement with the results of photon absorption experiments. The summed strength relative to the energy-weighted sum rule (EWSR) in the 10-35 MeV region in this analysis is 148 ± 21%, compared with 130% for the photon absorption data.[26] Significant monopole strength is not seen. Nonspin-flip quadrupole strength is concentrated in a broad resonance between 10 and 30 MeV; above 27 MeV the quadrupole strength is consistent with zero. It is found that 76 ± 12% of the EWSR is exhausted in the excitation energy range of 10-27 MeV. These measurements should not be considered definitive because of the limited angular range of the data; further measurements at the HRS have been approved to extend the angular range. Strong evidence has been provided that quantitative subtraction of spin-flip background under giant resonances is a sound approach to extraction reliable strengths.

A preliminary multipole analysis of the σ_1 spectra has been performed; simple particle-hole transitions have been assumed. A giant spin-flip dipole excitation is observed to extend from about 12 to 28 MeV with a centroid at approximately 18-20 MeV, as shown in Fig. 10. Roughly similar observations have been made at 334 MeV incident proton energy using a different approach.[27] Spin dipole strength above 30 MeV of excitation is consistent with zero. Evidence for a spin quadrupole excitation is seen above about 25 MeV, with an apparent centroid at 33-34 MeV of excitation. Definitive analysis of the spin quadrupole requires extension of the data to both larger angles and higher excitation energies.

CONCLUSIONS

Measurements of S_{nn} have been shown to yield a general picture of the relative spin response of the nucleus to intermediate energy protons. This is fundamental data about nuclei that would be interesting even if no unexpected behavior was revealed.

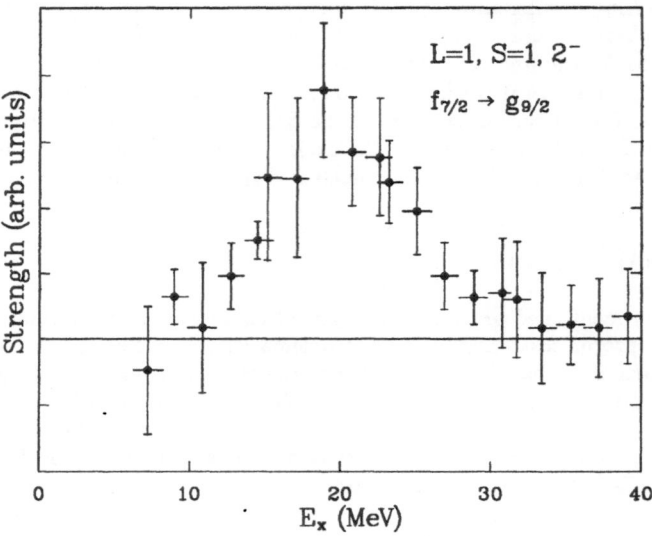

Fig. 10. Preliminary analysis of spin-dipole strength in ^{40}Ca$(p, p')^{40}$Ca* inelastic scattering at 319 MeV incident proton energy.

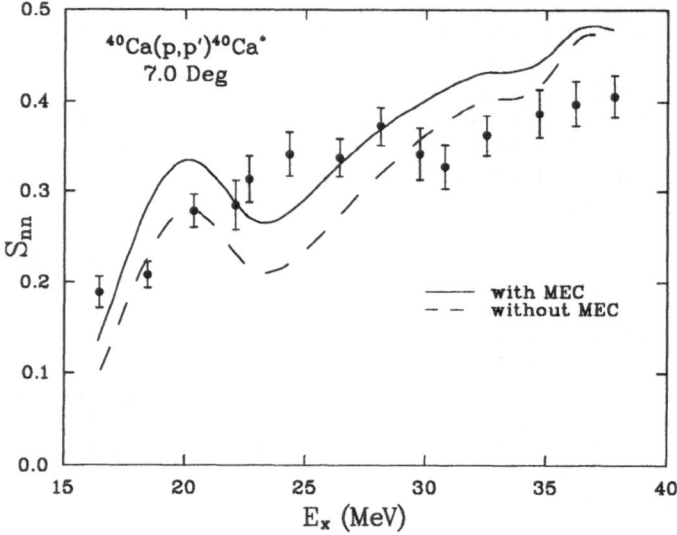

Fig. 11. Spin-flip probability S_{nn} for for inelastic scattering at 7° from ^{40}Ca at 319 MeV incident proton energy. The curves are schematic-model calculations by Boucher, et al.

The large enhancement of S_{nn} at high excitations in the 0.5 fm^{-1} momentum transfer region is not yet definitively explained. The ^{44}Ca data of Fig. 5 suggest that a relatively narrow spin resonance is not responsible. The preliminary 800-MeV data show it is not an artifact of 300-MeV scattering. The suggestion was made in Ref. 28 that exhaustion of $\Delta S = 0$ strength at lower excitation energies would imply an enhancement of relative $\Delta S = 1$ strength at high E_x above the free ratio. The slab model calculations indicated, however, that this was only a partial explanation. However, new RPA calculations of S_{nn} published recently by Boucher et al.^{29} lend support to this hypothesis (Fig. 11). The localization of $\Delta S = 0$ strength present in these RPA calculations but missing in the slab calculations may be the crucial ingredient. It will be important to determine whether improved versions of these RPA calculations can explain the lack of A dependence in the S_{nn} spectra, as well as the absolute cross sections for both $\Delta S = 0$ and $\Delta S = 1$ transfer. The separation of the longitudinal and transverse response will also be a useful experimental test of this theory. A consistent picture of $\Delta S = 0$ and $\Delta S = 1$ collectivity in nuclei seems to be emerging.

ACKNOWLEDGMENTS

A substantial program such as that described here is the result of the efforts of many participants. Major contributors to the work at LAMPF are F. T. Baker, L. Bimbot, R. W. Fergerson, C. M. Glashausser, A. A. Green, and S. K. Nanda. At TRIUMF the major participants, led by O. Hausser, include R. Abegg, R. Henderson, K. Hicks, K. P. Jackson, R. Jeppesen, J. Lisantti, C. A. Miller, R. Sawafta, M. Vetterli, and S. Yen. This work was supported in part by the U. S. Department of Energy, and by the National Science Foundation.

REFERENCES

1. R. S. Hicks and G. A. Peterson, The Search for $M1$ Strength, in: "Spin Excitations in Nuclei," F. Petrovich, et al., eds., Plenum Press, New York (1984).
2. G. M. Crawley, et al., Observation of $M1$ Strength in Medium-heavy Nuclei via the (p, p') Reaction, in: "Spin Excitations in Nuclei," F. Petrovich, et al., eds., Plenum Press, New York (1984).
3. N. Anantaraman, et al., *Phys. Rev. Lett.* 46, 1318 (1981).
4. S. K. Nanda, et al., *Phys. Rev. Lett.* 51, 1526 (1983).
5. S. K. Nanda, et al., *Phys. Rev.* C29, 660 (1984).
6. D. J. Horen, et al., *Phys. Rev.* C31, 2049 (1985).
7. S. K. Nanda, et al., *Phys. Lett.* 188B, 177 (1987).
8. R. S. Henderson, et al., *Nucl. Instr. and Meth.* A254, 61 (1987).
9. O. Hausser, et al., *Nucl. Instr. and Meth.* A254, 67 (1987).
10. J. B. McClelland, et al., "A Polarimeter for Analyzing Nuclear States in Proton-Nucleus Reactions Between 200 and 800 MeV," LA-UR-84-1671.
11. R. A. Arndt, *Phys. Rev.* D28, 97 (1983).
12. J. Lisantti, et al., *Phys. Lett.* 147B, 23 (1984).
13. D. K. McDaniels, et al., *Phys. Rev.* C33, 1943 (1986).
14. G. S. Adams, et al., *Phys. Rev.* C33, 2054 (1986).
15. F. E. Bertrand, et al., *Phys. Rev.* C34, 45 (1986).
16. S. J. Seestrom-Morris, et al., *Phys. Rev.* C26, 2131 (1982).
17. J. B. McClelland, et al., *Phys. Rev. Lett.* 52, 98 (1984).
18. G. Bertsch and O. Scholten, *Phys. Rev.* C25, 804 (1982).
19. H. Esbensen and G. Bertsch, *Ann. Phys.* 157, 255 (1984).
20. H. Esbensen and G. Bertsch, *Phys. Rev.* C34, 1419 (1986).
21. H. Esbensen, private communication.
22. R. D. Smith, private communication and contribution to these proceedings.
23. C. Horowitz and D. Murdock, preprint IUNTC-87-8.
24. L. Rees, private communication.
25. F. T. Baker, et al., *Phys. Rev.* C37, 1350 (1988), and private communication.

26. J. Ahrens, et al., *Nucl. Phys.* A251, 479 (1975).
27. D. J. Horen, et al., Submitted to *Phys. Rev.*
28. C. M. Glashausser, et al., *Phys. Rev. Lett.* 58, 2404 (1987).
29. P. M. Boucher, et al., *Phys. Rev.* C37, 906 (1988).

NUCLEON–NUCLEUS SCATTERING TO THE CONTINUUM

Richard D. Smith

Los Alamos National Laboratory
Los Alamos, NM 87545

ABSTRACT

Various aspects of inclusive nucleon-nucleus scattering are discussed which are important in understanding features of cross sections and spin observables in the continuum region. These include: 1) contributions from two-step processes, 2) damping of the response from 2p-2h excitations, 3) the "optimal" choice of frame in which to evaluate the nucleon-nucleon amplitudes, and 4) the effect of optical and spin-orbit distortions, which are included in a model based in the distorted wave impulse approximation.

INTRODUCTION

In recent years there has been a growing interest in intermediate-energy nucleon- and electron-nucleus scattering to the continuum region of quasi-elastic scattering. These reactions offer a means to study the fundamental two-body interactions in the nuclear medium, and to probe the structure of the nucleus by seeing how it responds to large energy, momentum, spin and isospin transfer.

The basic reaction mechanism in these processes is the single-step quasi-free scattering of the projectile off of a target nucleon. They can therefore be used to test models of the underlying projectile-nucleon interaction and how it is modified by the nuclear medium. For example, the relativistic Dirac theory predicts that the nucleon-nucleon (NN) amplitudes should be substantially altered compared to the free values due to the presence of large scalar and vector potentials in the nucleus,[1] and this leads to dramatic changes in the predicted quasi-elastic spin observables.

Continuum scattering can also be used to study the residual particle-hole interaction V_{ph}, which induces collective motion as the struck nucleon interacts with other target nucleons. Although signatures of shell structure, such as low-lying collective states and giant resonances, disappear at large excitation energies, the nucleus continues to respond collectively through the action of V_{ph}. This collectivity manifests itself not in sharp states or resonances, but in the gross features of the spectrum, such as shifts in the position of the quasi-elatic peak and deviations of the spin observables from the free values. Recent experiments at TRIUMF [2,3] and LAMPF [4] have clearly observed these effects in (p, p'), (p, n), and (n, p) reactions near 300 MeV. Quasi-elastic scattering therefore offers a means of studying the residual interaction in a region where it is currently not well known: at large energy and momentum transfer.

In order to extract detailed information on collectivity and two-body interactions from experimental spectra, it is clearly necessary to have a good theoretical handle on the important features of the reaction mechanism and the nuclear structure input. Traditional models such as the distorted wave impulse approximation (DWIA) combined with large-basis shell model or random phase approximation (RPA) calculations have proven very successful in understanding features of the spectrum at low excitation energies. However, these techniques become more difficult and sometimes impossible to implement numerically when there is a large transfer of momentum (q) and energy (ω) to the nucleus, so it is desirable to develop approximate techniques which include the essential physics, but at the same time involve manageable computations. Furthermore, the traditional methods should be re-examined to see if any of the approximations that were made (such as evaluating the NN amplitudes in the two-body laboratory frame) are no longer valid in the continuum region, or if any physical processes that were neglected at low q and ω (such as multi-step processes) must now be included.

In this talk several aspects of N-nucleus scattering to the continuum will be discussed, beginning with a brief review of what I call the Standard Reaction Model for quasi-elastic scattering, which is based on Glauber theory and uses the nuclear response funcion calculated in the Fermi-gas or the Slab model. Then I will discuss four features of N-nucleus scattering not included in the standard model: 1) the contribution of two-step processes – in (p, p') these are generally small relative to the one-step, although they are somewhat larger in charge exchange reactions at lower energies; 2) damping of the continuum response from 2p-2h excitations, which has a large effect in all channels involving spin and isospin transfer; 3) the "optimal" choice of frame for the NN amplitudes, which at large ω can be quite different from standard choices such as the Breit frame or the two-body laboratory frame; and 4) the effect of optical and spin-orbit distortions: the full spin-dependent distorted waves are included in a calculation of the continuum response using a technique based on the DWIA and the eikonal approximation, and the results are compared to experimental data on (p, p') spin observables, including the Los Alamos measurement of the ratio of longitudinal to transverse spin response at 500 MeV.[5]

THE STANDARD REACTION MODEL

In electron scattering, the inclusive cross section in the plane wave impulse approximation (PWIA) is proportional to the two-body (Mott) cross section times the nuclear response function $S(q, \omega)$. This model works very well for describing quasi-elastic (e, e'),[6] and has led to the *ansatz* that in the case of nucleon scattering a similar formula should hold for the cross section, modified only by a constant reduction factor to account for the attenuation of the beam due to the strong absorption of the probe. The N-nucleus cross section then has the simple factorized structure

$$\frac{d\sigma}{d\Omega dE} = N_{eff} \sum_{T,S} \frac{k'}{k} \, tr\{f_{TS}^{\dagger}(\vec{q}) f_{TS}(\vec{q})\} \, S_{TS}(q, \omega) \tag{1}$$

The sum is over spin (S) and isospin (T) transferred to the nucleus. S_{TS} is the response function in the T,S channel, and f_{TS} is the corresponding piece of the free NN amplitude. Eq. (1) assumes the nuclear ground state spin and isospin are both zero, although it can be modified to account for neutron excess (corrections are of order $(N-Z)/A$). The trace is over both projectile and target nucleon spins, and is normalized such that $tr(\sigma_i) = 0$, $tr(1) = 1$. Formulas for spin-dependent cross sections[7] are similar except that projection operators for the incident and final projectile spin are sandwiched between the NN amplitudes inside the trace. N_{eff} is the effective number of nucleons seen by the projectile, and is determined in Glauber theory from the in-medium total NN cross section σ

$$\begin{aligned} N_{eff} &= \int d^2b\, T(b)\, e^{-\sigma T(b)} \\ T(b) &= \int_{-\infty}^{\infty} dz\, \rho(r)\,, \quad \vec{r} = (\vec{b}, z) \end{aligned} \tag{2}$$

The thickness function $T(b)$ is the integral of the nuclear density along the projectile's path at impact parameter b. Thus N_{eff} is the probability, averaged over impact parameters, that the projectile will both find a target nucleon with which to interact and escape absorption on the remaining nucleons. In the limit of a short-range NN interaction σ can be related to the volume integral of the imaginary optical potential.[8] At intermediate energies it is typically 20 to 30 mb, and the reaction is strongly surface-peaked – the main contribution coming from impact parameters where the density is about one fourth the central value.

The Slab Model

If the response function is calculated in the PWIA it contains no information that the probe is surface-peaked. However, this is an important feature which can dramatically alter the response. In order to include it in a model which can easily be applied to quasi-elastic scattering, the slab model was developed by Bertsch, Scholten, and Esbensen.[9] In the slab model the surface region of the nucleus is modeled as a semi-infinite slab of nuclear matter, which is generated from a potential whose surface thickness and interior depth are similar to typical finite-nucleus potentials. The response function is constructed from the slab eigenstates and the probing field

$$\vartheta_{\vec{q}}(\vec{r}) = e^{i\vec{q}\cdot\vec{r}} e^{-\sigma T(b)/2} \tag{3}$$

which confines the scattering to the surface region. This field can be thought of as the product of incoming and outgoing eikonal distorted waves, which are generated using only the absorptive term in the optical potential.

The cross section is calculated from the slab response function using Eq. (1). This is deceptive, however, because it appears that the effect of distortion resides only in N_{eff}. In fact, the response is normalized such that, in the absence of Pauli-blocking, the integrated response is unity. This amounts to dividing the response by N_{eff}, so that the final result must again be multiplied by it, as in Eq. (1). So the slab response includes the most important effect of the distortion: the absorptive potential which

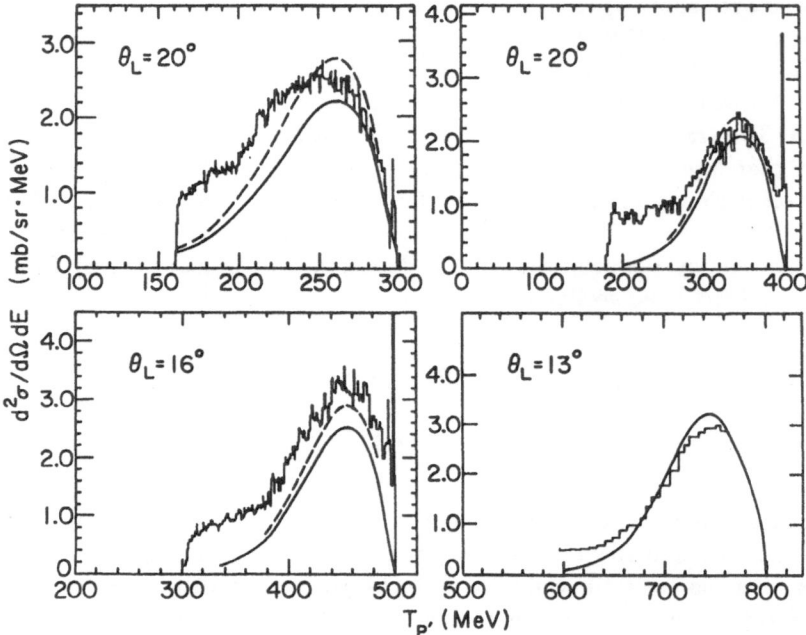

Figure 1. Slab-model calculations of ^{208}Pb(p, p') cross sections at 300, 400, 500, and 800 MeV.

confines the scattering to the surface region. Consequently, it is much more successful in describing quasi-elastic scattering than simpler models based on plane-wave probing fields (such as the Fermi-gas model).

Figure 1 shows some typical slab-model calculations by Esbensen and Bertsch [10] for proton scattering on ^{208}Pb at various angles and incident energies. The results agree well with both the position and magnitude of the quasi-elastic peak. An important feature of the slab response is that, in contrast to the Fermi-gas response, it has a long tail in ω. Later we will see in more detail how this results from including distortion in the probing field.

Figure 2 shows RPA slab-model calculations of the spin observables Ay and Snn compared to ^{54}Fe(p, p') data at 300 MeV from TRIUMF [3]. These were preliminary calculations based on the Lorentzian parameterization of the RPA slab response given in Ref. [9] (results using the exact slab response are essentially the same), and they include a small contribution from two-step processes, which are discussed in the next section. The curves in Fig. 2 deviate substantially from the free NN spin observables which are essentially flat as a function of excitation energy. This is because the response is dominated by the isoscalar (T,S=0,0) channel at low excitation, and by the spin and isospin channels at high excitation. Since the free NN amplitudes are different in each channel, the spin observables vary with ω. While the slab model cannot account for the sharp structures seen in the data at low excitation, which are associated with resonances of the finite system, the calculations are nevertheless in

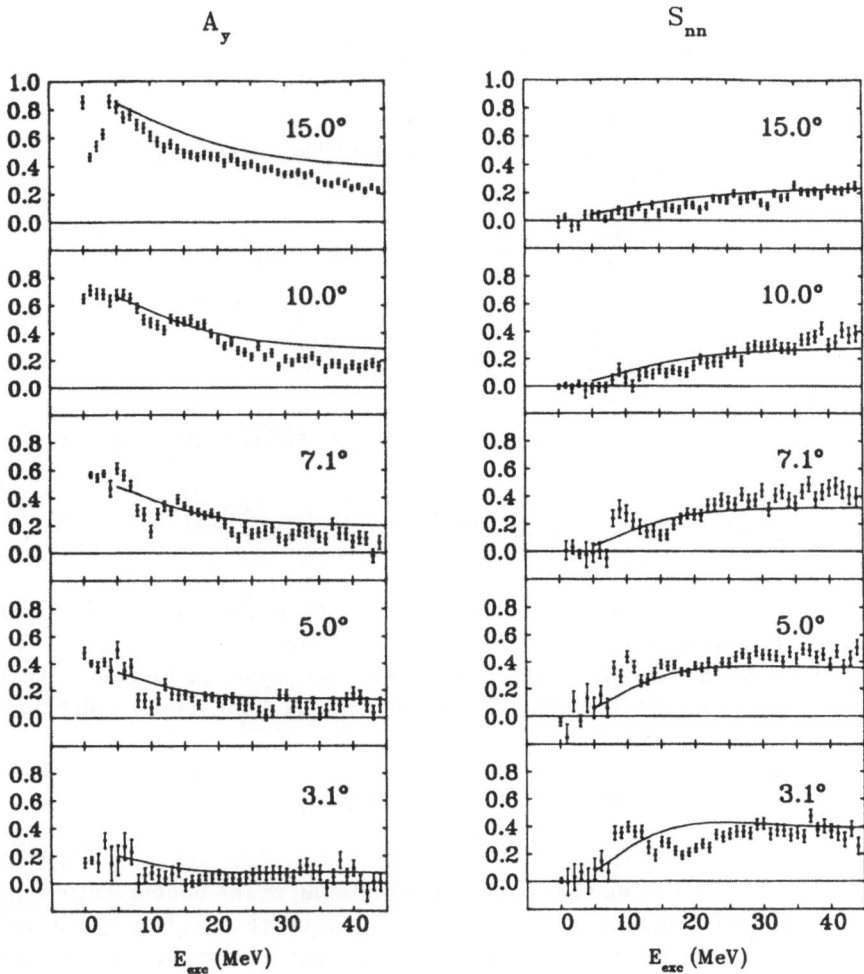

Figure 2. Analyzing power and spin-flip probability for ^{54}Fe(p, p') at 290 MeV, from Ref. [3].

reasonably good agreement with the gross features of the data, suggesting that we are indeed seeing the collectivity predicted by the RPA, even at the higher excitation energies.

CONTRIBUTION FROM MULTI-STEP PROCESSES

The standard reaction model can be extended using Glauber theory to include the contribution from multi-step processes in which the projectile has quasi-elastic collisions with more than one target nucleon. The full cross sections can be expressed as a sum over n-step processes [7,8] as follows

$$\frac{d\sigma}{d\Omega dE} = kk' \sum_{n=1}^{A} D_n W_n(q,\omega)$$

$$D_n = \binom{A}{n} \int d^2b \, [t(b)]^n \, [1 - \sigma t(b)]^{A-n} \;\;, \;\;\; t(b) = T(b)/A$$

$$\approx \int d^2b \frac{1}{n!} \, [T(b)]^n \, e^{-\sigma T(b)} \qquad (A \to \infty)$$

$$W_n(q,\omega) = \int \frac{d^2q_1}{k^2} \cdots \frac{d^2q_n}{k^2} \, \delta^2(\vec{q} - \sum_{i=1}^{n} \vec{q}_i)$$

$$\times \, tr\{ f^\dagger(\vec{q}_1) \cdots f^\dagger(\vec{q}_n) \, f(\vec{q}_n) \cdots f(\vec{q}_1) \} \, S_n(q_1, \ldots q_n, \omega)$$

$$S_n(q_1, \ldots q_n, \omega) = \int \prod_{i=1}^{n} [d\omega_i \, S_1(q_i, \omega_i)] \, \delta(\omega - \sum_{i=1}^{n} \omega_i) \qquad (4)$$

W_n describes the sequence of n hard collisions, and D_n is the distortion factor which accounts for absorption on the remaining $A - n$ nucleons. The $n = 1$ term is identical to Eq. (1) except that the spin-isospin sums have been supressed for simplicity. Note that $D_1 = N_{eff}$. This can be generalized to define an N_{eff} for the nth order term

$$N_{eff}^{(n)} = \binom{A}{n} D_n(\sigma)/D_n(0) \qquad (5)$$

Thus $N_{eff}^{(2)}$ is the effective number of pairs participating in the double scattering, etc. W_n involves the NN amplitudes for each hard collision (which must be properly ordered due to their spin dependence [7]) and the n-step response function S_n. Assuming there are no correlations between successive collisions, S_n is simply a convolution of n one-step response functions, as in the last line of Eq. (4). Given $S_1(q,\omega)$, Eq. (4) can easily be evaluated for $n \le 2$, while higher order terms require a Fourier transform. Fortunately, the series converges rapidly: successive terms are smaller by roughly an order of magnitude,[7] so it is usually sufficient to keep only the first two terms. In heavy nuclei one might expect the multi-step contributions to be larger, since there are more nucleons present. This is not the case, however, because the higher density is compensated by stronger absorption. The ratio of the multi-step cross sections to the single-step is determined primarily by σ, and is only weakly dependent on A.

An example of the two-step contribution is shown in Fig. 3 for the 290 MeV $^{54}Fe(p,p')$ reaction at 7.1°. The two-step produces a small background which rises up slowly from zero to about 0.2 mb at 60 MeV. In the region of the quasi-elastic peak it is only $\sim$5% of the single-step cross section, which is typical for (p,p') reactions. At much larger excitation energies (in this case above 60 MeV) the two-step begins to dominate.

In (p,n) and (n,p) reactions the ratio of the two-step to the one-step cross section can be somewhat larger than in (p,p'). This occurs for two reasons: first, the charge-exchange, which involves the isovector part of the NN amplitude, occurs on only one of the collisions in the two-step process, while the other involves the larger isoscalar amplitude; and second, there are two orderings of the charge-exchange and non-charge-exchange collisions which must both be included, yielding an extra factor of two. The two-step can also be larger at lower incident energies (say around 200 MeV), because Pauli-blocking is more effective in reducing the in-medium NN cross section σ, and the ratio $N_{eff}^{(2)}/N_{eff}^{(1)}$ increases as σ decreases.

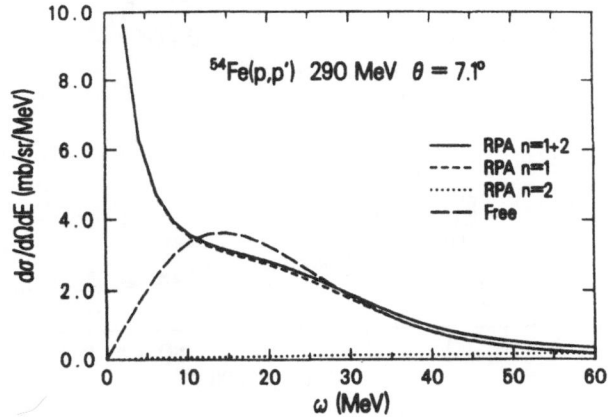

Figure 3. Slab-model calculation including two-step contribution.

Figure 4 shows the two-step contribution to the forward angle (p, n) and (n, p) reactions on ^{90}Zr at 200 MeV. Dotted curves show the two-step, and the full curves are the sum of one and two-step cross sections, all calculated using Eq. (4). The normalization of the cross sections is determined by the distortion factors D_1 and D_2 which were evaluated using $\sigma = 14.6$ mb. This value should be fairly accurate, since the magnitude of the single-step cross sections are in good agreement with the data in the low excitation region. The one-step curves employ a full finite-nucleus response function calculated by J. Wambach using the Second RPA formalism,[11] while the two-step calculations use the slab-model response. The slab response does not contain any shell structure, but it should be adequate for the two-step cross section, since any detailed structure in the single-step response will be smoothed out by the convolution over $\vec{q}$ and ω in Eq. (4). Comparing the calculations to the data, we see that in the (p, n) case the full spectrum is nicely explained by including the two-step. This calculation and its implications for the issue of quenching of Gamow-Teller strength are discussed in more detail in Ref. [8]. The (n, p) calculations, however, do not agree as well with the data: the cross section above 15 MeV is too low by about a factor of two. Part of this descrepancy may be due to the fact that single-step response was calculated using a plane-wave probing field. If the distortion is included in the response calculation, the results can be quite different at large excitation, since the momentum transferred on the hard collision is different from the external momentum transfer $\vec{q}$. We will return to this point in the last section of the talk.

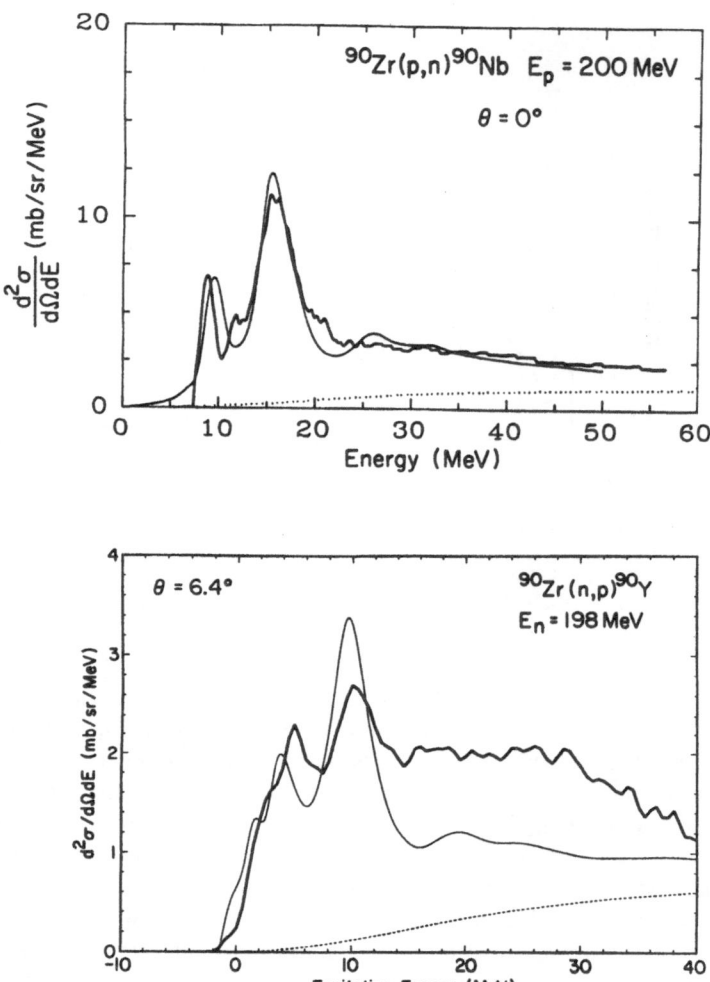

Figure 4. Forward-angle (p, n) and (n, p) cross sections on ^{90}Zr. The solid lines include one and two-step contributions. The one-step is calculated in the SRPA, and the two-step (dotted lines) is calculated in the slab model. The (p, n) data is from Ref. [12], and the (n, p) data is from Ref. [13].

The response function can be substantially altered at high excitation energy due to the collisional damping of single-particle motion. In the quantum many-body theory this corresponds to the coupling of 1p-1h to 2p-2h or higher configurations through the action of V_{ph}. The formalism for including this coupling in full finite-nucleus calculations, known as the Second RPA, has been developed and successfully applied to a variety of reactions.[11] The difficulty with this method is that it involves large-basis calculations which quickly become numerically infeasible as they are pushed to higher angles and excitation energies. In the region of giant resonances in medium and heavy nuclei, the density of 2p-2h states is already on the order of 10^2-10^3 states per MeV, and rapidly increases with energy. The problem is even worse at larger momentum transfers, where many more multipoles must be included to describe the full spectrum. In the region of the quasi-elastic peak, the SRPA can only be applied to light nuclei like ^{12}C up to momentum transfers of about 1 to 1.5 fm^{-1}.[14]

To overcome this problem we have developed an approximate method for including the effect of collisional damping which can easily be applied in regions of large energy and momentum transfer.[15] The basic idea is to replace the microscopic coupling to specific 2p-2h configurations with empirical information already contained in the low-energy phenomenological optical potentials and measured single-particle spreading widths. These are used to construct the self-energies of RPA p-h vibrations which, in a semi-classical limit, depend only on their energy, spin, and isospin. This simplification allows the full response including 2p-2h damping to be expressed directly as an integral over the RPA response.

To see how this comes about, consider a spectral representation of the response in terms of the eigenstates $|\alpha\rangle$ of the RPA Hamiltonian, which are collective states composed of 1p-1h components constructed from the Hartree-Fock orbitals. The RPA response function is

$$S_{rpa}(q,\omega) = \sum_{\alpha} |\langle \alpha | \vartheta_{\vec{q}} | 0 \rangle|^2 \, \tfrac{1}{\pi} \, \mathrm{Im} \left(\frac{1}{E_{\alpha} - \omega - i\eta} + \frac{1}{E_{\alpha} + \omega - i\eta} \right)$$

$$= \sum_{\alpha} |\langle \alpha | \vartheta_{\vec{q}} | 0 \rangle|^2 \, \delta(E_{\alpha} - \omega) \tag{6}$$

and the full response is

$$S(q,\omega) = \sum_{\alpha} |\langle \alpha | \vartheta_{\vec{q}} | 0 \rangle|^2$$

$$\times \tfrac{1}{\pi} \, \mathrm{Im} \left(\frac{1}{E_{\alpha} - \omega - \Sigma_{\alpha}(\omega) - i\eta} + \frac{1}{E_{\alpha} + \omega - \Sigma_{\alpha}(-\omega) - i\eta} \right) \tag{7}$$

where Σ_{α} is the complex self-energy of the state α due to its coupling to 2p-2h or higher configurations. In writing this expression it has been assumed that the 1p-1h excitations act as a doorway, and the direct transitions from the ground state to 2p-2h excitations, which are allowed because of the existence of 2p-2h ground-state correlations, have been neglected. However, it has recently been shown, using an extended version of the SRPA, that these transitions do not contribute significantly to the response.[16] If they are not included, then the full response has the same integrated strength as the RPA response.[15]

$$\int_0^\infty d\omega \, S(q,\omega) = \int_0^\infty d\omega \, S_{rpa}(q,\omega) \tag{8}$$

Therefore, the effect of collisional damping is merely to redistribute the strength as a function of ω, leaving the total strength unchanged.

The self-energy can be formally expressed in terms of the residual interaction and the nuclear Hamiltonian H as

$$\Sigma_\alpha(\omega) = \langle \alpha | V_{ph} Q \frac{1}{QHQ - \omega - i\eta} QV_{ph} | \alpha \rangle \tag{9}$$

where Q is a projection operator which projects out of the RPA subspace of 1p-1h states. The self-energy is assumed to be diagonal in the RPA basis, which should be a good approximation for highly collective states.

At this point, the full response cannot be expressed directly in terms of the RPA response, because the self-energy Σ_α depends on the quantum numbers of the state α. However, it can be shown [17] that in the classical limit of high angular momentum multipoles, which should be approximately valid in the continuum region, the self-energy depends only on the energy, spin and isospin of the excited state. In this case $\Sigma_\alpha(\omega) \to \Sigma_{TS}(\omega)$, and the response in each TS channel takes the simpler form

$$S(q,\omega) = \sum_\alpha |\langle \alpha | \vartheta_{\vec{q}} | 0 \rangle|^2 \, [\rho(E_\alpha, \omega) + \rho(E_\alpha, -\omega)]$$

$$\rho(E,\omega) = \frac{\frac{1}{\pi} \Gamma(\omega)/2}{[E - \omega - \Delta(\omega)]^2 + [\Gamma(\omega)/2]^2}$$

$$\Sigma(\omega) \equiv \Delta(\omega) + i\Gamma(\omega)/2 \tag{10}$$

where spin and isospin indices have been omitted for simplicity. The Lorentzian function $\rho(E,\omega)$ gives the RPA states an energy-dependent shift $\Delta(\omega)$ and a width $\Gamma(\omega)$ which correspond to the real and imaginary parts of $\Sigma(\omega)$. Summing over all quantum numbers except energy spin and isospin, the full response can now be expressed in terms of the RPA response

$$S(q,\omega) = \int_0^\infty dE \, S_{rpa}(q,E) \, [\rho(E,\omega) + \rho(E,-\omega)] \tag{11}$$

(here the integral may also include a sum over discreet states). Given the RPA response function, Eq. (11) provides a simple method for including the effect of collisional damping once the self-energy $\Sigma_{TS}(\omega)$ has been determined.

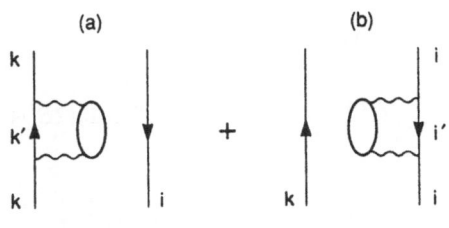

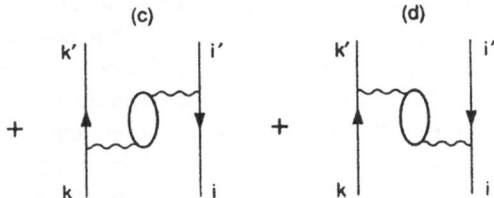

Figure 5. Particle-hole self-energy diagrams.

Self-Energy of p-h Vibrations

The dominant diagrams which contribute to the self-energy of a given p-h component of an RPA state α are shown in Fig. 5. Graphs (a) and (b) are essentially self-energy diagrams for the individual particles and holes. The imaginary parts of these graphs will be associated with the empirically determined widths of single-particle and hole states

$$\text{Im}\Sigma_{(a)}(\omega) = \gamma_p(\omega + \varepsilon_i)/2$$

$$\text{Im}\Sigma_{(b)}(\omega) = \gamma_h(\varepsilon_k - \omega)/2 \qquad (12)$$

where ε_i, ε_k are the energies of the spectator hole, particle in graphs (a), (b) (all energies are measured with respect to the Fermi energy $\varepsilon_F \equiv 0$.) Graphs (c) and (d) are interference diagrams which arise from the coherence of particle and hole decay amplitudes, and may interfere either constructively or destructively with the direct graphs (a) and (b), depending on the quantum numbers of the initial p-h pair. Graphs (c) and (d) can be related to (a) and (b) through recoupling coefficients which depend on the spin, isospin, and angular momentum of the incident and intermediate p-h pairs. In the semi-classical limit discussed above, this recoupling is independent of the angular momentum of the intermediate pair, so the ratio of the interference to the direct graphs depends only on their spin and isospin structure [15,17] and is given by

$$\text{Im}\Sigma_{(c,d)} = C_{TS}\,\text{Im}\Sigma_{(a,b)}$$

$$C_{TS} = \sum_{T',S'} V_{T'S'}^2\, C_{T'S'}^{TS} \left/ \sum_{T',S'} V_{T'S'}^2\,(2T'+1)(2S'+1)\right.$$

$$C_{T'S'}^{TS} = -(-)^{T+S+T'+S'}(2T'+1)(2S'+1)4\left\{\begin{matrix} \tfrac{1}{2} & \tfrac{1}{2} & T \\ \tfrac{1}{2} & \tfrac{1}{2} & T' \end{matrix}\right\}\left\{\begin{matrix} \tfrac{1}{2} & \tfrac{1}{2} & S \\ \tfrac{1}{2} & \tfrac{1}{2} & S' \end{matrix}\right\} \quad (13)$$

where S,T (S',T') are the spin and isospin of the incident (intermediate) p-h pair, and V_{TS} is the residual interaction in the TS channel. The coefficients C_{TS}, calculated using the residual interactions used in the slab model,[9] are given in Table 1. In the spin and isospin channels ($TS = 01, 10, 11$) the interference terms $\text{Im}\Sigma_{(c)}$ and $\text{Im}\Sigma_{(d)}$ are small, and the imaginary self-energy is approximately equal to the sum of particle and hole decay widths. This occurs because the coefficients $C_{T'S'}^{TS}$ have oscillating signs in different intermediate $T'S'$ channels, so they tend to cancel one another. In fact, C_{10}, C_{10}, and C_{11} would all be identically zero if the residual interaction were the same in all channels. In the 00 channel, on the other hand, the imaginary self-energy vanishes, because C_{00} is always -1, and the interference terms exactly cancel the direct terms. In full finite-nucleus calculations this cancellation is not exact, but the same features appear, namely, that the spin-isospin states have large widths, and the isoscalar non-spin states have very small widths.[17]

Table 1. Ratio of interference to direct graphs in each channel.

TS	00	10	01	11
V_{TS} (MeV fm^3)	-284.4	300	50	220
C_{TS}	-1.0	0.19	-0.25	-0.05

In order to evaluate the self-energy of a collective state, we must sum over its p-h components. To do this, we assume that the density of p-h states making up a given RPA state is uniform in energy, and that on the average the particle and hole energies satisfy $\varepsilon_k - \varepsilon_i \approx \omega$. The imaginary self-energy of the collective state is then given by

$$\text{Im}\Sigma_{TS}(\omega) \equiv \Gamma_{TS}(\omega)/2$$

$$\Gamma_{TS} = \frac{1}{\omega}\int_0^\omega d\varepsilon\,[\gamma_p(\varepsilon) + \gamma_h(\varepsilon - \omega)]\,\xi_{TS}$$

$$\xi_{TS} = 1 + C_{TS} \quad (14)$$

Thus, the full width is simply a classical average width for the particles plus holes, multiplied by a factor ξ_{TS} due to the quantum coherence of particle and hole decay amplitudes. The real self-energy can be obtained from the imaginary part using the dispersion relation

$$Re\Sigma_{TS}(\omega) \equiv \Delta_{TS}(\omega) = \frac{1}{2\pi} P \int_{-\infty}^{\infty} d\omega' \frac{\Gamma_{TS}(\omega')}{\omega' - \omega} \qquad (15)$$

To complete the evaluation of Σ_{TS}, it remains only to specify how the particle and hole widths γ_p and γ_h are determined empirically. Above the Fermi sea the single-particle widths can be deduced from matrix elements of the absorptive part of the low-energy optical potential, and below the Fermi sea they are determined from the measured spreading widths of hole states. Figure 6 shows a compilation of single-particle widths in medium-heavy nuclei by Mahaux and Ngô.[18] These are reasonably well described by the continuous parameterization (shown by the solid curve in Fig. 6)

$$\gamma(\varepsilon) = 21.5 \left(\frac{\varepsilon^2}{\varepsilon^2 + 18^2} \right) \left(\frac{110^2}{\varepsilon^2 + 110^2} \right) \text{ MeV} \qquad (16)$$

which is assumed to be symmetric about the Fermi energy, so that $\gamma(\varepsilon) = \gamma_p(\varepsilon) = \gamma_h(-\varepsilon)$. Using this parameterization, the real and imaginary parts of the full self-energy Σ_{TS} can be calculated from Eqs. (14) and (15). The results are shown in Fig. 7 for the spin and isospin channels (in the 00 channel both real and imaginary parts vanish due to the cancellation discussed above). In Ref. [15] this calculation was also performed for the specific case of ^{208}Pb using a parameterization of $\gamma(\varepsilon)$ based on the low-energy neutron optical potential from Ref. [19]. The results were in good agreement with both nuclear matter calculations and with full finite-nucleus calculations of the widths of isoscalar and isovector monopole states in ^{208}Pb. This gives us confidence that the semi-classical model provides a good approximation to more exact calculations, and can be reliably extended to higher momentum and energy transfers. It is interesting to note that in the medium-heavy nuclei the widths Γ_{TS} are much larger near the Fermi energy than in ^{208}Pb. This is due to the surface dynamics: the self-energy is enhanced in the surface region, which in lighter nuclei constitutes a larger percentage of the total volume.

Now that Σ_{TS} has been determined, we can return to Eq. (11) and calculate the effect of 2p-2h damping on the slab-model RPA response functions. Figure 8 shows the response in the various channels for 290 MeV nucleons scattered at 5° in the laboratory, calculated using $\sigma = 23$ mb in the probing field (Eq. 3). Dotted lines show the free slab response (no residual interaction), dashed lines are the RPA response, and solid lines are the RPA response with damping, calculated from Eqs. (11-16).‡ There is no damping in the 00 channel, but in the spin and isospin channels the response is quenched at the peak and enhanced at higher excitation energies. The effect of damping should therefore be large in charge-exchange and spin-flip reactions, but small in (p, p') which is dominated by isoscalar non-spin excitations.

---

‡The real self-energy Δ_{TS} was not included in these calculations, because the residual interactions, which were originally determined phenomenologically for 1p-1h calculations, already effectively contain the real part of the energy shift due to the 2p-2h coupling. The sum rule (Eq. 8) is violated if Δ_{TS} is not included, but in this case it causes only a few percent increase in the integrated strength. An alternative method which satisfies the sum rule, but yields essentially the same results, is to subtract a constant from the energy-dependent $\Delta_{TS}(\omega)$.

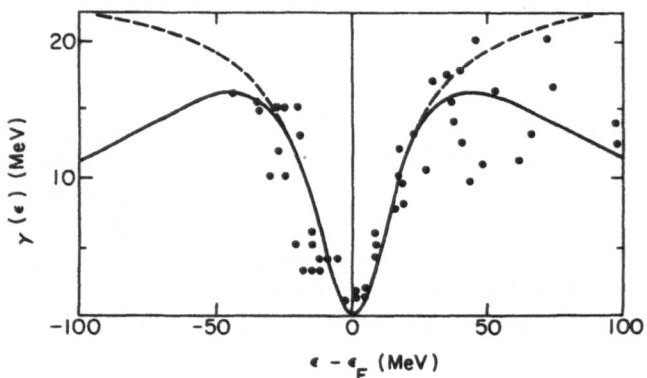

Figure 6. Single-particle widths for medium-heavy nuclei from Ref. [18].

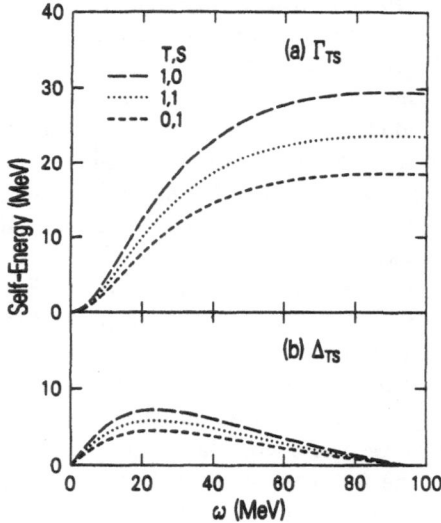

Figure 7. Real and Imaginary parts of the averaged p-h self-energy calculated from Eqs. (14-16).

Figure 9 shows cross sections, calculated using the slab response and Eq. (1), for the 300 MeV ^{54}Fe(n, p) reaction along with data at 5°, 8°, and 12° from Ref. [2]. The RPA (dashed) and RPA+2p-2h (solid) curves also include the contribution from double scattering (Eq. 4). Comparing the solid and dashed curves, we see that the effect of 2p-2h damping is to bring the 1p-1h RPA calculations into nice agreement with both the magnitude and overall shape of the experimental spectra. The magnitude is governed primarily by N_{eff}, which should be fairly accurately determined since it also yields the correct normalization for the (p, p') cross sections at 290 MeV.

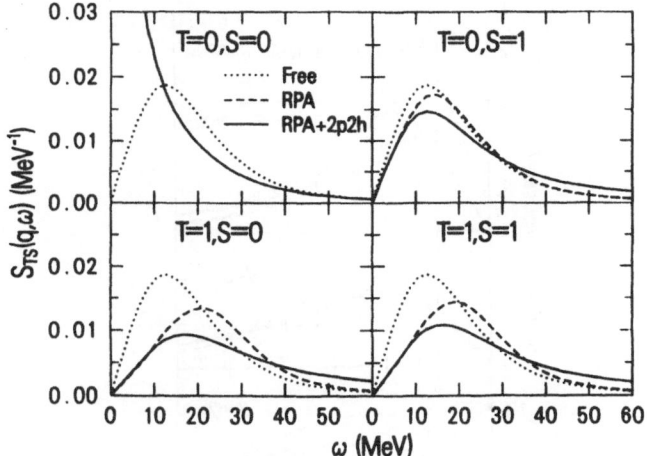

Figure 8. Slab-model response functions with 2p-2h damping for 290 MeV nucleons scattered at 5°.

It should be emphasized that, although these calculations use the slab response, the method can be applied to any RPA response function. Recently, Co' *et al.* [20] used it to estimate the effect of 2p-2h damping on the (e, e') charge response in ^{12}C and ^{40}Ca, which was treated in a full finite-nucleus continuum RPA. They found that the discrepancy previously observed between theory and experiment can be explained by a combination of the the 2p-2h damping and an effect due to the mean field non-locality.

Finally, I conclude this section with a word of caution addressed to people who calculate exclusive reactions like $(e, e'p)$ and $(p, 2p)$. To date, most calculations of these reactions include only the scattering of the ejected particle in the optical potential, leaving out the hole scattering and their interference. In our formalism this corresponds to keeping only diagram (a) in Fig. 5. By accident, this turns out to be approximately correct for the charge (longitudinal) response in (e, e'),[20] since the average width is given by

$$\Gamma_{ch} \approx (\Gamma_p + \Gamma_h) \frac{\xi_{00} + \xi_{10}}{2}$$

$$\approx \Gamma_p \qquad (17)$$

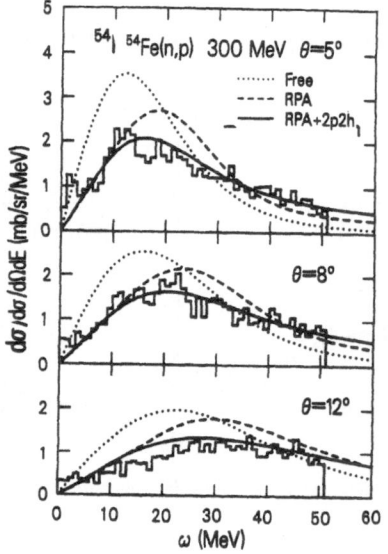

Figure 9. Charge-exchange cross sections compared to experimental data from Ref. [2].

where the last line follows because $\xi_{00} = 0$, $\xi_{10} \approx 1$, and $\Gamma_p \approx \Gamma_h$. The longitudinal response is a special case, however, since it is the average of the isoscalar and isovector response functions. Equation (17) does not apply to the transverse part, which is predominately spin response, or to any of the various response functions which can be measured in polarized $(e, e'p)$. In general, the final state interaction must include the coherent scattering of both particle and hole in the optical potential.

OPTIMAL FRAME FOR NN AMPLITUDES

In this section we turn to the question of how to evaluate the two-body amplitudes which describe a quasi-elastic collision. In most calculations they are associated with the free on-shell amplitudes derived from experimental phase shifts, but in principle they should be evaluated off-shell and should include medium modifications. Furthermore, Eq. (1) presumes that the NN amplitudes depend only on the incident energy and the momentum transfer $\vec{q}$, but in general they also depend on the momentum of the struck nucleon, which varies due to its Fermi motion. This dependence can be very important in continuum scattering and, as we will see, leads to a breakdown of the standard reaction model at high excitation energy.

In order to calculate quasi-elastic cross sections it is necessary to integrate over the struck nucleon's Fermi momentum. This problem is greatly simplified if the two-body amplitudes are factored out of the integration by evaluating them in a frame where the struck nucleon's momentum has a constant "optimal" value. Such an approximation is clearly required in order to derive a formula with the factorized structure of Eq. (1). The question of how best to choose this frame has been answered in the non-relativistic theory by Gurvitz and collaborators.[21] The result depends on both momentum transfer and excitation energy. In the case of elastic scattering ($\omega = 0$) the best choice is the Breit frame, in which the struck nucleon has momentum $\vec{p} = -\vec{q}/2 + \mathcal{O}(1/A)$. At the quasi-elastic peak ($\omega = q^2/2m$), the struck nucleon is on average at rest, and the optimal frame is the two-body laboratory frame where $\vec{p} = 0$. In the general case (arbitrary ω), the struck nucleon's momentum is determined by requiring that it satisfies energy conservation: $\omega = (\vec{p} + \vec{q})^2/2m - p^2/2m$, and that it lies along the only preferred direction: that of the momentum transfer $\vec{q}$. Then the optimal momentum is given by

$$\vec{p}_{opt} = -\frac{\vec{q}}{2}\left(1 - \frac{2m\omega}{q^2}\right) \qquad (nonrelativistic)$$

$$\vec{p}_{opt} = -\frac{\vec{q}}{2}\left(1 - \frac{\omega}{q}\sqrt{1 + \frac{4m^2}{q^2 - \omega^2}}\right) \quad (relativistic) \tag{18}$$

The second line gives the result using relativistic kinematics. Note that $\vec{p}_{opt}$ reduces to zero at the quasi-elastic peak and to the Breit-frame momentum at $\omega = 0$. A diagram of the optimal frame is shown in Fig. 10.

In Ref. [21] a detailed derivation of the optimal frame is given, in which the projectile-nucleus scattering operator is expanded in terms of the difference between the exact and an approximate Green's function for the system, then the optimal momentum and two-body energy are chosen such that the leading term in this expansion vanishes. Fortunately, the optimal energy turns out to be the same as the on-shell energy. This is because the difference between the on and off-shell energies is cancelled by the leading order correction coming from the final-state interaction of the struck nucleon with the nuclear binding potential. This cancellation gives justification to the use of on-shell amplitudes in nonrelativistic models at high energy, where the binding potentials are small compared to the incident energy. However, in relativistic models with large scalar and vector potentials, the on-shell approximation is probably not justified, and further research on this question is clearly needed.

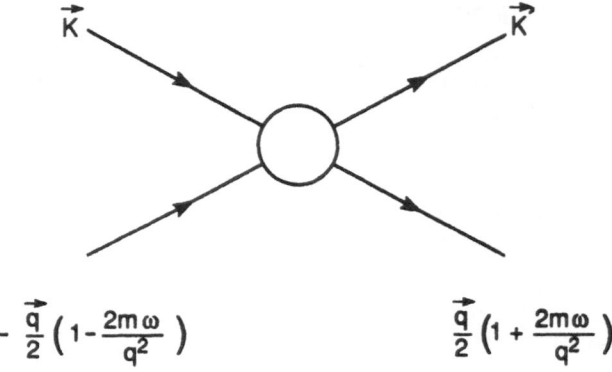

$$-\frac{\vec{q}}{2}\left(1-\frac{2m\omega}{q^2}\right) \qquad\qquad \frac{\vec{q}}{2}\left(1+\frac{2m\omega}{q^2}\right)$$

Figure 10. The optimal frame with nonrelativistic kinematics.

Since the experimentally determined amplitudes are usually given in the center-of-mass (c.m.) frame,[22] it is necessary to perform a Lorentz boost to the optimal frame. To do this we follow the method of Ref. [23], where the c.m. amplitudes are boosted to the Breit frame. The scattering amplitude in the c.m. can be parameterized in terms of five complex amplitudes

$$f_{cm} = A_{cm} + iC_{cm}(\sigma_{1q} + \sigma_{2q}) + B_{cm}\sigma_{1n}\sigma_{2n} + D_{cm}\sigma_{1q}\sigma_{2q} + E_{cm}\sigma_{1z}\sigma_{2z} \qquad (19)$$

where the Pauli spin matrices are projected onto the axes $\hat{n} \sim \vec{k}_{cm} \times \vec{k}'_{cm}$, $\hat{q} \sim \vec{k}_{cm} - \vec{k}'_{cm}$, and $\hat{z} = \hat{n} \times \hat{q}$ ($\sigma_{1n} \equiv \vec{\sigma}_1 \cdot \hat{n}$, etc). The c.m. amplitude can also be expressed in terms of the invariant Dirac matrices γ_μ and free Dirac spinors for the external lines

$$f_{cm} = \bar{u}_1(\vec{k}'_{cm})\bar{u}_2(-\vec{k}'_{cm})\,\mathcal{F}(s,t)\,u_2(-\vec{k}_{cm})u_1(\vec{k}_{cm})$$

$$\mathcal{F} = \mathcal{F}_S + \mathcal{F}_V\gamma_1 \cdot \gamma_2 + \mathcal{F}_P\gamma_1^5\gamma_2^5 + \mathcal{F}_A\gamma_1^5\gamma_1^\mu\gamma_2^5\gamma_{2\mu} + \mathcal{F}_T\sigma_1^{\mu\nu}\sigma_{2\mu\nu} \qquad (20)$$

where s and t are the invariant energy and momentum transfer, and $\mathcal{F}$ is normalized such that $tr\{f^\dagger_{cm}f_{cm}\} = d\sigma/d\Omega_{cm}$, the two-body c.m. cross section. The five Dirac amplitudes $\mathcal{F}_S, \ldots, \mathcal{F}_T$ can be derived from the c.m. amplitudes in Eq. (19) by expanding Eq. (20) in Pauli matrices and equating like terms. The optimal frame amplitudes are then obtained by sandwiching the Dirac amplitudes between free spinors in the optimal frame:

$$f_{opt} = \bar{u}_1(\vec{k}')\bar{u}_2(\vec{p}'_{opt})\,\mathcal{F}(s,t)\,u_2(\vec{p}_{opt})u_1(\vec{k})$$

$$= A + iC_1\sigma_{1n} + iC_2\sigma_{2n} + B\sigma_{1n}\sigma_{2n} + D\sigma_{1q}\sigma_{2q} + E\sigma_{1z}\sigma_{2z}$$

$$+ F\sigma_{1q}\sigma_{2z} + G\sigma_{1z}\sigma_{2q} \qquad (21)$$

where the spin axes are now given by $\hat{n} \sim \vec{k} \times \vec{k}'$, $\hat{q} \sim \vec{k} - \vec{k}'$, and $\hat{z} = \hat{n} \times \hat{q}$. Notice that in the optimal frame there are eight complex amplitudes compared to five in the c.m. frame. The spin-orbit terms C_1 and C_2 are no longer equal as they were in the c.m., and the two new terms F and G are non-vanishing at all energy transfers except $\omega = 0$, the Breit frame. Thus there are eight amplitudes even in the two-body laboratory frame. Other terms would also be present if the optimal momentum had components along $\hat{n}$ or $\hat{z}$. It should be mentioned that since Eqs. (19-21) represent a Lorentz boost between frames, it is necessary to use the relativistic version of the optimal momentum (Eq. 18), otherwise energy will not be properly conserved. The matrices which transform between the c.m. amplitudes, the Dirac amplitudes, and the optimal amplitudes are given for both πN and NN scattering in Ref. [24].

The inclusive projectile-nucleus cross section is expressed in terms of the optimal amplitudes by

$$\frac{d\sigma}{d\Omega dE} = N_{eff} \, C(k, k', q) \, tr\{f_{opt}^{\dagger} f_{opt}\} \, S(q, \omega) \qquad (22)$$

where $C(k, k', q)$ is a kinematic factor given by

$$C(k, k', q) = \frac{k'}{k} \left(\frac{s}{E_{p_{opt}} E_{p'_{opt}}} \right) \qquad (23)$$

where $E_p = \sqrt{p^2 + m^2}$. It approaches k'/k in the limit of nonrelativistic nucleon scattering. The transformation of the amplitudes between frames in Eqs. (19-21) is actually not necessary for simply calculating the two-body cross section, since $tr\{f_{opt}^{\dagger} f_{opt}\} = tr\{f_{cm}^{\dagger} f_{cm}\}$. However, it is necessary for calculating spin observables, which involve $tr\{\sigma_i f_{opt}^{\dagger} \sigma_j f_{opt}\}$, or for separating the cross section into contributions from different spin-isospin channels for use with the RPA response.

Perhaps the most important aspect of the optimal frame is that the invariant energy s varies rapidly with ω due to its dependence on $\vec{p}_{opt}$, and it can be quite different from the energy in the two-body laboratory frame. This can be seen by examining the effective laboratory kinetic energy defined by

$$T_L^{eff} \equiv (s - 4m^2)/2m = \frac{E_k E_{p_{opt}} - \vec{k} \cdot \vec{p}_{opt} - m^2}{m} \qquad (24)$$

which is plotted in Fig. 11(a) as a function of ω for 290 MeV nucleons scattered at various angles. At each angle the curves are only plotted in the regions of ω allowed in the Fermi-gas model, which restricts $|\vec{p}_{opt}(\omega)| \leq k_F = 268$ MeV/c. T_L^{eff} agrees with the true kinetic energy (shown by the straight line at 290 MeV) at the quasi-elastic points $\omega = \sqrt{q^2 + m^2} - m$, which correspond to $\vec{p}_{opt} = 0$ the two-body laboratory frame. However, away from these points it varies by as much as ± 100 MeV. Such variation will clearly have a large effect on the amplitudes in regions where they are strongly energy dependent.

To illustrate this point, and to show how well the optimal factorization works, consider π-nucleus scattering in the region of the Δ_{33} resonance, where the πN amplitudes vary rapidly with energy. Figure 11(b) shows the effective laboratory kinetic energy

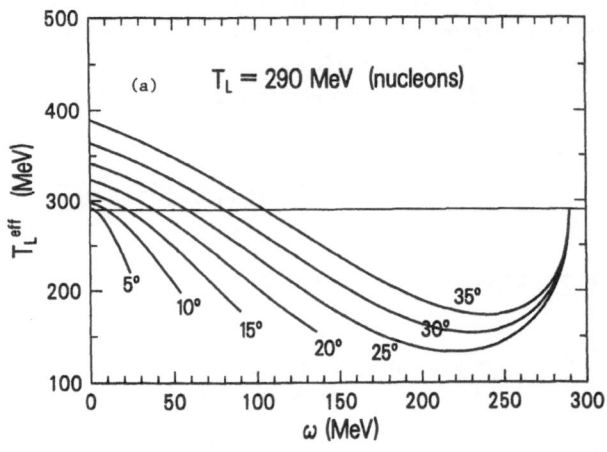

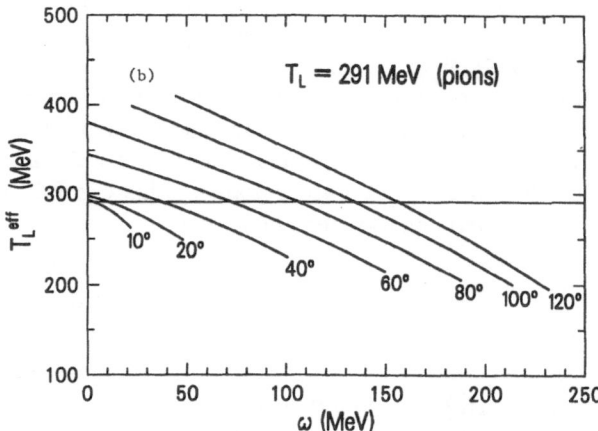

Figure 11. The effective laboratory kinetic energy in the optimal frame for pions and nucleons near 300 MeV.

in the optimal frame for pions incident at 291 MeV, and again we see a variation of $\sim \pm 100$ MeV over the allowed regions of ω. Figure 12 shows cross sections for the 291 MeV $\pi^{+}-^{40}$Ca reaction at 120° calculated using the relativistic Fermi-gas response. The dotted and dashed curves are evaluated with Eq. (22) using, respectively, the optimal and two-body-lab amplitudes.[‡] The two curves agree at the quasi-elastic point near ~ 155 MeV, but at lower ω the optimal curve is below the two-body-lab curve, while at higher ω it is much larger. This is because as ω increases, T_L^{eff} decreases, and the πN cross section rises as it nears the peak of the Δ_{33} resonance. The solid curve in Fig. 12 shows a "full" calculation with no factorization, where the on-shell πN amplitudes are included inside the integration over the nucleon's Fermi-momentum. Comparison with the dashed curve shows that the optimal factorization provides a very good approximation to the full result. Figure 13 shows the optimal and two-body-lab calculations using the slab response function, along with data from Ref. [25], and the optimal factorization is seen to improve agreement with the position and magnitude of the quasi-elastic peak.

The optimal frame has a smaller effect in intermediate energy N-nucleus scattering since the NN amplitudes are more slowly varying with energy. Figures 14 and 15 show optimal and two-body-lab calculations of the cross section and spin observables for the 290 MeV ^{54}Fe(p, p') reaction at 10°. The solid and long-dashed curves use the RPA slab response, and the short-dashed and dotted curves use the free Fermi-gas response. At this angle the difference between the optimal and two-body-lab calculations is negligible at low ω, but increases beyond the quasi-elastic point at ~ 10 MeV. This can be understood by looking at the 10° curve in Fig. 11(a) which shows that T_L^{eff} and T_L are fairly close at low ω, but deviate substantially above about ~ 25 MeV.

The curves in Figures 13-15 were calculated using slab model response functions in Eq. (22). This procedure should be applied with caution, however, because Eq. (22) is only valid within the region of the Fermi-gas response. It is meaningless to apply it at large ω beyond the point where the Fermi-gas response vanishes, because there the amplitudes would be evaluated with an optimal momentum which is greater than the Fermi momentum. Nucleons with such momenta do not exist in the nucleus, except for a very small percentage in the tail of the momentum distribution, and these make a negligible contribution to the response. Moreover, it is equally meaningless to use the two-body-lab amplitudes in this region, because the projectile cannot transfer such large energies to a nucleon at rest without violating energy conservation. The point is that the standard reaction model based on Eq. (1) or (22) cannot be consistently applied at high ω, because it assumes that all the momentum transfer occurs on the hard collision, and this makes the high excitation region kinematically inaccessible. This is especially troublesome at small scattering angles, where most if not all of the experimental spectrum falls outside the Fermi-gas region. As we will see in the next section, the way out of this problem is to explicitly include the momentum transferred in the distortion, which changes the momentum transfer on the hard collision, and thereby permits scattering at high ω without violating energy conservation.

[‡]In these calculations the magnitude is determined using a more complicated version of N_{eff} valid for large angle scattering. The simple version of N_{eff} (Eq. 2) is based on the assumption that in the eikonal limit the projectile traverses an essentially straight-line path through the nucleus. Since this is evidently not a good approximation for 120° scattering, N_{eff} is instead calculated by assuming straight-line propogation only in the initial and final states, and allowing the projectile to change direction on the hard collison as dictated by the scattering angle. Furthermore, the total πN cross section is evaluated in the final state at the lower energy $T_L - \omega$. Thus at large angles N_{eff} depends on both θ_{lab} and ω.

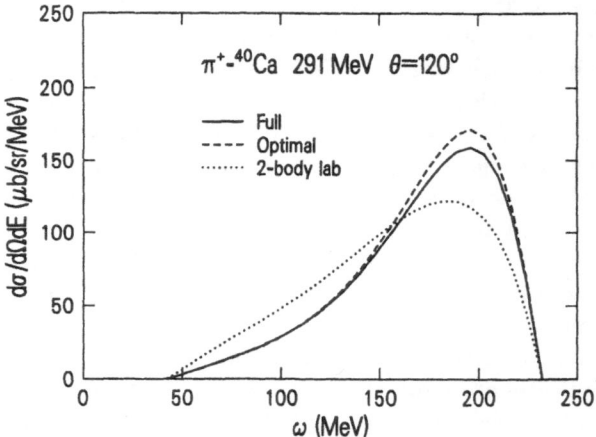

Figure 12. Pion inelasic cross sections based on Eq. (1) with the Fermi-gas response and various choices of frame for the πN amplitudes. In the solid curve, the amplitudes are inside the Fermi-momentum integration.

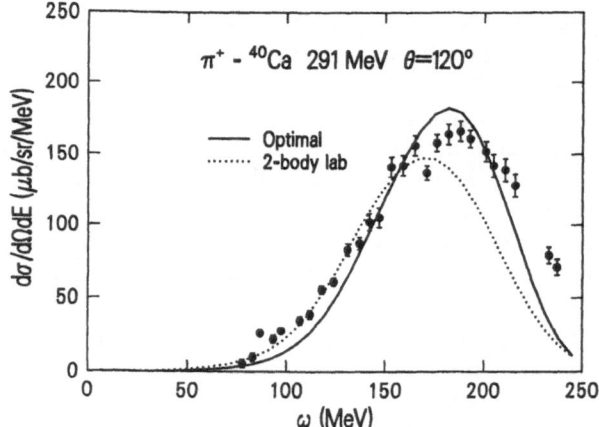

Figure 13. Slab-model calculations with optimal and two-body-lab amplitudes compared to data from Ref. [25].

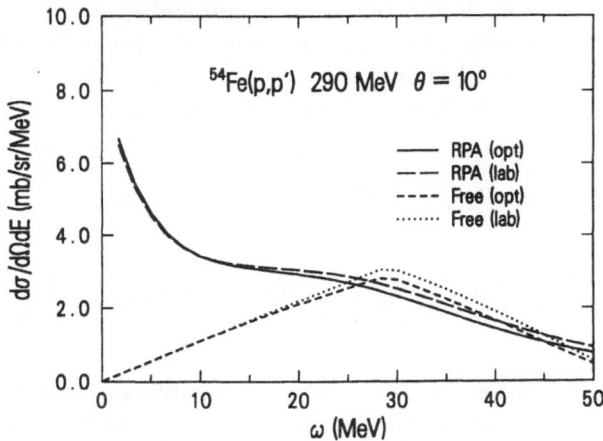

Figure 14. Cross section for proton scattering with optimal and two-body-lab amplitudes. RPA curves use the slab response, and Free curves use the Fermi-gas response.

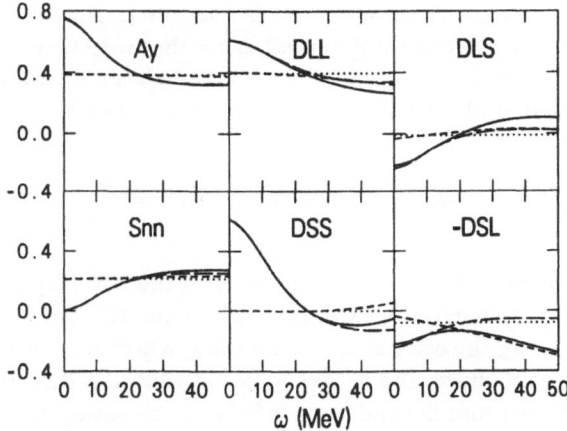

Figure 15. Spin observables for the same reaction as Fig. 14 with optimal and two-body-lab amplitudes.

RESPONSE CALCULATION WITH EIKONAL DISTORTED WAVES

In this section a method is developed for including the full distorted waves in a continuum response calculation based on the DWIA and the eikonal approximation. This will be used to examine the effects of optical and spin-orbit distortion on the cross sections and spin observables, and to address the problems with the standard reaction model raised in the last section.

In the DWIA the cross section can be expressed as

$$
\frac{d\sigma}{d\Omega dE} = S(q,\omega) = \sum_\alpha |\langle \alpha | \vartheta_{\vec{q}} | 0 \rangle|^2 \, \delta(\omega - E_\alpha)
$$

$$
\vartheta_{\vec{q}}(\vec{r}) = \int d^3r' \, \chi_{\vec{k}'}^{(-)*}(\vec{r}') \left[\int \frac{d^3q'}{(2\pi)^3} \, e^{i\vec{q}' \cdot (\vec{r} - \vec{r}')} f_{opt}(\vec{q}') \right] \chi_{\vec{k}}^{(+)}(\vec{r}') \tag{25}
$$

where α labels the excited states of the nuclear Hamiltonian. The probing field $\vartheta_q(\vec{r})$ is now more complicated than in Eq. (3), since it includes the full distorted waves $\chi^{(\pm)}$ and the two-body amplitude f_{opt}.[‡] The distorted waves can be generated by solving the Schrödinger equation for the projectile in the nuclear optical potential. In the eikonal limit they can be simply expressed as a modified plane wave

$$
\chi_{\vec{k}}^{(\pm)}(\vec{r}) = \exp\{i\vec{k} \cdot \vec{r} + iS_\pm(\vec{r})\}
$$

$$
S_\pm(\vec{r}) = \mp \frac{m}{k} \int_{\mp\infty}^{z} dz' \, \{V_c(b, z') + \vec{\sigma} \cdot \vec{b} \times \hat{z} \, kV_{so}(b, z')\} \tag{26}
$$

where $\vec{r} = (\vec{b}, z)$, and z is defined along the average momentum direction $\hat{z} \sim \vec{k} + \vec{k}'$ (note this is different than the $\hat{z}$ in Eq. (21) if $\omega > 0$). $S_\pm(\vec{r})$ is the eikonal phase, which is the integral of the optical potential along the trajectory at impact parameter $\vec{b}$ up to the point z, where the hard collision takes place. To simplify the numerical calculations the eikonal phase will be evaluated at $z = 0$ and written as

$$
S_\pm(\vec{r}) \rightarrow S_\pm(\vec{b}, 0) \equiv S(\vec{b})/2 \tag{27}
$$

which is exact if the distorted waves commute with the NN amplitudes, since $S_+ + S_-$ is independent of z, and corrections from the noncommuting spin-orbit term are small. The advantage of using the eikonal approximation, which is accurate for foward-angle scattering at energies above about 200 MeV, is that the distorted waves have no explicit angular momentum dependence, so it is not necessary to perform a multipole decomposition of the probing field.

Since the eikonal phase depends only on impact parameter, momentum is transferred in the distortion only in the two directions perpendicular to $\hat{z}$. Thus the distortion imposes a cylindrical geometry on the problem. The response in Eq. (25) can be expressed in a nonspectral representation in terms of the imaginary part of the p-h

---

[‡]In Eq. (25) a factor $C^{\frac{1}{2}}$ (see Eq. (23)) has been absorbed into the definition of f_{opt}.

propogator as follows

$$S(q,w) = \int d^2r'_\perp d^2r_\perp \, \vartheta^\dagger_{\vec{q}_\perp}(\vec{r}'_\perp) \, \frac{1}{\pi} \mathrm{Im}\, G_{ph}(\vec{r}'_\perp, \vec{r}_\perp, q_z, \omega) \, \vartheta_{\vec{q}_\perp}(\vec{r}_\perp)$$

$$\vartheta_{\vec{q}_\perp}(\vec{r}_\perp) = \int d^2b \; e^{i\vec{q}_\perp \cdot \vec{b}} \, e^{iS(\vec{b})} \int \frac{d^2q'_\perp}{(2\pi)^2} \, e^{i\vec{q}'_\perp \cdot (\vec{b}-\vec{r}_\perp)} f_{opt}(\vec{q}'_\perp, q_z) \tag{28}$$

For the moment, we have ignored the spin dependence of $S(\vec{b})$ and f_{opt} in order to simplify the equations. The $\perp$ subscripts indicate two-dimensional vectors perpendicular to $\hat{z}$, e.g., $\vec{q} = (\vec{q}_\perp, q_z)$. A factor $e^{iq_z(z-z')}$ has been extracted from the probing fields and used to Fourier transform G_{ph} to momentum space along the $\hat{z}$ direction. Note that all the momentum transfer in this direction occurs on the hard collison.

Distorted-Wave Fermi-Gas Model

So far the nuclear structure input to Eq. (25) or (28) has not been specified. It would be nice to use the slab model with RPA correlations to evaluate G_{ph}. However, this is not possible, because the slab cannot incorporate the cylindrical geometry of the distortion. If the slab response function is expressed in momentum space, it becomes apparent that momentum transfer in the distortion is allowed in only one direction: perpendicular to the slab surface. Fortunately, the most important ingredient in the slab model is the surface-peaked probe, not the use of slab wavefunctions vs. plane waves. In Ref. [9] it was shown that the slab response is well approximated by a Fermi-gas model with a surface-peaked probe, as long as k_F is evaluated at about one-third nuclear matter density. Following this method, the p-h propagator in Eq. (28) can be expressed as

$$\frac{1}{\pi}\mathrm{Im}\, G_{ph}(\vec{r}'_\perp, \vec{r}_\perp, q_z, \omega) = T^{\frac{1}{2}}(r'_\perp) T^{\frac{1}{2}}(r_\perp) \int \frac{d^2q'_\perp}{(2\pi)^2} \, e^{i\vec{q}'_\perp \cdot (\vec{r}_\perp - \vec{r}'_\perp)} R_{FG}(q', \omega) \tag{29}$$

where R_{FG} is the plane-wave Fermi-gas response evaluated at $q' = (q'^2_\perp + q^2_z)^{\frac{1}{2}}$. The $T^{\frac{1}{2}}$ factors are needed to relate the infinite to the finite geometry and to insure the response is properly normalized. They also appear in the slab model as a renormalization of the probing field. Inserting Eq. (29) into (28), and making one further approximation – that the NN amplitudes are short-ranged compared to the distance over which $T^{\frac{1}{2}}(r_\perp)$ varies significantly – we arrive at the following expression for the response

$$S(q,\omega) = \int \frac{d^2q'_\perp}{(2\pi)^2} \, R_{FG}(\vec{q}-\vec{q}'_\perp, \omega) \, |\Delta(\vec{q}'_\perp) f_{opt}(\vec{q}-\vec{q}'_\perp)|^2$$

$$\Delta(\vec{q}'_\perp) = \int d^2b \; e^{i\vec{q}'_\perp \cdot \vec{b}} \, e^{iS(\vec{b})} \, T^{\frac{1}{2}}(b) \tag{30}$$

where $\vec{q}'_\perp$ is now the momentum transferred in the distortion. If the spin and isospin dependence of f_{opt}, $S(\vec{b})$, and R_{FG} are included, then Eq. (30) is slightly more complicated and takes the form

$$S(q,\omega) = \sum_{TS} \int \frac{d^2 q'_\perp}{(2\pi)^2} \, R_{TS}(\vec{q}-\vec{q}'_\perp,\omega) \, tr\{F^\dagger_{TS}(\vec{q},\vec{q}'_\perp) F_{TS}(\vec{q},\vec{q}'_\perp)\}$$

$$F_{TS}(\vec{q},\vec{q}'_\perp) = \int d^2 b \, e^{i \vec{q}'_\perp \cdot \vec{b}} \left\{ e^{iS(\vec{b})/2} f_{TS}(\vec{q}-\vec{q}'_\perp) e^{iS(\vec{b})/2} \right\} T^{\frac{1}{2}}(b) \tag{31}$$

This equation is the central result of this section. The main new ingredients not present in the standard model of Eq. (1) are the optimal frame NN amplitudes and the use of the full optical potential in the distorted waves. Furthermore, the optimal amplitudes are now inside the integration over $\vec{q}'_\perp$, and are evaluated at the same momentum transfer as the Fermi-gas response.[‡] This insures that the optimal momentum is always less than the Fermi momentum, which resolves the problem raised in the last section. Also note that the momentum transfer is the same in f and $f^\dagger$, which is not generally true in a distorted-wave calculation. This is a consequence of the approximation that f is short-ranged compared to variations in $T^{\frac{1}{2}}$. One advantage of this approximation is that the optimal momentum given by Eq. (18) remains valid even with distortion present. If the momentum transfers in f and $f^\dagger$ are not the same, then there is more than one preferred direction, and the optimal momentum is a more complicated function which links together the integrals over the two momentum transfers, resulting in a much more difficult calculation.[24]

The simplest version of the standard model (Eq. 1) can be recovered from Eq. (30) by factoring f_{opt} and R_{FG} out of the integral and evaluating them at the external momentum transfer $\vec{q}$. Then, if the eikonal phase is evaluated using only the central absorptive potential, the distortion function Δ satisfies

$$\int \frac{d^2 q'_\perp}{(2\pi)^2} \, |\Delta(\vec{q}'_\perp)|^2 = N_{eff} \tag{32}$$

which follows from Eqs. (2), (30) and the replacement

$$\left| e^{iS(\vec{b})} \right|^2 \rightarrow e^{-2\frac{m}{k}\int_{-\infty}^{\infty} dz \, \mathrm{Im} V_o} \approx e^{-\sigma T(b)} \tag{33}$$

where the approximate equality follows from the impulse approximation in the limit of a zero-range NN interaction.

A less severe approximation is to only factor f_{opt} out of the integral and leave R_{FG} inside. Then we obtain a surface response function similar to the slab-model response

$$S(q,\omega) = \int \frac{d^2 q'_\perp}{(2\pi)^2} \, R_{FG}(\vec{q}-\vec{q}'_\perp,\omega) \frac{|\Delta(\vec{q}'_\perp)|^2}{N_{eff}} \tag{34}$$

where we have set $f_{opt} \rightarrow 1$ and divided by N_{eff} so that, as in the slab model, the

---

[‡]There is some ambiguity in how to evaluate the optimal amplitudes in Eq. (30) or (31), since the initial-state distortion changes the incident momentum for the hard collision. To completely specify the frame the amplitudes will always be evaluated by assuming half the momentum transfer occurs in the initial state and half in the final state.

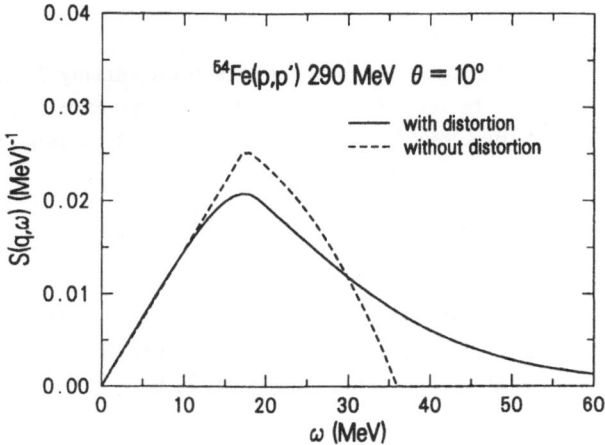

Figure 16. Surface response function calculated from Eq. (34). The dashed line is the plane-wave Fermi-gas response. In the tail region above ~36 MeV the momentum transfer on the hard collision is always greater than the external momentum transfer q.

ω-integrated response is normalized to unity in the absence of Pauli blocking. This formula is similar to the Fermi-gas version of the slab response in Ref. [9], except that it allows for two-dimensional rather than one-dimenional momentum transfer in the distortion. An example of the response calculated with Eq. (34) is shown in Fig. 16 for 290 MeV protons scattered at 10° using $\sigma = 23$ mb. For comparison, the plane-wave Fermi-gas response $R_{FG}(q, \omega)$ is shown by the dashed line. In both curves R_{FG} is evaluated at a density $\rho_o = 0.058$ fm^{-3}, which corresponds to $k_F = 0.95$ fm^{-1}. The main feature of this calculation is that the distortion produces a long tail in ω which is not present in the plane-wave response. The reason for this is clear from Eq. (34): the convolution integral spreads the plane-wave response over many values of $\vec{q}$. The same effect is responsible for the tail in the slab-model response. An important consequence is that beyond the region where the plane-wave response vanishes, there is no contribution to the integral if $|\vec{q} - \vec{q}''_{\perp}| < |\vec{q}|$, because R_{FG} continues to vanish at smaller momentum transfers. Therefore, in the tail region the momentum transfer on the hard collision is always *greater* than the external momentum transfer. As we will see, this can have a large effect on the observables if the two-body amplitudes vary rapidly with increasing $\vec{q}$.[‡]

[‡]Another consequence is that the convolution integral washes out the effect of Pauli blocking, which diminishes with increasing q. This may help explain why the ratio of the forward-angle (n, p) to (p, n) strength in Fig. (4) is larger than predicted by the theoretical curves. There the ratio is small largely because of Pauli blocking, which enhances (p, n) and reduces (n, p) if $N > Z$. The convolution integral will increase the ratio toward Z/N, which is the value obtained in the absence of Pauli blocking.

Interacting Fermi-Gas

RPA correlations can be introduced by using the interacting Fermi-gas model to evaluate R_{TS} in Eq. (31). The RPA Green's function can be expressed in terms of the residual interaction and the Green's function for the noninteracting system by

$$G_{rpa} = G_o \left(1 + V_{ph}G_o\right)^{-1} \tag{35}$$

If the interaction is local this reduces to a simple algebraic equation in the Fermi-gas model, since momentum is a good quantum number. The RPA response can be expressed in terms of the real and imaginary parts of G_o by

$$G_o(q,\omega) = \rho_o \left[P_o(q,\omega) + i\pi R_o(q,\omega)\right]$$

$$R_{TS}(q,\omega) = \frac{1}{\pi\rho_o} \text{Im}\, G_{TS}(q,\omega)$$

$$= \frac{R_o(q,\omega)}{[1 + \kappa_{TS}(q)P_o(q,\omega)]^2 + [\kappa_{TS}(q)\pi R_o(q,\omega)]^2} \tag{36}$$

where R_o and P_o are given by the Linhard functions [26], and $\kappa_{TS} = \rho_o V_{TS}$. In the isoscalar ($TS = 00$) channel the residual interaction is given by

$$\kappa_{00}(0) = -1/P_o(0,0)$$

$$\kappa_{00}(q) = \kappa_{00}(0)\frac{\Lambda}{q^2 + \Lambda} \tag{37}$$

The value at $q = 0$ is determined from a self-consistency requirement which insures the system saturates at the correct density ρ_o. It places a pole in the isoscalar response at zero frequency which corresponds to the collective mode associated with a static displacement of the entire system. If $\kappa_{00}(0)$ were larger, the pole would move to negative frequency, and the system would collapse. At finite q, κ_{00} is multiplied by a form factor with cutoff $\Lambda = 2.5$ fm^{-2} to account for the finite range of the interaction.

In the spin and isospin channels the values of V_{TS} given in Table 1 will be used. In these channels there are two branches which contribute to the response. One is from the region $R_o \neq 0$, and the other is a collective zero-sound mode which occurs when $R_0 = 0$ and $1 + \kappa_{TS}P_o = 0$. If the interaction is repulsive, these conditions can be satisfied at small values of q where P_o is negative. However, this branch will not be included in the present calculations, because it makes a very small contribution to the surface response. It falls off rapidly with increasing q and is washed out by the convolution integral in Eq. (34). The only noticeable effect of the zero-sound mode is in the isovector ($TS = 10$) channel where the residual interaction is largest, but even there the only contribution is at very small q and lies entirely below $\sim$11 MeV excitation.

Figure 17 shows a calculation of the RPA surface response based on Eq. (34) with R_{FG} replaced by R_{TS} from Eq. (36) and $\rho_o = 0.058\ \mathrm{fm}^{-3}$. For comparison with the slab response, the calculation was performed at the same energy and angle as in Fig. (8), again using $\sigma=23$ mb. The results are very similar to the slab response: in the spin-isospin channels the strength is pushed to higher excitation energy, and in the 00 channel the response is enhanced at low excitation, reflecting the pole at $q, \omega = 0$.

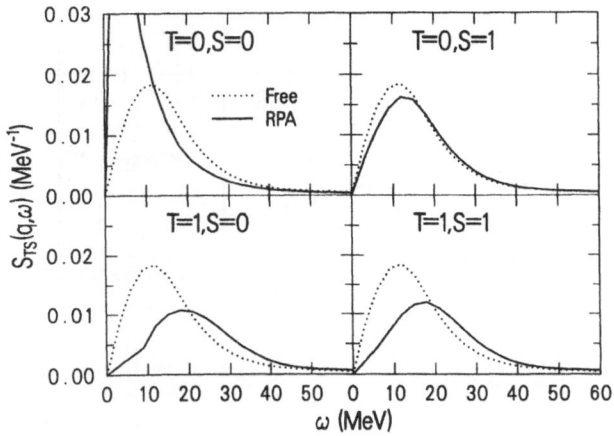

Figure 17. RPA surface response in the distorted wave Fermi-gas model (Eqs. 34,36).

Returning to Eq. (31), we are now in a position to calculate cross sections and spin observables with the effects of distortion and RPA correlations included. Figure 18 shows sample calculations of spin observables for the (p, p') reaction near 300 MeV. The dotted curves are calculated using only the optimal NN amplitudes (that is, they are based on Eq. (1) with the free response function, and since the spin observables are ratios of spin-dependent to spin-averaged cross sections, they are independent of both the response and N_{eff}), the dashed curves are based on Eq. (31) using the free response $R_{FG} = R_o$, and the solid curves are the full results using the RPA response. Figure 18(a) shows results for ^{54}Fe at 20°. The central and spin-orbit potentials in this case were calculated from the impulse approximation using the Breit-frame amplitudes, and the imaginary terms were reduced by ~25% to account for the effect of Pauli blocking. Comparing the dotted and dashed lines, we see that at this large angle the effects of the distortion are small, while the solid lines show the RPA correlations have a sizeable effect on the analyzing power and the spin-flip probability Snn.

The effects are much larger at smaller angles, as seen in Fig. 18(b) which shows the results for ^{40}Ca at 7° using the phenomelological optical potential from Ref. [27]. In this case the difference between dotted and dashed curves is mostly due to the

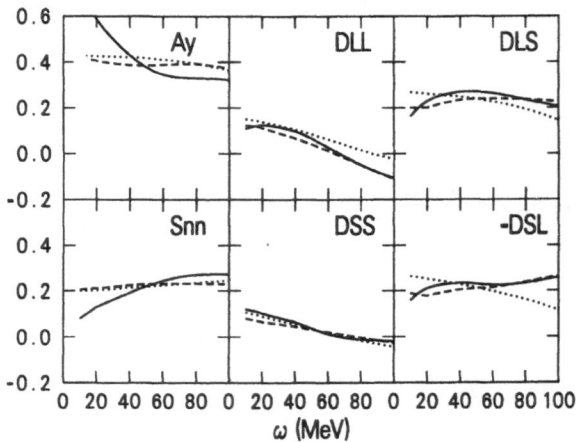

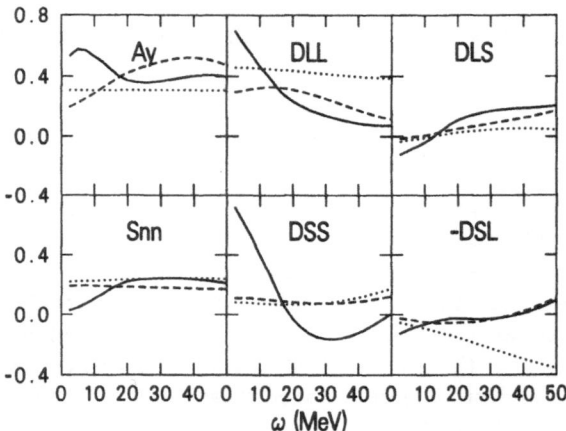

Figure 18. (p,p') spin observables for ^{54}Fe at 20° (a) and ^{40}Ca at 7° (b) calculated with Eqs. (31) and (36). Dotted lines are the free NN results, dashed lines use the full distortion with the free response, and solid lines use the RPA response.

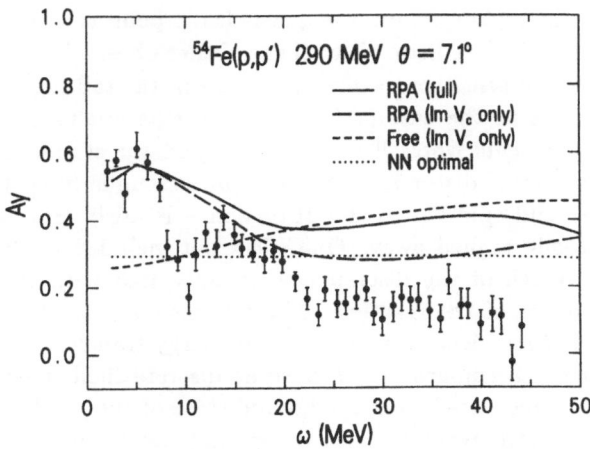

Figure 19. Effect of the distortion and RPA correlations on the forward-angle analyzing power.

momentum transfer in the distortion, which, as discussed above, causes the NN amplitudes to be evaluated on the average at a higher momentum transfer at large ω. This effect is more prominent at forward angles, because the amplitudes are more rapidly varying with q, and because the onset of the "tail" region occurs at lower ω. Some spin observables, such as the analyzing power, are also sensitive to the spin-flip in the distortion induced by the spin-orbit potential, while others, such as the cross section and the spin-flip probability, are not.

To see the magnitude of the the various effects, consider the analyzing power calculation in more detail. Figure 19 shows the analyzing power for the 290 MeV ^{54}Fe(p, p') reaction at 7.1° along with data from Ref. [3]. The dotted line again shows the free NN analyzing power using the optimal amplitudes. The short-dashed line is based on Eq. (31) using the free response and only the imaginary central potential in the distorted waves. The short-dashed and dotted curves differ substantially at large ω, because the distortion causes the free analyzing power to be evaluated at larger momentum transfers in the tail region. This effect is especially large in this case, because the free analyzing power increases linearly with q at small angles. Unfortunately it causes the calculation to move away from the data. The situation is improved if the RPA correlations are turned on, as shown by the long-dashed curve, where again only the absorptive potential is used in the distortion. The solid curve shows the RPA result with the full optical potential. Comparing the solid and long-dashed curves, we that the effect of spin-flip in the distortion causes the analyzing power to increase even further at large ω. This effect occurs mainly in the tail region, and is associated with the fact that the momentum transfer in the distortion is oriented *opposite* to the external momentum transfer in this region. On the other hand there is no restriction on this direction in the region around the quasi-elastic point $\omega = q^2/2m$ near 5 MeV, so the spin-orbit potential has little effect there. Other observables, such as the cross section and Snn, are not sensitive to this effect even in the tail region.

It appears, at least in the case of the forward-angle analyzing power, that the description of the data gets worse if we improve upon the standard reaction model by including the effects of distortion. Yet the physics underlying the effects – the momentum transfer and spin-flip in the distortion – is fairly transparent, and cannot, I believe, be easily argued away. One effect not included in these calculations, which may remove much of the discrepancy, is associated with the Q-value of the spin-transfer component of the reaction. The response is actually a function of the excitation energy of the nucleus, not the external energy transfer ω. In simple models such as the slab or the Fermi-gas, which have no discrete shell structure, the excitation energy should be measured from the ground state of the *final* nucleus. If spin or isospin is transferred in the reaction, then the ground-state energy of the final nucleus is higher than that of the initial nucleus, and the response in each channel should be shifted by the difference, which is minus the Q-value for that channel. The response is therefore zero for $\omega < -Q$. The Q-value for the spin-transfer reactions can be estimated from the data for Snn, which appears to vanish below about 5 MeV in ^{54}Fe (see Fig. 2). In the RPA, the analyzing power at large ω is smaller than the free value because the spin response functions dominate over the isoscalar ($TS = 00$) response. If they were shifted out by 5 MeV, they would dominate even more at large ω, and the analyzing power would drop further. This effect could also explain the observed enhancement of Snn seen in experiments on both ^{54}Fe [3] and ^{40}Ca [4] compared to slab-model calculations.

Relativistic m^* Effect

The distorted-wave model presented here can easily be extended to include modifications of the NN amplitudes due to the relativistic effective mass m^* as proposed by Horowitz and Iqbal.[1] They incorporated the effect of the large scalar potential predicted in the Dirac formalism by employing Dirac spinors with enhanced lower components characterized by the effective mass $m^*(r) = m + S(r)$, where the scalar potential $S(r)$ is evaluated at an average radius appropriate for the surface-peaked reaction. To include this effect we need only calculate the optimal amplitudes using spinors which depend on m^* rather than m. That is, Eq. (21) is replaced by

$$ f_{opt} = \bar{u}_1(\vec{k}', m^*)\bar{u}_2(\vec{p}'_{opt}, m^*)\,\mathcal{F}(s,t)\,u_2(\vec{p}_{opt}, m^*)u_1(\vec{k}, m^*) \qquad (38) $$

Following the prescription given in Ref. [1], the invariant amplitude $\mathcal{F}$ is evaluated with s and t determined from the external kinematics ($m^* = m$), while the Fermi-gas response functions R_{TS} are evaluated for nucleons of mass m^*.

Figure 20 shows calculations for the 290 MeV ^{54}Fe(p, p') reaction, again at 20°, along with data for the complete set of spin observables which were recently measured at TRIUMF.[28] The dashed lines are based on Eqs. (31) and (38) with $m^* = 0.85m$,[28] and the solid lines are the $m^* = m$ results which were also shown in Fig. 18(a). We see that the m^* effect improves the agreement with data for Ay, DLL and DSS, it leaves Snn unchanged, and although it does not really improve the agreement with DLS and DSL, it at least moves the curves in the right direction. However, it would be premature to say this is clear evidence in favor of the relativistic model, in view of the fact that at 500 MeV the spin observables, with the exception of Ay, are either unaffected or are in worse agreement with the data compared to nonrelativistic predictions.[1] More theoretical work and experimental data is needed before this question can be finally settled.

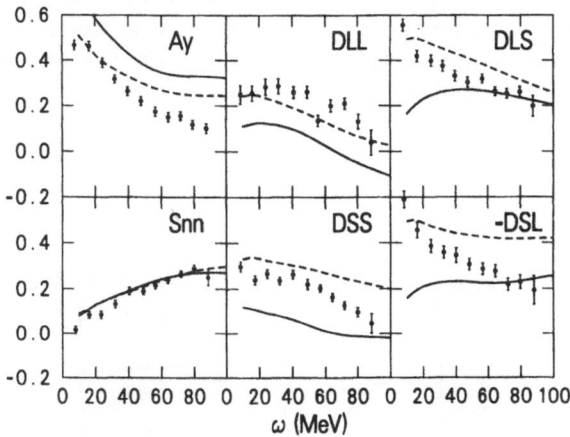

Figure 20. Spin observables with and without the relativistic m^* effect compared to recent data from TRIUMF [28] for 290 MeV ^{54}Fe(p, p') at 20°. Solid curves use $m^* = m$ and dashed curves use $m^* = 0.85m$.

Longitudinal and Transverse Spin Response

The final topic I wish to address using the distorted-wave Fermi-gas model is the question of collectivity in the longitudinal vs. transverse isovector spin response, which has been a subject of some controversy in the last few years.

Theoretical models based on $\pi + \rho$ exchange predict that at moderate momentum transfers V_{ph} should be substantially different in the longitudinal $(\vec{\sigma} \cdot \vec{q})$ and transverse $(\vec{\sigma} \times \vec{q})$ channels.[29] At momentum transfers in the region $q \sim 1 - 2$ fm^{-1} the longitudinal interaction is attractive, because the short-range repulsion is overcome by long-range attraction due to π exchange. In the same region the transverse interaction is repulsive, because it is governed by ρ exchange, and the long-range attraction does not dominate until much higher q due to the larger mass of the rho meson. The theory therefore predicts that between 1 and 2 fm^{-1} the response should be very different in these two channels. Calculations in infinite nuclear matter show that, depending on ω, the longitudinal response S_L can be $\sim 2 - 10$ times as large as the transverse response S_T. On the other hand, recent measurements at LAMPF [5] of the complete

(p, p') spin observables on ^{208}Pb at $q = 1.75$ fm^{-1} have shown that the ratio S_L/S_T is consistent with unity throughout the region of the quasi-elastic peak.

This problem has been addressed by various authors [30,31] and there are essentially three effects which have been employed to explain the discrepancy without abandoning the $\pi + \rho$ model; these are: 1) mixing of the longitudinal and transverse channels, which occurs in a finite nucleus where momentum is not a good quantum number; 2) the use of surface rather than volume response functions for the (p, p') reaction; and 3) incorporating the effect of the isoscalar ($TS = 01$) background, which in principle is present in the experimental ratio. The last two effects can be investigated using the distorted-wave model of Eq. (31). To do this we use the same interaction in the longitudinal and transverse channels as was used in Refs. [29,30] (e.g. $g' = 0.7$). Furthermore, we include in the Fermi-gas response ΔN^{-1} as well as NN^{-1} excitations in the same way as described in these references.

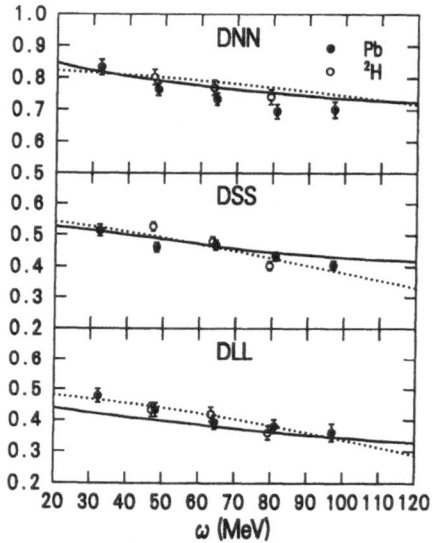

Figure 21. Spin observables at 500 MeV 18.5° which are needed to extract the ratio S_L/S_T. Data are from Ref. [5].

Figure 21 shows calculations of the spin observables DNN, DSS and DLL (which are needed to extract the ratio S_L/S_T), along with data from Ref. [5]. The observables depend individually on the non-spin response functions, and in these channels V_{ph} is taken from Table 1. In the $TS = 01$ channel the residual interaction is small but poorly known, and is taken to be zero as in Ref. [30]. The distorted waves were

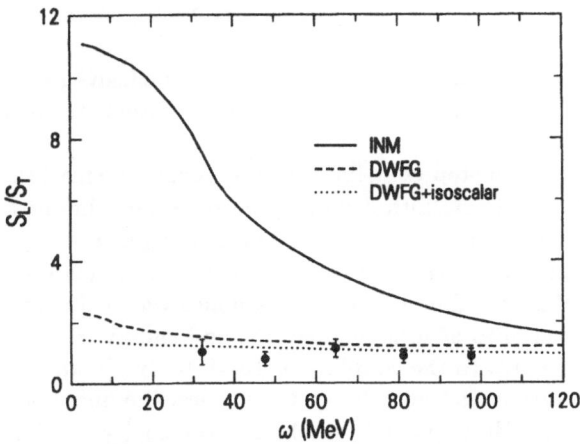

Figure 22. Ratio of longitudinal to transverse spin response along with data from Ref. [5]. The solid line is the infinite nuclear matter result, and the dashed and dotted lines are calculated in the distorted-wave Fermi-gas model (Eq. 31) as described in the text.

calculated using the ^{208}Pb optical potential from Ref. [32]. Dotted lines show the free NN spin observables in the optimal frame. Note that the ω-dependence of the observables is nicely predicted by the optimal-frame amplitudes. The solid lines are the full calculation with the effects of distortion and RPA correlations included. Neither of these effects drastically alters the results, so it is clear that the ratio σ_L/σ_T of the longitudinal to transverse cross sections will be close to the free value.

Figure 22 shows the ratio S_L/S_T. The solid line is the pure ratio calculated in infinite nuclear matter, and the dashed and dotted curves are determined from

$$\frac{S_L}{S_T} = \frac{\left(\sigma_L/\sigma_T\right)^{^{208}\mathrm{Pb}}}{\left(\sigma_L/\sigma_T\right)^{\mathrm{NN}_{opt}}} \tag{39}$$

using Eq. (31) to calculate σ_L and σ_T for ^{208}Pb. In the dashed line the isoscalar ($TS = 01$) response is set to zero, and the dotted line is the full result including the isoscalar background. We see that the effect of distortion is to completely wash out the differences in S_L and S_T predicted in nuclear matter. Most of this effect is from the absorptive term in the optical potential, although the other terms also play a role in reducing the ratio, and part of it is due to the fact that the amplitudes are inside the convolution integral in Eq. (31). It is interesting to note that these results are almost identical if the relativistic m^* effect is included. The reason is that while DSS and DLL are increased by the m^* effect, σ_L and σ_T depend on the difference DSS−DLL, which is not strongly effected by m^*.

CONCLUSION

In this talk we have investigated several aspects of inclusive projectile-nucleus scattering to the continuum, which are extentions or corrections to the standard reaction model of Eq. (1).

It was found that two-step contributions are generally small in the region of the single-step quasi-elastic peak, although they become more important at larger excitation energies. They are also more important in charge-exchange reactions, and at lower incident energies where the absorptive potential is smaller due to Pauli blocking.

The effect of collisional damping of the response was included in a simple model which can be easily applied at large momentum and energy transfer. Microscopically, this damping corresponds to the coupling of 1p-1h to 2p-2h states. Due to the quantum coherence of particle and hole decay amplitudes, the final state interaction must include the scattering of both particle and hole in the nuclear optical potential, as well as their interference. In the continuum region this interference depends on the energy, spin and isospin of the excited p-h pair, and as a result the damping is different in each channel. In channels involving spin or isospin transfer the damping is relatively large, but it is small in the isoscalar non-spin channel due to a cancellation between direct and interference diagrams. The model was applied to the forward-angle $^{54}Fe(n,p)$ reaction, and the effect of damping was to bring the RPA slab-model calculations into good agreement with the data.

We also investigated the question of how best to evaluate the two-body amplitudes which describe a quasi-elastic collision. To date, most continuum calculations based on the PWIA or the DWIA have used either the c.m. amplitudes, the two-body-lab amplitudes or the Breit-frame amplitudes. However, these choices violate energy conservation at the two-body level and cannnot be consistently applied at arbitrary energy transfers. The "optimal" choice of frame is determined by by choosing a value for the struck nucleon's momentum which satisfies energy conservation and lies along the direction of momentum transfer $\vec{q}$. The main effect of the optimal frame is that the two-body energy varies rapidly with energy transfer. This can have a large effect on the observables if the two-body amplitudes have a strong energy-dependence.

Finally, a method was developed for including the effects of spin-dependent distortion in a continuum response calculation based on the eikonal approximation and the DWIA. The main feature of the model is that the momentum transfer in the distortion is explicitly included, and this changes the momentum transfer on the hard collision. It is essential to include this effect in the tail region beyond the plane-wave response, which would be kinematically inaccessible if all the momentum transfer occured on the hard collision. In general, the effects of momentum transfer and spin flip in the distortion were found to be largest at forward angles and high excitation energies. The standard reaction model therefore cannot be applied in these regions. It is only valid at large $\vec{q}$ (above about ~ 1.5 fm) at excitation energies near the quasi-elastic point $\omega = q^2/2m$. In this region all the effects we have discussed are relatively small, including RPA correlations, which, as was seen in the calculations of longitudinal and transverse spin response, are washed out by the effects of the distortion.

The numerical results shown here have have for simplicity employed either the slab or the Fermi-gas model for the nuclear structure. It should be emphasized, however, that the techniques are general and can be used with more sophisticated structure input if desired. In particular, the method for including 2p-2h damping based on Eqs. (11-16), and the distorted-wave model of the continuum response in Eq. (31) can be applied to any version of the plane-wave RPA response. One need only calculate the plane-wave response on a grid in q or ω, and then perform the necessary convolution integral, as in Eq. (11) or (31). It is also easy to merge the 2p-2h damping and distorted-wave effects into one calculation by applying the damping integral to the response R_{TS} before inserting it into the distortion integral. When these effects are combined, the result is a simple yet comprehensive model for continuum scattering, which includes the spin-dependent distortion of the projectile, and the effects of RPA correlations and collisional damping which describe the struck nucleon's final state interaction.

REFERENCES

[1]   C. J. Horowitz and M. J. Iqbal, Phys. Rev. **C33** (1986) 2059;
      C. J. Horowitz and D. P. Murdock, preprint.
[2]   M. C. Vetterli, *et al.*, Phys. ReV. Lett. **59** (1987) 439.
[3]   O. Häusser, Can. J. Phys. **65** (1987) 691.
[4]   C. Glashausser, *et al.*, Phys. Rev. Lett. **58** (1987) 2404.
[5]   T. A. Carey, *et al.*, Phys. Rev. Lett. **53** (1984) 144.
[6]   E. J. Moniz, *et al.*, Phys. Rev. Lett. **26** (1971) 445.
[7]   R. D. Smith and S. J. Wallace, Phys. Rev. **C32** (1985) 1654.
[8]   R. D. Smith and J. Wambach, Phys. Rev. **C36** (1987) 2704.
[9]   H. Esbensen and G. F. Bertsch, Ann. Phys. **157** (1984) 255.
[10] H. Esbensen and G. F. Bertsch, Phys. Rev. **C34** (1986) 1419.
[11] S. Drożdż, V.Klempt, J. Speth, J. Wambach, Nucl. Phys. **A451** (1986) 11.
[12] C. Gaarde, *et al.*, Nucl. Phys. **A365** (1981) 258.
[13] S. Yen, *et al.*, to be published.
[14] S. Drożdż, G. Co', J. Wambach, J. Speth, J. Lett **B185** (1987) 287.
[15] R. D. Smith, J. Wambach, to be published in Phys. Rev. C.
[16] S. Nishizaki, S. Drożdż, J. Speth, J. Wambach, to be published.
[17] G. F. Bertsch, P. F. Bortignon, R. A. Broglia, Rev. Mod. Phys. **55** (1983) 287.
[18] C. Mahaux, H. Ngô, Phys. Lett. **B100** (1981) 285.
[19] C. H. Johnson, P. J. Horen, C. Mahaux, Phys. Rev. **C36** (1987) 2252.
[20] G. Co', K. F. Quader, R. D. Smith, J. Wambach, University of Illinois preprint.
[21] S. A. Gurvitz, Phys. Rev. **C33** (1986) 422.
[22] R. A. Arndt, *et al.*, Phys. Rev. **D28** (1983) 97.
[23] J. A. McNeil, L. Ray, S. J. Wallace, Phys. Rev. **C27** (1983) 2123.
[24] R. D. Smith and E. R. Siciliano, to be published.
[25] G. R. Burleson, *et al.*, Phys. Rev. **C21** (1980) 1452.
[26] J. Linhard, Mat.-Fys. Medd. Danske Vid. Selsk. **28** (8) (1954).
[27] E. D. Cooper, *et al.*, Phys. Rev. **C36** (1987) 2170.
[28] O. Häusser, *et al.*, submitted to Phys. Rev. Lett.
[29] W. M. Alberico, M. Ericson, A. Molinari, Nucl. Phys. **A379** (1982) 429.
[30] H. Esbensen, H. Toki, G. F. Bertsch, Phys. Rev. **C31** (1985) 1816.
[31] W. M. Alberico, *et al.*, Phys. Lett. **B183** (1987) 135;
      see also talks in these proceedings by M. Ichimura, W. Alberico,
      and G. Chanfray.
[32] A. M. Kobos, *et al.*, Nucl. Phys. **A445** (1985) 605.

RELATIVISTIC CALCULATIONS FOR QUASI-ELASTIC PROTON SCATTERING

M. J. Iqbal

TRIUMF
4004 Wesbrook Mall
Vancouver, BC, Canada V6T 2A3

Quasi-elastic scattering on nuclei implies scattering process that has strong relationship to free nucleon-nucleon scattering. Thus, it is one of the cleanest reaction to investigate both the single-particle properties (shell structure) of a nucleus and the nuclear medium effects on the free nucleon-nucleon interaction. In this article we will present calculations for spin observables for quasi-elastic proton scattering, calculated in a simple relativistic model.[1]

In a relativistic approach the NN interaction depends upon four component Dirac wave functions. In the presence of strong scalar (S) and time-like vector (V) nucleon-nucleus optical potentials[2] the lower components of these wave functions are enhanced. This enhancement of the lower components of the wave functions can be characterised by an effective mass m^* which is smaller than the free value m. The effective mass m^* can be calculated as follows.

Dirac equation for the incident proton, under the influence of the scalar and time-like vector optical potentials, can be written as,

$$\left[-i\vec{\alpha} \cdot \vec{\nabla} + \beta \left(m + S(r,E) \right) - \left(E - V(r,E) \right) \right] \psi_{\vec{K},S}^{\pm}(r) = 0 . \tag{1}$$

Here superscripts ($\pm$) correspond to ($^{\text{incoming}}_{\text{outgoing}}$) distorted waves. Coulomb interaction has been ignored. In the eikonal approximation the wave function $\psi_{\vec{K},S}^{\pm}(r)$ can be written as[3]

$$\psi_{\vec{K},S}^{\pm}(\vec{r}) = \left(\frac{E+m}{2m} \right)^{1/2} \left(\begin{array}{c} 1 \\ \dfrac{\vec{\sigma} \cdot \vec{K}}{(E - V(r) + m + S(r))} \end{array} \right) e^{i\vec{K}\cdot\vec{r}} \, e^{iS^{\pm}(\vec{r})} \chi_S , \tag{2}$$

where the eikonal phase factor $S^{\pm}(\vec{r})$ can be written in an integral form as

$$S^{\pm}(\vec{b},z) = -\frac{m}{K} \int_{\pm\infty}^{z} dz' \left\{ V_c(\vec{b},z') + V_{s.o}(\vec{b},z')(\vec{\sigma} \cdot \vec{b} \times \vec{K} \, - iKz') \right\} , \tag{3}$$

where $\vec{K}$ is the average momentum defined in terms of initial ($\vec{k}$) and initial ($\vec{k}'$) momenta as

$$\vec{K} = \frac{1}{2}(\vec{k} + \vec{k}') . \tag{4}$$

The central (V_c) and spin-orbit ($V_{s.o.}$) potentials are defined in terms of S and V as

$$V_c(r) = S(r) + \frac{E}{m} V(r) + \frac{1}{2m}(S^2(r) - V^2(r))$$

$$V_{s.o.}(r) = -\frac{1}{2mr} \frac{\frac{\partial}{\partial r}(S - V)}{(E - V) + m + S}$$

$$r = \sqrt{b^2 + z^2} . \tag{5}$$

The transmission probability for going through the nucleus at an average impact parameter b is given in terms of eikonal phase factor $S(b, z)$ as

$$T(b) = \left| e^{iS^+(b,\infty)} \right|^2 . \tag{6}$$

Ignoring that spin-orbit interaction (effect of spin-orbit distributions on spin-observables is discussed in Ref. 1) we can write

$$T(b) = \exp\left(\frac{4m}{k} \int_0^\infty dz\, Im\, V_c(b, z) \right) . \tag{7}$$

Given the transmission probability $T(b)$ one can construct the average density for forward angle scattering from a given nucleus by integrating over all the impact parameters as,

$$\langle \rho \rangle = \frac{\int b\, db\, T(b) \int dz\, \rho_B^2(b, z)}{\int b\, db\, T(b) \int dz\, \rho_B(b, z)} . \tag{8}$$

Here $\rho_B(b, z)$ is the baryon density for a given nucleus.

If we assume that the nuclear scalar and vector optical potential scale as nuclear density i.e.,

$$S(r, E) = S_0(E) \frac{\rho(r)}{\rho(0)}$$

$$V(r, E) = V_0(E) \frac{\rho(r)}{\rho(0)} , \tag{9}$$

where $S_0(E)(V_0(E))$ is the strength of the scalar (vector) optical potential at origin, then the scalar and vector optical potential at the average density $\langle \rho \rangle$ can be defined as

$$\bar{S}(E) = S_0(E) \frac{\langle \rho \rangle}{\rho(0)} \tag{10}$$

$$\bar{V}(E) = V_0(E) \frac{\langle \rho \rangle}{\rho(0)} , \tag{11}$$

we then can define an effective mass m_1^* for the projectile

$$m_1^* = m + \bar{S}(E) , \tag{12}$$

and effective energy

$$E_1^* = E - \bar{V}(E) = (p_1^2 + m_1^{*2})^{1/2} . \tag{13}$$

Here p_1 is the asymptotic momentum of the projectile. For the target nucleon one can use an effective scalar interaction from Walecka's mean field theory,[4] and define an effective mass m_2^* as

$$m_2^* = m + S_{mft} \frac{\langle \rho \rangle}{\rho(0)} , \tag{14}$$

where $S_{mft} = -0.44$ m at $\rho(0)$. The corresponding energy E_2^* is defined as

$$E_2^* = (p_2^2 + m_2^{*2})^{1/2} . \tag{15}$$

The Dirac wave function for nucleon of mass m^* and energy E^* is

$$u(\vec{p}, s) = \left(\frac{E^* + m^*}{2m^*}\right)^{1/2} \left(\begin{array}{c} 1 \\ \dfrac{\vec{\sigma} \cdot \vec{p}}{E^* + m^*} \end{array}\right) \chi_s . \tag{16}$$

For the scattering of a projectile from a target nucleon, the invariant matrix element $\mathcal{M}$ can be written as (Fig. 1)

$$\mathcal{M} = \sum_{i=s}^{t} \bar{u}_1(k_1, s_f) \, \lambda^i \, u(p_1, s_i) \, t_i(q, T_{eff})$$

$$\bar{u}(k_2, s_2') \lambda^i \, u_2(p_2, s_2) , \tag{17}$$

where Bjorken and Drell conventions are used. The details of calculations can be found in Ref. 5. Here we compare our calculations with the most recent data on spin observables from TRIUMF.

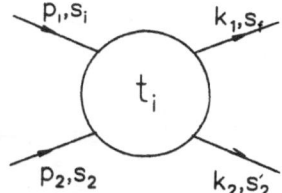

Fig. 1. Schematic diagram for nucleon-nucleon scattering

Figure 2 shows a complete spin observables for inclusive quasielastic scattering of 290 MeV protons from ^{54}Fe at a momentum transfer of 1.36 fm^{-1}, taken at TRIUMF. The energy transfer (ω) varies from 0-90 MeV. Details about the experimental procedure are given elsewhere.[6] Here we will concentrate on the comparison of the experimental data with the calculations of Smith[7] (long dashed curves) based upon non-relativistic impulse approximation (NIRA) and that of Horowitz, Iqbal and Murdock (solid lines) based upon the relativistic impulse approximation (RIA). The quasielastic peak for the above kinematics is around $\omega = 40$ MeV. An interesting feature of all the spin-observables is their energy transfer dependence. The slope for the polarization (P) is correctly described by the NRIA calculations which use an infinite slab model[8] to calculate the nuclear response. However, the magnitude of P does not agree with the experimental data. At the quasifree value NRIA calculations predict a value for P which is close to the free value (dashed curves). The RIA calculations predict a considerable ($\sim 40\%$) decrease in P compared to the free value and agree well at quasielastic peak. The relativistic calculations use Fermi-gas model for the nuclear response. The energy transfer dependence of spin observables, in RIA, arises due to Fermi motion of the target nucleons. It is clear that a model of nuclear response based upon the Fermi gas model cannot produce the correct energy transfer dependence. The reason may be that the spin observables P, A_y and D_{nn} depend sharply upon spin-scalar isospin-scalar response. The non-relativistic RPA calculations based upon the infinite slab model push the strength of $S = 0$ $T = 0$ response at lower energies, hence causing an enhancement (for these observables) at lower ω. A Fermi-gas model response does not distinguish between various spin-isospin parts of the nuclear response. So by comparing the experimental data for P with NRIA and RIA calculations we can make two observations. Firstly, P decreases considerably (compared to free values) because of the nucleon medium effects and that the RIA calculations predict this decrease in P very successfully. Secondly the ω dependence of P cannot be explained due to Fermi-motion effects alone. One needs a realistic model (which includes RPA correlations) for the nucleon response function. The NRIA calculations which use such a model can explain the ω dependence for P, A_y and D_{nn} very nicely.

The spin observable D_{nn} is explained quite well both in NRIA and RIA calculations.

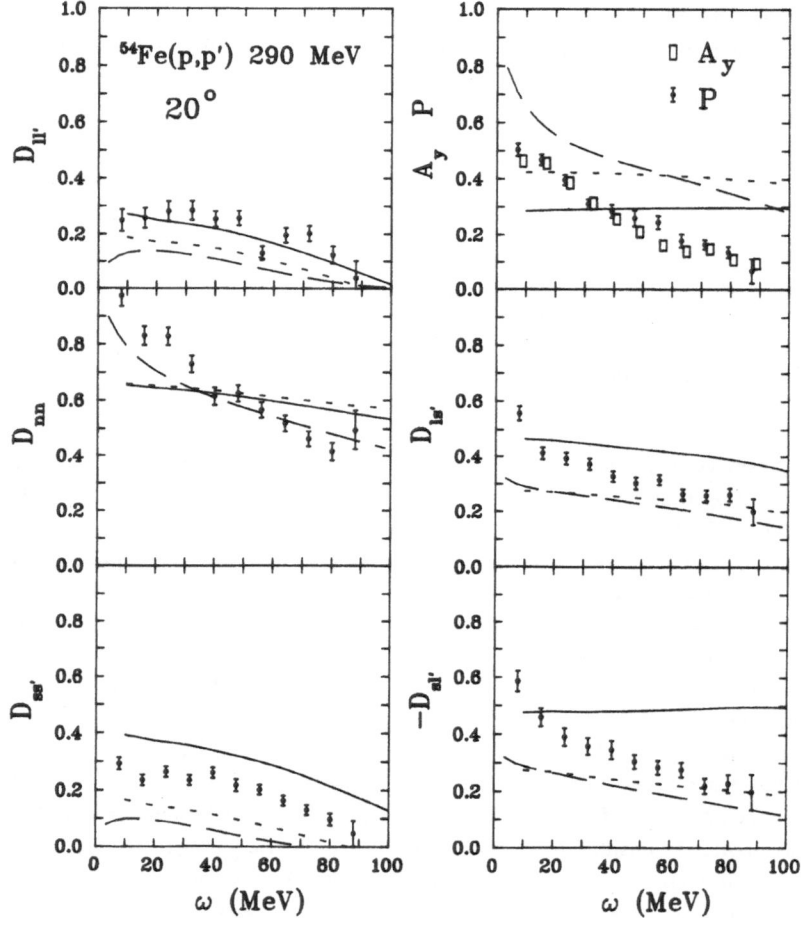

Fig. 2. Spin observables for inclusive proton scattering of 290 MeV protons from ^{54}Fe at 20°. The theoretical curves represent relativistic (solid lines) and non-relativistic (dashed lines) calculations explained in the text.

The slope of D_{nn} is again due to enhancement of isoscalar-scalar nucleon response at low ω. Fermi-gas model predicts an almost flat energy dependence. In this case the medium effects do not cause any change compared to the free values. At top of quasielastic peak RIA calculations agree well with the experiment.

Now we come to observables D_{LL}, D_{SS}, D_{SL} and D_{LS}. The slopes of these observables, both for the RIA and NRIA calculations of Smith, agree well with the experimental slopes. For these observables there is no enhancement of $T = 0$ $S = 0$ nuclear response for small ω. The Fermi averaging predicts the slopes correctly for all of these observables. In RIA calculation D_{LL} is predicted to be slightly (10–15%) enhanced compared to the free values. This is in quite a good agreement with the experimental results. The values for D_{SS} are slightly enhanced due to the nuclear medium effects. The RIA calculations predict D_{SS} to be considerably enhanced compared to the free values. The experimental data does show quite as strong an enhancement. There is a quantitative disagreement between experiment and theory. This is also true for the spin observables D_{SL} and D_{LS}. In both these cases, relativistic calculations predict an enhancement, compared to the free values. The experimental data does show this enhancement but not quite as much.

Thus we see that the experimental data provides a strong evidence for the nuclear medium effects which in the RIA calculations are due to strong scalar (S) and vector (V) potentials. In all the cases, RIA calculations predict the change from the free values in the correct direction. There is an excellent agreement between RIA calculations and experiment at the quasielastic peak for P, D_{nn} (which is predicted not to change) and D_{LL}. The enhancements for D_{SS}, D_{LS} and D_{SL} are qualitatively correct although there still are quantitative disagreements.

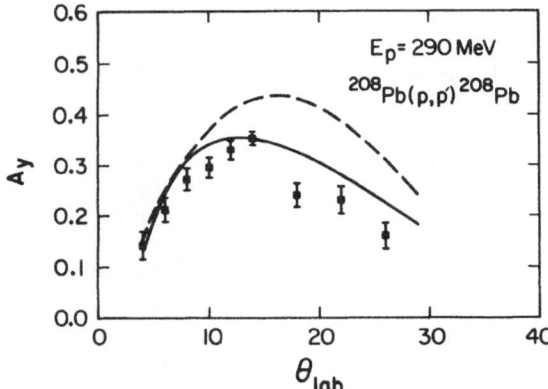

Fig. 3. Analyzing powers for the quasifree peak for 290 MeV protons scattered from ^{208}Pb. Solid curves correspond to RIA predictions for $m^* = 0.83m$ and dashed curves are the free values ($m^* = m$).

The agreement between theory and experiment for P is again observed by comparing RIA calculations with the experimental data of Chen et $al.$[8] They measured P for quasielastic scattering of 290 MeV protons for ^{208}Pb. In this case cross sections were measured over an angular range of $4° - 26°$ and for excitation energies up to 160 MeV.[1] Figure 3 shows the comparison of the experimental data with RIA calculations (solid). Also shown are the free values (dash curve) for $m^* = m$. Again we see a good agreement between experiment and RIA predictions.

We conclude that there is a strong evidence for the nuclear medium effects in quasifree nucleon-nucleon scattering. The quantitative differences between theory and experiment call for a more realistic input for the nuclear response in the relativistic calculations.

This work has been supported by the Natural Sciences and Engineering Research Council, Canada.

REFERENCES

1. C.J. Horowitz and M.J. Iqbal, Phys. Rev. C **33**, 2059 (1986)
2. See, for example, B.C. Clark, S. Hamma and R.L. Mercer, AIP Conf. Proc. No. 97, (260) 1982.
3. R. Amado *et al.*, Phys. Rev. C **28**, 1663 (1983).
4. B.D. Serot and J.D. Walecka, Advances in Nuclear Physics, Vol. 16, Plenum Press, N.Y., 1987.
5. C.J. Horowitz and D. Murdock, Phys. Rev. C **35**, 1442 (1987).
6. O. Haüsser *et al.*, (to appear in Phys. Rev. Lett.)
7. See the talk by R.D. Smith in these proceedings.
8. G.F. Bertsch and O. Scholten, Phys. Rev. C **25**, 804 (1982).
9. X.Y. Chen *et al.*, Phys. Lett. **205B**, 434 (1988).

---

[1] Thus the quasi-elastic peak varies over a range of values.

SPIN EXCITATIONS IN A SCHEMATIC MODEL

P. M. Boucher [1], B. Castel [1], Y. Okuhara [1], I. P. Johnstone [1], J. Wambach [2], and
T. Suzuki [3]

[1]Department of Physics
Queen's University
Kingston K7L 3N6 Canada

[2]Department of Physics
University of Illinois at Urbana-Champaign
Urbana-Champaign IL 61801

[3]Research Institute for Theoretical Physics
Kyoto University
Kyoto 606, Japan

Several recent results of proton inelastic scattering, some of them presented at this conference, have shed a new light on the character of the nuclear spin response [1,2]. In particular, in a recent experiment using the proton beam at LAMPF, Glashauser et al.[3] have shown that spin-flip states in ^{40}Ca are strongly excited at excitation energies above 10 MeV and that nuclear spin excitations appear *relatively suppressed at low excitation energies* but surprisingly *enhanced at higher excitation.*

The purpose of this paper is to present a simple theoretical context for these results and discuss some possible directions for future experiments. The model we will use is inspired by earlier schematic models based on energy weighted sum rules.

We first relate the nucleon-nucleon scattering amplitude to the cross section for proton-nucleus scattering. The nucleon-nucleon scattering amplitude is written as

$$M(q) = A(q) + B(q)\,(\vec{\sigma}^i\cdot\hat{n})\,(\vec{\sigma}^p\cdot\hat{n}) + C(q)\,(\vec{\sigma}^i + \vec{\sigma}^p)\cdot\hat{n} + E(q)\,(\vec{\sigma}^i\cdot\hat{q})\,(\vec{\sigma}^p\cdot\hat{q}) + F(q)\,(\vec{\sigma}^i\cdot\hat{p})\,(\vec{\sigma}^p\cdot\hat{p}), \qquad (1)$$

where the superscripts p and i denote the projectile and target nucleons, respectively, and we have used the abbreviations

$$A(q) = A_0(q) + A_1(q)\vec{\tau}^i\cdot\vec{\tau}^p, \text{ etc.} \qquad (2)$$

The choice of coordinates is as follows

$$\hat{q} = \frac{\vec{k}_i - \vec{k}_f}{|\vec{k}_i - \vec{k}_f|}, \quad \hat{n} = \frac{\vec{k}_i \times \vec{k}_f}{|\vec{k}_i \times \vec{k}_f|}, \quad \hat{p} = \hat{q} \times \hat{n}, \qquad (3)$$

---

[1] Talk presented by J. Wambach

which form an orthogonal set of axes $(\hat{x}, \hat{y}, \hat{z})$ under the assumption that $k_f \approx k_i$. The coefficients of Eq. (1) are calculated according to the prescription of Love and Franey [4]. Using their t-matrix at $E_p = 325$ MeV, we obtain

$$|A_0|^2 = 43793.9 \, fm^2 \, MeV^6 \text{ and } |B_1|^2 = |F_1|^2 = 16190.0 \, fm^2 \, MeV^6 \tag{4}$$

which correspond to the experimental conditions $E_p = 319$ MeV and $\theta = 7°$. All other coefficients are found to be negligible when compared with the above values. This fact implies that the cross section for proton-nucleus scattering is dominated by the $T = S = 0$ and $T = S = 1$ channels, under the present experimental conditions. Then, the spin-flip cross section, σS_{nn}, defined by Glashauser et al. can be written in natural units as

$$\sigma S_{nn} = \frac{m^2}{4\pi^2} |F_1|^2 \sum_n |<n|\sum_i \vec{\sigma}^i \cdot \hat{p} \tau_z^i e^{i\vec{q} \cdot \vec{r}_i}|0>|^2 \delta(\omega - \omega_n), \tag{5}$$

where $q = 0.5 \, fm^{-1}$. Here ω and ω_n denote the energy transfer and the excitation energy of the nuclear state $|n>$, respectively. Note that σ on the left hand side represents the sum of spin-flip and non-flip cross sections, while S_{nn} denotes the spin-flip probability [3]. The total cross section, σ, is given by

$$\sigma = \frac{m^2}{4\pi^2} \sum_n \left\{ |A_0|^2 |<n|\sum_i e^{i\vec{q} \cdot \vec{r}_i}|0>|^2 + |B_1|^2 |<n|\sum_i \vec{\sigma}^i \cdot \hat{n} \tau_z^i e^{i\vec{q} \cdot \vec{r}_i}|0>|^2 \right.$$

$$\left. + |F_1|^2 |<n|\sum_i \vec{\sigma}^i \cdot \hat{p} \tau_z^i e^{i\vec{q} \cdot \vec{r}_i}|0>^2| \right\} \delta(\omega - \omega_n) \tag{6}$$

We now briefly discuss our model. It is based on the assumption that a single collective state exhausts the classical sum rule value,

$$S_1(r^\lambda Y_{\lambda 0}) = \sum_\nu (E_\nu - E_0)| <\nu|r^\lambda Y_{\lambda 0}|0> |^2 = \frac{A}{2m} \frac{(2\lambda + 1)\lambda}{4\pi} < r^{2\lambda - 2} > . \tag{7}$$

As in the classical schematic model [5], all of the states which correspond to $\Delta N = 1$ and $\Delta N = 2$ states are assumed degenerate at the observed peak energies. Thus, in the $S = T = 0$ channel, the states due to the operators $r^2 Y_{20}(2^+)$ and $r^2(0^+)$ are considered. The energy of the 0^+ state is taken as 23 MeV from the empirical relationship $E^{0+} \simeq 80A^{-1/3}$ (Ref 7) and the energy of the 2^+ state is taken as 16 MeV, in order to correspond with the proton scattering result being considered. The total strength is then calculated by invoking the energy-weighted density sum rule, where the density is denoted by $\rho(\vec{r})$,

$$S_\rho(r^\lambda Y_{\lambda 0}) = \sum_\nu (E_\nu - E_0) < 0|\hat{\rho}(\vec{r})|\nu> <\nu|\sum_i r_i^\lambda Y_{\lambda 0}(r_i)|0> = -\frac{1}{2m}\nabla(\rho(\vec{r})\nabla(r^\lambda Y_{\lambda 0}(r))). \tag{8}$$

With these assumptions, the total strength in the $S = T = 0$ and $S = T = 1$ channels can be evaluated.

We now consider meson exchanges and their effect on the strength [8]. The nuclear force is assumed to be given by those one-boson exchange potentials which are spin dependent, since the spin independent part of the potential makes a vanishing contribution [9]. The form of the π and ρ exchange potentials has been taken as

$$V_\pi(r) = f_\pi^2 m_\pi \vec{\tau}_j \vec{\tau}_l \left\{ [\frac{1}{3m_\pi r} + \frac{1}{(m_\pi r)^2} + \frac{1}{(m_\pi r)^3}] e^{-m_\pi r} S_{jl} + \frac{1}{3}\vec{\sigma}_j \vec{\sigma}_l \frac{e^{-m_\pi r}}{\cdot m_\pi r} \right\} \tag{9}$$

$$V_\rho(r) = f_\rho^2 m_\rho \vec{\tau}_j \vec{\tau}_l \left\{ -[\frac{1}{3m_\rho r} + \frac{1}{(m_\rho r)^2} + \frac{1}{(m_\rho r)^3}] e^{-m_\rho r} S_{jl} + \frac{2}{3}\vec{\sigma}_j \vec{\sigma}_l \frac{e^{-m_\rho r}}{m_\rho r} \right\} \tag{10}$$

The following values have been assumed for the interactions [10]: $f_\pi^2 = 0.08$, $f_\rho^2 = 5.38$, $m_\pi = 138$ MeV, $m_\rho = 770$ MeV. This contribution includes all angular momentum values, but is appropriate at this small momentum transfer since the higher L effects are negligible. Using these values within the context of the Fermi Gas Model, we find that the additional contribution to the $S = T = 1$ sum rule value is

$$\frac{24A}{\pi \omega} k_F j_\pi^2 m_\pi \int dr j_1^2(k_F r) e^{-m_\pi r} \left(\frac{2}{3m_\pi r} [1 - j_0(qr)] - [\frac{1}{3m_\pi r} + \frac{1}{(m_\pi r)^2} + \frac{1}{(m_\pi r)^3}] j_2(qr) \right)$$

$$+ \frac{24A}{\pi \omega} k_F j_\rho^2 m_\rho \int dr j_1^2(k_F r) e^{-m_\rho r} \left(\frac{4}{3m_\rho r} [1 - j_0(qr)] + [\frac{1}{3m_\rho r} + \frac{1}{(m_\rho r)^2} + \frac{1}{(m_\rho r)^3}] j_2(qr) \right). \quad (11)$$

In order to evaluate the strength, a nuclear density with a Fermi dependence is chosen:[5]

$$\rho(r) = \rho_o \left[1 + \exp(\frac{r - R}{a}) \right]^{-1}, \quad (12)$$

where the radius and diffuseness parameters are

$R = 1.12A^{1/3} - 0.86A^{-1/3} = 3.58 \ fm$ and $a = 0.54 \ fm$.

With these parameters, the evaluation of the rescaling radial integral at $q = 0.5 \ fm^{-1}$ gives an overall enhancement of the strength in the $S = T = 1$ channel equal to 56%.

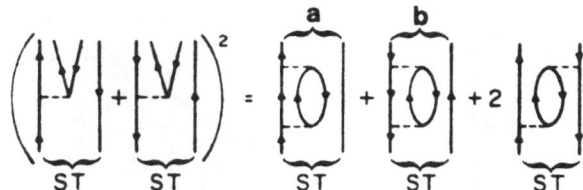

Figure 1 The 2p-2h diagrams contribution to the spreading width of the resonances

We now briefly discuss the origin of the asymmetric width obtained for the states in both channels as a result of the coupling to 2p-2h background [8].
Microscopically, the 2p-2h spreading width originates from the processes shown in Fig. 1. The total width Γ^{ST} can be obtained from the imaginary part of the optical potential and the hole widths. Assuming that the optical potential is symmetric about the Fermi energy and denoting the (a) and (b) parts of diagrams (1) and (2) by Γ_p and Γ_h we obtain the total width

$$\Gamma^{ST}(\omega) = \frac{1}{\omega} \int_0^\omega de \{ \Gamma_p(e) + \Gamma_h(e - \omega) + 2C_{ST}[\Gamma_p(e)\Gamma_h(e - \omega)]^{1/2} \}. \quad (13)$$

Thus, Γ^{ST} is completely specified by single particle data once C_{ST} is calculated, thus producing asymmetric widths [8]. The asymmetry is important in order to give a reasonable contribution at high excitation energy. With a symmetric width and a Lorentzian distribution, the high energy cross section is severely underestimated.

Several features of the results are now worthy of discussion.
i) Fig. 2 shows the calculated cross section in the $S = T = 1$ channel. It is evident that the rescaling due to meson exchange is instrumental in improving agreement with experiment, especially in the region between 15 and 25 MeV. The model underestimates the spin flip strength above 30 MeV Clearly, absorption effects and two-step processes could play a role in that region.
ii) In Fig. 3 the spin-flip probability has been plotted and compared with experiment. The enhancement of the relative response at high excitation energies is clearly a feature of the model. The origin of the oscillation can be attributed to the inclusion of the 0^+ state in the $S = T = 0$ channel. Although this state is anticipated theoretically, experimental efforts to detect it have been inconclusive.

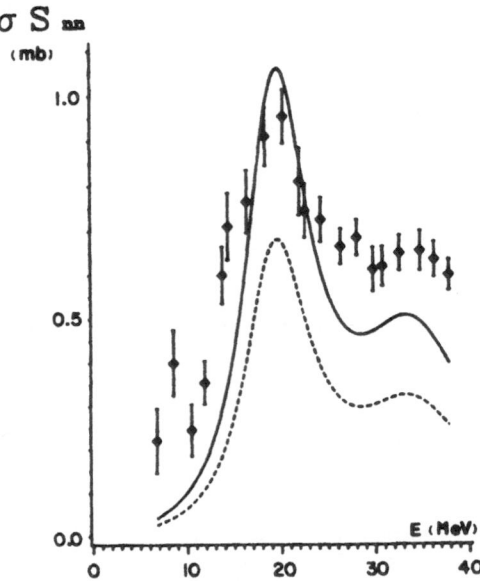

Figure 2 Comparison of the calculated cross section in the $S = T = 1$ channel with the experimental values. The dashed curve represents the cross section *without* meson exchange.

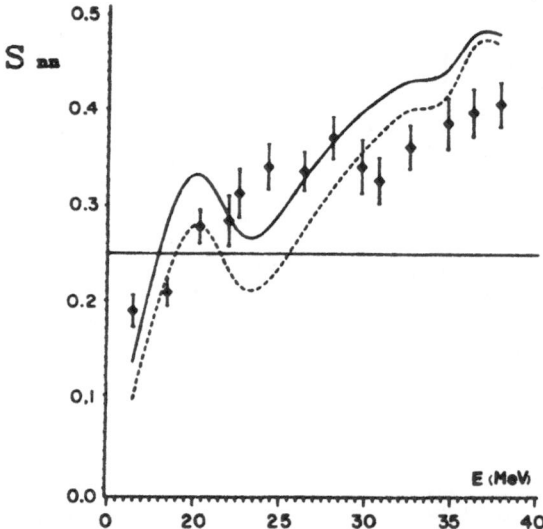

Figure 3 Calculated and experimental spin-flip probabilities displaying an increase with excitation energy. The horizontal straight line represents the prediction from free nucleon-nucleon scattering (Ref. 3) and the dashed curve represents the probability calculated *without* meson exchange.

In conclusion, we should remark that two features of the model are crucial in determining the overall shape and scale of the spectrum. The first is the inclusion of the *2p-2h* background, producing an asymmetric width which is needed to explain the high degree of strength found at higher excitation energy. The second is the inclusion of meson exchange effects without which the observed scale would not be obtained.

It would be interesting to extend the range of experimental investigation in order to observe if the increase in cross section persists beyond 40 MeV. The model *predicts an increase in the height of the second peak* (at 35 MeV) *at higher* $\vec{q}$ *values* (or at higher scattering angles). Such experiments are clearly needed as tests of sum rule models like the one above as well as of more microscopic calculations which we are currently undertaking.

Acknowledgments

The authors are grateful to G. E. Brown, H. Toki and L. Zamick for helpful discussions and to C. Glashauser for communicating his results. The work was supported by N.S.E.R.C. (Canada) and the N.S.F. (U.S.A.).

References

1. C. Djalali, G.M. Crawley, B.A. Brown, V. Rotberg, G. Caskey, A. Galonsky, N. Marty, M. Morlet and A. Willis, Phys. Rev. **C35,** 1201 (1987).
2. C. Djalali, N. Marty, M. Morlet, A. Willis, J.-C. Jourdain, N. Anantaraman, G.M. Crawley and A. Galonsky, Phys. Rev. **C31,** 758 (1985).
3. C. Glashauser, K. Jones, F.T. Baker, L. Bimbot, H. Esbensen, R.W. Ferguson, A. Green, S. Nanda and R.D. Smith, Phys. Rev. Lett. **58,** 2407 (1987).
4. W. G. Love and M.A. Franey, Phys. Rev. **C24,** 1073 (1981).
5. A. Bohr and B. Mottelson, *Nuclear Structure, Vol. I,* W.A. Benjamin, N.Y. (1976).
6. T. Suzuki, Phys. Lett. **83B,** 147 (1979);
7. J. Speth and A. Van der Woude, Rep. Prog. Phys. **44,** 719 (1981); T. Suzuki, Ann. Phys. (Fr) **9,** 535 (1984). Benjamin, New York, 1969) p.161.
8. T. Suzuki, Phys. Lett. **101B,** 298 (1981).
9. S.-O. Backman, G. E. Brown and J. A. Niskanen, Phys. Rep. **124,** 1 (1985).
10. J. Wambach and D. Smith, preprint (to be published); S. Drozdz, F. Osterfeld, J. Speth and J. Wambach, Phys. Lett. **189B,** 271 (1987).

SPIN OBSERVABLES IN ELECTRON SCATTERING FROM NUCLEI*

T. W. Donnelly

Center for Theoretical Physics
Massachusetts Institute of Technology
Cambridge, Massachusetts 02139   U.S.A.

I.   INTRODUCTION

This talk is intended to serve as a general introduction to the subject of spin observables in electron scattering from nuclei. The archetypical reaction,

$$\vec{A}(\vec{e}, e'\vec{x}_1 x_2 \ldots)X \ ,$$

is characterized at least by having an electron scattered, (ee'), with the initial electron possibly polarized; the target A can also possibly be polarized. Furthermore, as labeled by the x_i, one or more particles (and possibly their polarizations as well) may be detected in coincidence with the scattered electron, leaving the rest, X, unobserved. In what follows we begin with the simplest inclusive, unpolarized reaction and then progressively add the new features of polarization and/or coincidence.

The kinematics for the various cases discussed are illustrated in Fig. 1. In each an electron with 3-momentum $\vec{k}$ and energy ϵ is scattered through an angle θ_e to be detected with 3-momentum $\vec{k}'$ and energy ϵ'. The 3-momentum transfer is $\vec{q} = \vec{k} - \vec{k}'$ with magnitude $q = |\vec{q}|$ and the energy transfer is $\omega = \epsilon - \epsilon'$. The 4-momentum transfer $q_\mu q^\mu = \omega^2 - q^2$ is space-like (≤ 0). In general (see Fig. 1a), we may consider coordinate systems fixed by the electron momenta so that $\vec{u}_L$ is along $\vec{k}$, $\vec{u}_N$ is normal to the electron scattering plane and $\vec{u}_S = \vec{u}_N \times \vec{u}_L$, with similar labels L', N', S' referred to the scattered electron. Here $L \leftarrow$ longitudinal, $N \leftarrow$ normal, $S \leftarrow$ sideways (as used in hadron scattering). The cross section may be broken down into specific spin projections, $\sigma^{P'P}$, where $P = L$, N or S and $P' = L'$, N' or S'. We are generally only interested in the Extreme Relativistic Limit (ERL) in which $\gamma = \epsilon/m_e >> 1$ and $\gamma' = \epsilon'/m_e >> 1$ and so where terms of order γ^{-1} or γ'^{-1} can safety be neglected (*i.e.* in all but a few very specific circumstances such as when $\theta_e < \gamma^{-1}$). In the ERL we find that[1]

$$\sigma^{P'P} \sim \begin{cases} O(1) & \text{for } P'P = L'L \\ O(\gamma^{-1} \text{ or } \gamma'^{-1}) & \text{for } P'P \neq L'L \ , \end{cases}$$

* This work is supported in part by funds provided by the U. S. Department of Energy (D.O.E.) under contract #DE-AC02-76ER03069.

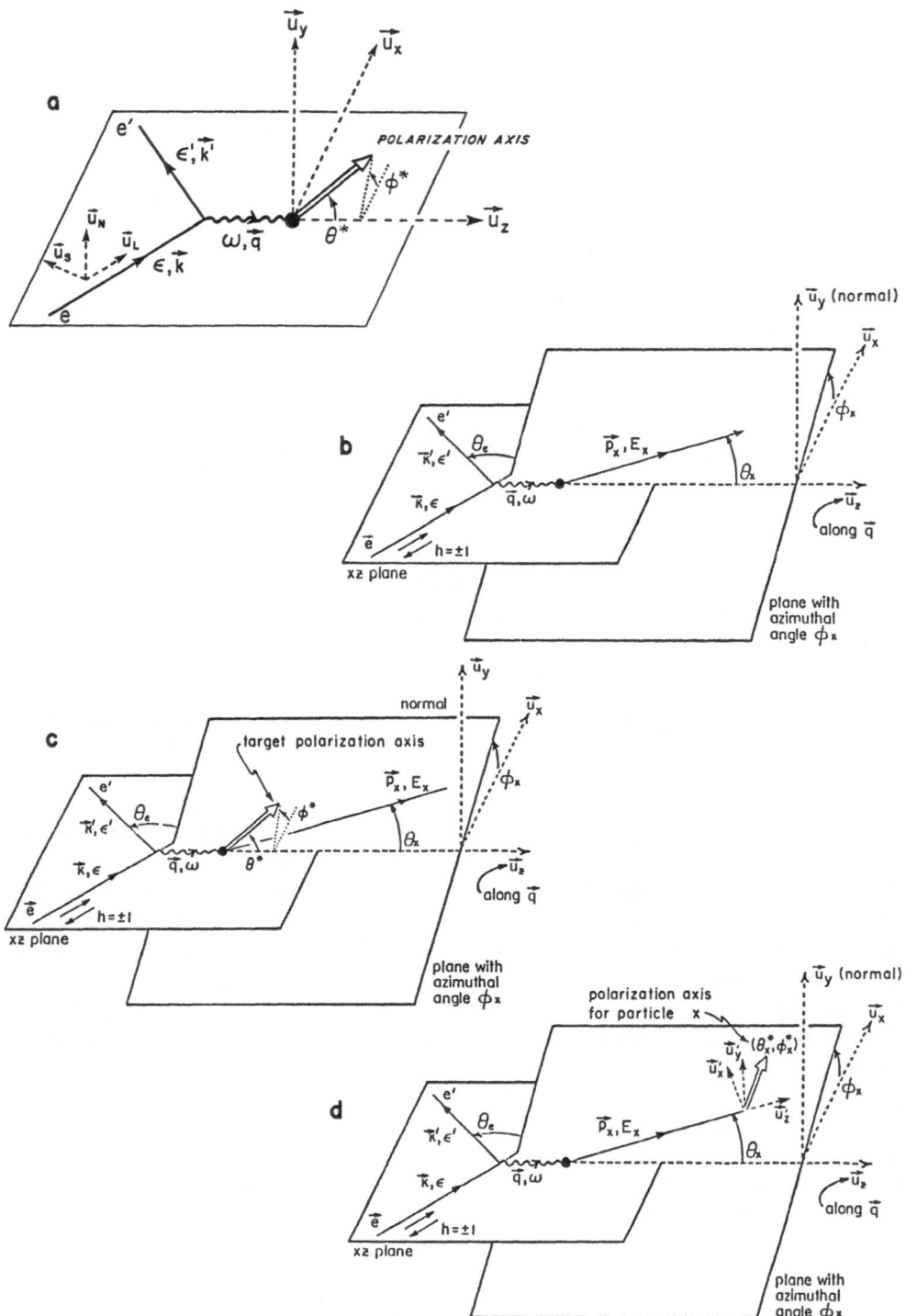

Fig. 1.   Kinematics for electron scattering from nuclei (see text for discussion of individual processes).

and so practically we only need to consider *longitudinally polarized electrons* having helicities $h = \pm 1$ and $h' = \pm 1$. Furthermore, the scattering process is *helicity conserving* to $O(\gamma^{-1}$ or $\gamma'^{-1})$, $h' = h$ and so the information obtained using an incident longitudinally polarized electron beam is the same as that obtained by measuring the longitudinal polarization of the scattered electron. We only consider the former as they are trivially related (see Figs. 1b – d).

Thus we have the first major way of characterizing the responses in electron scattering. Using the electron spin dependence the cross section can be decomposed into two contributions denoted Σ and Δ:

$$\sigma^h = \Sigma + h\Delta \ , \tag{1}$$

where the helicity-averaged cross section

$$\Sigma = \frac{1}{2}\left(\sigma^{+1} + \sigma^{-1}\right) \tag{2a}$$

is obtained using unpolarized beams and where determination of the helicity-difference cross section

$$\Delta = \frac{1}{2}\left(\sigma^{+1} - \sigma^{-1}\right) \tag{2b}$$

requires the use of longitudinally polarized electrons.

II. CLASSES OF ELECTROMAGNETIC RESPONSE FUNCTIONS: THE "SUPER–ROSENBLUTH" DECOMPOSITION

A. *Single-Arm Scattering, No Polarizations: $A(e, e')X$*

This is the simplest and most studied electron scattering reaction. It involves only the unpolarized cross section (the Σ piece introduced above) and contains two contributions labelled L (longitudinal) and T (transverse) for projections with respect to the virtual photon direction (the z-axis in Fig. 1):

$$\Sigma \sim v_L W^L + v_T W^T \ . \tag{3}$$

Here v_L and v_T are two members of a complete set of six electron kinematical factors,[1]

$$
\begin{aligned}
v_L &= \lambda^2 \\
v_T &= \frac{1}{2}\lambda + \tan^2 \frac{\theta_e}{2} \\
v_{TL} &= -\frac{1}{2}\lambda\sqrt{\lambda + \tan^2 \frac{\theta_e}{2}} \\
v_{TT} &= -\frac{1}{2}\lambda \\
v_{T'} &= \sqrt{\lambda + \tan^2 \frac{\theta_e}{2}}\ \tan \frac{\theta_e}{2} \\
v_{TL'} &= -\frac{1}{\sqrt{2}}\lambda \tan \frac{\theta_e}{2} \ ,
\end{aligned}
\tag{4}
$$

where $\lambda \equiv -q_\mu q^\mu / q^2 = 1 - \left(\frac{\omega}{q}\right)^2$ so that $0 \leq \lambda \leq 1$. By fixing q and ω and varying the scattering angle θ_e, the longitudinal kinematical factor can be kept fixed and the transverse factor varied. Since the responses W^L and W^T are functions only of

q and ω, but not θ_e, they may then be separated. Thus the familiar Rosenbluth decomposition yields two response functions each of which is a function of two variables. It should be recognized that even in rather simple situations there can be rich structure in the nuclear electromagnetic currents which are present that is inseparably locked into the two accessible responses. For instance, suppose that the electron scattering excites a nucleus with ground state $\frac{3}{2}^+$ to a state having spin-parity $\frac{5}{2}^-$. The six electromagnetic multipoles permitted are $C1/E1$, $M2$, $C3/E3$ and $M4$. The two responses which can be separated for the class of reactions under discussion involve only the incoherent sums, $(C1)^2 + (C3)^2$ for longitudinal and $(E1)^2 + (M2)^2 + (E3)^2 + (M4)^2$ for transverse, whereas the individual multipoles cannot be extracted in such experiments (see, however, subsection C, below).

B. *Single-Arm Scattering, Electron Polarized: $A(\vec{e}, e')X$*

With longitudinally polarized electrons it becomes possible to separate the cross section into its Σ and Δ contributions as in Eq. (1), where Σ is the unpolarized cross section discussed above and where Δ has a similar generalized Rosenbluth form:

$$\Delta \sim v_L W_{AV}^P + v_T W_{AV}^T + v_{T'} W_{VA}^{T'} \ , \tag{5}$$

in which $v_{T'}$ is another of the electron kinematical factors in Eqs. (4). This particular single-arm helicity-difference cross section comes entirely from parity-violating effects, for instance, involving the interference of the photon-exchange electromagnetic amplitude with the Z^0-exchange weak neutral current amplitude. The labels "AV" denote axial-vector leptonic and vector hadronic currents, whereas the labels "VA" denote the reverse situation.

This constitutes an important, very special case. For such single-arm experiments using longitudinally polarized electrons it is possible to isolate the very small parity-violating effects. Indeed the maximum asymmetry $|\Delta/\Sigma|$ is expected to range from $\sim 10^{-6}$ at medium energies (100's of MeV) up to $\sim 10^{-4}$ at higher energies (a few GeV). In contrast, for all other reactions discussed below, the helicity-difference cross section is in general non-zero even when only parity-conserving electromagnetic interactions are considered. Typically the asymmetries found in those cases lie in the range 0.1 – 1.0 and completely obscure the parity-violating effects which should also be present. Thus, except to indicate where parity-violating effects fall in the general classification of electron scattering response functions as we have in the present subsection, we consider only parity-conserving electromagnetic interactions in the following.

C. *Single-Arm Scattering, Target Polarized, Electron Possibly Polarized: $\vec{A}(e, e')X$, $\vec{A}(\vec{e}, e')X$*

The kinematics for this class of reactions are illustrated in Fig. 1a. In addition to the electron variables, which are treated exactly as in the previous two subsections, we now have to specify an axis along which the target polarization is given; two angles, θ^* and ϕ^*, characterize this axis of quantization. The target polarization is then given in terms of the magnetic substate population probabilities or equivalently[1] the Fano tensors $\left\{ f_I^{(i)} \right\}$ referred to this axis. For example, in a simple case a target having spin $\frac{3}{2}$ might be completely polarized with projection $+\frac{3}{2}$ along the axis of quantization. The angles (θ^*, ϕ^*) are assumed to be at our disposal to vary as we wish. The Σ and Δ parts of the cross section can, as usual, be separated by forming

68

the electron helicity sum and difference (Eqs. (2)), and each in turn can be expressed in a "Super-Rosenbluth" decomposition:

$$\Sigma \sim v_L W^L(\theta^*) + v_T W^T(\theta^*) + v_{TL} \cos \phi^* W^{TL}(\theta^*) + v_{TT} \cos 2\phi^* W^{TT}(\theta^*) \quad (6a)$$

$$\Delta \sim v_{T'} W^{T'}(\theta^*) + v_{TL'} \cos \phi^* W^{TL'}(\theta^*) \quad , \quad (6b)$$

now using all six electron kinematical factors given in Eqs. (4). The dependence on ϕ^* is explicit in Eqs. (6) and so, using the electron helicity dependence, the ϕ^*-dependence and the θ_e-dependence (in the v's), it is possible to isolate six classes of responses. The resulting W's are functions of q and ω (as in subsections A and B above), as well as of the polar angle θ^*. This last dependence can also be made explicit (see Ref. [1] for details):

$$W^L(\theta^*) = W^L + \sum_{\substack{I \le 2 \\ \text{even}}} P_I(\cos \theta^*) W_I^L \quad (7a)$$

$$W^T(\theta^*) = W^T + \sum_{\substack{I \le 2 \\ \text{even}}} P_I(\cos \theta^*) W_I^T \quad (7b)$$

$$W^{TL}(\theta^*) = \sum_{\substack{I \le 2 \\ \text{even}}} P_I^1(\cos \theta^*) W_I^{TL} \quad (7c)$$

$$W^{TT}(\theta^*) = \sum_{\substack{I \le 2 \\ \text{even}}} P_I^2(\cos \theta^*) W_I^{TT} \quad (7d)$$

$$W^{T'}(\theta^*) = \sum_{\substack{I \le 1 \\ \text{odd}}} P_I(\cos \theta^*) W_I^{T'} \quad (7e)$$

$$W^{TL'}(\theta^*) = \sum_{\substack{I \le 1 \\ \text{odd}}} P_I^1(\cos \theta^*) W_I^{TL'} \quad . \quad (7f)$$

Here the quantities W^L and W^T in Eqs. (7a,b) (with no arguments) are the unpolarized responses discussed in subsection A. The new reduced responses, W_I^K, $K = L, T, \ldots$, which can be studied with polarized targets fall into two classes, those accessible without having polarized electrons (L, T, TL and TT) involving only even-rank tensors ($I = $ even, $I \ge 2$) and those accessible only with polarized electrons (T' and TL') involving only odd-rank tensors ($I = $ odd, $I \ge 1$). Furthermore, the tensor ranks are restricted to satisfy $I \le 2J_i$, where J_i is the target spin. All of the responses in Eqs. (7) written without arguments are functions only of q and ω.

There are immediate consequences of these simple statements: (1) For $J_i = 0$, only W^L and W^T survive, as expected since there is nothing that can be polarized in this case; (2) For $J_i = \frac{1}{2}$ and for unpolarized electrons, only W^L and W^T survive again. Thus, considering a polarized spin-$\frac{1}{2}$ target without having polarized electrons yields only the unpolarized cross section. On the other hand when such a target is polarized and when the helicity-difference cross section is measured, two new responses become accessible, $W_1^{T'}$ and $W_1^{TL'}$. We return to these shortly.

The new polarization observables W_I^K, $K = L, T, \ldots$, $I > 0$, contain the nuclear electromagnetic current matrix elements, just as the unpolarized responses W^L and W^T do, however in new *interfering* combinations. For example returning to the $\frac{3}{2}^+$ to $\frac{5}{2}^-$ transition discussed in subsection A, now in addition to the incoherent sums which occur in the unpolarized responses, there are CJ/EJ', CJ/MJ' and EJ/MJ' interferences. In fact we can see that there are two new $I = 1$ responses, four new

$I = 2$ responses and two new $I = 3$ responses, for a total of 10 observables. It can be shown[1] that this provides enough information to determine all six multipole form factors (including their relative signs) for such a spin-parity transition.

Rather than discussing the general problem, let us consider a few specific examples to illustrate the potential contained in such polarized target studies. Let us begin with electron scattering from a polarized spin-1 target, such as $^2\vec{H}$. For instance, for elastic scattering there are three basic form factors, $C0$, $M1$ and $C2$ (see Refs. [1] and [2]). The unpolarized cross section involves the longitudinal and transverse form factors,

$$W^L = F_L^2 = F_{C0}^2 + F_{C2}^2$$
$$W^T = F_T^2 = F_{M1}^2 \ ,$$

but the $C0$ and $C2$ contributions are summed incoherently and no relative phase information is available. The additional reduced response functions which are accessible with polarized targets are (see Refs. [1–3] and the talk by R. Holt at this conference)

$$W_2^L = -2\sqrt{3}F_{C2}\left(F_{C0} + \frac{1}{2\sqrt{2}}F_{C2}\right)$$

$$W_2^T = -\frac{1}{2}\sqrt{\frac{3}{2}}F_{M1}^2$$

$$W_2^{TL} = \frac{3}{\sqrt{2}}F_{M1}F_{C2}$$

$$W_2^{TT} = \frac{1}{4}\sqrt{\frac{3}{2}}F_{M1}^2 \ .$$

Clearly with this polarization information it is possible to separate the individual multipole form factors. In particular, the first of these, W_2^L, forms the interesting part of the tensor polarization t_{20}. Having this together with F_L^2 permits the extraction of the individual F_{C0} and F_{C2} multipoles.

In passing, a special circumstance should be mentioned: for elastic scattering and light nuclei it is possible to obtain the same information with polarized targets or by measurement of the final-state recoil polarization. An example is provided by recent experiments at Bates involving measurements of the recoil tensor polarization in elastic scattering from deuterium (see the talk by M. Garçon at this conference). It should be remarked, however, that inelastic excitations are not generally accessible with the final-state polarization measurements (since the final states generally decay too fast, although the reaction $(e, e'\gamma)$ can be a powerful alternative tool here and can be related directly to the present polarization discussions[4]) and that all but the lightest targets are probably impractical (since the slow recoil is usually too hard to handle).

Next let us consider the scattering of longitudinally polarized electrons from polarized spin-$\frac{1}{2}$ targets. From Eqs. (6) and (7) we have

$$\Sigma \sim v_L W^L + v_T W^T \tag{8a}$$

$$\Delta \sim v_{T'}\cos\theta^* W_1^{T'} + v_{TL'}\sin\theta^*\cos\phi^* W_1^{TL'} \ . \tag{8b}$$

Clearly to emphasize the T' contributions the polarization axis should be chosen along $\vec{q}$, whereas to emphasize the TL' contributions it should be chosen perpendicular to $\vec{q}$ and in the electron scattering plane (i.e., with $\phi^* = 0°$ or $180°$.) First, consider elastic scattering in which F_{C0} and F_{M1} form factors occur (equivalently, we can use G_E and G_M for the nucleon). The unpolarized cross section in Eq. (8a) involves $W_L = F_{C0}^2$ and $W^T = F_{M1}^2$ which can in principle be separated by making a Rosenbluth decomposition. In practice, however, one may be dominant and it may be very difficult to extract the smaller from the larger. For example, at all but the lowest values of q the present information on G_{E_n} comes from unpolarized electron scattering using deuterium as perhaps the simplest target containing neutrons. But at low-to-intermediate values of q, $|G_{E_n}| << |G_{M_n}|$ and the separation is very poorly defined. Now suppose this polarized electron/polarized target information is added. We have[1-3]

$$W_1^{T'} = -\sqrt{2}F_{M1}^2$$
$$W_1^{TL'} = -2\sqrt{2}F_{C0}F_{M1} \ .$$

The former just involves F_T^2 again, whereas the latter is the one of interest for the present purposes: it involves the *interference* between the two form factors and, when one is small in magnitude and the other large, it provides a much more sensitive way to extract one from the other. The specific measurements of relevance here are $\vec{p}(\vec{e}, e')p$ (to extract G_{E_p} from G_{M_p}; even this is interesting for some values of momentum transfer) and $^2\vec{H}(\vec{e}, e')$ or $^3\vec{H}e(\vec{e}, e')$ in the region where the process corresponds best to quasi-free scattering from a nucleon (to extract G_{E_n} from G_{M_n} and to check the approximations involved by extracting G_{E_p} and G_{M_p} as well).

As another example, consider inelastic scattering for the transition $\frac{1}{2}^{\pm} \rightarrow \frac{3}{2}^{\pm}$ in which F_{M1}, F_{C2} and F_{E2} form factors occur. Again using polarized electron scattering from a polarized target, the four accessible responses here are[1]

$$W^L = F_{C2}^2$$
$$W^T = F_{M1}^2 + F_{E2}^2$$
$$W_1^{T'} = \frac{1}{\sqrt{2}}\left(F_{M1}^2 - F_{E2}^2 - 2\sqrt{3}F_{M1}F_{E2}\right)$$
$$W_1^{TL'} = -\sqrt{2}F_{C2}\left(F_{M1} + \sqrt{3}F_{E2}\right) \ .$$

A specific situation is the $N \rightarrow \Delta$ transition, say in $\vec{p}(\vec{e}, e')\Delta$. To the extent that channels other than the $\frac{3}{2}^+$ final state can be neglected, we have the above responses. For this transition the $M1$ contribution is dominant and the $C2/E2$ pieces, which reflect the baryon deformations, are small. The polarization responses, in contrast to the unpolarized responses, involve interferences and especially the $W_1^{TL'}$ contribution is interesting, since it can only be non-zero when $F_{C2} \neq 0$.

These few simple examples are indicative of what is found more generally for higher spin situations, for elastic or inelastic scattering (see Ref. [1] for more complete discussions). We can see that the prime use of polarization in single-arm scattering is to provide a "Multipole Meter" by making accessible new *interference information* in the polarization observables.

D. *Two-Arm Coincidence, Electron Possibly Polarized: $A(e, e'x)X$, $A(\vec{e}, e'x)X$*

The kinematics for this class of reactions are shown in Fig. 1b. As before, the electron scattering process involves the variables q, ω and θ_e, together with the electron's helicity h, if polarized electrons are considered. The fact that particle x is detected in coincidence with the scattered electron means that we have two angles θ_x and ϕ_x to specify the direction in which particle x is going, as well as its energy E_x.

The two major classes of response (again, separated using the electron helicity sum and difference) can as before be expressed in a "Super-Rosenbluth" decomposition:

$$\Sigma \sim v_L W_{(x)}^L + v_T W_{(x)}^T + v_{TL} \cos \phi_x W_{(x)}^{TL} + v_{TT} \cos 2\phi_x W_{(x)}^{TT} \qquad (9a)$$

$$\Delta \sim v_{TL'} \sin \phi_x W_{(x)}^{TL'} \; , \qquad (9b)$$

where the notation "(x)" is used to indicate that particle x is detected in coincidence with the scattered electron. The responses here depend on q, ω, E_x and θ_x, but not on ϕ_x; all of the ϕ_x-dependence is explicit in Eqs. (9) (as is the θ_e-dependence through the v's). Thus by varying the azimuthal angles ϕ_x and using the electron kinematical factors it is possible to isolate five responses, four when the electron beam is unpolarized plus a fifth when polarized electrons are available. Note that the fifth response can only be observed when particle x is detected out of the electron scattering plane. Perhaps not so obvious is the fact that, as well, the L and TT contributions can only be separated using out-of-plane measurements: the factor $\cos 2\phi_x$ is $+1$ in the electron scattering plane (for $\phi_x = 0°$ or $180°$); furthermore, the factors v_L and v_T are the only ones which contain no dependence on $\tan \frac{\theta_e}{2}$ (see Eqs. (4)) and so are fixed when q and ω are fixed. Therefore, the only way to make relative changes in the factors multiplying the L and TT responses in attempting to separate them is to perform at least one measurement with $\phi_x \neq 0°$ or $180°$.

To get some feeling for how these various response functions reflect different aspects of the dynamics underlying a specific reaction, let us briefly examine the general behavior of the fifth response. As noted above in subsection B, in contrast to single-arm scattering with polarized electrons but unpolarized targets, this fifth response function is non-zero in general even when parity is conserved. Thus only the helicity-difference cross section in $A(\vec{e}, e')$ is likely to be practical for studies of electroweak parity violating effects: such effects would usually be overwhelmed by the non-zero parity conserving asymmetries. The TL' fifth response function and the usual TL response have similar structures:

$$W_{(x)}^{TL} \sim \mathrm{Re}(T^* L)$$
$$W_{(x)}^{TL'} \sim \mathrm{Im}(T^* L) \; ,$$

where $T^* L$ represents the appropriate (*i.e.* determined by the dynamics of the specific problem of interest) bilinear combination of (transverse)* × (longitudinal) matrix elements. The *same* combinations occur in the two responses; the only difference is that one has the real part and the other the imaginary part. Now, if the reaction proceeds through a channel in which a single phase dominates for all projections of the current ($T \sim |T| e^{i\delta}$, $L \sim |L| e^{i\delta}$, with the same δ), then $T^* L$ is real and, while $W_{(x)}^{TL}$ is non-zero in general, $W_{(x)}^{TL'}$ vanishes. Moreover, it happens that $W_{(x)}^{TL'}$ also vanishes in the absence of final-state interactions. On the other hand, if $W_{(x)}^{TL'} \neq 0$ then interesting effects must be coming into play. For example, in the Δ-region coincidence electron scattering will be driven to a large degree by the 33-amplitude with a single phase, δ_{33}, and, while $W_{(x)}^{L, T, TL, TT}$ may all be non-zero, $W_{(x)}^{TL'}$ may be expected to vanish. To the extent that it does not vanish, we will be able to access information concerning interferences of the 33-amplitude with amplitudes for other channels which are usually too weak to be studied directly.

The L, T, TL and TT (unpolarized) responses on the one hand and the TL' (polarized electrons, but otherwise unpolarized) response on the other may be characterized by their time-reversal properties, even and odd, respectively. Time-reversal even responses are always real parts of bilinear products involving the currents, while time-reversal odd responses involve imaginary parts (as for TL and TL' above, respectively). As we shall see in the following subsections, this characterization can be generalized to include situations when target and/or final-state polarizations are specified and where wider classes of response becomes accessible.

In passing let us remark that three-arm, four-arm, ... coincidences can be catalogued in a similar way to the two-arm situation discussed here (see Ref. [5]). For instance, for reactions $A(e, e'xy)X$ and $A(\vec{e}, e'xy)X$, the structure becomes

$$
\begin{aligned}
\Sigma \sim\; & v_L W^L_{(xy)} + v_T W^T_{(xy)} \\
& + v_{TL} \left(\cos \Phi_{xy} W^{TL}_{(xy)} + \sin \Phi_{xy} \tilde{W}^{TL}_{(xy)} \right)) \\
& + v_{TT} \left(\cos 2\Phi_{xy} W^{TT}_{(xy)} + \sin 2\Phi_{xy} \tilde{W}^{TT}_{(xy)} \right)
\end{aligned}
\tag{10a}
$$

$$
\Delta \sim v_{T'} \tilde{W}^{T'}_{(xy)} + v_{TL'} \left(\sin \Phi_{xy} W^{TL'}_{(xy)} + \cos \Phi_{xy} \tilde{W}^{TL'}_{(xy)} \right) \;,
\tag{10b}
$$

for a (maximal) complement of nine classes of response. Here $\Phi_{xy} = \frac{1}{2}(\phi_x + \phi_y)$ is the *average* azimuthal angle and each response is a function of seven variables, q, ω, E_x, θ_x, E_y, θ_y and $\varphi_{xy} = \phi_x - \phi_y$. Rather than pursuing this as a topic of discussion let us turn to the simpler two-arm coincidence reactions, but now with polarizations specified.

E. *Two-Arm Coincidence, Target Polarized, Electron Possibly Polarized:*
$\vec{A}(e, e'x)X,\ \vec{A}(\vec{e}, e'x)X$

The kinematics for this class of reactions are indicated in Fig. 1c. Now, as in the discussions of the past two subsections, polarization angles (θ^*, ϕ^*) are needed together with angles (θ_x, ϕ_x) which specify the direction in which particle x is detected. The helicity sum and difference cross sections can be written in the forms

$$
\begin{aligned}
\Sigma \sim\; & v_L W^L_{(x)}(\theta^*, \phi^*) + v_T W^T_{(x)}(\theta^*, \phi^*) \\
& + v_{TL} \left(\cos \phi_x W^{TL}_{(x)}(\theta^*, \phi^*) + \sin \phi_x \tilde{W}^{TL}_{(x)}(\theta^*, \phi^*) \right) \\
& + v_{TL} \left(\cos 2\phi_x W^{TT}_{(x)}(\theta^*, \phi^*) + \sin 2\phi_x \tilde{W}^{TT}_{(x)}(\theta^*, \phi^*) \right)
\end{aligned}
\tag{11a}
$$

$$
\begin{aligned}
\Delta \sim\; & v_{T'} \tilde{W}^{T'}_{(x)}(\theta^*, \phi^*) \\
& + v_{TL'} \left(\sin \phi_x W^{TL'}_{(x)}(\theta^*, \phi^*) + \cos \phi_x \tilde{W}^{TL'}_{(x)}(\theta^*, \phi^*) \right) \;,
\end{aligned}
\tag{11b}
$$

where each response depends on q, ω, E_x and θ_x as before, together with new dependences on the target polarization (for emphasis, the dependences on θ^* and ϕ^* are displayed here). Note the similarity to Eqs. (10) above. For an unpolarized situation, the responses with tildas vanish and the ones without revert to the responses discussed above (see Eqs. (9):

$$
\tilde{W}^K_{(x)}(\theta^*, \phi^*) \xrightarrow[\text{unpol.}]{} 0
$$

$$
W^K_{(x)}(\theta^*, \phi^*) \xrightarrow[\text{unpol.}]{} W^K_{(x)} \;.
$$

The target polarization information may be organized into spherical tensors characterized by rank I, where I may be even or odd, with $I = 0$ corresponding to

73

the unpolarized cross sections in the previous subsection. When target polarizations are considered, this is the same type of tensor decomposition encountered earlier for single-arm scattering in subsection C. The general break-down into time-reversal even and odd responses is as follows:[5]

		$I = $ even	$I = $ odd
Σ	L	TR E	TR O
$\left\{ \begin{array}{c} \text{electron} \\ \text{unpolarized} \end{array} \right\}$	T	TR E	TR O
	TL	TR E	TR O
	TT	TR E	TR O
Δ	T'	TR O	TR E
$\left\{ \begin{array}{c} \text{electron} \\ \text{polarized} \end{array} \right\}$	TL'	TR O	TR E

where TR E (TR O) refers to time-reversal even (odd). As in the more limited situation without target polarizations discussed above in subsection D, the responses in time-reversal even or odd sectors can have rather different sensitivities to the nature of the final state.

To help clarify these ideas a little, let us consider a specific spinology (in fact, this is the one we return to in the next major section). We take the simplest non-trivial case of a spin-$\frac{1}{2}$ target which may be unpolarized ($I = 0$) or vector polarized ($I = 1$). Let us characterize the polarizations by projections along three orthogonal directions: (ℓ) along $\vec{q}$, (n) in the direction $\vec{q} \times \vec{p}_x$, and (s) in the direction $\vec{u}_n \times \vec{u}_\ell$. The ϕ_x dependences of the complete set of 18 responses are the following:

		Target Unpolarized	Target Polarized ℓ	s	n
Electron Unpolarized	L	1	0	0	1
	T	1	0	0	1
	TL	$\cos\phi_x$	$\sin\phi_x$	$\sin\phi_x$	$\cos\phi_x$
	TT	$\cos 2\phi_x$	$\sin 2\phi_x$	$\sin 2\phi_x$	$\cos 2\phi_x$
Electron Polarized	T'	0	1	1	0
	TL'	$\sin\phi_x$	$\cos\phi_x$	$\cos\phi_x$	$\sin\phi_x$

With electron and target either unpolarized or both polarized, the responses are TR E; with only one or the other polarized, they are TR O. Note that even in the absence of polarized electrons there are new responses for polarized spin-$\frac{1}{2}$ targets when coincidence reactions are being considered (in contrast to single-arm scattering, see above).

Before returning to one specific example of coincidence–plus–target–polarization, let us place in context a final class of reactions.

F. *Two-Arm Coincidence, Polarization of x Measured, Electron Possibly Polarized:* $A(e, e'\vec{x})X$, $A(\vec{e}, e'\vec{x})X$

The kinematics for this class are shown in Fig. 1d and now involve angles (θ_x, ϕ_x) for the direction in which x is detected together with (θ_x^*, ϕ_x^*) which are used to specify the polarization axis for outgoing particle x in a frame of reference fixed by the kinematics of that particle and the momentum transfer $\vec{q}$. The general characterization discussed above is still true in form, but now the responses of Eqs. (11) are replaced by new ones which depend on (θ_x^*, ϕ_x^*). Note that the responses in the two cases are in general *different* and that complementary information is obtained using the two classes of reactions. For example, even in a case as simple as pion electroproduction from the proton,[6,7]

$$p(e, e'p)\pi^0 \ , \qquad p(\vec{e}, e'p)\pi^0$$
$$\vec{p}(e, e'p)\pi^0 \ , \qquad \vec{p}(\vec{e}, e'p)\pi^0$$
$$p(e, e'\vec{p})\pi^0 \ , \qquad p(\vec{e}, e'\vec{p})\pi^0 \ ,$$

the polarized target responses are not all the same as the final-state polarization responses. In general it is desirable to have some information from both classes as specific examples such as this pion electroproduction reaction suggest that the differences can be advantageously used to extract the basic underlying information of interest.

The subject of final-state proton polarizations in $(e, e'p)$ reactions is discussed in more detail by W. Van Orden in another talk at this conference. Specific results are given there for TR E and TR O responses in studies of the $^{16}O(e, e'p)$ reaction. Here, instead of continuing along the same lines, we leave this general overview of the subject and return to one particularly simple example of the class of reactions introduced in subsection E.

III. SPECIFIC EXAMPLE: THE REACTION $^3\vec{H}e(\vec{e}, e'\pi)$

In Ref. [8] we considered an application of the model of Lipkin and Lee[9] to electroproduction of pions from polarized ^{3}He using longitudinally polarized electrons. Their model is based on the following assumed properties of the Δ: (1) that it has isospin-$\frac{3}{2}$; (2) that isospin is conserved in strong interactions which mix the Δ into the nuclear wave function for the ground state of ^{3}He; (3) that the amplitude for absorbing a virtual photon on a pre-existing Δ is proportional to the Δ charge. Final-state interactions and non-resonant contributions are neglected. Lipkin and Lee find that the relative rates for the different pion charge states depend on whether a Δ is electroproduced (p) and decays or whether a pre-existing Δ is struck by the virtual photon, knocked out (k) of ^{3}He and decays:

$$\pi^+ : \pi^0 : \pi^- = \begin{cases} \frac{2}{9} : \frac{6}{9} : \frac{1}{9} & \text{``}p\text{''} \\ \frac{19}{21} : \frac{2}{21} : 0 & \text{``}k\text{''} \end{cases} .$$

Using the ideas which have been summarized here in subsection II.E (specifically with $x = \pi$), this model was taken in Ref. [8] to represent the reaction mechanism for pion electroproduction in the region of the Δ. Referring to the table for scattering from polarized spin-$\frac{1}{2}$ targets in subsection II.E, we see that in general there are 18 responses to consider. Let us reduce the number to a small set of particularly interesting responses by selecting the kinematics in the following ways:

(1) Choose coplanar kinematics — then with $\phi_\pi = 0°$ or $180°$ we have four unpolarized responses (L, T, TL, TT), four responses involving n-polarized ^{3}He (L, T, TL, TT), two responses involving ℓ-polarized ^{3}He (T', TL') and two responses involving s-polarized ^{3}He (T', TL'). The remaining six cases cannot contribute as they are multiplied by explicit factors $\sin \phi_\pi$ or $\sin 2\phi_\pi$.

(2) Choose the target spin in the scattering plane — then the n-polarized ^{3}He cases are absent.

(3) Moreover, choose the target spin perpendicular to $\vec{q}$ — then the ℓ-polarized ^{3}He responses are also absent (the s-polarized ^{3}He cases still remain).

75

(4) Choose to detect the pions along $\vec{q}$ — from general symmetry arguments the T' responses are then absent. The cross section consequently has the form

$$\sigma \sim \{\text{unpolarized responses}\} + p_e p_A v_{TL'} \tilde{W}^{TL'}_{(\pi)}(s) \ , \tag{12}$$

where p_e (p_A) is the electron (helium) polarization. By flipping either of these spins the polarization response can be separated from the unpolarized responses to form the polarization asymmetries for $\pi^{\pm}$-production, $A(\pi^{\pm})$. Using the Lipkin and Lee model the π^+/π^- asymmetry ratio is

$$\frac{A(\pi^+)}{A(\pi^-)} = 1 + \frac{57}{14} \frac{\Gamma_k}{\Gamma_p} \ , \tag{13}$$

where Γ_k/Γ_p is the ratio of knockout to production in the longitudinal-transverse interference terms which survive. We see that in the approximation where final-state interactions and non-resonant contributions are neglected the ratio of charged pion asymmetries deviates from unity only if there is a contribution from knockout of Δ components in the ground state of ^{3}He. We can write the ratio of longitudinal-transverse interference terms as

$$\frac{\Gamma_k}{\Gamma_p} = \frac{P_\Delta G^{\Delta\Delta}_{CO} G^{\Delta\Delta}_{M1}}{G^{N\Delta}_{C2} G^{N\Delta}_{M1}} \tag{14}$$

where $G^{N\Delta}_{C2}$ and $G^{N\Delta}_{M1}$ are the C2 and M1 Δ-production form-factors respectively, $G^{\Delta\Delta}_{CO}$ and $G^{\Delta\Delta}_{M1}$ are the CO and M1 $e - \Delta$ elastic scattering form factors respectively, and P_Δ is the probability to find a Δ in the ground state of ^{3}He. Now at low $Q^2 \approx 0.1 (\text{GeV}/c)^2$, we have

$$G^{\Delta\Delta}_{CO}, \ G^{\Delta\Delta}_{M1}, \ G^{N\Delta}_{M1} \sim 1 \quad \text{and} \quad G^{N\Delta}_{C2} \sim 0.10$$

and so $\frac{A(\pi^+)}{A(\pi^-)}$ is very sensitive to P_Δ. If P_Δ is of order 2% in ^{3}He, as estimated from fits to the electromagnetic form factors,[10] then the ratio of charged pion asymmetries will differ from unity by a factor of two. This large sensitivity results from the use of the spin degree of freedom to suppress the large transverse production cross-section, leaving a relatively small longitudinal-transverse production contribution.

It is clear that more sophisticated models could be brought to bear in treatments of reactions such as this. Nevertheless, such a simple model helps to point the way as a guide to which of the many responses that will become accessible in coincidence studies with polarizations specified are likely to be especially sensitive to some feature of the underlying reaction mechanism.

IV. SUMMARY

What has been presented here is a brief overview of some of the highlights of "spin physics" in electron scattering in which interference effects become accessible when polarization degrees of freedom can be controlled. For single-arm scattering the general case requires that polarized targets be available; for coincidence reactions there are interesting processes to explore both with targets polarized and when a specific particle in the final state is detected together with its polarization. In many cases, it is important (or essential, such as with single-arm scattering from polarized spin-$\frac{1}{2}$ targets) to have longitudinally polarized electrons available.

The practical implications are severe. For detection of final-state polarizations, a polarimeter is required and these are usually devices with limited efficiencies. For polarized target studies the problem is to obtain significant enough luminosities to have feasible experiments: to be practical for nuclear physics studies it must be possible in general to obtain luminosities above, say, 10^{30} cm^{-2} s^{-1} (and frequently considerably above this). To have good resolution capability in general requires that the target not be too thick. Typical external cryogenic polarized targets cannot withstand more than about 1 nAmp before depolarizing and so to reach the desired range of luminosity requires a very thick target. In fact, with such targets the degree of polarization is usually rather low and so the effective luminosity is actually quite a bit smaller than the nominal value. With internal targets using a circulating electron beam the current can be very high (for example, Bates[11] is designed for 80 m Amp internal current). In general for such studies with adequate liminosities it will then be necessary to have internal polarized target densities lying above about 10^{13} nuclei/cm^2. Even given these technical challenges, it is now clear that the new facilities being built with which spin observables in electron scattering from nuclei can be studied will open up an exciting new area of electromagnetic nuclear physics.

REFERENCES

1. T. W. Donnelly and A. S. Raskin, *Ann. Phys.* **169**, 247 (1986).

2. T. W. Donnelly and I. Sick, *Rev. Mod. Phys.* **56**, 461 (1984).

3. R. G. Arnold, C. E. Carlson and F. Gross, *Phys. Rev.* **C23**, 363 (1981).

4. T. W. Donnelly, A. S. Raskin and J. F. Dubach, *Nucl. Phys.* **A474**, 307 (1987).

5. T. W. Donnelly, "Polarization Degrees of Freedom in Electron Scattering from Nuclei," at the Summer School *New Vistas in Electro-Nuclear Physics*, Banff, Canada (1985).

6. A. S. Raskin, Ph.D. Thesis (M.I.T., 1987, unpublished).

7. A. S. Raskin and T. W. Donnelly (to be published).

8. R. G. Milner and T. W. Donnelly, *Phys. Rev.* **C37**, 870 (1988).

9. H. J. Lipkin and T.-S. H. Lee, *Phys. Lett.* **B183**, 22 (1987).

10. L. Heller, S. Kumano, J. C. Martinez and E. J. Moniz, *Phys. Rev.* **C35**, 718 (1987).

11. Proposal for a CW Upgrade of the William H. Bates Linear Accelerator Center (June, 1984).

SPIN OBSERVABLES IN EXCLUSIVE ELECTRON SCATTERING

J. W. Van Orden

Continuous Electron Beam Accelerator Facility
Newport News, Virginia 23606

and

Department of Physics and Astronomy
University of Maryland
College Park, Maryland 20742

INTRODUCTION

Experimental studies of both inclusive and exclusive quasielastic electron scattering have yielded results which have, so far, defied explanation within the context of traditional nuclear theory. In the case of the inclusive reaction, the difficulties lie in the experimentally observed suppression of the longitudinal response and in the region of the "dip" between the quasielastic and resonance peaks in the transverse response.[1] In the case of the exclusive reactions, which have so far been limited to the (e,e'p) reaction using certain restricted choices of kinematics, the difficulty is the observation of insufficient spectral response[2] and in the description of the missing energy spectrum of the longitudinal and transverse response functions.[3] These difficulties have led to a considerable amount of theoretical activity in improving the application of traditional many-body methods to the description of these reactions, and in the application of more exotic models involving relativistic dynamics[4,5,6] or quark degrees of freedom.[7,8,9] So far these efforts have resulted in varying degrees of success and continued work in these areas will be required for the foreseeable future.

The difficulty in using the available experimental results to discriminate between these various models is related to the intrinsic properties of the quasielastic response. Since the quasielastic response is in large part determined by the nuclear momentum distribution and the phase space available to the reactions, the quasielastic response functions tend to be smooth broad peaks. As a result, given the range of assumptions which can reasonably be made in constructing models of these processes, it is quite easy to fit any limited set of quasielastic data by the adjustment of a small number of parameters. At present, the data are not capable of distinguishing unambiguously among the various classes of models. It is, therefore, necessary to determine classes of experiments which have the potential for discriminating among the various models.

The focus of this paper is on the $(\vec{e}, e'\vec{N})$ reaction where polarized electrons are used to eject polarized nucleons from an unpolarized nucleus. This reaction has several advantages as a means for increasing the available information necessary to constrain theory. The additional measurable quantities are discrete spin degrees of freedom which can be accessed by providing a polarized electron beam and/or using a polarimeter for the ejected nucleons. Both of these elements exist and the advent of the coming generation of high duty factor electron accelerators should make possible the simultaneous use of these elements in coincidence experiments. The discreteness of the spin degrees of freedom can also be used to minimize systematic errors by allowing all of the continuous kinematical variables to be fixed while the spin of the beam is flipped. While this is also true of coincidence experiments using polarized targets, the measurement of ejectile spin circumvents the difficulties of producing polarized targets which can be used in a high current electron beam. From the theoretical standpoint, the $(\vec{e}, e'\vec{N})$ reaction provides additional access to the spin response of the nuclear system. This is, of course, of considerable importance since the strong interactions of the nuclear system are explicitly spin dependent as is the electromagnetic interaction of the electrons with the hadrons of the nucleus. There is, by inference from elastic proton scattering[10,11,12] and from the electrodisintegration of the deuteron,[13] reason to believe that the addition of these spin observables will considerably constrain the various elements of models of quasielastic electron scattering.

In a recent paper,[14] a formal framework was presented for the discussion of the $(\vec{e}, e'\vec{N})$ reaction which was constructed to explicitly display the dependence of the differential cross section on the polarization of the ejected nucleon in a manner which provides a direct generalization of the usual description of the unpolarized reaction and which treats the spin of the ejected nucleon in a manner consistent with that used in elastic proton scattering. The constraints placed on the eighteen response functions (thirteen of which depend on ejectile spin) by various symmetries were also discussed. In the present work, this formalism is reviewed and results showing final state interaction, relativistic and off-shell effects are presented.

REVIEW OF FORMALISM

For light to medium heavy nuclei, the electron scattering cross section can be described to a good approximation by the exchange of a single photon as illustrated schematically in Fig. 1. Here, an incident electron with

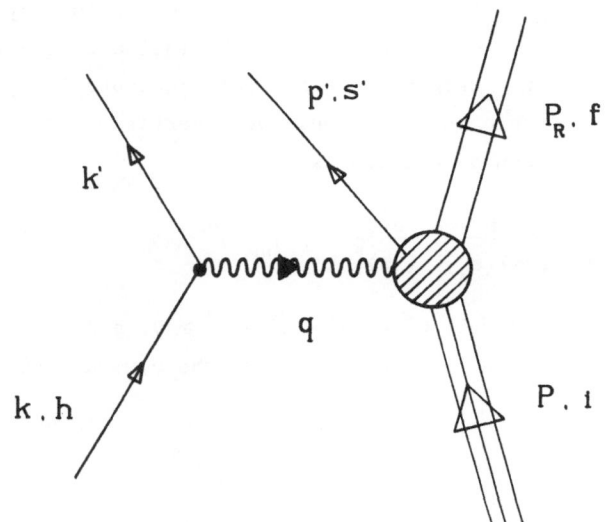

Fig. 1. Schematic diagram of the $(\vec{e}, e'\vec{p})$ reaction.

four-momentum $k = (\epsilon_k, \vec{k})$ and helicity h scatters into a final state with four-momentum $k' = (\epsilon_{k'}, \vec{k}')$ by emitting a virtual photon carrying a four-momentum $q = (\omega, \vec{q})$. The virtual photon is absorbed by the target nucleus with four-momentum P and intrinsic quantum numbers i, ejecting a nucleon with four-momentum $p' = (E', \vec{p}')$ and spin s', leaving a residual nuclear system with four-momentum P_R and intrinsic quantum numbers f. The transition matrix for this process is proportional to

$$j_\alpha \frac{g^{\alpha\beta}}{q^2} J_\beta$$

where

$$j_\alpha = \bar{u}(k',s'_e) \; \gamma_\alpha \; u(k,s_e)$$

is the electromagnetic current density for the electron, $g^{\alpha\beta}/q^2$ is the photon propagator and

$$J_\beta = <p',s'(-);P_R,f|\hat{J}_\beta(q)|P,i>$$

is the electromagnetic transition current density for the nuclear system. Electromagnetic current conservation requires that both the electron and nuclear currents satisfy the continuity equation which in momentum space requires $q^\alpha j_\alpha = 0$ and $q^\beta J_\beta = 0$. Current conservation implies that there are only three independent components for each current. For conceptual reasons it is convenient to associate these with coupling to the three helicity states of the virtual photon. The electron therefore emits a virtual photon which is in a superposition of helicity states determined by the electron scattering kinematics and the helicity of the incident electron. This photon is then absorbed by the target nucleus. This conceptual separation of the reaction into a leptonic and a hadronic part carries over to the differential cross section which can be written as

$$d\sigma = \frac{2\pi}{|\vec{k}|} \frac{e^4}{Q^4} \frac{d^3k}{(2\pi)^3 \epsilon_{k'}} \frac{d^3p'}{(2\pi)^3} \frac{m}{E'} \; \eta_{\mu\nu} \; W^{\mu\nu}$$

where e is the unit charge, $Q^2 = -q^2 = \vec{q}^2 - \omega^2 \geq 0$, and m is the nucleon mass. The cross section is proportional to the contraction of an electron tensor

$$\eta_{\mu\nu} = m_e^2 \sum_{s'_e} j_\mu^\dagger \, j_\nu$$

and a nuclear response tensor

$$W^{\mu\nu} = \overline{\sum_i} \sum_F \int d^3P' \; \delta^4(q + P - P' - p') \; J^{\mu\dagger} \, J^\nu.$$

where m_e is the electron mass and s'_e is the spin of the scattered electron.

By use of general symmetries of the reaction, the differential cross section for the $(\vec{e},e'\vec{N})$, when the residual nucleon is left in its ground state or some discrete excited state, can be written as[14]

$$\left[\frac{d^5\sigma}{d\epsilon_k, d\Omega_k, d\Omega_{p'}}\right]_{s',h} = \frac{m|\vec{p}'|}{2(2\pi)^3}\left[\frac{d\sigma}{d\Omega_{k'}}\right]_{Mott}$$

$$\times \left\{\left[\frac{Q^4}{\vec{q}^4}\right](R_L + R_L^n\,\hat{n}\cdot\hat{s}_R') + \left[\frac{Q^2}{2\vec{q}^2} + \tan^2\frac{\vartheta}{2}\right](R_T + R_T^n\,\hat{n}\cdot\hat{s}_R')\right.$$

$$+ \frac{Q^2}{2\vec{q}^2}\left[(R_{TT} + R_{TT}^n\,\hat{n}\cdot\hat{s}_R')\cos(2\beta) + (R_{TT}^t\,\hat{t}\cdot\hat{s}_R' + R_{TT}^l\,\hat{l}\cdot\hat{s}_R')\sin(2\beta)\right]$$

$$+ \frac{Q^2}{\vec{q}^2}\left[\frac{Q^2}{\vec{q}^2} + \tan^2\frac{\vartheta}{2}\right]^{1/2}\left[(R_{LT} + R_{LT}^n\,\hat{n}\cdot\hat{s}_R')\cos\beta + (R_{LT}^t\,\hat{t}\cdot\hat{s}_R' + R_{LT}^l\,\hat{l}\cdot\hat{s}_R')\sin\beta\right]$$

$$+ h\frac{Q^2}{\vec{q}^2}\tan\frac{\vartheta}{2}\left[(R_{LT'} + R_{LT'}^n\,\hat{n}\cdot\hat{s}_R')\sin\beta + (R_{LT'}^t\,\hat{t}\cdot\hat{s}_R' + R_{LT'}^l\,\hat{l}\cdot\hat{s}_R')\cos\beta\right]$$

$$+ h\tan\frac{\vartheta}{2}\left[\frac{Q^2}{\vec{q}^2} + \tan^2\frac{\vartheta}{2}\right]^{1/2}(R_{TT'}^t\,\hat{t}\cdot\hat{s}_R' + R_{TT'}^l\,\hat{l}\cdot\hat{s}_R')\right\}$$

where ϑ is the electron and the Mott cross section is given by

$$\left[\frac{d\sigma}{d\Omega_{k'}}\right]_{Mott} = \left[\frac{\alpha\,\cos(\vartheta/2)}{2\,k\,\sin^2(\vartheta/2)}\right]^2 = \left[\frac{2\,\alpha\,k'\cos(\vartheta/2)}{Q^2}\right]^2.$$

The direction of ejectile spin quantization in the ejectile rest frame is given by the unit vector $\hat{s}_R'$. The direction of $\vec{p}'$ is given by the polar and azimuthal angles α and β relative to the coordinate system shown in Fig. 2. The unit vectors $\hat{n}$, $\hat{l}$ and $\hat{t}$, shown in Fig. 2, form a right-handed coordinate system defined such that $\hat{l}$ lies along $\vec{p}'$, $\hat{n} = (\vec{q}\times\hat{l})/|\vec{q}\times\hat{l}|$, and $\hat{t} = \hat{n}\times\hat{l}$. The plane containing $\vec{p}'$ and $\vec{q}$ (or, alternately, $\hat{l}$ and $\hat{t}$) contains the momentum of the ejectile $\vec{p}'$ and the recoil momentum of the residual nuclear system $\vec{P}_R = \vec{q} - \vec{p}'$, and plays the same role in this case as does the scattering plane in elastic proton scattering. For convenience we will refer to this as the hadronic plane. The nuclear response functions R_j^i are related to the nuclear response tensor $W^{\mu\nu}$ through

$$\frac{1}{2}(R_L + R_L^n\,\hat{n}\cdot\hat{s}_R') = \int_{line} dE'\, W^{00}(\hat{s}_R')$$

$$\frac{1}{2}(R_T + R_T^n\,\hat{n}\cdot\hat{s}_R') = \int_{line} dE'\, [W^{11}(\hat{s}_R') + W^{22}(\hat{s}_R')]$$

$$\frac{1}{2}\left[\left(R_{TT} + R_{TT}^{n}\ \hat{n}\cdot\hat{s}_{R}'\right)\cos(2\beta) + \left(R_{TT}^{t}\ \hat{t}\cdot\hat{s}_{R}' + R_{TT}^{1}\ \hat{1}\cdot\hat{s}_{R}'\right)\sin(2\beta)\right]$$

$$= \int_{line} dE' \left[W^{11}(\hat{s}_{R}') - W^{22}(\hat{s}_{R}')\right]$$

$$\frac{1}{2}\left[\left(R_{LT} + R_{LT}^{n}\ \hat{n}\cdot\hat{s}_{R}'\right)\cos\beta + \left(R_{LT}^{t}\ \hat{t}\cdot\hat{s}_{R}' + R_{LT}^{1}\ \hat{1}\cdot\hat{s}_{R}'\right)\sin\beta\right]$$

$$= - \int_{line} dE' \left[W^{01}(\hat{s}_{R}') + W^{10}(\hat{s}_{R}')\right]$$

$$\frac{1}{2}\left[\left(R_{LT'} + R_{LT'}^{n}\ \hat{n}\cdot\hat{s}_{R}'\right)\sin\beta + \left(R_{LT'}^{t}\ \hat{t}\cdot\hat{s}_{R}' + R_{LT'}^{1}\ \hat{1}\cdot\hat{s}_{R}'\right)\cos\beta\right]$$

$$= i \int_{line} dE' \left[W^{20}(\hat{s}_{R}') - W^{02}(\hat{s}_{R}')\right]$$

$$\frac{1}{2}\left(R_{TT'}^{t}\ \hat{t}\cdot\hat{s}_{R}' + R_{TT'}^{1}\ \hat{1}\cdot\hat{s}_{R}'\right) = i \int_{line} dE' \left[W^{12}(\hat{s}_{R}') - W^{21}(\hat{s}_{R}')\right]$$

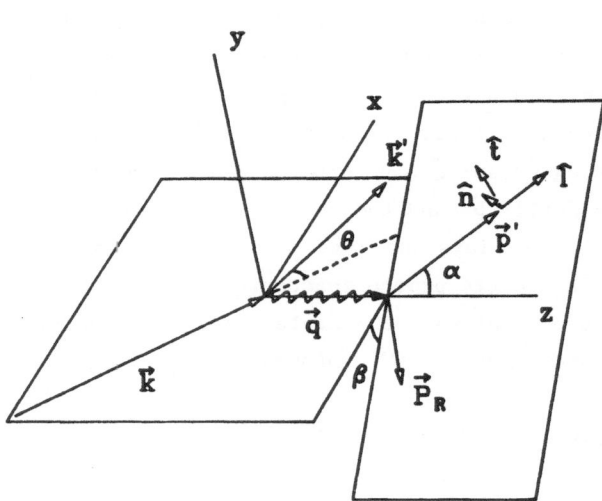

Fig. 2.   Coordinate system used to in deriving the cross section.

Table 1 contains a summary of some of the properties of the eighteen response functions. The first of these is the behavior of the response functions under time reversal and parity transformations. Since the cross section will be invariant under a combination of time reversal and when only one channel is open, this symmetry is good only in cases when only one channel is physically open or in models which ignore the additional channels such as the plane wave impulse approximation (PWIA). In the column headed "TP", response functions which survive the imposition of this symmetry are labeled as even, while those which do not are labeled odd.

Table 1. Response Function Characteristics

Response Function	TP	In Plane	Parallel Kinematics	Electron Polarization	Symmetry
R_L	even	yes	yes	no	even
R_L^n	odd	yes	no	no	even
R_T	even	yes	yes	no	even
R_T^n	odd	yes	no	no	even
R_{TT}	even	yes	no	no	even
R_{TT}^n	odd	yes	no	no	even
R_{TT}^l	odd	no	no	no	odd
R_{TT}^t	odd	no	no	no	odd
R_{LT}	even	yes	no	no	odd
R_{LT}^n	odd	yes	yes	no	odd
R_{LT}^l	odd	no	no	no	even
R_{LT}^t	odd	no	yes	no	even
$R_{LT'}$	odd	no	no	yes	even
$R_{LT'}^n$	even	no	yes	yes	even
$R_{LT'}^l$	even	yes	no	yes	odd
$R_{LT'}^t$	even	yes	yes	yes	odd
$R_{TT'}^l$	even	yes	yes	yes	even
$R_{TT'}^t$	even	yes	no	yes	even

Additional properties of the response functions are of interest from the experimental viewpoint. Since it is difficult to measure ejected nucleons which have momenta lying out of the electron scattering plane, it is useful to know which of the response functions can be detected in the electron scattering plane ($\beta = 0$ or $\beta = \pi$) and which require going out of plane. The column of Table 1 labeled "in plane" indicates which response functions contribute in the electron scattering plane.

Coincidence experiments are often performed with the ejectile momentum is parallel ($\alpha = 0$) or antiparallel ($\alpha = \pi$) to the momentum transfer. This is the so-called parallel-antiparallel kinematics.[15] The column labeled "parallel kinematics" indicates whether the response functions contribute to the cross section under these kinematical conditions. In order to obtain this information, the unit vector $\hat{s}'_R$ is assumed to point in an arbitrary direction. The limit is taken $\alpha \rightarrow 0, \pi$ and the independence of the cross section of azimuthal angle β in this limit is used to determine which of the response functions are constrained by this requirement. Those response functions denoted by "no" in this column of Table 1 are constrained to be zero. The response functions R_L, R_T and R^l_{TT}, are not constrained in any way. The remaining response functions must satisfy the constraints $R^n_{LT} \mp R^t_{LT} \rightarrow 0$ and $R^n_{LT'} \pm R^t_{LT'} \rightarrow 0$ where the upper (lower) sign refers to the limit $\alpha \rightarrow 0$ ($\alpha \rightarrow \pi$). These limits provide a useful check on numerical calculations of the response functions. A careful examination of the cross section once these constraints have been imposed shows that the LT response contribution to the cross section is proportional to the x-component of the spin vector while the LT' contribution is proportional to the y-component of the spin vector. This implies that there is a natural choice for the spin coordinate system in this limit. The unit vector $\hat{l}$ is chosen to point along $\vec{p}'$, $\hat{n}$ points in the positive y-direction, and $\hat{t} = \hat{n} \times \hat{l}$. This corresponds to the limiting process of first taking the limit $\beta \rightarrow 0$ and then $\alpha \rightarrow 0, \pi$.

Since the polarization of existing electron beams is limited to about 40 percent and such beams have limited current, coincidence experiments which do not require such a beam can be performed more rapidly. The column labeled "electron polarization" indicates which of the response functions contribute only when the beam is polarized.

Finally, terms contributing to the cross section in the electron scattering plane with a contribution to the cross section which changes sign under the change in kinematics $\beta = 0 \rightarrow \beta = \pi$ can be easily separated from the total cross section by making the change in azimuthal angle and subtracting cross sections. The last column of Table 1, labeled "Symmetry" gives the symmetry of the contribution of the term in the cross section associated with each of the response functions under the reflection of the ejectile momentum $\vec{p}'$ through the plane containing $\vec{q}$ and normal to the hadronic plane. The terms of interest are those with even symmetry which contribute in the electron scattering plane.

DWIA CALCULATIONS OF $(\vec{e},e'\vec{p})$

Before discussing the results it is useful to make some comment on the likely use of the information contained in the $(\vec{e},e'\vec{p})$ cross section. It is clearly a formidable task to undertake the separation of all eighteen response functions. For simple systems such a the deuteron it may be possible for a comprehensive program of measurements, including target and ejected nucleon polarization, to completely determine the transition current densities up to an overall phase. In this case, the greater degree of accuracy with which various dynamical models may be applied to the two-nucleon system may justify the effort inherent in a comprehensive separation of response functions. For many-body systems, the greater degree of uncertainty in our understanding of reactions in systems where many reaction channels may be open, militates against such an ambitious approach. A more modest and realistic approach would be to select response functions which show a high degree of sensitivity to the constituents of a given model of the reaction, or for which various models give disparate predictions. The choice of response functions would also be greatly influenced by the degree of difficulty required to separate them from the cross section. From the theoretical standpoint, it is useful to study the variations of the response functions caused by various dynamical ingredients or models.

Figures 3-5 represent an extension of the calculations of Ref. 16 to include the thirteen response functions which depend on the spin of the ejected proton. Results are presented for the ejection of a 135 MeV proton from the $1p_{1/2}$ shell of ^{16}O at a constant momentum transfer of 2.641 fm^{-1}. The response functions are plotted as a function of the magnitude of the recoil momentum $|\vec{p}'-\vec{q}|$. Four different calculations are presented for each of the eighteen response functions.

The solid lines represent the Dirac DWIA calculations as described in Ref. 16. These use Dirac optical model scattering wave functions for the ejected nucleon,[12] Dirac Hartree independent particle bound state wave functions[17] and the free Dirac current operator as given by

$$\Gamma^{\mu}(q) = F_1(Q^2) + \frac{F_2(Q^2)}{2m} i\, \sigma^{\mu\alpha} q_{\alpha} \;.$$

The dotted lines represent the Dirac PWIA calculation. The equivalent nonrelativistic DWIA is represented by the dashed lines. A calculation, which for convenience we will refer to as "on shell", is represented by the

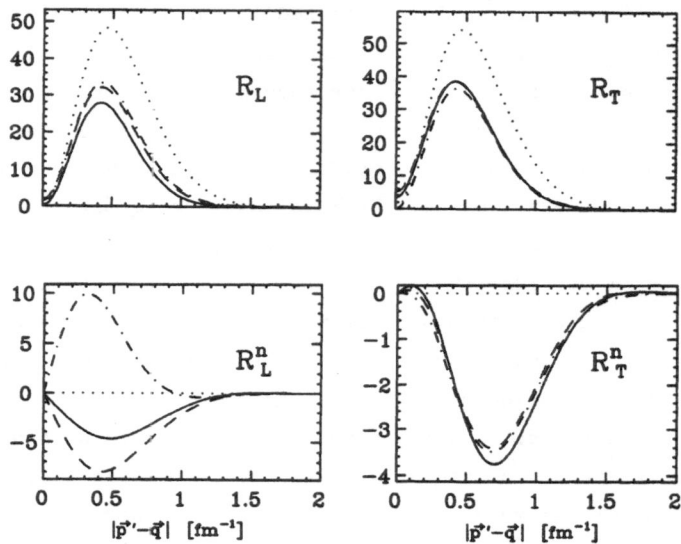

Fig. 3. Response functions for the knockout of 135 MeV protons from the $1p_{1/2}$ shell of ^{16}O at a constant momentum transfer of 2.64 fm^{-1} as a function of the recoil momentum. A description of the various calculations is included in the text.

dot-dashed lines. In this calculation, only the pole part of the propagator which appears in the Møller operator for the scattering wave function is kept. This forces the t matrix which appears in this Møller operator to be on shell.

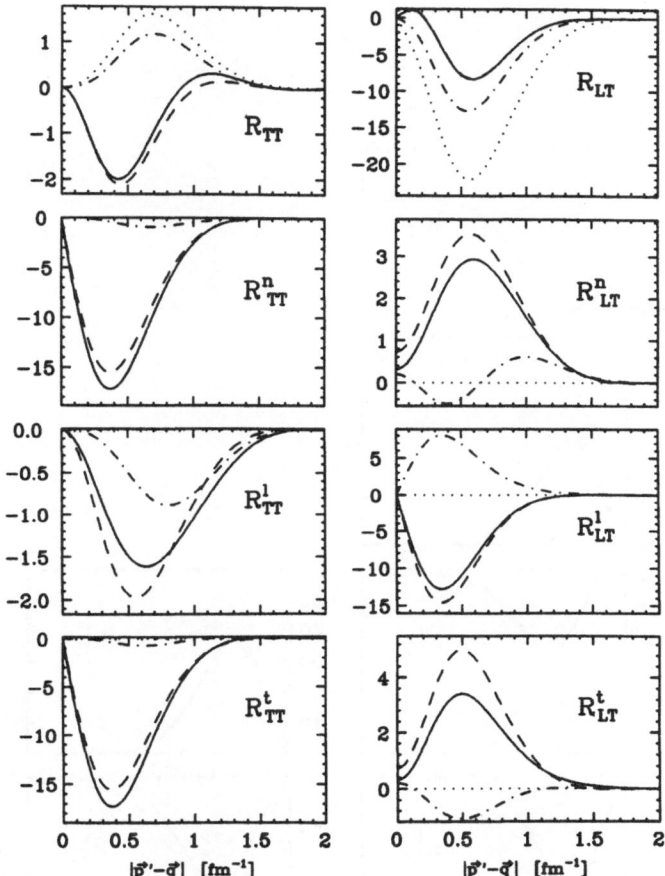

Fig. 4.   Response functions for the knockout of 135 MeV protons from the $1p_{1/2}$ shell of ^{16}O at a constant momentum transfer of 2.64 fm^{-1} as a function of the recoil momentum.   A description of the various calculations is included in the text.

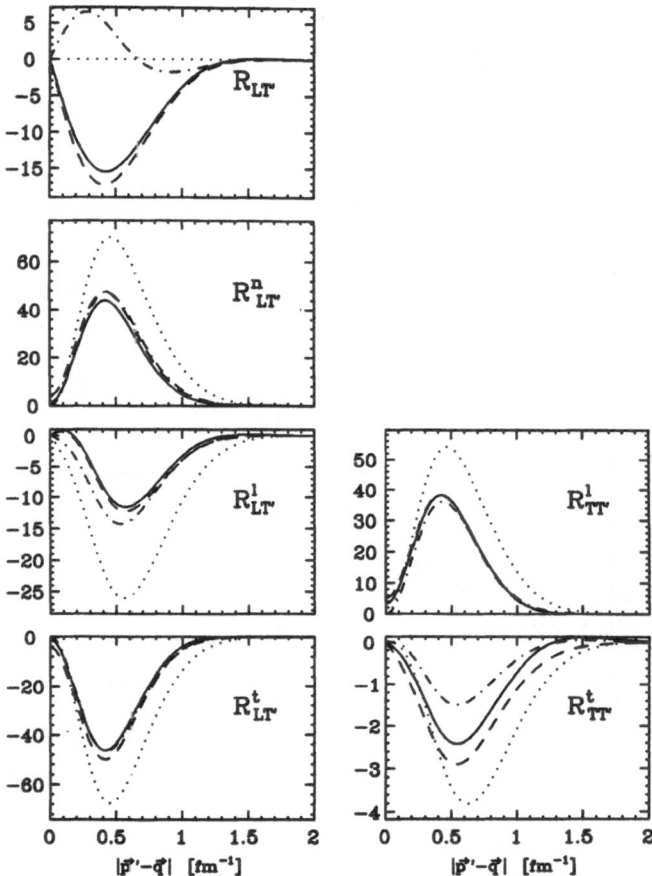

Fig. 5. Response functions for the knockout of 135 MeV protons from the
$1p_{1/2}$ shell of ^{16}O at a constant momentum transfer of 2.64 fm^{-1} as
a function of the recoil momentum. A description of the various
calculations is included in the text.

This calculation can be used as a rough measure of the sensitivity of the DWIA calculations to off-mass-shell components of the t matrix which are not highly constrained by elastic proton scattering. An important characteristic of these off-shell contributions to the scattering wave functions is that, since such contributions can not propagate to infinity, they are nonzero only in the nuclear interaction volume. Therefore, response functions which are very sensitive to the inclusion of the off-shell contributions are sensitive to the detailed characteristics of the scattering wave function in the nuclear interior. This has implications beyond the DWIA since it suggests that these response functions also have the potential to be sensitive to many-body effects, such as exchange currents and correlations, which modify the effective current operator in the nuclear interior.

A careful examination of the eighteen response functions shown in Figs. 3-5 shows that there is no consistent relationship among the various approximate calculations which holds for all of the response functions. This is not surprising since seven of the eighteen response functions cannot contribute in the PWIA, but do so in the various distorted wave calculations. This diversity suggests that it is indeed possible that a selective separation of response functions may be used to constrain various models of this reaction. Although there seems to be no global relationship between the four calculations, some interesting patterns appear in Figs. 3-5.

For the large response functions the dynamical differences between the relativistic and nonrelativistic DWIA calculations result in differences in size on the order of 5 to 10 percent with the longitudinal response function R_L showing an effect of 15 to 20 percent. For the smaller response functions the dynamical effects of relativity are on the order of 5 to 10 percent with the exceptions of R_{LT}^t and R_{TT}^l where the effects are from 25 to 35 percent, and R_L^n where the effect is 75 percent. In general, small response functions are more likely to be very sensitive to variations in the contents of calculations since they are small because of the cancellations of the leading order contributions which dominate the large response functions. Anything which can disturb these cancellations can result in large changes in these small response functions.

With the exception of R_T^n, all of the TP even response functions (those which vanish in PWIA) are very sensitive to the off-shell components of the scattering wave function. The transverse-transverse response function R_{TT} is also sensitive to off-shell components by virtue of the sensitive cancellation between the squares of the two transverse components of the transition current density.

Finally, it should be noted that a number of the polarization response functions are large. While many of these are also large in the PWIA, the two transverse-transverse response function R_{TT}^n and R_{TT}^t do not contribute in PWIA. The fact that several of the polarization response functions are large provides encouragement for efforts to measure the polarization response functions.

CONCLUSIONS

The sensitivities of the response functions to the various approximations presented here suggests that the measurement of response functions depending on ejected nucleon and/or electron polarizations may be able to discriminate between dynamical models of the (e,e'N) reaction. For example the response functions show a sensitivity to relativistic dynamics which vary from a few percent to 75 percent. Many of the response functions show a considerable sensitivity to the off-shell components of the scattering wave function. It is clear that possible advantages to be gained by measuring these new response functions merits an investment in studies of the feasibility of separating some or all of these response functions from the cross section and in efforts to develop any new experimental techniques which may be necessary to achieve this goal.

ACKNOWLEDGEMENT

The author would like to acknowledge the support of this work by the U. S. Department of Energy.

REFERENCES

1. R. Altemus et al., Phys. Rev. Lett. $\underline{44}$,965 (1980); P. Barreau et al., Nucl. Phys. $\underline{A402}$, 515 (1983); M. Deady et al., Phys. Rev. $\underline{C28}$, 631 (1984); Z.-E. Meziani et al., Phys. Rev. Lett. $\underline{52}$, 2130 (1984); M. Deady et al., Phys. Rev. $\underline{C33}$, 1897 (1986); C. C. Blatchley et al., Phys. Rev. $\underline{C34}$, 1243 (1986).
2. S. Frullani and J. Mougey, Adv. Nucl. Phys. $\underline{14}$,1 (1984).
3. P. E. Ulmer et al., Phys. Rev. Lett. $\underline{59}$, 1375 (1987).
4. T. de Forest, Jr., Phys. Rev. Lett. $\underline{53}$, 895 (1984).
5. G. Do Dang and N. Van Giai, Phys. Rev. $\underline{C30}$, 731 (1984).
6. C. R. Chinn and J. W. Van Orden, Bull. Am Phys. Soc. $\underline{32}$, 1030 (1987).
7. J. V. Nobel, Phys. Rev. Lett. $\underline{46}$, 412 (1981); Phys. Lett. $\underline{B178}$, 285 (1986).
8. L. S. Celenza, A. Rosenthal, and C. M. Shakin, Phys. Rev. Lett $\underline{53}$, 892 (1984); Phys. Rev. $\underline{C31}$, 232 (1985).
9. P. J. Mulders, Phys. Rev. Lett $\underline{54}$, 2560 (1985); Nucl. Phys. $\underline{A459}$, 525 (1986).

10. J. A. McNeil, J. R. Shepard, and S. J. Wallace, Phys. Rev. Lett. <u>50</u>, 1429 (1983); J. R. Shepard, J. A. McNeil and S. J. Wallace, Phys. Rev. Lett. <u>50</u>, 1443 (1983); B. C. Clark, S. Hama, R. L. Mercer, L. Ray, and B. D. Serot, Phys. Rev. Lett. <u>50</u>, 1644 (1983).

11. A. Picklesimer, P. C. Tandy, R. M. Thaler, and D. H. Wolfe, Phys. Rev. <u>C29</u>, 1582 (1984); <u>30</u>, 1861 (1984).

12. M. V. Hynes, A. Picklesimer, P. C. Tandy, and R. M. Thaler, Phys. Rev. Lett. <u>52</u>, 978 (1984); Phys. Rev. <u>C31</u>, 1438 (1985).

13. W. Fabian and H. Arenhovel, Nucl. Phys. <u>A314</u>, 253 (1979).

14. A. Picklesimer and J. W. Van Orden, Phys. Rev. <u>C35</u>, 266 (1987).

15. The term "parallel-antiparallel kinematics" is conventionally used to denote the relative orientation of the recoil momentum $\vec{P}_R$ with respect to the momentum transfer $\vec{q}$.

16. A. Picklesimer, J. W. Van Orden and S. J. Wallace, Phys. Rev <u>C32</u>, 1312 (1985).

17. C. J. Horowitz and D. D. Serot, Nucl. Phys. <u>A368</u>, 503 (1981).

RELATIVISTIC TREATMENT OF MESONIC CONTRIBUTIONS TO QUASIELASTIC (e,é)

P. G. Blunden and M. N. Butler

TRIUMF
4004 Wesbrook Mall
Vancouver, BC, Canada V6T 2A3

It is clear now that meson exchange currents play an important role in the description of observables in electron scattering. This has been shown most dramatically in the description of the high momentum components of the elastic form factors of $A=3$ nuclei.[1] For quasielastic scattering, two calculations exist, one which examines the effects at the quasielastic peak for 1p–1h exchange currents,[2] and one which examines 2p–2h contributions.[3] (It should be pointed out that the results of Ref. [2] are somewhat confusing, since they show a suppression of the quasielastic cross section when exchange currents are included, rather than the expected enhancement.) All of these calculations have been performed with the approximation that the system is non-relativistic. The validity of this may be called into question by the fact that the exchange-current effects are most significant at high momentum transfers and often are coupled to high momentum components of the nuclear wavefunction. It can also be unclear which exchange current processes should be included and which are already included in the bound state structure of the nucleus.

Another problem is that current conservation, evaluated using Siegert's theorem, is preserved only to $\mathcal{O}(q/M)$ (q is the momentum transfer and M is the nucleon mass). Relativistic models for the bound state structure of the nucleus contain the mesons as explicit degrees of freedom, providing a framework for calculating their effects, and virtually all other dynamical effects, in a completely self–consistent manner. Since the mesons and nucleons are coupled to the electromagnetic field in a gauge invariant way, it is possible to write down an electromagnetic current for the nucleons and mesons that is conserved both in theory and in practice. We use a relativistic model with a pseudovector pion coupling to study the exchange current contributions, with emphasis on quasielastic kinematics.

We begin with a Lagrangian for nucleons interacting with scalar and vector mesons, together with a pseudovector coupling to pions:

$$
\begin{aligned}
\mathcal{L} = {}& \bar{\psi}(i\slashed{\partial} - M)\psi - \frac{1}{4}F_{\mu\nu}F^{\mu\nu} - \frac{1}{4}G_{\mu\nu}G^{\mu\nu} + \frac{1}{2}(\partial_\mu\phi)^2 + \frac{1}{2}(\mathcal{D}^\mu_{ij}\pi^j)(\mathcal{D}^{ik}_\mu\pi_k) \\
& - \frac{1}{2}m_s^2\phi^2 - \frac{1}{2}m_\pi^2\bar{\pi}^2 - \frac{1}{2}m_v^2V_\mu^2 \\
& - g_s\phi\bar{\psi}\psi + e\bar{\psi}\Gamma_\mu A^\mu\psi + g_v\gamma_\mu V^\mu\bar{\psi}\psi - i\frac{g_\pi}{2M}\gamma_5\bar{\psi}\gamma_\mu\tau^i\mathcal{D}^\mu_{ij}\pi^j\psi,
\end{aligned}
\tag{1}
$$

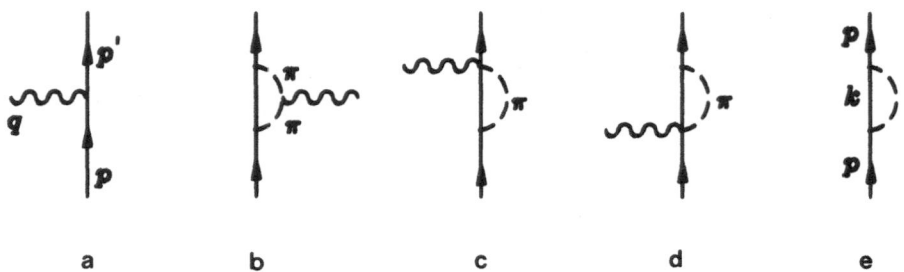

Fig. 1. Irreducible diagrams contributing to the one-body current (a–d) and the pionic contribution to the Hartree-Fock self energy (e).

where

$$\mathcal{D}^{\mu}_{ij} = \delta_{ij}\partial^{\mu} - i\frac{e}{2}A^{\mu}\epsilon_{ij3} , \quad \Gamma_{\mu} = F_1(Q^2)\gamma_{\mu} + i\frac{F_2(Q^2)}{2M}\sigma_{\mu\nu}q^{\nu} ,$$

and

$$F_{\mu\nu} = \partial_{\mu}A_{\nu} - \partial_{\nu}A_{\mu} , \quad G_{\mu\nu} = \partial_{\mu}V_{\nu} - \partial_{\nu}V_{\mu} .$$

The bound state problem using this model has been studied in detail elsewhere.[4] Here, we use this Lagrangian to derive the one and two-body electromagnetic currents. In doing so, we can use Ward–Takahashi Identities, relating the Hartree-Fock self-energy to the vertex function, to verify our results.[5]

Fig. 1 shows the irreducible diagrams that could contribute to the one-body electromagnetic current up to first order in g_{π}^2. Diagram (a) represents the standard impulse approximation, and diagrams (b–d) represent the meson-exchange current corrections. Diagram (e) gives the Hartree-Fock self energy. In a Fermi gas, the expressions for these diagrams are given by

$$\Lambda^a_{\mu} = \left(F_1 + \frac{M^*}{M}F_2\right)\gamma_{\mu} - \frac{F_2}{2M}(p + p')_{\mu} ,$$

$$\Lambda^b_{\mu} = i\left(\frac{g_{\pi}}{2M}\right)^2 2\tau_3 F_{\pi} \int \frac{d^4k}{(2\pi)^4} \frac{\gamma_5(\slashed{p}' - \slashed{k})G_D(k^*)\gamma_5(\slashed{p} - \slashed{k})}{((p-k)^2 - m_{\pi}^2)((p'-k)^2 - m_{\pi}^2)}(p' + p - 2k)_{\mu} ,$$

$$\Lambda^{c+d}_{\mu} = -i\left(\frac{g_{\pi}}{2M}\right)^2 2\tau_3 F_{\pi} \int \frac{d^4k}{(2\pi)^4} \left(\frac{\gamma_5\gamma_{\mu}G_D(k^*)\gamma_5(\slashed{p} - \slashed{k})}{((p-k)^2 - m_{\pi}^2)} + \frac{\gamma_5(\slashed{p}' - \slashed{k})G_D(k^*)\gamma_5\gamma_{\mu}}{((p'-k)^2 - m_{\pi}^2)}\right) ,$$

$$\Sigma^{HF} = i\left(\frac{g_{\pi}}{2M}\right)^2 \vec{\tau}\cdot\vec{\tau} \int \frac{d^4k}{(2\pi)^4} \frac{\gamma_5(\slashed{p} - \slashed{k})G_D(k^*)\gamma_5(\slashed{p} - \slashed{k})}{((p-k)^2 - m_{\pi}^2)} .$$

Here M^* is the effective mass and $G_D(k^*) = 2\pi i(\slashed{k}^* + M^*)\delta(k^{*2} - M^{*2})\theta(k_F - k)\theta(k_0)$ is the propagator for an occupied single particle state. F_1 and F_2 are parameterizations of the nucleon form factors given by Preston and Bhaduri,[6] and we use $F_{\pi}(Q^2) = F_1^p - F_1^n$ so that the electromagnetic current is conserved. In order to simplify the calculation, we neglect any momentum dependence in the self energies and approximate $k_{\mu}^* = (k_0 + \Sigma_0, \mathbf{k})$. This is an excellent approximation justified by relativistic Hartree-Fock calculations. One can now replace $p_{\mu} - k_{\mu}$ by $p_{\mu}^* - k_{\mu}^*$ in the above expressions. We can then evaluate the effective one-body current between on-shell spinors as

$$J_{\mu} = \bar{u}(p'^*)[\Lambda_{\mu}^{1a} + \Lambda_{\mu}^{1b} + \Lambda_{\mu}^{1c} + \Lambda_{\mu}^{1d}]u(p^*) .$$

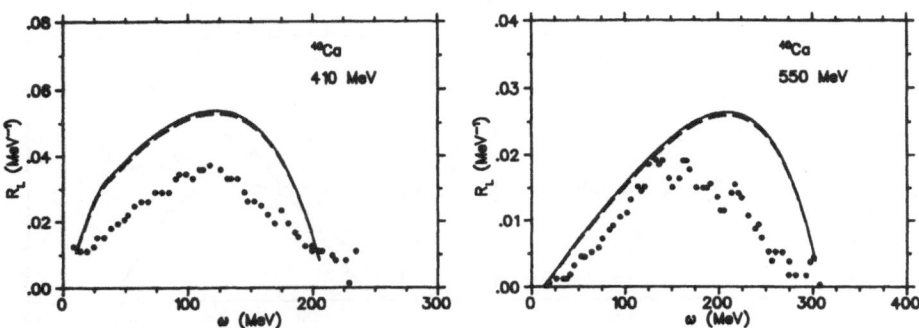

Fig. 2. $R_L(q,\omega)$ for ^{40}Ca at q=410 and 550 MeV. The solid curve represents the relativistic impulse approximation and the dashed curve the result including the mesonic effects. Data from Ref. [7].

Making the replacement $\not{p}^* u(p^*) = M^* u(p^*)$, one finds general forms for the time–like and space–like parts of the current, given by

$$J_0 = \bar{u}(p'^*)[A_1\gamma_0 + A_2]u(p^*)\,,$$
$$J_i = \bar{u}(p'^*)[B_1\gamma_i + B_2 p_i + B_3\gamma_i\gamma_0 + B_4\gamma_0 p_i]u(p^*)\,.$$

The factors A_j and B_j can be loosely thought of as medium form factors and are functions of q, ω, k_F and p. The lack of manifest Lorentz invariance is due to the fact we have evaluated the expressions for the vertex functions in a specific frame. In general, the current is indeed Lorentz invariant.

We can now use these expressions for the current to calculate the longitudinal (R_L) and transverse (R_T) pieces of the quasielastic cross section for various nuclei and kinematics. These are given as

$$R_L = V\sum_{\text{isospin}}\int\frac{d^3p}{(2\pi)^3}\frac{\text{Tr}[J_0^\dagger J_0]}{4E_p^* E_{p+q}^*}\theta(k_F - p)\theta(|\vec{p}+\vec{q}| - k_F)\delta(\omega + E_p - E_{p+q})\,,$$

$$R_T = V\sum_{\text{isospin}}\int\frac{d^3p}{(2\pi)^3}\left(\delta^{ij} - \frac{q^i q^j}{q^2}\right)\frac{\text{Tr}[J_i^\dagger J_j]}{4E_p^* E_{p+q}^*}\theta(k_F - p)\theta(|\vec{p}+\vec{q}| - k_F)\delta(\omega + E_p - E_{p+q})\,.$$

M^* and k_F are parameters which we use to fit the position of the quasielastic peak. In the future, we would like to input these from nuclear structure calculations of the initial and final states.

As an example, let us consider the Saclay data for ^{40}Ca at q=410 and 550 MeV.[7,8] The results of our calculation for R_L in the relativistic impulse approximation, and with the inclusion of exchange currents can be seen in Fig. 2 (the parameters used are $M^* = 0.8M$ and $k_F = 1.3$ fm^{-1}). It is clear that the exchange currents have a negligible effect in this channel, as would be expected. On the other hand, in Fig. 3 we see the results for R_T and here the effects are more significant. At both energies, the location of the quasielastic peak is shifted to lower energy transfer, and there is an enhancement in overall strength. This effect comes in large part from a correct treatment of the relativistic kinematics, as the retardation in the pion propagators turns out to be a fairly small enhancement and contributes mainly to the low ω region of the structure function. Lastly, in Fig. 4, we show

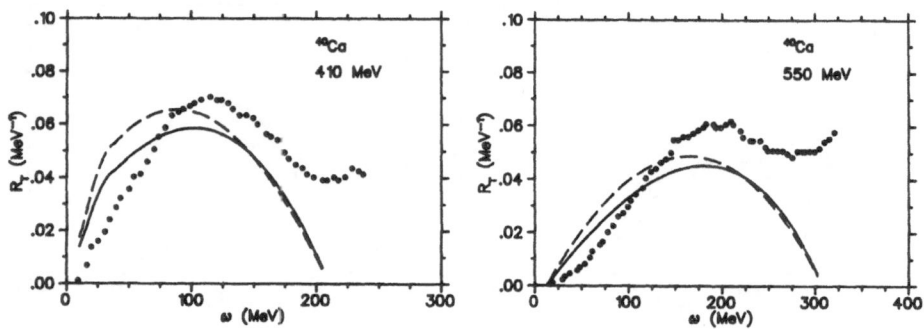

Fig. 3. Same as Fig. 2, but for R_T. Data from Ref. [8].

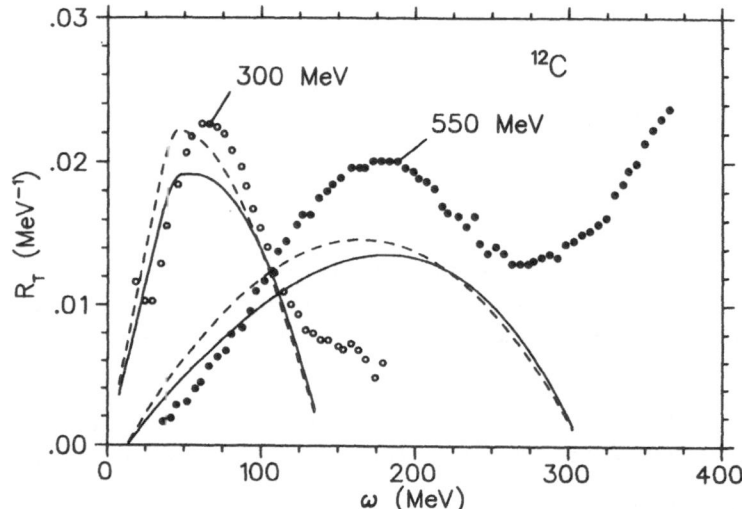

Fig. 4. The structure function R_T for ^{12}C at q=300 and 550 MeV. The solid curve represents the impulse approximation result, and the dashed line the result including exchange currents. Data from Ref. [9].

results for the transverse structure function of ^{12}C at q=300 and 550 MeV (data from Ref. [9]). We see the same sort of behaviour here that we saw in the case of ^{40}Ca.

We can also apply our model to the question of the ratio of R_T/R_L in quasielastic $(e, e'p)$ for parallel kinematics. Specifically, we study

$$R_G = \left(2M^2 \frac{q^2}{Q^4} \frac{R_T}{R_L} \right)^{1/2} .$$

Significant deviations from the expected value of $R_G = G_M^p/G_E^p$ were found at NIKHEF[10] for ^{12}C$(e, e'p)^{11}$Be, and in Saclay data[7,8] on the inclusive reaction ^{12}C(e, e'). Several attempts have been made to explain this, among them studies of the final-state interaction of the struck proton with the residual nucleus.[11] We have found that, in our model, the inclusion of exchange currents into R_T gives a non-negligible contribution to the enhancement of R_G seen experimentally (Fig. 5).

In conclusion, we have found that the effects of meson-exchange currents are important in a relativistic model, particularly with respect to the location of the quasielastic peak. However, there are a few points which need to be pursued further. It should be noted that the Lagrangian of Eq. (1) does not contain explicit information about the Δ, which will be needed for both completeness and a reasonable description of pion electroproduction. Another point is that we do not include RPA corrections in either R_L or R_T. These corrections might be balanced, however, were we to include the delta. The contribution to the structure functions of two-nucleon knockout due to exchange currents is also being studied, especially with regard to the so-called "dip" region, and we hope to have results on that soon.

This work has been supported by the Natural Sciences and Engineering Research Council, Canada.

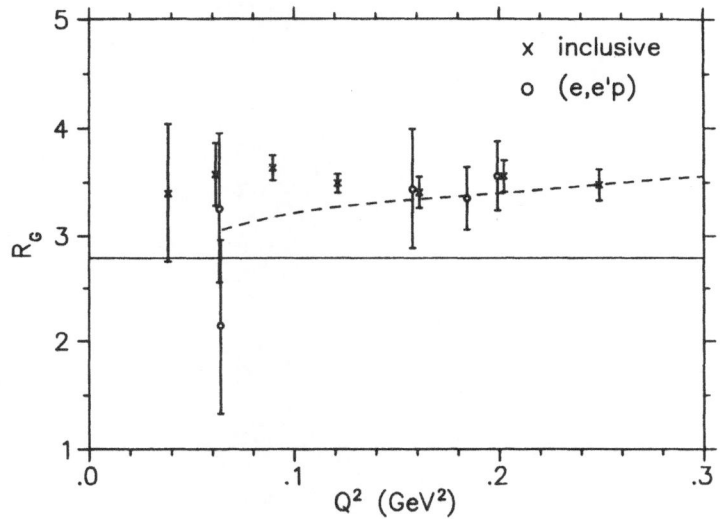

Fig. 5. The ratio R_G. The solid curve is the free proton result, and the dashed curve includes exchange currents in R_T. Data from Ref. [10]

REFERENCES

[1] See, for example, J.L. Friar, *New Vistas in Electro-Nuclear Physics*, Nato ASI Series (Plenum, New York NY, 1986) p213.
[2] M. Kohno and N. Ohtsuka, Phys. Lett. **98B**, 335 (1981).
[3] J.W. Van Orden and T.W. Donnelly, Ann. Phys. **131**, 451 (1981).
[4] B.D. Serot and J.D. Walecka, Adv. in Nucl. Phys. **16**, 1 (1986).
[5] W. Bentz et al., Nucl. Phys. **A436**, 593 (1985).
[6] M.A. Preston and R.K. Bhaduri, *The Structure of the Nucleus* (Addison-Wesley, Reading, Mass., 1975).
[7] Z.E. Meziani et al., Phys. Rev. Lett. **52**, 2130 (1984).
[8] Z.E. Meziani et al., Phys. Rev. Lett. **54**, 1233 (1985).
[9] P. Barreau et al., Nucl. Phys. **A402**, 515 (1983).
[10] G. van der Steenhoven et al., Phys. Rev. Lett. **57**, 182 (1986).
[11] T.D. Cohen, J.W. Van Orden and A. Picklesimer, Phys. Rev. Lett. **59**, 1267 (1987), and T. Suzuki, Phys. Rev. **C37**, 549 (1988).

THE (p,2p) REACTION IN THE DIRAC IMPULSE APPROXIMATION

E.D. Cooper

Dept. of Physics, McGill University, 3600 University Street
Montreal, Quebec, Canada H3A 2T8
and
TRIUMF, 4004 Wesbrook Mall, UBC Campus
Vancouver, B.C., Canada T6G 2N5

O.V. Maxwell

Dept. of Physics, Florida International University
Tamiami Trail, Miami, Florida, 33199, USA
and
TRIUMF, 4004 Wesbrook Mall, UBC Campus
Vancouver, B.C., Canada T6G 2N5

ABSTRACT

In this paper we present a formalism and preliminary results for the (p,2p) reaction on ^{40}Ca at incident bombarding energy of 300 MeV. The calculation is essentially that of an impulse approximation using a relativistic Love Franey t-matrix. The proton distortions are however obtained from a global optical potential. Exchange graphs are treated on an equal footing with direct graphs, and finite range effects are exactly included. The agreement with data is seen to be qualitative. Possible reasons for the disagreement are given.

1. INTRODUCTION

Recently, proton knockout reactions have become the focus of renewed interest, both experimental and theoretical. A historical review is contained in ref [1], whereas a typical modern calculation can be found in ref [2] and subsequent references. Much of this interest lies in the possibility that (p,2p) data and that of related reactions can be used to directly map out the momentum distributions of single proton hole states in nuclei. Within the impulse approximation picture, the principle is clear: the incident proton ejects a second proton from the target nucleus, which is then left in a single hole state. By adjusting the kinematics of the reaction so as to select different nuclear recoil momenta, the desired information concerning the momentum components of the single hole state can be extracted.

In practice the situation is not so simple. In light nuclear systems, explicit two step processes can be expected to make significant contributions to the t-matrix, thereby masking the impulse approximation probe of single nucleon momentum components. In heavy nuclei, initial and final state interactions come into play, requiring the use of distorted waves. In the past the lack of reliable distorted waves has proven to be a serious source of ambiguity in (p,2p) calculations, which in view of the order of magnitude drop in the cross sections resulting when distorted waves are used instead of plane waves, clearly require a careful treatment of distortions. In this respect, (e,e'p) reactions enjoy an advantage over (p,2p) reactions in that only one of the three particles involved experiences nuclear distortions. On the other hand, (p,2p) cross sections are much larger than (e,e'p) cross sections so that with a reliable treatment of distortions, (p,2p) data, especially that obtained with a dual arm spectrometer system, has the possibility of providing nuclear structure information more readily than (e,e'p) data.

The ability to reliably calculate nuclear distorted waves has been boosted considerably in the last several years by the development of the Dirac impulse approximation (DIA) [3] [4] [5]. The DIA has been used to generate an optical potential which has provided a good description of elastic scattering for a wide range of incident energies and target nuclei. Both the real and imaginary parts of the optical potential are essential to obtain agreement with the elastic data. Since the imaginary part includes contributions to elastic scattering from the (p,2p) channel and since the DIA optical potential reproduces the elastic data so successfully, it does not seem unreasonable to expect the DIA to describe the (p,2p) reaction directly. This observation has motivated us to undertake a study of (p,2p) reactions in which nuclear distortion effects are treated completely relativistically in the DIA. We note that our study is not unique in this regard. There now exist zero range relativistic calculations [6]. Our calculations are the first we know of to include both relativistic and finite range effects simultaneously.

Dynamical (p,2p) calculations require the NN t-matrix, both on and off-shell. Since, strictly speaking, only the on-shell part is known phenomenologically, some off-shell ansatz has to be adopted. The assumption of maximum locality[3], has proven quite successful in the description of nucleon-nucleus elastic scattering, especially at energies exceeding 300 MeV. Below this it expected that exchange effects and their associated non-localities will become important. A simple parameterisation of the NN t-matrix as a sum of Yukawa terms has become available [7] in which the direct and exchange contributions to the t-matrix are considered separately. After incorporating medium effects (Pauli blocking) via an effective G-matrix, this t-matrix and the impulse approximation have yielded [5]. a good description of the proton elastic scattering data all the way down to 150 MeV. Since in (p,2p) reactions one of the protons can be quite low in energy, we expect that probably explicit exchange effects and Pauli Blocking effects on the t-matrix we are using will also be important. However the preliminary calculations we present here include just the separate treatment of exchange effects: the effect of the Pauli Blocking factors will be discussed elsewhere. For completeness, a brief description of the t-matrix is presented in section 2.

In a completely consistent treatment it would be desirable to use the same NN t-matrix to describe nuclear distortions as drives the (p,2p) reaction. Unfortunately the t-matrix we use, even corrected for Pauli Blocking effects, is not sufficiently reliable at very low energies (less than 100 MeV) to make its use tenable for obtaining the distorted wavefunctions. Instead, we have generated distorted waves using two different global optical potentials, which have been fit over a wide range of energies. These are described in section 3.

In section 4 we introduce the Dirac distorted wave formalism which has been employed to obtain the (p,2p) matrix elements in the present study. This formalism combines the NN t-matrix discussed in section 2 with the distorted waves discussed in section 3 and a bound state solution of the Dirac equation containing vector and scalar potentials. Our philosophy here has been to make as few approximations as possible in order the test the physics of the model adopted with as little ambiguity as possible. In particular, in contrast with most previous (p,2p) calculations, we have performed a full finite range calculation. This allows us to distinguish relativistic and distortion effects in our results from effects that are just artifacts of the zero-range approximation. Preliminary results obtained with the model are presented in section 5, while section 6 contains a summary, some conclusions, and an outline of future work.

2. HOROWITZ-LOVE-FRANEY t-MATRIX

The nucleon-nucleon (NN) t-matrix employed in this work is based upon a relativistic Love-Franey formalism that is described in detail in ref. [7]. The major advantage of this formalism over previous analyses is that the direct and exchange contributions to the NN t-matrix are each treated explicitly, so that scattering processes at relatively low energy can be considered with some degree of reliability. The direct part of the t-matrix, expressed as a function of the total centre of mass energy s and the squared momentum transfer variable t, is given by

$$T_d(s,t) \;\; = \;\; \sum_{i=1}^{5} T_i(s,t) O_i^1 \cdot O_i^2 \tag{1}$$

where O_i are the Lorentz invariant amplitudes. The real and imaginary parts of the complex coefficients $T_i(s,t)$ are separately parameterised by expressions of the form

$$Re(T_i(s,t)) \;\; = \;\; \frac{g_i(s)^2}{m_i(s)^2 - t} \{ \frac{\Lambda_i(s)^2}{(\Lambda_i(s)^2 - t)} \}^2 \tag{2}$$

for the real parts and corresponding expressions for the imaginary parts, where the various parameters are fixed by fitting the NN amplitudes at each value of s.

The exchange contributions to the NN t-matrix are functions of s and the mandelstam variable u, so that the complete t-matrix is a function of all three mandelstam variables, as it should be, in contrast with the original impulse approximation, which employs a t-matrix dependent on only s and t. How well the particular parameterisation given by eq. 2 represents the off-shell properties of the NN t-matrix is difficult to determine. On the one hand, it works quite well for elastic scattering; on the other hand, it is well known that this type of parameterisation does not guarantee unitarity off shell [8].

One of the major ambiguities in the off-shell properties of the NN t-matrix concerns the choice of vertex in the pseudoscalar amplitude. The simplest choice is a pseudoscalar vertex; however, a pseudovector vertex is more compatible with the large vector and scalar fields which are required in the formalism by considerations involving pion production[9]. Also, in studies based on effective chiral theories, the convergence properties of the pseudovector lagrangian have been found superior to those of the pseudoscalar lagrangian [10]. Hence, we adopt the pseudovector form, rather than the simpler pseudoscalar one.

At low energies, medium effects in the nucleus arising from the Pauli blocking of intermediate states can be important. One means by which such effects can be taken into account is to replace the NN t-matrix by a density dependent g-matrix. In the case of elastic scattering, a local density approximation was employed [5] to estimate the influence of medium effects. As expected, they are non-negligible at the lower end of the energy spectrum. In principle such effects can also be incorporated in the (p,2p) formalism without major difficulties; however, they have not been included in the results reported in this paper, which are confined to incident lab energies of 300 MeV. It is expected that the influence of medium effects on the (p,2p) matrix elements will be studied in future work.

103

3. GLOBAL OPTICAL POTENTIALS (GOP'S)

To evaluate the (p,2p) matrix elements of the t-matrix discussed above, one requires distorted wave functions in both the incoming and outgoing channels. These are generated through solution of the Dirac equation with both vector and scalar distorting potentials. In principle, the distorting potentials should be obtained at each energy from a microscopic calculation using the same NN t-matrix as drives the (p,2p) calculation. Unfortunately no impulse calculation to date has proved itself trustworthy below 135 MeV, whereas (p,2p) calculations often require distorting potentials as low as 60 MeV.

An alternative procedure, adopted here, is to employ a global optical potential (GOP). The principles underlying these are discussed in detail in refs [11] and [12]. In essence, the idea is to use a single optical potential with energy-dependent parameters that have been fit to the elastic data over a wide range of energies. Not only does this procedure avoid the difficulties encountered with interpolating energy independent potentials, but it is also convenient from a computational point of view since the GOP can be hard wired into the program and distortions calculated without the need for additional input.

In the present work, two different GOPs are employed. Neither of these GOPs represents a superb fit to the data of the quality found in refs [11] and [12], however they have the property of spanning, in one parameterisation, the desired energy range of 70 to 300 Mev. The two GOPs do provide reasonable fits to the data in this desired range. At the time of writing, better GOPs are now available in this range. The effect of using these GOPs will be studied in more detail later. All GOPs presently consist of vector and scalar potentials that are sums of symmetric Woods-Saxon forms with energy-dependent strengths. In particular,

$$V_v(E,r) = V_v(E)f(r,R_{v_1},a_{v_1}) + iW_v(E)f(r,R_{v_2},a_{v_2})$$
$$+ iX_v(E)f'(r,R_{v_3},a_{v_3}) \tag{3}$$

$$V_s(E,r) = V_s(E)f(r,R_{s_1},a_{s_1}) + iW_s(E)f(r,R_{s_2},a_{s_2})$$
$$+ iX_s(E)f'(r,R_{s_3},a_{s_3}) \tag{4}$$

The shapes of these potentials are assumed to be independent of energy, except that the imaginary part of the potential GOP1 passes smoothly from a fixed surface peaked shape at low energy to a volume shape at high energy. A symmetric Woods-Saxon shape is adopted in preference to the ordinary one, since the symmetric form is better behaved at short distances and can be easily Fourier transformed to momentum space, if so desired. In particular, we have

$$\int f(r,R,a)d^3r\,e^{i\vec{q}\cdot\vec{r}} = \frac{\pi a}{2q}\frac{exp(R/a)}{cosh(R/a)}\frac{\partial}{\partial q}\{\frac{sin(qR)}{sinh(q\pi a)}\} \tag{5}$$

with

$$f(r,R,a) = \frac{1}{(1+exp(\frac{r-R}{a}))(1+exp(\frac{-(r+R)}{a}))} \tag{6}$$

It should be noted that for calcium, the ratio R/a is so large that the second factor in the denominator above can be effectively replaced by 1; i.e., the results obtained with the symmetric and ordinary Woods-Saxon forms are nearly identical.

The GOP strengths are expressed as polynomials in the energy. For the vector potentials we have

$$V_v(E) = \sum_{n=0}^{3} V_v^{(n)}(E/400)^n \tag{7}$$

$$W_v(E) = \sum_{n=0}^{3} W_v^{(n)}(E/400)^n \tag{8}$$

$$X_v(E) = X_v^{(0)} e^{-X_v^{(1)}(E/400)} \tag{9}$$

The scalar potentials are parameterised in a similar manner. As mentioned above, (p,2p) calculations have been carried out with two different GOPs in the present study. This permits us to estimate the sensitivity of the results to the particular form of the distorting potentials. One of the potentials, GOP1, fits the elastic scattering data on ^{40}Ca from 21 MeV through 1040 MeV (19 energies) and incorporates surface absorption. The other, GOP2, was tailored explicitly for use in the (p,2p) calculations presented here. It spans the energy range from 80 through 300 MeV (6 energies) and consists of just symmetric Woods-Saxon forms (no surface absorption). The strength and radial parameters for both potentials are listed in Table 1.

4. DISTORTED WAVE FORMALISM

In the mean field approximation, the lowest order contribution to the *direct* part of the (p,2p) t-matrix is given, with spin labels suppressed, by the expression

$$T(p_i, E_i, p_{f_1}, p_{f_2}) = \int \int d^4x\, d^4x'\, \bar{\Psi}(p_{f_1}, x)_1 \bar{\Psi}(p_{f_2}, x')_2$$

$$g(x) t_d(s, x - x') g(x') \Psi(p_i, x)_1 \Psi(E_i, x')_2$$

$$\tag{10}$$

which contains, in an obvious notation, one incoming and two outgoing distorted waves, a single nucleon overlap factor (bound state) with binding energy E_i, and the Fourier transform of the t-matrix in eq. 2 with respect to the mandelstam variable t. The three distorted waves are obtained in a standard way from one of the two GOPs discussed in the preceding section. For the single nucleon overlap factor, we employ a bound state solution of the Dirac equation containing vector and scalar bound state potentials. These bound state potentials are obtained by extrapolating GOP1 down in energy from 21 MeV and correcting the resulting strengths to get proton separation energies of 6.7 and 15.2 MeV for the $1D_{3/2}$ and $1D_{5/2}$ levels respectively. With the geometrical parameters listed in table 1, this yields vector and scalar strengths of 331 MeV and -424 MeV. The factors g(x) in eq. 10 are the Pauli blocking factors discussed in section 2. They have been set equal to unity for all the results presented here.

The exchange part of the (p,2p) t-matrix is obtained from the direct part, eq. 10, by the interchange of the two final proton states. The complete t-matrix is then the difference between the direct part and the exchange part.

The evaluation of the expression in eq. 10 involves two four dimensional integrals. The time parts of these integrals can be performed analytically in the usual manner, yielding the energy conserving delta-function and fixing the energy transfer between the incident nucleons in the t-matrix. The remaining integrals have to be performed numerically and with the aid of partial wave expansions. In order to test the physics of the interaction model with as little uncertainty as possible, we have constructed a numerical code that evaluates these integrals without approximations. With minor modifications this code can also be used to study (e,e'p) reactions and to study (e,2e) reactions, which are of interest in atomic physics.

The integral over $\vec{r'}$ is actually a sum of folding integrals involving Yukawa exchanges. The radial parts of these are performed by transforming the integrals into equivalent differential equations, which are then solved by a stabilized marching method. The remaining three dimensional integral over $\vec{r}$ could be accomplished by means of a partial wave expansion; however, because of the large number of intermediate angular momentum sums involved and the relatively slow convergence of the partial wave series, this technique proves to be very inefficient. An alternative technique, first employed in zero-range calculations in ref [13] and adopted in the finite range calculations presented in this work, is to carry out the final three dimensional integral directly using a three dimensional Gaussian grid.

For the incident and outgoing energies considered here, it happens that the energy transfer q_0 can exceed the masses of some of the quanta exchanged in the NN t-matrix. When this occurs, the range parameter $\mu = \sqrt{mass^2 - q_0^2}$ is imaginary so that the corresponding Yukawa function is oscillatory rather than exponentially decaying. This feature is unique to the relativistic approach. For such terms the radial part of the final three dimensional integral is too slowly convergent to make a straight forward evaluation along the real axis practical. To handle these terms, we employ a procedure developed in ref [14] where the integration contour is rotated from the real axis into the complex plane, thereby transforming an oscillatory integral into one which is exponentially convergent. This procedure also provides a numerical test of the program in that the calculated integral must be independent of the 'turning radius' where the integration contour first turns away from the real axis. Since the integration in the complex plane is carried out separately from that along the real axis and moreover, involves the nuclear phase shifts, the calculated integral will be independent of the turning radius only if the distortions are correctly incorporated in the code. We have confirmed that the numerical code indeed satisfies this criterion. We have also compared the results obtained with the code in the absence of distortions with those obtained from an independent momentum space pwba code. The two sets of results agree reasonably well for all energies and angles tested.

5. RESULTS AND CONCLUSIONS

We present here only results for the reactions $^{40}Ca(p, 2p)^{39}K(3/2^+, 5/2^+)$. In figures 1 and 2 we show the results for the cross section and analysing power to the $1D_{5/2}$ and the $1D_{3/2}$ states respectively, plotted as a function of the energy difference of the two protons coming out at 30° and −55°. Figure 3 is an enlargement of scale of figure 1. The data is that of Kitching et al[15] multiplied by 2 as these authors give the 5-fold differential cross section with respect to e1-e2, whereas the calculations are for cross sections with respect to e1. All calculated curves assume full occupancy of the shells.

The dashed curves are plane wave calculations. Immediately apparent is that the PWBA severely overestimates the cross sections. The solid curves are the result of what we call our full calculation. This is using the GOP_1 for the 3 distortions, using the bound state wavefunction obtained from this potential. By considering the difference between the PWIA and DWIA calculations for the *cross-sections* one may argue that the effect of distortions is to just reduce the cross-sections uniformly, presumably due to flux absorption. Whilst this is almost true (the peak is shifted towards the position of the peak in the data also), it is apparent from a comparison of DWIA and PWIA analysing powers that more is going on. This is most dramatic for the 1d5/2 state, where the analysing powers change sign in the vicinity of the quasi-elastic peak. What is the physical origin of this change of sign? We note that the non-relativistic calculations of Kudo et al predict a positive analysing power here. Ergo, it seems that this sign change may be a relativistic effect. Indeed Kudo's zero-range relativistic calculations do show such a sign change. However, we reserve final comment on this point until more calculations have been done.

The full circles in figure 2, are the results of doing the calculation with GOP2, but keeping the same bound state. The differences between the GOP1 and GOP2 calculations are much smaller than between either and the PWIA calculations, but it is interesting to see that the differences between the curves from the two global potentials is comparable to that between the calculations and the data. This implies that if one wants to improve the calculations, the first thing to try is to recalculate

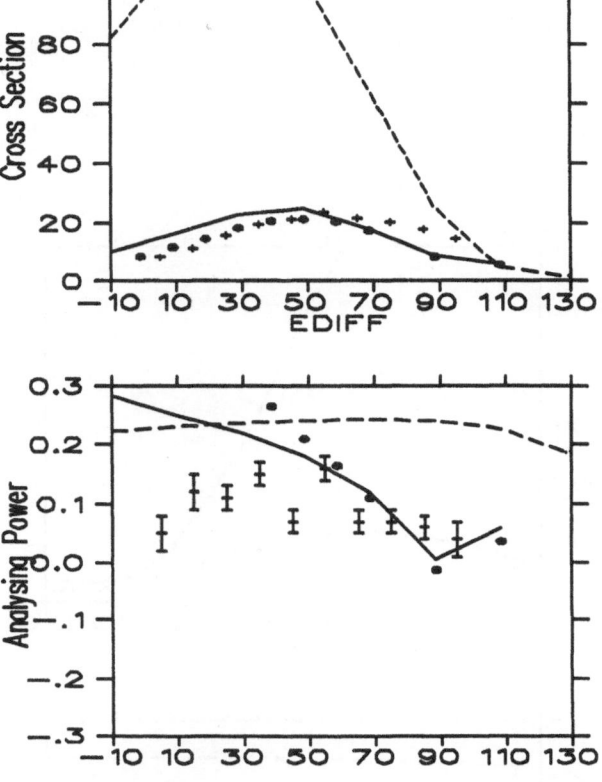

Figure 1. Data and results for $1D_{3/2}$ state. Dashed line is PWBA, solid line and dots are DWBA calculations using GOP1 and GOP2 respectively. Crosses are data of ref [15].

with an improved pair of GOPs. To our amusement we find that the curves with the two Gops come together when the energy difference becomes large and positive. This is when the proton going off at the smallest angle, 30°, has the highest energy. We have no explanation for this at present.

There are several other things that must be explored before the 'final' DWIA answer can be given in this model. We have already touched on the point of sharpening up the GOPs. There is also the question of modifying the off-shell behaviour of the t-matrix used [17] (it is not difficult to impose off shell unitarity to first order in these calculations). The question of how sensitive we are to the difference between using a pseudovector or pseudoscalar coupling for the pion is interesting. A quick calculation has shown us that the results do not depend critically on this choice, but this needs to be confirmed. Also the choice of the bound state wavefunction used must be examined. In (p,π^+) calculations it seems that the use of a wavefunction obtained from a Dirac Hartree calculation or that from the recipe used here results in no appreciable differences, so that we expect the same to be true for (p,2p). Lastly, but perhaps not least, we must consider the effects of Pauli-Blocking which we have so far not included. This causes no more problems in the calculation as it has been set up.

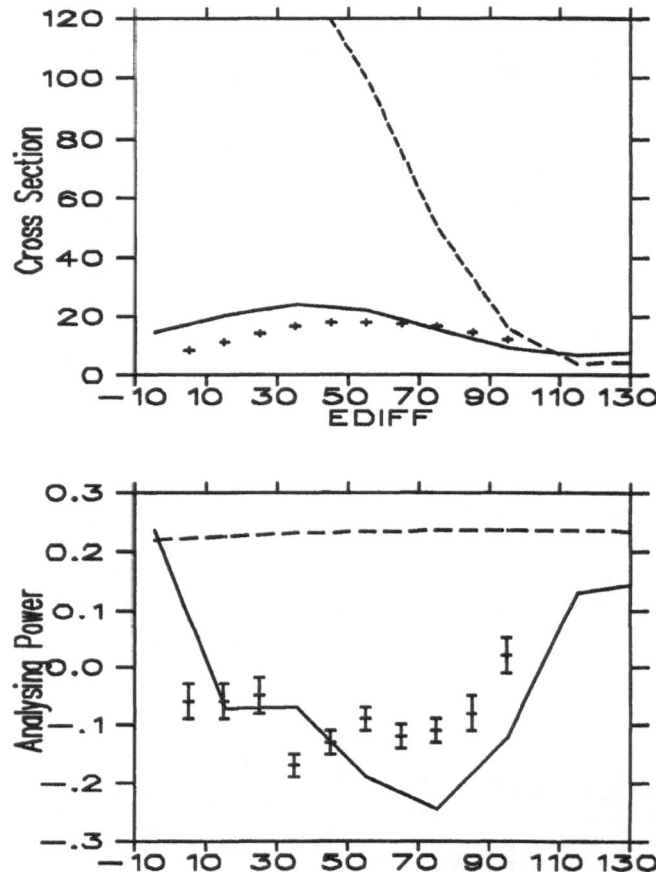

Figure 2. Data and results for $1D_{5/2}$ state. Dashed line is PWBA, solid line is DWBA calculation using GOP1. Crosses are data of ref [15].

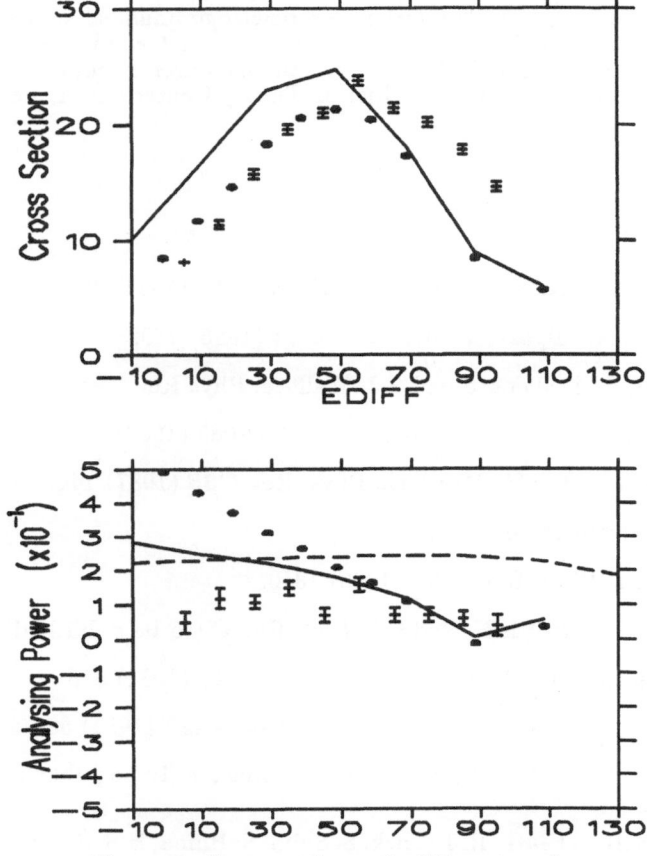

Figure 3. Same as figure 1, different scale.

Table 1 The Global Optical Model Parameters Used

	V_v^n	W_v^n	X_v^n	V_s^n	W_s^n	X_s^n
			GOP_1			
n=0	319.7897	-34.0381	-69.8580	-405.395	46.24880	109.4527
n=1	-85.2109	-61.5664	1.689790	89.28785	-8.68631	1.187799
n=2	11.27108	40.37400		-47.3242	40.37400	
n=3	-1.45800	-19.1239		13.64275	-19.1239	
n=4	-0.06563	3.415059		-0.06562	3.415059	
R	3.538693	3.888458	3.893386	3.538693	3.933882	3.762518
a	0.608548	0.505919	0.4372	0.639027	0.468648	0.6212
			GOP_2			
n=0	380.6916	-35.1995		-441.741	36.71259	
n=1	-479.229	-10.6658		355.9410	-51.0801	
n=2	985.3442	16.63852		-896.998	16.63852	
n=3	-773.973	2.620688		803.1932	2.620688	
n=4	0.0	0.0		0.0	0.0	
R	3.604181	3.747939		3.577595	3.854686	
a	0.607233	0.606344		0.636644	0.440297	

Acknowledgements -- The authors acknowledge financial support from the Natural Sciences and Engineering Research Council of Canada. B.K. Jennings, A. Miller, N. DeTakacsy and C. Horowitz are thanked useful comments. We would like to thank the Indiana University Nuclear Theory Center and Triumf for hospitality during part of this work.

References

[1] P. Kitching et al. Advances in Nuclear Physics **15** (1985) 43.

[2] Y. Kudo and K. Miyazaki, Phys. Rev. **C34** (1986) 1192

[3] J.A. McNeil, J.M. Shepard and S.J. Wallace. Phys Rev Lett **50** (1983) 1443

[4] J.A. Tjon and S.J. Wallace. Phys. Rev. **C36** 1085 (1987).

[5] C.J. Horowitz and D.M. Murdock. Phys. Rev **C35** (1987) 1442.

[6] Y. Kudo. Private Communication.

[7] C.J. Horowitz, Phys. Rev. **C31** (1985) 1340.

[8] M.H. Macfarlane and E.F. Reddish. Phys Rev C (to be published).

[9] E.D. Cooper and H.S. Sherif. Phys. Rev. Lett. **47** (1981) 818.

[10] B.K. Jennings and O.V. Maxwell, Nucl. Phys. **A422** (1984) 589.

[11] E.D. Cooper, B.C. Clark, S. Hama and R.L. Mercer. To be published in Physics Letters.

[12] E.D. Cooper,B.C. Clark, R. Kozack, S. Shim, S. Hama, H.S. Sherif, R.L. Mercer and B.D. Serot. Phys Rev **C36** (1987) 2170.

[13] N.S. Chant and P.G. Roos Phys. Rev **C27** (1983) 1060.

[14] E.D. Cooper,K. Hicks and B.K. Jennings. Nucl. Phys. **A470** (1987) 523

[15] P.Kitching et al. AIP Conference **97** p232.

[16] A. Miller, Private Communication.

[17] B.K. Jennings and C.J. Horowitz. In Progress

[18] E. Rost and J. Shepard. Phys Rev **C35** (1987) 681.

SPIN-OBSERVABLES FOR THE $(\vec{p}, \vec{p}\, \gamma)$ REACTIONS AT 400 MeV

K.H. Hicks

TRIUMF
4004 Wesbrook Mall
Vancouver, BC, Canada V6T 2A3

INTRODUCTION

The success of the relativistic impulse approximation (RIA) for elastic scattering[1] has suggested that relativistic effects may be important for intermediate-energy proton scattering. Relativistic effects are features such as the large vector and scalar potentials (which could be responsible for medium effects such as an effective mass[2] for a nucleon in the nucleus) or cancellation of short-range structure in the particle propagator due to antiparticles[3]. However, the non-relativistic impulse approximation (NRIA) is not to be discounted; indeed, it does as well as the RIA (or in some cases even better) for elastic scattering[4,5] and inelastic scattering to the first excited 2^+ states for light even-even nuclei[6] at incident proton energies between 130 and 400 MeV.

It is natural to extend these studies to inelastic proton scattering (for excited states other than "collective" states), which has a richer set of eight independent observables compared to the three independent ones for elastic scattering. Several recent studies have compared new or existing (p,p') data with RIA and NRIA calculations[7,8,9,10] and often the cross sections are reasonably well reproduced, but predictions for the spin-observables miss the data by several standard deviations.

The present study is a new application of a technique[11] used in low energy nuclear physics, which gives observables that are sensitive to the spin of the recoil nucleus. The hope is that these new observables will lead to improvements in the RIA and NRIA, and possibly to identify the importance of relativistic effects for inelastic scattering.

THEORY OF THE $(\vec{p}, p'\gamma)$ OBSERVABLES

The form of the (p,p') transition amplitude is restricted by the requirements of invariance under space-inversion (parity), rotations, and time-reversal. Assuming the initial nuclear state to have spin-parity $J_i^{\pi_i} = 0^+$, we may write the inelastic scattering amplitude to a final state of spin-parity J^π and spin projection M as

$$T_M = < J^\pi, M; \vec{p}_f, m_f|T|0^+; \vec{p}_i, m_i >, \qquad (1)$$

where we have explicitly included reference to the initial and final spin projections of the projectile, m_i and m_f, respectively. Assuming the interaction T to be invariant with respect to the parity transformation and rotations, we find that the general form for a 0^+ to 1^+ transition amplitude is given by

$$
\begin{aligned}
T_{M=q} &= C_q \sigma_p + D_q \sigma_q \\
T_{M=p} &= C_p \sigma_p + D_p \sigma_q \\
T_{M=n} &= A_n \mathbf{1} + B_n \sigma_n
\end{aligned}
\tag{2}
$$

where $\vec{q} = \vec{k}_f - \vec{k}_i$, $\vec{p} = \frac{1}{2}(\vec{k}_i + \vec{k}_f)$, and $\vec{n} = \vec{p} \times \vec{q}$ in terms of the initial and final projectile momenta $\vec{k}_i$ and $\vec{k}_f$. The T_i are operators in the projectile spin space for residual nucleus substates with projection $M = 0$ along the direction $i = \hat{q}$, $\hat{p}$, or $\hat{n}$. The projectile spin operators are $\sigma_q = \vec{\sigma} \cdot \hat{q}$, etc. in terms of the Pauli spin matricies, and we quantize along the $\hat{q}$ direction.

The spin-observables from measuring the in-going and out-going spin of the scattered proton, where the spin substate, M, of the target is *unrestricted*, are related to the transition amplitude by

$$
ID_{ij} = \frac{\text{Tr}}{2} \sum_M \vartheta_i T_M \vartheta_j T_M^\dagger
\tag{3}
$$

where $\vartheta_i = \{1, \sigma_n, \sigma_p, \sigma_q\}$ for $i = \{0, n, p, q\}$ and where $I = \frac{\text{Tr}}{2} \sum_M T_M T_M^\dagger = d\sigma/d\Omega$ is the cross section. The indicated trace is over the spin states of the projectile. Imposing parity and rotational invariance leave only *eight* of the D_{ij} to be non-zero. Hence, the sum over the target spin substate M prevents a measurement of the complete set of D_{ij} from uniquely determining all six complex numbers given by equation (2).

The values of M can be restricted by measuring the γ-ray given off by the 1^+ to 0^+ deexcitation of the residual nucleus, assuming the γ-ray lifetime is short enough to be unambiguously associated with the associated inelastic scattering event. The γ-ray carries off 1 unit of angular momentum in the direction of its motion (i. e., it has helicity $\lambda = \pm 1$). This means that $M = 0$ in the direction perpendicular to the γ-ray direction, giving a constraint on the transition amplitudes in equation (2) that can contribute. Details of the connection between the transition amplitudes and the $(\vec{p}, p'\gamma)$ observables is given in Ref. 12.

The present experiment does not measure the outgoing spin of the scattered proton, preventing a complete determination of all parts of the transition amplitude. Rather, different combinations of the transition amplitude are determined as compared to measurements of the D_{ij} spin-observables. This new information may point to strengths or deficiencies in some combinations of the transition amplitudes, and could lead to identifying problems (and hopefully make improvements) in the RIA or NRIA models.

DESCRIPTION OF THE EXPERIMENT

Polarized protons of 400 MeV incident energy from the TRIUMF cyclotron were scattered from a 94.7 mg/cm^2 ^{12}C target and detected in the medium resolution spectrometer[13] (MRS) at laboratory angles of 6, 7, 9, 11, and 13 degrees. The MRS has a large solid angle, and the position at the entrance to the MRS is measured with a front-end chamber, which allows an overlapping measurement between successive angles. The cross sections for these overlapping regions were usually within statistical errors, and always within 2σ. The data presented here were obtained by averaging the overlapping angles. The data acquisition livetime was measured by comparing the number of electronically generated psuedo-events sent out to the number recorded by the computer. The polarization of the proton beam was measured from p-p scattering using a previously calibrated in-beam polarimeter.

Eight Bismuth Germinate (BGO) detectors were placed at angles between 90° and 270° in the scattering plane at a distance of 35.5 cm from the target. In addition, two BaF_2 detectors were placed above and below the target, normal to the scattering plane at a distance of 10.2 and 17.7 cm. All detectors were shielded by cylinders of 7.5 cm thick lead to eliminate room background. A threshold of approximately 1.2 MeV (as determined from a ^{60}Co source) were set on the pulse height of the anode signal from the photomultiplier tube base. A timing resolution of about 1 nanosecond was achieved with the use of constant-fraction discriminators for the γ-ray detectors, with respect to the timing of the scattered proton. The high duty-cycle of the TRIUMF cyclotron, with beam pulses every 43 nanoseconds, gave a clean coincidence with a reals:randoms ratio of 3:1.

To calibrate the detectors, a ^{6}Li target was put in the beam and the isotropic γ-ray from deexcitation of the 3.56 MeV 0^+ to ground state transition was observed in coincidence with protons scattered in the MRS. The ratio of coincidence cross section to singles cross section for this reaction gives the efficiency folded with the solid angle of the γ-ray detector. Tagged singles data (with a known prescale factor) were taken simultaneously with the coincidence data, thus eliminating the need for comparing data from two separate runs. The known solid angles was used to extract the γ-ray detection efficiency for this γ-ray energy. The normalized spectra were then compared with Monte-Carlo calculations[14] and are shown in the top of Fig. 1 for both the BGO and BaF_2 detectors. The excellent agreement between calculation and data gives confidence for using the Monte-Carlo detector efficiencies for the 15.1 MeV γ-ray from deexcitation of the 1^+ $T=1$ state in ^{12}C , where no direct measurement of the γ-ray efficiency was available. The shape for this calculation are shown in the bottom of Fig. 1 which is also in good agreement with the data.

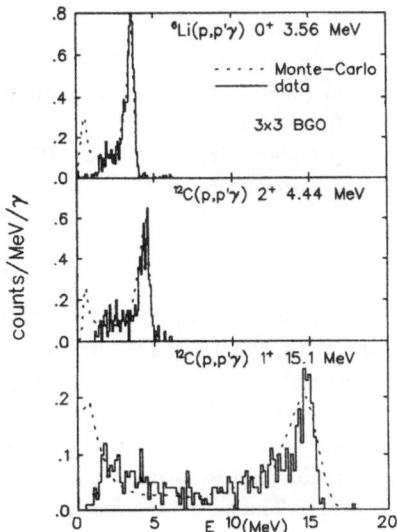
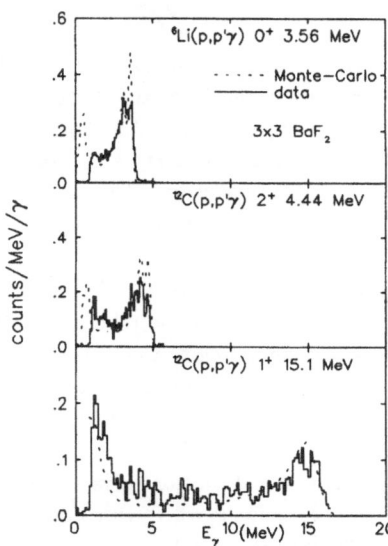

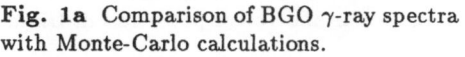

Fig. 1a Comparison of BGO γ-ray spectra with Monte-Carlo calculations.

Fig. 1b Comparison of BaF_2 γ-ray spectra with Monte-Carlo calculations.

RESULTS AND ANALYSIS

The singles cross section for inelastic scattering to the 15.1 MeV 1^+ T=1 state in ^{12}C is shown in Fig. 2, along with the data of Jones[15] for comparison. Also shown in Fig. 2 are calculations using the codes DREX[7,16] (for the RIA model) and DW81[17] (for the NRIA model). The details of the inputs to these calculations are the same as for Ref. 10. Both calculations give a good prediction of the inelastic cross section and analyzing power in the range of angles considered here. No preference for either model is found for the singles data.

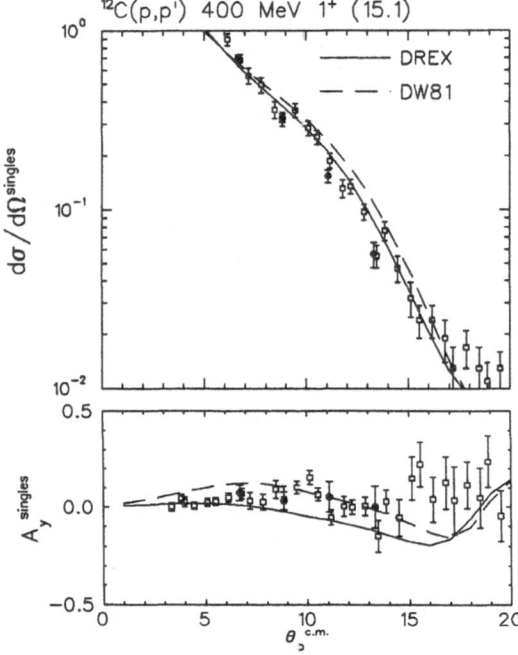

Fig. 2 Singles cross sections and analyzing powers. The open squares are previous data from Ref. 15 and the solid circles are from the present measurement.

The results of the in-plane γ-ray angular correlations are shown in Fig. 3 along with calculations from the RIA and NRIA models for several (center-of-mass) angles of the scattered proton, θ_p. The coincidence cross sections and calculations have been divided by their respective singles cross sections to minimize any ambiguities associated with an overall normalization error. The data exhibit the shape expected for a 1^+ to 0^+ γ-ray transition, which has the general form[18]

$$\frac{d\sigma}{d\Omega_p d\Omega_\gamma} = A(\theta_p) + B(\theta_p)cos(2\theta_\gamma) + C(\theta_p)sin(2\theta_\gamma) \qquad (4)$$

where θ_γ is measured in the laboratory frame with respect to the beam axis. The three constants corresponding to the isotropic, symmetric, and antisymmetric terms can be predicted from the impulse approximation[12,19]. The smaller (proton) angle data do not show a preference for either model, but the two larger angles show a definite preference for the relativistic model. The basis for this behaviour has been traced to the distortions, because plane-wave calculations give a shape at all proton angles that is very similar to that shown in Fig. 3 for θ_p=6.7° (for both DREX and DW81). The superior prediction for the RIA model may also be due to the enhanced lower-components of the nuclear structure wave-functions, which is a signature of the large scalar and vector potentials in Dirac phenomenology[7].

The in-plane analyzing power angular correlations are presented in Fig. 4. These data again show no preference for either model for the smaller proton angles. At the larger proton angles, there is a slight preference for the DREX calculation, although the agreement is not as dramatic as for the cross section angular correlations. Similarly, the plane wave calculations show the same shape for all angles, which looks like that shown at 6.7° in Fig. 4.

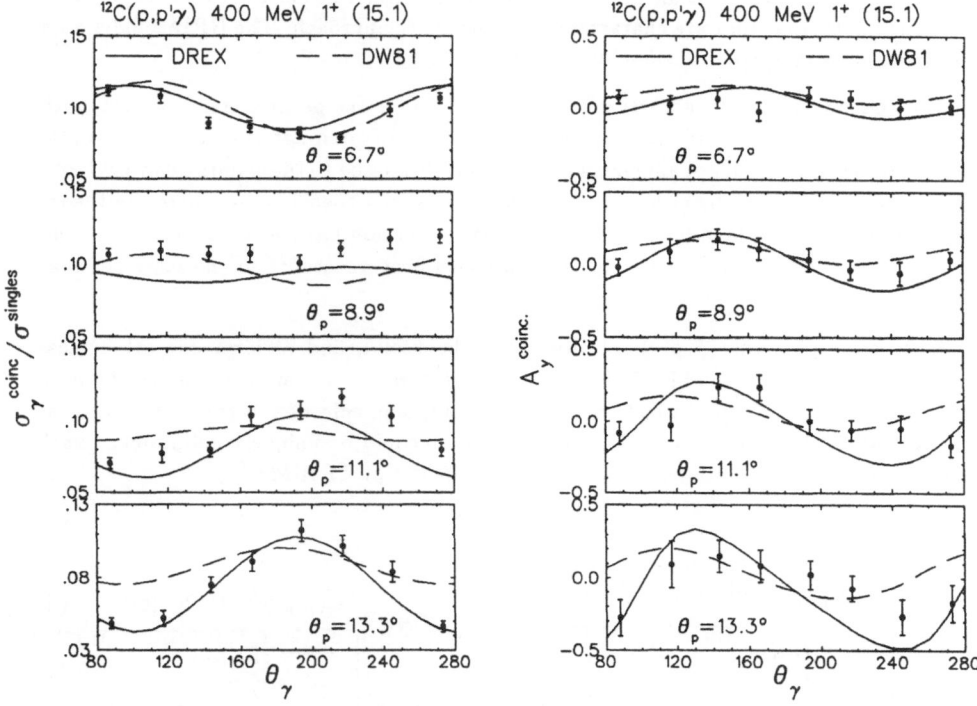

Fig. 3 In-plane γ-ray cross section angular correlations.

Fig. 4 In-plane γ-ray analyzing power angular correlations.

The out-of-plane cross sections have been known for some time[11] to be proportional to the singles cross section times the spin-flip probability S_{nn} . In addition, there exist previous measurements of S_{nn} for ^{12}C at 400 MeV taken at LAMPF[20]. The results of these data and those obtained from the present experiment are shown in Fig. 5. Both sets of data show a clear preference for the relativistic model. Unlike the angular correlations, the calculations for S_{nn} are affected very little when plane-waves are used. This has also been noted for calculations of S_{nn} for quasi-elastic scattering[21].

The out-of-plane coincidence analyzing power data are shown in Fig. 6. This observable is related to the singles observable $P-A_y$ by

$$A_y^{coinc.} = -\frac{1}{2}\frac{P-A_y}{S_{nn}} \qquad (5)$$

which is sensitive to non-local (or momentum-dependent) terms in proton-nucleus interaction. A previous measurement of $P-A_y$ for ^{12}C at 400 MeV has been modified according to the above relation and plotted in Fig. 6. The data are in strong disagreement with either the RIA or NRIA models. This implies that the non-local nature of the interaction needs to be improved in both models.

CONCLUSIONS

The $(\vec{p}, p'\gamma)$ reaction has proven useful in identifying combinations of the transition amplitude that do and do not provide agreement with measurements. Specifically, the coincidence cross section data (both in- and out-of-plane) indicate that the RIA model gives a good prediction for the parts of the transition amplitude that make up these observables. The source of the agreement for the in-plane angular correlations is due to the large scalar and vector potentials that are characteristic of the Dirac approach. The NRIA model does not give a good prediction of these observables.

The coincidence in-plane analyzing power data follow the general trend of both with RIA and NRIA models, although neither calculation is in good agreement with the data. The out-of-plane coincidence analyzing power, which is related to the singles quantity $P-A_y$ is very poorly predicted by either theory. This points to a specific area where the theory needs improvement, which is the representation of the non-local features of the proton-nucleus interaction. This same conclusion has also been reached[22] for measurements at incident proton energies of 150 MeV.

More data from the $(\vec{p}, p'\gamma)$ reaction would be useful, and at least one other measurement will be done at TRIUMF in the near future. Even more useful would be a measurement of the $(\vec{p}, \vec{p}'\gamma)$ reaction, where the outgoing spin of the scattered proton is also measured; then all six quantities of equation (2) could be uniquely determined. Such a measurement would be possible with existing facilities at TRIUMF.

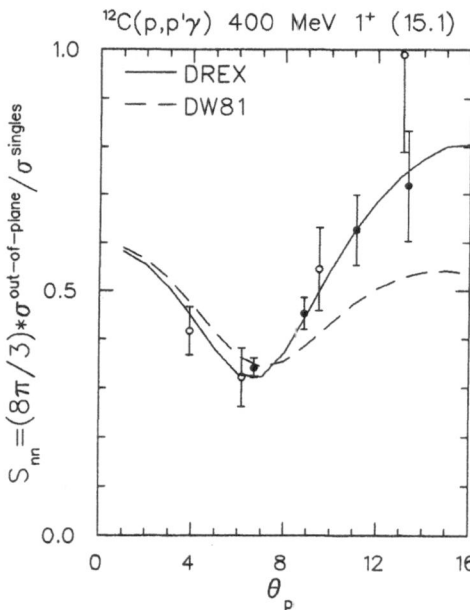

Fig. 5 Out-of-plane γ-ray coincident cross section. The open points are from Ref. 20 and the solid points are from the present measurement.

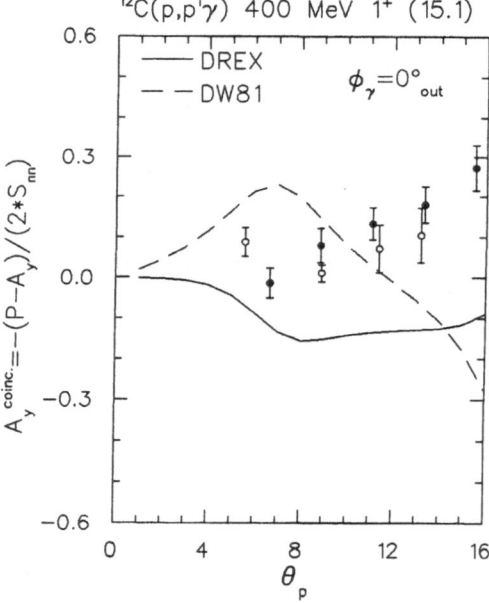

Fig. 6 Out-of-plane γ-ray coincident analyzing powers. The open points are from Ref. 10 and the solid points are from the present measurement.

ACKNOWLEDGEMENTS

These measurements were done in collaboration with members from TRIUMF, Los Alamos National Laboratory, Oak Ridge National Laboratory, and the University of Kentucky. The BaF$_2$ detectors were provided by LANL/ORNL and the BGO detectors were provided by both LANL/ORNL and the University of Kentucky. The calculations were provided by J. Shepard of the University of Colorado. This work was supported in part by a grant from the National Sciences and Engineering Research Council of Canada.

REFERENCES

[1] J.R. Shepard, J.A. McNeil, and S.J. Wallace, Phys. Rev. Lett. **50**, 1443 (1983).

[2] C.J. Horowitz and M.J. Iqbal, Phys. Rev. **C33**, 2059 (1986); J. Iqbal, contribution to these proceedings.

[3] E.D. Cooper and B.K. Jennings, Nucl. Phys. **A458**, 717 (1986).

[4] O. Häusser, et al., Phys. Lett. **184B**, 316 (1987).

[5] Ma Ji, M.Sc. Thesis, Simon Fraser University, 1987, unpublished.

[6] K.H. Hicks, et al., accepted in Phys. Rev. C.

[7] E. Rost and J.R. Shepard, Phys. Rev. **C35**, 681 (1987).

[8] W. Bauhoff, et al., Nucl. Phys. **A410**, 180 (1983).

[9] J.R. Shepard and E. Rost, Phys. Rev. **C33**, 634 (1986).

[10] K.H. Hicks, et al., Phys. Lett. **201B**, 29 (1987).

[11] R.H. Howell, et al., Phys. Rev. **C21**, 1153 (1980); R.N. Boyd, et al., Phys. Rev. Lett. **29**, 955 (1972).

[12] J.R. Shepard, to be published; J.R. Shepard and E. Rost, University of Colorado Annual Report, 1986, p. 120.

[13] C.A. Miller, in *Studying Nuclei with Medium Energy Protons*, Edmonton (1983), TRIUMF Proc. TRI-83-3, p. 339.

[14] S.A. Wender, et al., NIM **258**, 225 (1987).

[15] K. Jones, Ph.D. thesis, LAMPF report LA-10064-T, 1984.

[16] E. Rost, code DREX, unpublished.

[17] R. Schaeffer and J. Raynal, code DWBA70, unpublished; extended version DW81, J. Comfort, unpublished.

[18] A.J. Ferguson, "Angular Correlation Methods in Gamma-Ray Spectroscopy", North-Holland, Amsterdam, 1965; A.A. Debenham and G.R. Satchler, Part. Nucl. **3**, 117 (1972); F. Rybicki, T. Tamura, and G.R. Satchler, Nucl. Phys. **A146**, 659 (1970)

[19] N. Mobed, S.S.M. Wong, and X. Zhu, Nucl. Phys. **A456**, 644 (1986).

[20] S.J. Seestrom-Morris, et al., Phys. Rev. **C26**, 2131 (1982).

[21] K. Jones contribution in these proceedings.

[22] M.A. Kovash, et al., submitted to Phys. Rev. C.

PROTON INDUCED DELTA PRODUCTION

B. K. Jain

Nuclear Physics Division
Bhabha Atomic Research Centre
Bombay 400 085, India

Pions, in addition to neutrons and protons, form the integral part of the nucleus. Associated with them are the Δ-isobars which get excited easily in real state, when pions of enough energy couple to nucleons. The coupling constant for $\pi N \rightarrow \Delta$ is about two times that for $\pi N \rightarrow N$. Due to this there is much interest in the propagation of pions and deltas in the nuclear medium. Specifically, one wants to learn about the behaviour of the self energy, $\Pi(\omega, q)$, of pions, which, in general, is complex, and the modification in the coupling constant, $f^*(f)$, and the mass and decay width of the Δ in the nucleus. In addition, it is also of interest to understand the mechanism of the reactions type (p, Δ^{++}), because, they involve the transfer of spin-isospin of one or two units, and momentum transfer beyond ~ 300 MeV/c. This provides a potentially powerful tool to explore the spin-isospin excitations in nuclei in high momentum domain. This is unlike the intermediate energy (p, n) reaction which explores these excitations in low momentum region. The reaction mechanism for (p, Δ^{++}) reaction may be simple, because, here, unlike normal nuclear reactions, the high momentum transfer (≥ 300 MeV/c) occurs in one step when N (938 MeV) $\rightarrow \Delta$ (1232 MeV).

(p, Δ^{++}) REACTION

Experimentally the (p, Δ^{++}) reaction is difficult. So far, using proton beams, only one experiment has been done.[1] It uses ^{6}Li target, 1.04 GeV proton beam from Saturne and records the events corresponding to different kinetic energies and emission angles of the recoiling nucleus ^{6}He. Missing mass spectrum and four momentum transfer distribution are constructed from them. Another experiment, $p(^3\text{He}, t)\Delta^{++}$, which has also been done, uses ^{3}He beam.[2] Both these data distinctly exhibit bumps around 1232 MeV missing mass with width around that of the free Δ isobar (FWHM $\sim$90 MeV). In these measurements, apart from tagging the events, the decay products of Δ^{++} are not seen. Therefore every data point in missing mass spectrum corresponds to inclusive $p\pi^+$ events. Due to this, in theoretical description, the Δ^{++} can be considered as an elementary particle which simplifies the formalism considerably. Theoretically, these data and other data of ^{3}He projectile have been analysed by us.[3] These studies establish that the Δ-production reactions mainly proceed in one step and, quantitatively, they can be well described by the distorted wave Born approximation (DWBA). For ^{6}Li $(p, \Delta^{++})^6$He and $p(^3\text{He}, t)\Delta^{++}$ reactions, where the transition densities for ^{6}Li $\rightarrow ^6$He

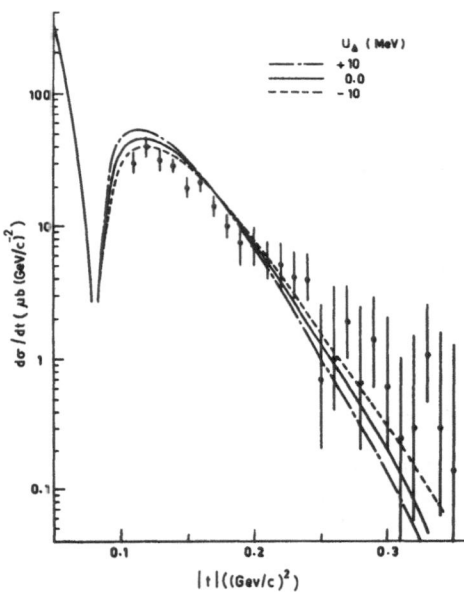

Fig. 1. Differential cross section $d\sigma/dt$ vs. $|t|$ for ^{6}Li$(p, \Delta^{++})^6$He for various values of the real part (U_Δ) of the delta optical potential. The experimental points are from Ref. 1.

and ^{3}He $\rightarrow$ t can be well estimated, it has been found that the data can be understood using one pion plus one rho exchange interaction for $V_{\sigma\tau}$ (pp $\rightarrow$ n Δ^{++}) (Figs. 1-2).

If only one pion-exchange (without any form factor) with an additional Migdal type spin-isospin interaction is used these data tells that the Migdal parameter, g'_Δ, is less than 0.5 (Fig. 3).

In the above studies it has also been observed that the interaction $V_{\sigma\tau}$, more or less, remains constant over the momentum range of interest. The shape of the four momentum transfer distribution is mainly determined by the variation of the transition density, $\rho_{\sigma\tau}$ $V_{\sigma\tau}$ determines the magnitude. Therefore, once $V_{\sigma\tau}$ is known, the (p,Δ^{++}) reaction provides a powerful tool to explore the spin-isospin densities in large momentum domain in nuclei.

However, as mentioned earlier, the measurements on the recoiling nucleus is difficult. Consequently the utility of these experiments gets limited to light nuclei. To overcome this limitation one can, alternatively, detect the decay products, p' and π^+, of the Δ^{++}. This of course requires the detection of two energetic particles in coincidence, which also, in general, is not simple. But with the current medium-energy machines it should not be too difficult.[4]

$(p,p'\pi^+)$ REACTION

Theoretically, as shown in Fig. 4, the situation for $(p,p'\pi^+)$ reaction does become less clear. Contribution to the measured events can come from four amplitudes. However, we can see by inspection that these four amplitudes can have quite different behaviour. One way to see this is to examine the energy carried by the virtual meson and by the isobar in the four amplitudes. In Fig. 4A, the energy of the

120

intermediate meson is given by the difference between the initial and final binding energies. This energy for the excitation of low-lying nuclear states, is quite small, and can be approximated by zero. The energy of the isobar in this amplitude is given by $T_p + M_p$, where T_p is the kinetic energy of the incident nucleon. This is a large energy, and it is independent of the kinematics of the final particles. In Fig. 4B, the energy of the intermediate meson is T_p. The energy of the isobar is $T_p + M_p$, same as in Fig. 4A.

In Fig. 4C, the energy of the intermediate meson is T'_p, the kinetic energy of the final proton. The energy of the isobar is $E_\pi + M_p$, where E_π is the total energy of the final pion. This energy is considerably smaller than the energy of the isobar corresponding to the amplitudes of Figs. 4A and 4B; also both the meson and isobar energies vary as we change the kinematics of the outgoing proton and pion. In Fig. 4D, the energy of the intermediate meson is $T_p - T'_p$; the energy of the isobar is $E_\pi + M_p$, same as for Fig. 4C.

Thus in the four amplitudes of Fig. 4, there is a tremendous variation in the energy of the virtual meson, from zero to the incident kinetic energy. As a consequence, the behaviour of the four amplitudes could be quite different, and this might be exploited to separate the various amplitudes by varying the final-state kinematics. If these amplitudes do get separated, then this reaction can provide a powerful handle on the investigation of self energy of pion and isobar in nuclear medium. In the remaining part of the talk we examine the sensitivity of the $(p,p'\pi^+)$ reaction to different amplitudes.

Since, at this stage, we are interested only in simple estimates of the relative size of the various reaction amplitudes, and in order of magnitude estimates of medium effects on intermediate mesons and isobars, the first calculations are done with simple forms for the nuclear wave functions and transition densities, and with very simple approximations for the nucleon and pion wave functions. Due to these simplifying assumptions, the results should be considered exploratory.

For the πNN and $\pi N\Delta$ coupling we have used the monopole form factor with the same value for the cut off parameter, $\Lambda(=1.2 \text{ GeV})$. The "bare" coupling constants are taken as $f_\Delta^2/4\pi = 0.224$. The parameters of the πNN is $f^2/4\pi = 0.08$. The ρ coupling parameters, which were used for calculating the (p,Δ^{++}) reaction, are

$$\Lambda_\rho = 2 GeV; \ f_\rho^2/4\pi = 4.86; \ f_\rho^*/4\pi = 1.85 f_\rho$$

The amplitude of Fig. 4A corresponds to emission of a virtual meson from a target nucleon, excitation of the incident proton to a Δ (1232), with subsequent decay to a continuum proton and pion; thus, we refer to this amplitude as the 'projectile excitation' term. For 'amplitude A' (referring to the diagram of Fig. 4A), the T-matrix expression has the form:

$$T_A = \sqrt{2}\rho_{\sigma\tau}(\vec{q}_A)\frac{ff_\Delta^2}{m_\pi^3}F(q_A)\frac{\langle \vec{k}' \mid \vec{S}^+ \cdot \vec{k}_\pi S \cdot q_A\rangle}{D_\Delta(q_A^\Delta, \omega_A^\Delta)D_\pi(q_A, \omega_A)} \tag{1}$$

where

$$\vec{q}_A = \vec{k}_\pi + \vec{k}' - \vec{k}; \qquad \omega_A = \epsilon_f - \epsilon_i \approx 0$$

$$\vec{q}_A^\Delta = \vec{k}_\pi + \vec{k}'; \qquad \omega_A^\Delta = E' + E_\pi. \tag{2}$$

The subscript A refers to the amplitude in question.

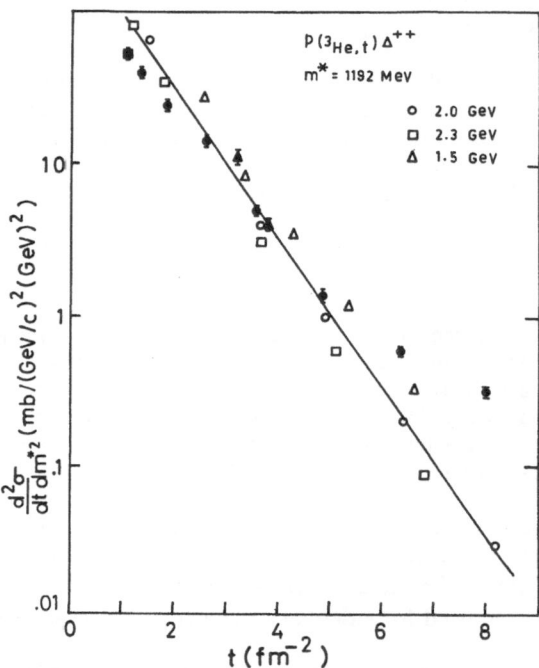

Fig. 2. Four-momentum transfer (t) distribution for $p(^3He,t)\Delta^{++}$ constructed from calculations at ^3He incident kinetic energies 1.5(Δ), 2.0(0) and 2.3( ) GeV for a delta mass, m^*, of 1192 MeV. Filled points are measured values from Ref. 2. Continuous curve is for visual guidance.

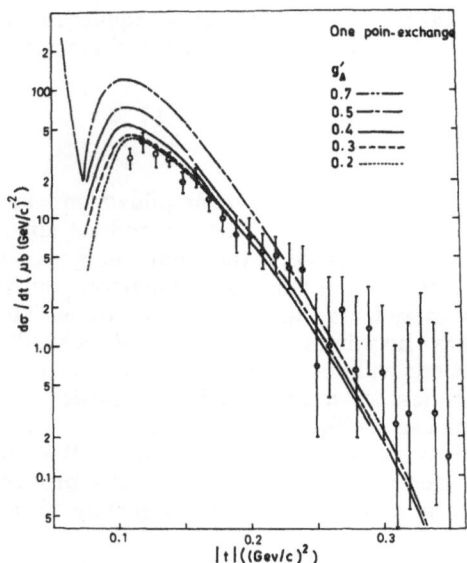

Fig. 3. Sensitivity of the ^{6}Li(p,Δ^{++})^{6}He differential cross section to g'_Δ, the Landau-Migdal parameter for NN $\rightarrow$ NΔ interaction, using the one pion-exchange potential (without the form factor) only.

In Eq. (1), D_Δ represents the propagator for the intermediate Δ, and D_π is the propagator for the meson exchanged between the projectile and target nucleons (for the time being, we will take this as a pion). The incident and final proton states are denoted by $\bar{k}$ and $\bar{k'}$, and the initial bound proton and final bound neutron are denoted by i and f, respectively. As mentioned earlier, the fourth component of the virtual meson momentum, ω_A, is given by the difference in binding energy of the initial proton and the final neutron. In comparison with the three-momentum q_A, ω_A can be set to zero.

The spin-isospin matrix $\rho_{\sigma\tau}$ in Eq. (1) is given by

$$\rho_{\sigma\tau}(\vec{q}) = \langle f \mid \Sigma_j \vec{\sigma}_j \cdot \vec{q}\tau_j \mid i \rangle \tag{3}$$

where the sum is over all target nucleons. Here we consider only those cases where the final state is a single neutron particle-proton hole state relative to the initial state, and we assume that the core is inert. Eq. (3) can then be written as

$$\rho_{\sigma\tau}(\vec{q}) = \langle \phi_{n,f} \mid \vec{\sigma} \cdot \vec{q} \mid \phi_{p,i} \rangle \tag{4}$$

In Eq. (4), only a single 'active' nucleon contributes to the transition density.

Since our aim here is to explore the general features of the (p,p'π^+) reaction, we have used simple Gaussian forms for the single-particle transition densities and wave functions.

$$\phi(p) = \left(\frac{4\pi}{p_0^2}\right)^{3/4} \exp(-p^2/2p_0^2) \tag{5}$$

$$\rho(p) = \exp(-p^2/4p_0^2) \tag{6}$$

The model functions depend upon a single momentum scale, p_0. This parameter has been fixed by requiring that it reproduces the mean squared single-particle momentum seen in electron scattering data. In terms of the Fermi momentum, p_0 then obeys the relation

$$p_0^2 = \frac{2}{3}\langle p^2 \rangle = \frac{2}{5}k_F^2 \qquad (7)$$

These nuclear wave functions are oversimplified and lack the dependence on orbital and spin degrees of freedom. They reproduce single-particle momentum distributions for the nucleus as a whole, rather than being typical of a particular shell. Also, they fall off too fast for large momentum. Therefore, our resulting cross sections will decrease too rapidly away from the peak cross sections. These wave functions are also structureless; they have no minima, for example. As a result, effects due to minima in the nuclear wave functions will not be present in our calculations. However, we have deliberately chosen featureless wave functions for this initial survey, as it is well known that PW approximations can overestimate the effects of structure (and, particularly minima) in single-particle wave functions: for example, a PW approximation involving a bound-state wave function with a zero will frequently produce a deep minimum in a cross section, whereas the minimum may be completely "filled in" when distortions are included.

The pion and Δ propagators are given by

$$D_\pi(\vec{q}, \omega) = [\omega^2 - \vec{q}^2 - m_\pi^2 - \Pi(q, \omega) + i\epsilon]^{-1} \qquad (8)$$

for the pion, where Π is the self-energy of the pion with momentum q and energy ω; similarly, the Δ propagator is given by

$$D_\Delta(\vec{q}, \omega) = [\omega - T_\Delta - M_\Delta - \nu_\Delta(q, \omega) + i\Gamma/2 + i\epsilon]^{-1} \qquad (9)$$

where ν represents the real part of the nuclear potential for a Δ with energy ω and momentum q·T_Δ is the kinetic energy and Γ the width of the isobar. In this form the (non- local) isobar-nucleus interaction has been approximated by a local effective potential.

The second amplitude considered corresponds to emission of a meson by the projectile proton, as shown in Fig. 4B. This meson excites a target nucleon to a Δ, which then decays to the final continuum proton and pion. We refer to this amplitude as the 'target excitation' term (or 'amplitude B'). This amplitude has the form:

$$T_B = \frac{\sqrt{2}f f_\Delta^2}{m_\pi^3} \int \frac{d\vec{q}F(q)\phi_f^+(\vec{k} - \vec{q})\phi_i(\vec{k}' + k_\pi - \vec{q})}{(2\pi)^3 D_\Delta(q_B^\Delta, \omega_B^\Delta)D_\pi(q_B, \omega_B)} \qquad (10)$$

$$\times \langle f \mid \vec{\sigma} \cdot \vec{q} \mid k \rangle \langle k' \mid \vec{S}^+ \cdot \vec{k}_\pi \vec{S} \cdot \vec{q} \mid i \rangle$$

where

$$\omega_B = T_p + \epsilon_f \approx T_p$$

$$\vec{q}_B^\Delta = \vec{k}' + \vec{k}_\pi = \vec{q}_A^\Delta \qquad (11)$$

$$\omega_B^\Delta = E' + E_\pi = \omega_A^\Delta.$$

This amplitude is proportional to the product of two single-particle wavefunctions, integrated over the momentum q. As this term is non-local, it cannot be written

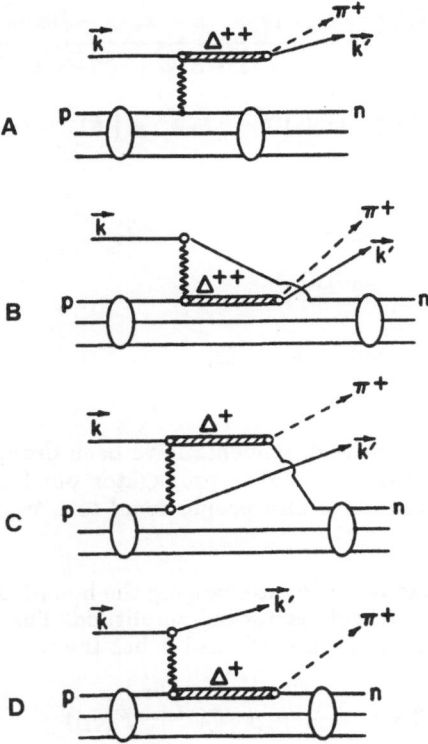

Fig. 4. Various diagrams considered for the $(p,p'\pi^+)$ reaction. (A) 'Projectile excitation' amplitude; (B) 'Target excitation' amplitude. Diagrams (C) and (D) represent the processes where the nucleon originating from the decay of the delta is captured by the nucleus and another nucleon is emitted.

in terms of a local spin-isospin transition density, but it requires a non-local density matrix. In this term, we have neglected the contribution of the single-particle binding energy to the energy carried by the virtual meson. The 'target excitation' amplitude is obtained from the 'projectile excitation' term by interchanging the projectile and target nucleon lines in Fig. 4A. All of the additional terms in our amplitude are obtained from the amplitude of Fig. 4A by interchanging either the initial or final-state nucleon lines.

The third amplitude, given in Fig. 4C, is obtained from 'amplitude A' by exchanging the continuum and bound nucleon lines in the final state. We refer to this term as 'amplitude C'; it has the form

$$T_C = -\frac{\sqrt{2}ff_\Delta^2}{3m_\pi^3} \int \frac{d\vec{q}\,F(q)\phi_f^+(\vec{k} - \vec{k}_\pi - -\vec{q})\phi_i(\vec{k}' - \vec{q})}{(2\pi)^3 D_\Delta(q_c^\Delta, \omega_c^\Delta)D_\pi(q, \omega_c)}$$

$$\times \langle k' \mid \vec{\sigma} \cdot \vec{q} \mid i\rangle\langle f \mid \vec{S}^+ \cdot \vec{k}_\pi \vec{S} \cdot \vec{q} \mid k\rangle \tag{12}$$

where

$$\omega_C = -\epsilon_i - T_{p'} \approx -T_{p'}$$

$$\vec{q}_C^\Delta \cong \vec{k}_\pi \tag{13}$$

$$\omega_C^\Delta \approx E_\pi + M_p$$

Small bound-state energies and momenta have been dropped in Eq. (13). This simplification allows us to take the isobar propagator outside the integral over q. Like amplitude B, this amplitude is also proportional to a transition density matrix operator.

The final amplitude is obtained by exchanging the bound and continuum nucleon lines in the final state of the 'target excitation' amplitude. This term is shown in Fig. 4D. We refer to this term as 'amplitude D', and it has the form:

$$T_D = \frac{-\sqrt{2}}{3}\tilde{\rho}_{fi}(-\vec{q}_A)\frac{ff_\Delta^2}{m_\pi^3}F(q_D)$$

$$\times \frac{\langle k' \mid \vec{\sigma} \cdot \vec{q}_D \mid k\rangle}{D_\Delta(q_D^\Delta, \omega_D^\Delta)D_\pi(q_D, \omega_D)} \tag{14}$$

where

$$\vec{q}_D = \vec{k}' - \vec{k}$$

$$\omega_D = T_{p'} - T_p$$

$$\vec{q}_D^\Delta \cong \vec{k}_\pi \tag{15}$$

$$\omega_D^\Delta \approx E_\pi + M_p$$

In Eq. (15), the same approximations were made, as in Eq. (13). This amplitude, like the "projectile excitation" amplitude, is proportional to a nuclear transition density;

126

the transition density $\rho_{fi}(q)$, which appears in Eq. (14), is a combination of spin-isospin and isospin densities.

$$\tilde{\rho}_{fi}(\vec{q}) = \frac{2}{3}\vec{k}_\pi \cdot \vec{q}\langle f \mid \Sigma_j \tau_j^- \mid i\rangle$$

$$- \frac{i}{3}(\vec{k}_\pi \times \vec{q})\langle f \mid \Sigma_j \vec{\sigma}_j \tau_j^- \mid i\rangle \tag{16}$$

In general, the transition densities and transition density matrix operators will have considerable dependence upon the spin and angular momentum of the active nucleons. With the simple forms which were chosen for the nuclear wave functions, this sensitivity and selectivity has been lost.

The c.m. differential cross section is given in terms of the above amplitudes as

$$\frac{d^3\sigma}{d\Omega' d\Omega_\pi dE'} = \frac{10^4 k' k_\pi E' E}{2(2\pi)^5 k(1 + E/E_A)} \mid T_{fi} \mid^2 \tag{17}$$

In Eq. (17), the units are $\mu b/sr^2/MeV$, and $\mid T_{fi} \mid^2$ is obtained by squaring the sum of the amplitudes of Eqs. (1), (10), (12) and (14) and averaging over initial spins and summing over final spins. With our approximations for the momentum dependence of the Δ propagator, and with the simple Gaussian wave functions and transition densities which we have employed, the PW cross sections can be reduced to a sum of terms requiring no more than a single integral. (For details see Ref. 5).

We show the calculated cross sections for the $(p,p'\pi^+)$ reaction for coplanar geometry; i.e., where the incident and outgoing particles all lie in a plane in the c.m. system. As the spin dependence of this interaction is not investigated, there is little additional information obtained by going out of plane. In the calculated cross sections, the c.m. angle between the incident and outgoing protons is kept fixed at -10^0 relative to the incident direction and the pion and isobar propagators are taken those of free particles.

In Fig. 5, the cross sections are shown as a function of the outgoing pion angle, for fixed outgoing proton kinetic energy. Fig. 5a is for incident energy 450 MeV and outgoing energy 100 MeV, and Fig. 5b is for incident energy 800 MeV and outgoing energy 200 MeV. For each incident energy, the outgoing proton energies are chosen to correspond to the overall peak in the cross section.

The solid curve gives the full cross section; the other curves show the cross section for each amplitude separately. While the contribution from amplitude 'C' is smaller than the others, the remaining amplitudes contribute about equally to the cross section. The cross section has a broad peak for a small positive value of θ_π corresponding to the minimum momentum transfer available in the reaction, and the cross section decreases monotonically as the angle is varied with respect to the maximum. The width and shape of the cross section peak are determined by the nuclear wave functions and transition densities. The maximum cross section increases somewhat as the proton energy increases. At 450 MeV, the peak cross section is about 30 $\mu b/sr^2/MeV$, whereas at 800 MeV, the peak cross section is roughly 60 $\mu b/sr^2/MeV$.

Fig. 6 shows the outgoing proton energy spectrum for fixed pion angle of $+10^0$, and incident proton energies 450 and 800 MeV. We find that, as a function of T_p, the energy distributions for amplitudes A and B differ from those of C and D; compared to C and D, the former pair peaks at a considerably smaller value of T_p, and the

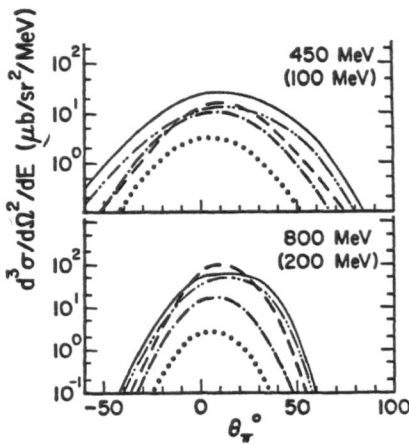

Fig. 5. Pion angular distribution for the reaction $(p,p'\pi^+)$ leading to a neutron particle-proton hole relative to an inert core for zero pion-self energy $(\Pi = 0)$. The proton angle is fixed at -10^0. The cross-sections, in $\mu b/sr^2/MeV$, are given for proton incident energies 450 MeV and 800 MeV. Number written in the brackets represents the energy of the outgoing proton. Solid curve: full differential cross-section; dashed curve: projectile excitation (amplitude A of Fig. 4); dot-dot-dashed curve: target excitation (amplitude B of Fig. 4); dotted curve: amplitude C; dot-dashed curve: amplitude D.

separation between peaks increases with incident energy T_p. This feature helps to separate contributions from two classes of diagrams.

Fig. 6 is plotted for a pion angle very near the peak of the cross sections; for such an angle, the nuclear momentum transfer does not change very much as the outgoing proton energy is varied. As a result, the nuclear transition densities are relatively constant, hence the quantity which varies most rapidly with outgoing proton energy is the propagator. For amplitudes C and D, the Δ propagator has the form

$$D_\Delta \approx [\delta M + T_\Delta - (T_p - T_{p'}) + i\Gamma/2]^{-1} \qquad (18)$$

where $\delta M \equiv M_\Delta - M_N$. The maximum in this term occurs when the real part of the denominator has a minimun, i.e., for $T_p' \approx T_p - M_.$. In Fig. 6, amplitudes C and D peak at this value T_p'. For amplitudes C and D, the Δ propagator is

$$D_\Delta \approx [\delta M + (\vec{k}' + \vec{k}_\pi)^2/2M_\Delta - T_p + i\Gamma/2]^{-1} \qquad (19)$$

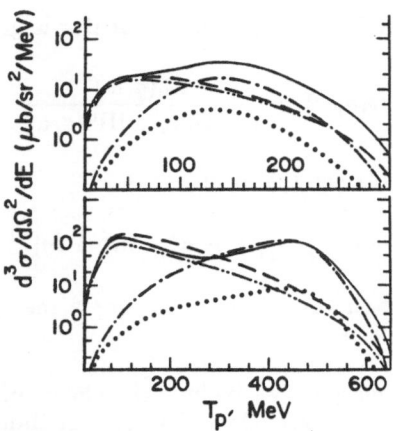

Fig. 6. Proton energy spectrum for the reaction $(p, p'\pi^+)$ with proton and pion angles fixed at -10.0 and +10.0 degrees respectively and no pion-self energy. Length scale, p_0, for the nucleon wave function and transition densities is taken 158 MeV/c. Identification of the curves is the same as given in Fig. 5. Upper and lower figures are for incident energies 450 and 800 MeV respectively.

Again, the maximum in this term occurs when the real part of the denominator is smallest. In contrast to Eq. (18), this occurs at a smaller value of T'_p. For proton incident energies of 450 and 800 MeV, the real part of Eq. (19) has a minimum at $T'_p = 50$ MeV, and 100 MeV, respectively. From Fig. 6, this is just where amplitudes A and B reach their maximum values. This analysis also explains the increased separation in the two peaks as the incident energy increases.

The two ways we have chosen to view the cross sections give us quite different information about the reaction. Fixing the outgoing kinetic energy and varying the pion angle gives direct information regarding the shape of the nuclear transition densities. On the other hand, if the pion angle is fixed around the maximum of the cross sections, the resulting proton energy spectrum allows a separation of the cross section into different reaction amplitudes.

MEDIUM EFFECTS ON THE PION

Next we calculate the effects arising from modification of the pion propagation due to the nuclear medium. The full meson propagator is given by

$$G(q, \omega) = \frac{G_0(q, \omega)}{1 - G_0(q, \omega)\Pi^0(q, \omega)} \tag{20}$$

where the free propagator G_0 has the form

$$G_0(q, \omega) = [\omega^2 - q^2 - m^2 + i\eta]^{-1} \tag{21}$$

In Eq. (20), Π^0 is the self-energy of the pion. For a real pion Π^0 is related to the optical potential through $\Pi^0 = 2\omega \, V_{opt}$.

To calculate the self-energies, we include the effects of particle-hole and Δ-hole excitations. We approximate the nuclear density distribution by a Fermi Gas, for which the expressions are well known.[6] The Fermi momentum has been chosen as k_F = 210 MeV/c, and the nucleon effective mass is taken as $M^* = 0.7 \, M_N$. Inserting the self-energies Π^0 into the propagator of Eq. (20) sums the ring diagrams for pion propagation in the nuclear medium.

In general, the self-energies Π^0 (q, ω) are complex. However, for $\omega = 0$, as is the case for amplitude A, the self- energy is real. As a result, the self-energy can produce a pole in the pion propagator for amplitude A, while for all other amplitudes it produces both a real and an imaginary part. In Fig. 7, the cross section at 450 MeV is plotted for pion c.m. angle + 10^0. In addition to the summed cross section, contribution from each amplitude is plotted separately; it is clear that the cross section shows a spike due to the pole amplitude A, which completely dominates the contribution from all other amplitudes.

While the self-energy effects will give a considerable increase in the calculated cross sections, the poles in amplitude A are unphysical. These poles are just the "pion condensate" seen in early calculations of the pion polarization potential in nuclei.[7] As is well known, this condensate occurs because the particle-hole and isobar-hole interactions in this approximation are strongly attractive at short distances. In reality, we should also include the strong short-ranged repulsion which arises from exchange of heavier mesons and other many-body effects; when this is done, the spurious poles in the full Green function disappear. The distortion of the continuum particles, which has been neglected here, would also 'soften' the poles which appear in the PW limit.

In order to estimate the effects of the short-ranged repulsion, we introduce the Migdal parameters, which are supposed to mock up the complicated density-dependent effective interaction between particles and holes in the nuclear medium. The additional short-range piece has the form

$$W_{ph} = \frac{f^2}{m_\pi^2} g' \vec{\sigma}_1 \cdot \vec{\sigma}_2 \vec{\tau}_1 \cdot \tau_2 + \frac{f f_\Delta}{m_\pi^2} g'_\Delta \vec{S}_1 \cdot \vec{\sigma}_2 \vec{T}_1 \cdot \vec{\tau}_2 \tag{22}$$

When all rings with short-ranged correlations in intermediate states are summed, the self-energies are renormalized. If the self-energies are written in terms of the 'susceptibilities' as

$$\Pi_N \equiv -q^2 \chi_N$$

$$\Pi_\Delta \equiv -q^2 \chi_\Delta \tag{23}$$

130

then the susceptibilities get renormalized to

$$\chi_N \rightarrow \frac{1 + \chi_\Delta(g'_\Delta - g')}{(1 + g'\chi_\Delta)(1 + g'\chi_N) - g'^2_\Delta \chi_N \chi_\Delta} \chi_N(q, \omega)$$

$$\chi_\Delta \rightarrow \frac{1 + \chi_N(g'_\Delta - g')}{(1 + g'\chi_\Delta)(1 + g'\chi_N) - g'^2_\Delta \chi_N \chi_\Delta} \chi_\Delta(q, \omega)$$

(24)

For our calculations, we have chosen $g'_\Delta = g'$, in which case the renormalization equations become

$$\chi_N(q, \omega) \rightarrow \frac{\chi_N(q, \omega)}{1 + g'(\chi_N + \chi_\Delta)}$$

$$\chi_\Delta(q, \omega) \rightarrow \frac{\chi_\Delta(q, \omega)}{1 + g'(\chi_N + \chi_\Delta)}$$

(25)

Most estimates of these quantities give a Migdal parameter between $0.5 \leq g' \leq 1$ for the πNN coupling. For the $\pi N\Delta$ case, there is much less certainty; for example, Arima et al.,[8] obtained a value $g'_\Delta \approx 0.3$, which value is also preferred in our analysis of the ^{6}Li $(p,\Delta^{++})^6$He reaction. However, given the uncertainty in this quantity, $g'_\Delta = g'$ is a common choice.

Fig. 7 shows the effect of the short-range repulsion in the particle-hole and isobar-hole terms. The cross sections are plotted for pion angle $+ 10^0$ vs. T'_p, the outgoing proton kinetic energy. Figs. 7a and b show the cross section for incident energy 450 MeV and Migdal parameters $g' = 0$ and $g' = 0.7$, respectively. For $g' = 0.7$, it is clear that the pole in the cross section which dominates Fig. 7a has completely disappeared. However, inclusion of the renormalized self-energy increases amplitude A relative to the other amplitudes, and it also increases the overall cross sections relative to the calculation without the pion self-energy. In our calculation, the self-energy Π^0 (q,ω) is enhanced considerably in the region $\omega \approx 0$, and for $q \approx 1$-3 fm^{-1}; thus inclusion of the self-energy produces a much larger effect on amplitude A than on any other amplitude.

From Fig. 7b, for low values of the outgoing kinetic energy, $T'_p \leq 150$ MeV, the contribution from amplitude A (the dashed curve in Fig. 7) dominates the cross sections. Fig. 7c shows the cross sections for incident energy 800 MeV and $g' = 0.7$; for outgoing kinetic energy $T'_p \leq 350$ MeV, the cross sections are dominated by amplitude A; above this energy, the amplitudes are dominated by amplitude D (the dot-dashed curve).

Fig. 8 shows the effects of the renormalized pion self-energy on the proton energy spectrum. Fig. 8a displays the energy spectrum for proton incident energy 450 MeV and pion angle $+10^0$. The dot-dashed curve is the result with no self-energy, the dashed and solid curves are the results with Migdal parameters $g' = 0.7$, $g' = 0.5$, respectively. In Fig. 8b the same results are shown for proton incident energy 800 MeV. First, the cross section is enhanced by a large amount relative to the case with no self-energy; there is an increase of a factor of five at 450 MeV and about eight at 800 MeV (for proton kinetic energy below 400 MeV). Second, as mentioned earlier the self-energy effect primarily enhances the "projectile excitation" amplitude: for large outgoing proton kinetic energy, where the contribution from Fig. 4D is important,

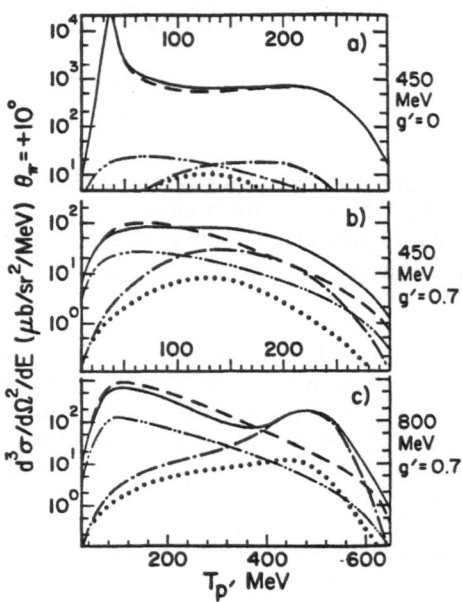

Fig. 7. Same as Fig. 6 except that the pion self energy (Π) is non-zero, (a) incident 450 MeV and $g' = 0$; (b) incident energy 450 MeV and $g' = 0.7$; (c) incident energy 800 MeV and $g' = 0.7$.

the self-energy produces only a slight increase in the cross section; this is especially noticeable for the 800 MeV results. The two curves in Fig. 8 corresponding to different values of g' differ by about 60% at the peak cross sections, and they differ by about 20% at the highest values of T'_p. This gives a measure of the variation in the cross-sections arising from the uncertainty in the precise value for g'.

MEDIUM EFFECTS ON THE Δ

In the previous section, nuclear medium effects were found to produce a significant change in the pion propagator. It is likely that isobar propagation will also be altered considerably by the nuclear medium. There has been many estimates of nuclear effects on intermediate isobars. The most prominent of these has been the isobar-hole approach, which has been applied extensively to production and propagation of isobars in pion elastic and inelastic scattering, and pion-induced total cross sections. The estimates presented in this paper are considerably less rigorous, as we wish to get only a rough idea of the size of the effects in the $(p,p'\pi^+)$ reaction. The qualitative results from more detailed calculations will be used to estimate the isobar medium effects.

The qualitative modifications of the Δ are expected to be of two types: first, we expect a shift in the position of the Δ peak. Second, we expect a modification of the isobar width in the medium. Changes in the isobar width arise from two competing mechanisms. First, Pauli effects in intermediate transitions inhibit the process $\Delta \to N + \pi$ as some of the nucleon states are occupied. Such effects tend to decrease the isobar width in the medium relative to the free width $\Gamma_0 = 116$ MeV. The other contribution to the width arises from nuclear interactions of the isobar. Such "spreading" transitions[9] produce both a shift in the position of the Δ peak, and a change in the isobar width in the medium. In the vicinity of the isobar resonance, a reasonable local estimate for the real part of the "spreading potential" is a 35 MeV attraction. The imaginary part of the spreading potential tends to increase the width of the isobar, at least in the vicinity of the isobar resonance. The two effects on the width differ somewhat, as the Pauli effect goes roughly as the nuclear density and the "spreading" effects go crudely as the square of the density; nevertheless, there is considerable cancellation between the two effects.

For the Pauli effects, a distinction has been made between the amplitudes A and B, and the amplitudes C and D. In the former case, the isobar carries considerably more energy (the sum of the final proton and pion *total* energies) than in the latter case, when the isobar carries the sum of the pion energy and proton *rest* mass. Changes in the isobar width in the medium have been parametrized by replacing the free width in the isobar denominator by

$$\Gamma_0 \to P\Gamma_0 + \Gamma_A \qquad (26)$$

where P represents the suppression of isobar decays in the medium due to Pauli effects, and Γ_A gives the "absorption width" due to isobar-nucleus collisions. The Pauli suppression should be considerably larger for amplitudes C and D, relative to amplitudes A and B. We have chosen P = 1 for amplitudes A and B (i.e., Pauli effects are neglected for these two amplitudes), and P = 0.7 for amplitudes C and D.

The absorption width has been calculated through the spreading width $\Gamma_A = g\Gamma_S$, where the spreading width is chosen as $\Gamma_S = 70$ MeV. This value is somewhat higher than the values deduced from pionic atoms,[10] and slightly lower than the spreading width deduced from pion scattering. The factor g is added to account for the fact that in our calculations the isobar is produced preferentially near the nuclear surface. This factor represents the fraction of the central nuclear density in the region where the isobar is formed. We have chosen g' = 0.7 for our calculations.

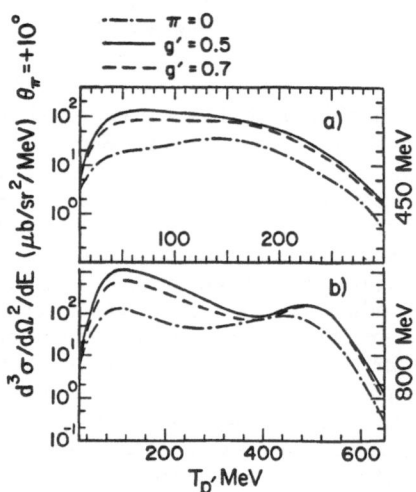

Fig. 8. Sensitivity of the $(p,p'\pi^+)$ energy spectrum to the value of the parameter g' at (a) 450 MeV and (b) 800 MeV incident energies.

Fig. 9 shows the results obtained when the isobar propagator is modified. The curves correspond to a proton incident energy of 450 MeV. The pion self-energy has been included with Migdal parameter $g' = 0.7$. The solid curve corresponds to Pauli effects only, i.e., both the real and imaginary parts of the spreading potential are zero. The dashed curve includes in addition a real spreading potential of -35 MeV, and the dot-dashed curve includes an absorption with $g\Gamma_S = 49$ MeV. Relative to the case with Pauli effects only, inclusion of the real spreading produces roughly a factor of two increase in the cross section, while including the spreading width results in a decrease of approximately 60% in the cross section. These effects are smaller than the estimated pion self-energy corrections, and the net result from both Pauli corrections and the spreading potential is less than a factor of two change in the overall cross sections.

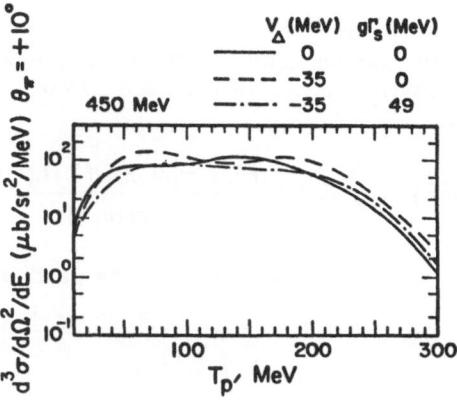

Fig. 9. Sensitivity of the proton energy spectrum in the $(p,p'\pi^+)$ reaction at 450 MeV to the parameters of the Δ-nucleus interaction in the intermediate state. V_Δ: real part of the Δ-nucleus potential; Γ_s: spreading width; g: density dependent factor representing the peripheral nature of the reaction.

These are qualitative estimates of the size and energy dependence of the isobar effects, which could be expected from a more quantitative calculations of the propagation of a Δ in the nuclear environment. Nevertheless, it is clear that the net result of all isobar medium effects is only a small change, and much smaller than the effects due to medium polarization of intermediate pion.

DISTORTION EFFECTS

Thus far we have calculated the $(p,p'\pi)$ cross section in a plane-wave approximation, which neglects the proton-nucleus and pion-nucleus interactions. The nuclear interactions generally produce both absorptive and dispersive effects, and both can be expected to be important. However, for the large incident energies treated in this calculation, the most important effects are likely to be the absorptive effects, and qualitative estimates of these effects are as follows.

We use the eikonal approximation to estimate the attenuation factor for a particle travelling through a medium. First, we associate a refractive index $n(r,E)$ with the nuclear medium, and we factorize the energy and radial dependence as

$$n(r, E) = n(E)\frac{\rho(r)}{\rho_0} \tag{27}$$

where ρ is the nuclear density distribution, and

$$n(E) = \frac{\kappa(E)}{k(E)} \tag{28}$$

Here k is the external wave number, and κ is the wave number in the medium. The imaginary part of this refractive index is defined as n_0. In terms of n_0, the attenuation factor can be written as

$$\eta(E) = \int d\vec{b}dz \frac{\rho(\vec{b}, z) \exp[-kn_0(E)L(b)]}{\int d\vec{b}dz\rho(\vec{b}, z)} \tag{29}$$

where $L(b)$ is the length of the path travelled by the particle in the medium. This length is given by

$$L(b) = \int\limits_0^\infty \frac{\rho(r)}{\rho_0}dz \tag{30}$$

The integration can be done analytically if the nucleon density is approximated by a Gaussian $\rho(r) = \rho_0 \exp(-r^2/\alpha^2)$, in which case the attenuation factor has the form

$$\eta(E) = \frac{1 - \exp[-\sqrt{\pi}\alpha kn_0(E)]}{\sqrt{\pi}\alpha kn_0(E)} \tag{31}$$

Once the value of n_0 is known, the attenuation due to the medium can be calculated. For protons, $n_0(E)$ is obtained from the imaginary part, W_0, of the optical potential,

$$n_0(E) = \frac{E}{k^2}W_0(E) \tag{32}$$

For pions, n_0 is calculated using the method of Ericson and Hufner.[11] κ is determined from the dispersion relation

$$\kappa^2 = k^2 + 4\pi\rho f_{\pi N}(\kappa, E) \tag{33}$$

where $f_{\pi N}$ is the π - N scattering amplitude for the effective pion-nucleon interaction.

Above dispersion relation, as shown by Tandy et al,[12] contains knock-out as primary reactive channel. However, in addition to nucleon knock-out, pion flux is also lost through real absorption in the medium. One term which contributes to real absorption in the final-state interaction is

$$\pi^+ + (N + (A - 1)) \rightarrow \Delta + (A - 1)^* \rightarrow N + (A - 1) \tag{34}$$

We have estimated the real absorption by adding to n_0 of Eq. (33) an additional term, n_A due to real absorption. It is estimated from the spreading width as

$$n_A = \frac{E_\pi}{2k_\pi^2} g\Gamma_s \tag{35}$$

where the value of the absorption width is taken as $g\Gamma_S = 49$ MeV.

In order to determine the total reduction factor for the $(p,p'\pi^+)$ reaction, the total attenuation due to both the protons and the pion has been estimated by replacing the factor kn in Eq. (31) by

$$kn \rightarrow k_\pi n_0(E_\pi) + k_p n_0(E_p) + k_{p'} n_0(E_{p'}) \tag{36}$$

We find that the overall distortion effects decrease the peak cross sections relative to the PW calculations by a factor 10-30. Consequently these calculations predict that with the distortion included the peak cross sections would be of the order of 1-50 microbarns per steradian squared per MeV.

CONCLUSIONS

We find that the nuclear spin-isospin transition densities in the large momentum domain can be reliably studied using the (p,Δ^{++}) reaction, where the measurements are done on the recoiling nucleus. For exploring the propagation of pions in the nuclear medium, the $(p,p'\pi^+)$ reaction, where the decay products of the Δ^{++} are seen in coincidence, seems richer. The estimated cross-section for the reaction, in the incident energy range 400-800 MeV, is around 1-50 microbarns per steradian squared per MeV. This should be measurable with the present detectors and beam intensities.

The work on the $(p,p'\pi^+)$ reaction was done in collaboration with J. T. Londergan and G. E. Walker of Indiana University, Bloomington, Indiana, U.S.A.

REFERENCES

1. T. Hennino et al., Phys. Rev. Lett. **48** (1982) 997.

2. V. Dmitriev, O. Sushkov and C. Gaarde, Nucl. Phys. **A459** (1986) 503; C. Ellegaard et al., Phys. Rev. Lett. **50** (1983) 1745; Phys. Lett. **154B** (1985) 110.

3. B. K. Jain, Phys. Rev. Lett. **50** (1983) 815; Phys. Rev. **C29** (1984) 1396; ibid **C32** (1985) 1253; H. Hasan and B. K. Jain, Phys. Rev. **C35** (1986) 1020; A. B. Santra and B. K. Jain, J. Phys. **G13** (1987) 745.

4. Indiana University Cyclotron Facility Scientific and Technical Report (1986).

5. B. K. Jain, J. T. Londergan and G. E. Walker, Phys. Rev. C **37** (1988) 1564.

6. G. E. Brown and W. Weise, Phys. Rep. 22 (1975) 22; A. Fetter and J. D. Walecka, *Quantum Theory of Many-Particle Systems* (McGraw Hill, NY, 1971); E. Oset and A. Palanques-Mestre, Nucl. Phys. **A359** (1981) 289.

7. E. Oset, H. Toki and W. Weise, Phys. Rep. **83** (1982) 282; A. B. Migdal, Rev. Mod. Phys. **50** (1978) 107; G. E. Brown and W. Weise, Phys. Rep. **27** (1976) 1.

8. A. Arima et al., Phys. Lett. **122B** (1983) 126.

9. M. Hirata, F. Lenz and K. Yazaki, Ann. Phys. **108** (1977) 116; F. Lenz, E. J. Moniz and K. Yazaki, Ann. Phys. **129** (1980) 84.

10. K. Stricker, H. McManus and J. A. Carr, Phys. Rev. **C19** (1979) 929.

11. T. E. O. Ericson and J. Hufner, Phys. Lett. **33B** (1970) 601.

12. P. C. Tandy, E. F. Redish and D. Bolle, Phys. Rev. **C16** (1977) 1924.

THE (n,p) REACTION AS A PROBE OF NUCLEAR STRUCTURE

K.P. Jackson
TRIUMF
4004 Wesbrook Mall
Vancouver, B.C., Canada V6T 2A3

A. Celler
Simon Fraser University
Burnaby, B.C., Canada and
University of Western Ontario
London, Ontario, Canada

INTRODUCTION

What follows is a brief account of some of the results of studies of the (n, p) reaction on nuclear targets initiated at TRIUMF in December 1985. The close connection between studies of nucleon charge exchange reactions and spin excitations in nuclei has been a central theme of earlier conferences held here at Telluride. The emphasis at this conference is on the information uniquely accessible through measurements of spin observables in nuclear reactions at intermediate energies. The (n, p) reaction, inducing spin-flip transitions in isospin space, appears to exhibit a unique sensitivity to certain aspects of nuclear structure.

Early experiments involving nucleon scattering as a probe of nuclear structure were carried out without reference to the concept of isospin. In fact the first direct evidence that isospin is a useful quantum number in medium weight or heavy nuclei was provided by Anderson and Wong[1] utilizing the (p, n) reaction at energies high enough to involve a single scattering mechanism and with resolution sufficient to identify the isobaric analogue state in the final nucleus. This pioneering study of isospin-flip transitions involving the Fermi component ($L=0$, $S=0$, $T=1$) of the effective nucleon-nucleon interaction was followed by a period of intense theoretical and experimental activity[2]. In many respects history was repeated more recently when the same reaction was used at higher energies to identify the Gamow-Teller resonance ($L=0$, $S=1$, $T=1$) as a simple mode of nuclear excitation prominent with targets from ^{2}H to ^{208}Pb.[3-7]

In the context of studies of the (p, n) reaction, the (n, p) reaction can be viewed as inducing "reverse" spin-flip transitions in isospin space. For a variety of reasons there are important distinctions to be drawn between the (p, n) and (n, p) reactions as probes of nuclear structure.

The most obvious difference in studies of the two charge exchange reactions is that the one is well established while the other, (n, p), is still in its infancy. The pioneering studies of the (n, p) reaction at 152 MeV on nuclear targets by Measday and Palmieri[8] were limited by low counting rates and an energy resolution of 6 MeV. More recently Brady et al.[9] have developed an (n, p) facility utilizing a collimated neutron beam at energies up to 65 MeV. For most applications involving the distorted wave impulse approximation (DWIA) higher energies are preferred. The TRIUMF (n, p) facility is the first in operation to exploit the (n, p) reaction as a detailed probe of nuclear structure at energies above 65 MeV.[10-14]

In terms of nuclear structure the initial experiments suggest a role for the (n, p) reaction significantly different from that associated with the (p, n) studies. In isospin space all targets are 100% polarized and the majority exhibit a neutron excess with the consequence that $T_z = (N - Z)/2 > 0$. Thus for all targets but ^{1}H and ^{3}He the operator associated with the (n, p) reaction is uniquely isospin raising ($T_i = T_z, T_f = T_z + 1$). The obvious consequence of this restriction is the absence of Fermi transitions but, as is illustrated below, the impact on Gamow-Teller and other collective transitions is dramatic.

Finally there are some specific nuclear transition matrix elements which can be estimated on the basis of (n, p) measurements. Detailed discussion of these important applications of the probe lies beyond the scope of this article but brief reference will be made to two examples in the context of a specific experimental result and in the conclusions.

THE TRIUMF (n, p) FACILITY

The TRIUMF charge exchange facility utilizes the medium resolution spectrometer (MRS) in very similar configurations for the study of both the (p, n) and (n, p) reactions at forward angles in the energy range between 190 and 500 MeV. The initial (p, n) results were reported by Alford et al.[15] and are also discussed in the following paper.[16] Brief descriptions of the (p, n) mode of operation are included in Refs. 11, 15 and 16. Figure 3 of Ref. 16 illustrates many of the features common to both modes of operation. The proton beam extracted from the cyclotron is incident on the primary target located at the entrance of a dipole magnet used to deflect the emerging protons 20° to a locally shielded beam dump. For (n, p) studies the secondary beam is produced by (p, n) reactions in a ^{7}Li primary target. The nearly monoenergetic neutrons emitted at 0° are incident on the (n, p) target located 92 cm away, just outside the clearing magnet as is illustrated in Fig. 1. The protons emerging from the secondary target are detected in the MRS.

Two essential differences in the operation of the facility in the (n, p) mode are the use of a segmented secondary target chamber[17] in place of a hydrogenous recoil scintillator and the location of the axis of rotation of the MRS near the centre of this (n, p) target chamber. This chamber, filled with counter gas, consists of alternating layers of wire chamber and target foil. The particular foil in which the (n, p) reaction takes place is identified by requiring all and only those wire chambers downstream to register the event. Correction can therefore be made for the proton energy loss in the subsequent targets, permitting the use of more secondary target material while maintaining the required energy resolution. Other advantages ensuing from the use of this design such as background reduction and the precision with which cross sections

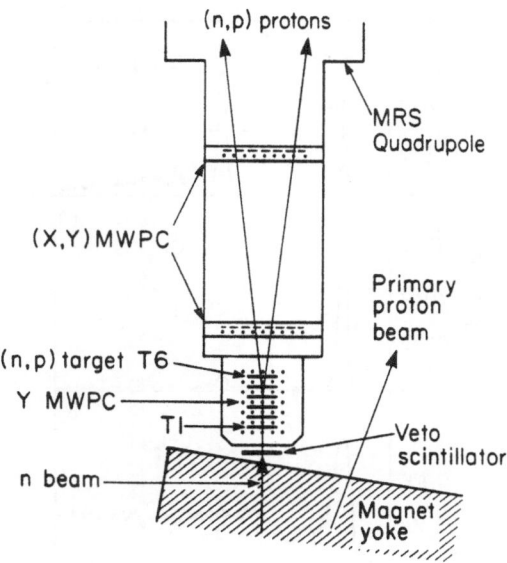

Fig. 1. A schematic diagram of the experimental configuration in the region of the secondary target for (n,p) experiments at TRIUMF.

can be normalized are discussed in detail in Refs. 13 and 17.

Figure 2(a) illustrates the spectrum obtained for the $^{12}C(n,p)^{12}B$ reaction at $0°$ with $E_n = 198$ MeV in the first experiment performed with this facility. Natural carbon targets each 5 cm high, 2 cm wide and 92 mg/cm^2 thick were mounted in positions T2 to T5 in the segmented chamber. The spectrum shown is the sum of all four targets with suitable energy loss corrections. The spectrum is dominated by the peak resulting from the (n,p) reaction populating the ground state of ^{12}B ($Q_0 = -12.59$ MeV). The observed energy resolution results from comparable contributions from

1) the energy spread of the incident proton beam,

2) energy loss in the 110 mg/cm^2 ^{7}Li target,

3) secondary neutrons populating both the g.s. and 0.43 MeV states in ^{7}Be, and

4) energy loss and straggling in the carbon targets.

The smaller peak at $\omega=0$ is associated with the H(n,p) reaction primarily on hydrogen in the mylar foils of the wire chambers and in the counter gas (50% isobutane, 50% argon in this case). This background and a very small contribution at higher excitation energies can adequately be accounted for with an "empty spectrum" recorded with the carbon targets removed.

Figure 2(b) is an (n,p) spectrum recorded *simultaneously* with that in 2(a) for CH$_2$ targets located in positions T1 and T6. The $^{12}C(n,p)$ cross section is measured relative to that for the H(n,p) peak that dominates this spectrum. This ratio is completely insensitive to the incident neutron flux, to computer dead time and to

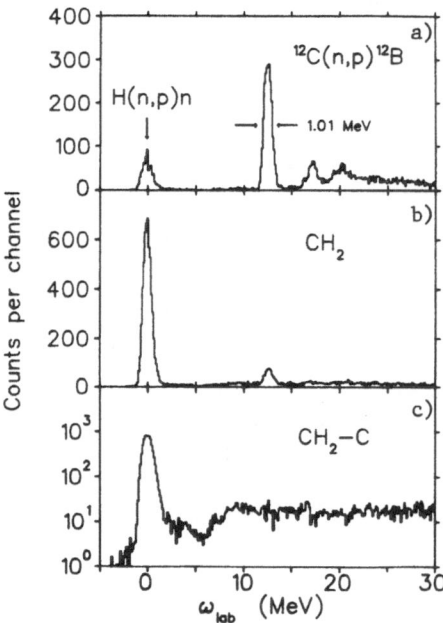

Fig. 2. (a) The (n, p) spectrum at 0° with 4 carbon targets and $E_n = 198$ MeV; ω is the kinetic energy lost by the nucleon, (b) the (n, p) spectrum recorded simultaneously with (a) for two CH_2 targets, (c) the spectrum used to infer the distribution of incident neutron energies [(a) and (b) are reprinted from Ref. 13, (c) from Ref. 11].

the efficiencies of all wire chambers except those in the target chamber. Any small differences associated with individual target locations are monitored by a separate measurement with CH_2 located in all target positions. Such spectra recorded with different settings of the MRS magnetic field are also used to measure the acceptance of the spectrometer as a function of the coordinate on the focal plane.

A novel feature of this (n, p) facility is the short uncollimated neutron beam. Figure 2(c) shows the spectrum obtained by subtracting the contribution of the C from a CH_2 spectrum and correcting for the variation with energy of the spectrometer acceptance. The peak arises from the $H(n, p)$ reaction induced by neutrons from the $^7Li(p, n)$ reaction populating the lowest two states in ^7Be. Lower-energy neutrons resulting from the population of unbound levels in ^7Be will give rise to a continuum at $\omega > 3$ MeV. Data from the time-of-flight facility at IUCF for the $^7Li(p, n)$ reaction at 200 MeV[18] indicate that this source accounts for the continuum present in Fig. 2(c). Estimates of the contribution of this continuum are subtracted from the (n, p) spectra recorded with the targets of interest by performing a simple deconvolution of the neutron line shape.

EXPERIMENTAL RESULTS

Below is a sample of the results acquired to date with the TRIUMF (n, p) facility. These are generally ordered by increasing target mass and are chosen to illustrate the trends in the data. Detailed discussion is avoided; in most cases reference is made to more complete accounts of these results.

$^6\mathrm{Li}(n,p)^6\mathrm{He}$

Figures 3(a) and 3(b) illustrate spectra at two angles recorded with neutrons incident at 198 MeV on $^6\mathrm{Li}$. The value $\theta_{\text{c.m.}} = 2.1°$ in Fig. 3(a) is an estimate of the rms scattering angle accounting for the finite acceptance of the MRS, which was centred at $\theta = 0°$. The dominant feature of these data is the strong transition to the $^6\mathrm{He_{g.s.}}(J^\pi=0^+)$ at forward angles. At 2.1° the strength of the transition to the 2^+ state at 1.80 MeV is <2% of that to the 0^+ despite the possible contribution of Gamow-Teller transitions from the $^6\mathrm{Li_{g.s.}}(1^+\ T=0)$ in both cases.

Figure 3(a) is a simple illustration of the sensitivity of the (n,p) reaction to the influence of space symmetry on nuclear wave functions. If the nuclear Hamiltonian was independent of both total spin S and isospin T, the energy levels would be eigenfunctions of SU_4:

$$\psi_n\{[f](\lambda\mu\ldots)LSJT\}$$

with eigenvalues:

$$E_n\{[f](\lambda\mu\ldots)L\}$$

exhibiting the $(2S+1) \times (2T+1)$ degeneracy of a Wigner supermultiplet.[19] In this classification $[f]$ specifies the spatial permutation symmetry and $\lambda\mu\ldots$ any additional quantum numbers. In general the attractive nature of the long-range part of the nucleon-nucleon interaction tends to lead to states of maximum orbital symmetry dominating the low-lying eigenfunctions. The operator responsible for Gamow-Teller transitions is particularly sensitive to the influence of space symmetry (SU_4) since it connects only states of the same $[f](\lambda\mu\ldots)L$ (i.e., states within the same supermultiplet).

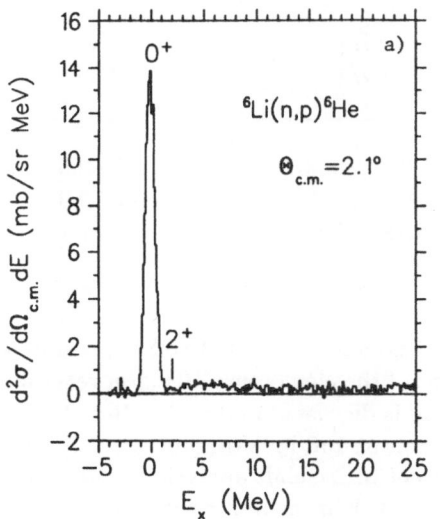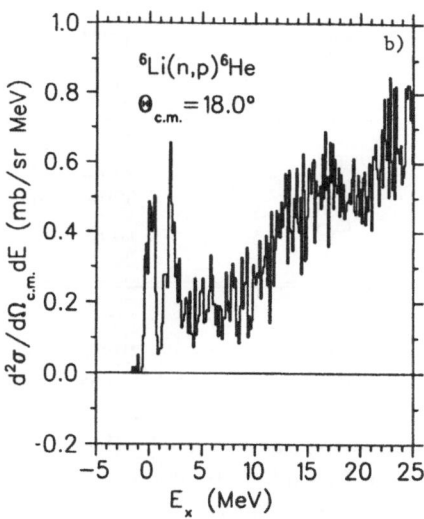

Fig. 3. (a) The cross section deduced from the observation of the (n,p) reaction on $^6\mathrm{Li}$ with $E_n = 198$ MeV and θ (MRS) = 0°. (b) The corresponding result with θ_{lab} (MRS) = 15°.

SU_4 is at best an approximate nuclear symmetry broken by, amongst other terms, a one-body spin-orbit interaction. The low-lying levels of the $A=6$ nuclei may exhibit the best examples of the influence of space symmetry in which the ground states of ^{6}Li and ^{6}He are predominantly $^3S_1$ and $^1S_0$, respectively, while the ^{6}He (1.80 MeV) is largely $^1D_2$.[20-24] The strong GT transition to the ^{6}He$_{g.s.}$ and the almost pure $L=2$ transition to the 2^+ state seen in Fig. 3 reflect these assignments. The complete (n,p) data set (including data at other angles) is being analyzed in terms of detailed nuclear wave functions and the abundant data available for the $A=6$ nuclei from other probes of nuclear structure.

Calibration of the Probe: $\hat{\sigma}_{GT}^+$

An essential first step in the quantitative analysis of the (p,n) data was a systematic series of measurements of the cross sections at $0°$ for transitions for which the Gamow-Teller strength, B_{GT}, was already known from β decay.[4] The result of a recent analysis by Taddeucci et al.[25] of $L=0$ spin-flip transitions within the framework of the DWIA is expressed in terms of $\hat{\sigma}_{GT}^-$, the cross section per unit GT strength extrapolated to $q = \omega = 0$.

An important objective in taking the data shown in Figs. 2, 3, and 4 was to obtain a similar calibration of the (n,p) reaction as a probe of B_{GT}^+, the distribution of Gamow-Teller strength for isospin-raising transitions. The accurately known ft values for the β^- decay of ^{6}He, ^{12}B and ^{13}B to the ground states of the daughter nuclei provide ideal opportunities for this task. The incident neutron energy, $E_n = 198$ MeV, was chosen to permit direct comparison with the most recent results for the corresponding (p,n) analogue transitions at $E_p = 200$ MeV.[25] The details of this investigation have recently been published.[13] The results are summarized in Table I.

Table I. A summary of the cross sections (c.m.) per unit Gamow-Teller strength for analogous (n,p) and (p,n) transitions at 200 MeV. For definitions of the symbols please see the text (reprinted from Ref. 13).

Target	B_{GT}^+	$\hat{\sigma}_{GT}^+$ mb/sr	$\hat{\sigma}_{GT}^-$ mb/sr
^{6}Li	1.593±0.007	9.90±0.36	9.1±0.5
^{12}C	0.999±0.005	9.42±0.31	9.2±0.9
^{13}C	0.759±0.018	10.97±0.56	14.7±1.1

For the (n,p) measurements the use of the segmented target chamber provides a natural normalization for cross sections in terms of the elementary H$(n,p)n$ reaction. This point was made with reference to Fig. 2 and is discussed in detail in Ref. 13. To derive the values for $\hat{\sigma}_{GT}^+$ the H(n,p) cross section at 198 MeV and $0°$ was assumed to be 12.5 mb/sr (c.m.), the value given in a recent phase-shift analysis.[26] The values for $\hat{\sigma}_{GT}^-$ were derived[25] from (p,n) cross sections at $0°$ measured relative to that for the ^{7}Li$(p,n)^7$Be reaction. The uncertainties quoted for $\hat{\sigma}_{GT}^+$ and $\hat{\sigma}_{GT}^-$ do not include

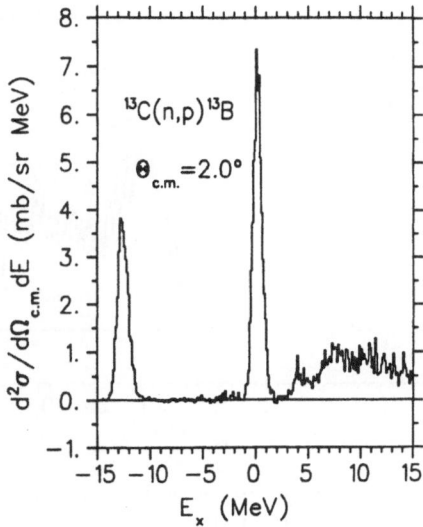

Fig. 4. The (n, p) cross section for ^{13}C with $E_n = 198$ MeV and θ (MRS)=0°. The prominent peak at "$E_x = -12.6$ MeV" results primarily from hydrogen in the mylar windows of the ^{13}C target container.

in either case the uncertainty in the cross section used for normalization.

As is discussed in Ref. 13, the values for the cross sections per unit strength given in Table I for A=6 and 12 are in reasonable accord with expectations. In contrast to these results there is a substantial discrepancy between the measured values of $\hat{\sigma}_{GT}^+$ and $\hat{\sigma}_{GT}^-$ for A=13. The latter value is based on the measured $\sigma_{pn}(0°)$ for the transition to the lowest $T = 3/2$ state at 15.1 MeV in ^{13}N and an estimate of B_{GT}^- = 0.23±0.01.[25] Considering the results for A=6 and 12, the 34% discrepancy in the two values for A=13 cannot reasonably be attributed to uncertainties in the overall normalizations. It should be pointed out that, in contrast to the (p, n) case, the (n, p) result involves a substantially stronger transition to the well-isolated ground state of ^{13}B (see Fig. 4).

^{9}Be$(n, p)^{9}$Li

Figure 5 presents the ^{9}Be$(n, p)^{9}$Li spectrum measured at $E_n = 198$ MeV with the MRS centred at 0°. The peak near $\omega = 0$ corresponds to the incomplete subtraction of the contribution of hydrogen in the segmented target chamber. Two relatively weak peaks are identified with transitions to the lowest known states in ^{9}Li.[27] The measured angular distribution for the transition to the ^{9}Li$_{g.s.}$ is consistent with DWIA calculations[28] based on the wave functions of Cohen and Kurath.[22] These calculations indicate that even for this weak transition $\sigma_{np}(0°)$ is dominated by the L=0 GT component. The value $B_{GT}^+ = 0.019$ is known from the decay of ^{9}Li,[27] and when combined with the measured $\sigma_{np}(0°)$ gives $\hat{\sigma}_{GT}^+(^{9}$Li$_{g.s.}) = 16$ mb/sr. The deviation between this value and the generally more consistent values of $\hat{\sigma}_{GT}^+$ given in Table I can probably be attributed to the fact that the A=9 transition is nearly

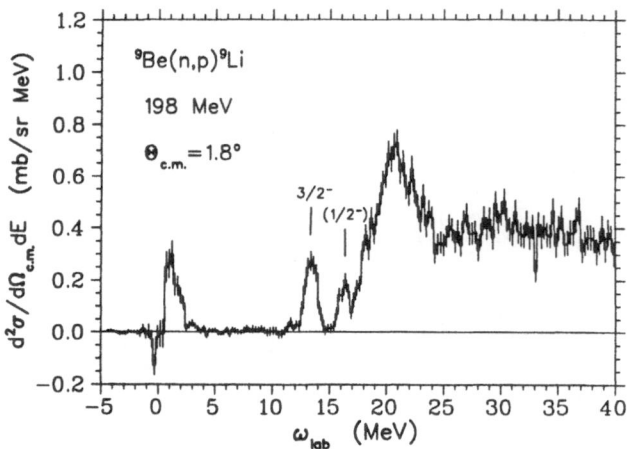

Fig. 5. The (n, p) cross section for ^{9}Be. Also shown are the expected locations of the 2 peaks corresponding to the lowest known states in ^{9}Li.

two orders of magnitude weaker than those chosen for the calibration of the probe.

A comparison between the (n, p) spectra recorded at 0° for ^{13}C (Fig. 4) and ^{9}Be (Fig. 5) reveals much more Gamow-Teller strength concentrated in the ^{13}B$_{g.s.}$ than is seen at low excitation energy in the final nucleus ^{9}Li. Reasonable wave functions calculated[22] for these nuclei account for this difference. It should be noted that if SU$_4$ is an exact symmetry for these nuclei there is no $T=3/2$ state of the same $[f]$ as the maximum orbital symmetry of the $T=1/2$ ground state. In general, in the SU$_4$ limit, $B_{GT}^+ = 0$ for any target with a neutron excess. Figure 5, like Fig. 3(a), dramatically illustrates the influence of spatial symmetry on these light nuclei. The much stronger B_{GT}^+ noted for ^{13}C can be attributed to the larger effect of the one-body spin-orbit interaction near the end of the $1p$ shell.

^{54}Fe$(n, p)^{54}$Mn

Results of measurements of the ^{54}Fe$(n, p)^{54}$Mn cross sections at 298 MeV have been reported by Vetterli et al.[10] and are shown in Fig. 6. There is a strongly forward peaked concentration of strength at excitation energies below 10 MeV. The shell model calculations of Bloom and Fuller[29] predict a distribution of GT strength with a shape shown for comparison with the 0° data. It is important to note that, although the predicted shape coincides reasonably well with the data for $E_x < 8$ MeV, the predicted magnitude of the cross section exceeds the data by more than a factor of 2.

To confirm the $L=0$ assignment for the peak at 0° in Fig. 6, Vetterli et al. used the measured angular distributions to perform a limited multipole decomposition of the data involving orbital angular momentum transfer $L=0$, 1 and 2. The results of this analysis for the 0° data are shown in Fig. 7. The measured cross section at 0° integrated to $E_x = 10$ MeV is 16.9 mb/sr of which 12.9 mb/sr is identified as $L=0$.

The most prominent feature at higher excitations in Fig. 7 is the broad spin dipole resonance ($L=1$, $S=1$, $T=1$). The analysis of Fig. 7 suggests an $L=0$ contribution of 8.5 mb/sr in the range $10 < E_x < 40$ MeV. It is not clear, however,

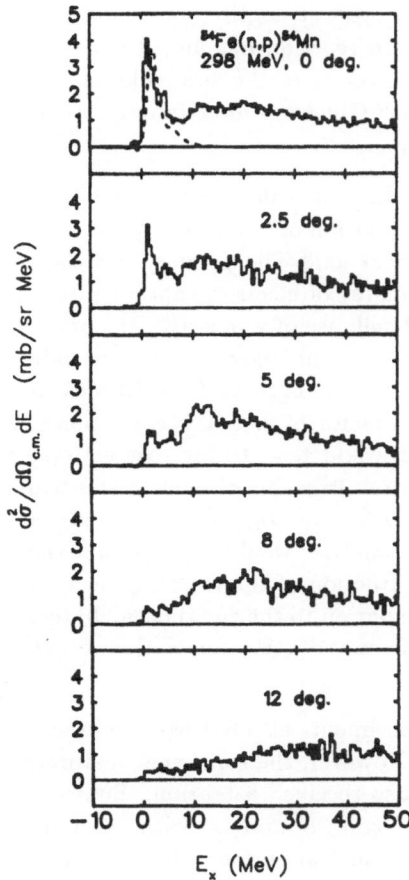

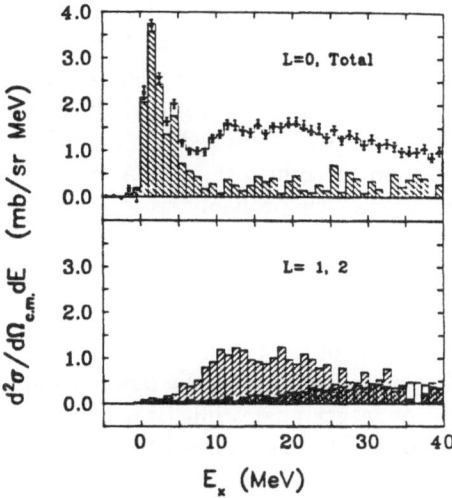

Fig. 6. Cross sections for the (n,p) reaction on ^{54}Fe at 298 MeV for five angles between 0° and 12°. The dashed curve in the 0° spectrum is the calculation of Bloom and Fuller (Ref. 29) normalized by a factor of 0.45 (reprinted from Ref. 10).

Fig. 7. A multipole decomposition of the ^{54}Fe$(n,p)^{54}$Mn spectrum at 0°. The top graph is a plot of the data with the total fitted curve, as well as the $L=0$ component shown hatched. The bottom graph shows the $L=1$ component (cross-hatched) and the $L=2$ (shaded) (reprinted from Ref. 10).

what fraction of this strength can properly be attributed to B_{GT}^+. As is mentioned in Ref. 10, this result is very sensitive to a possible uncertainty in the DWIA estimate of the angular distribution of the $L=1$ strength. Moreover, Auerbach et al.[30] have identified this region of excitation as the location of a spin isovector monopole resonance with the same quantum numbers as the GT resonance but involving a radial $(2\hbar\omega)$ excitation.

One of the important objectives of this study of the ^{54}Fe(n,p) reaction was to test the Gamow-Teller sum rule relating the strengths B_{GT}^- and B_{GT}^+ integrated over all excitation energies:

$$S_{\mathrm{GT}}^- - S_{\mathrm{GT}}^+ = 3(N-Z) . \tag{1}$$

As is mentioned above, there is significant uncertainty associated with possible B_{GT}^+

strength at $E_x > 10$ MeV. Also, the value of the total strength S_{GT}^- estimated by Rapaport $et\ al.$[31] from studies of the ^{54}Fe$(p,n)^{54}$Co reaction has an estimated uncertainty of 25%. Detailed analysis of the data in terms of the sum rule has been reported both by Vetterli $et\ al.$[10] and by Auerbach.[32] Given the uncertainties, no firm conclusions can be drawn in this regard at this time.

Despite the absence of a definitive test of the sum rule there are important implications of the comparison of the detailed distribution of strength B_{GT}^+ predicted by Bloom and Fuller[29] and measured by Vetterli $et\ al.$[10] The primary motivation of the theoretical calculations was to obtain estimates of electron capture rates on nuclei in this mass range during the gravitational collapse of a massive star prior to a supernova explosion. The results of the (n,p) experiment suggest that these calculations accurately give the relative distribution of the strength at $E_x < 10$ MeV but overestimate the total strength in this region by a factor of 2.4. The corresponding reduction in the predicted electron capture rates would tend to increase estimates of the electron/baryon ratio Y_e in the stellar interior. In terms of nuclear structure, Auerbach[32] has discussed the quenching of both B_{GT}^+ and B_{GT}^- for ^{54}Fe, with emphasis on the influence of RPA correlations. This analysis would suggest that these correlations are probably responsible for a major fraction of the rescaling necessary to fit the results of a limited-space shell model calculation to the (n,p) data. In terms of fractional changes in the total strengths, the influence of RPA correlations on B_{GT}^+ is much greater than on B_{GT}^-.

Much of the emphasis in the early (n,p) experiments at TRIUMF has been on B_{GT}^+ and hence data at the most forward angles. However, the somewhat featureless data at larger angles illustrated in Fig. 6 have also received attention. Smith and Wambach[33] have compared the ^{54}Fe(n,p) data at $\theta = 5°$, $8°$ and $12°$ with predictions of a semi-infinite slab model of the response of the nuclear surface. The predictions including significant $2p$-$2h$ damping of the $T{=}1$, $S{=}1$ response agree well with the data.

^{90}Zr$(n,p)^{90}$Y

Studies of the ^{90}Zr$(p,n)^{90}$Nb reaction have played a very prominent role in the consideration of isovector excitations in nuclei. Figure 8 illustrates both the data of Gaarde $et\ al.$[6] at $E_p = 200$ MeV, $\theta = 0°$ and one of the most recent calculations of this cross section by Smith and Wambach.[34] This calculation, based on the eikonal approximation, explicitly includes an estimate in the second random phase approximation of the $2p$-$2h$ spreading of the single scattering response and an estimate of the two-step contribution to the cross section. The results of this calculation are similar to earlier ones by Osterfeld $et\ al.$[35] and Klein $et\ al.$[36] The general conclusion based on these calculations is that for the (p,n) reaction at these energies the nuclear response can be accounted for in detail over a wide range of q and ω without invoking isobar-hole mixing in the nuclear wave functions.

The ^{90}Zr$(n,p)^{90}$Y reaction provides the opportunity for a critical test of the conclusions based on the analysis of the (p,n) data. The B_{GT}^- strength dominates the data of Fig. 8 at $\omega < 20$ MeV. However, to the extent that $Z{=}40$ is a closed proton shell in the ^{90}Zr$_{g.s.}$, the B_{GT}^+ strength is totally Pauli blocked. It is therefore very important to see whether the theoretical formalism apparently successful in accounting for (p,n) data can also correctly predict a dramatically different (n,p)

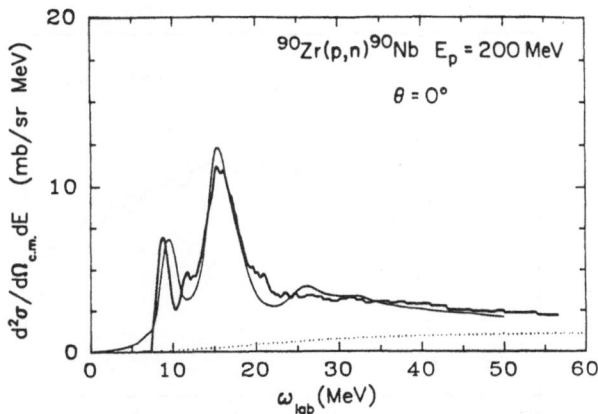

Fig. 8. The full 0° cross section calculated by Smith and Wambach (Ref. 34) for the (p,n) reaction on ^{90}Zr (thinner smooth line) compared to the data of Gaarde *et al.* (Ref. 6, heavier irregular line). The calculated two-step contribution is also given separately by the dotted line (reprinted from Ref. 34).

response. Results of studies of the ^{90}Zr$(n,p)^{90}$Y reaction at $E_n = 198$ MeV have recently been reported by Yen *et al.*[14]

Figure 9 is a comparison of the (n,p) data[14] at 6.4° and the results of a calculation by Smith and Wambach[34] carried out with the same formalism used to explain the (p,n) data in Fig. 8. The calculation gives a reasonable account of the location of the two peaks in the data which are attributed to concentrations of isovector spin dipole strength. The quantitative agreement between data and prediction exhibited in Fig. 8 is, however, notably absent in Fig. 9.

Other predictions of the ^{90}Zr(n,p) strength exist[36,37] but none appears to account in detail for the data. Until the discrepancy between theory and experiment

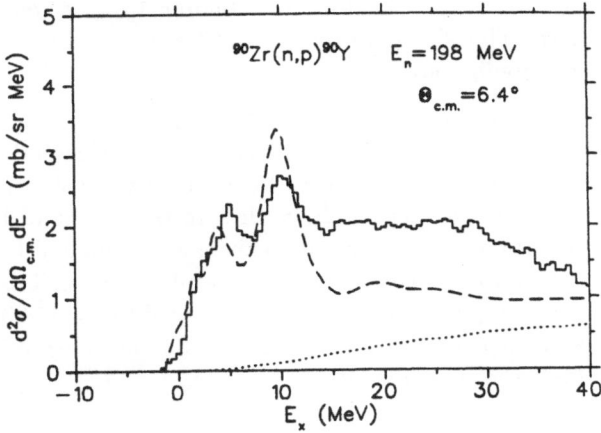

Fig. 9. The cross section for the (n,p) reaction on ^{90}Zr at 6.4° (solid histogram, Ref. 14) is compared with the full calculations of Smith and Wambach (dashed line, Ref. 34). This full calculation includes the contribution of the two-step processes also shown separately by the dotted line (reprinted from Ref. 14).

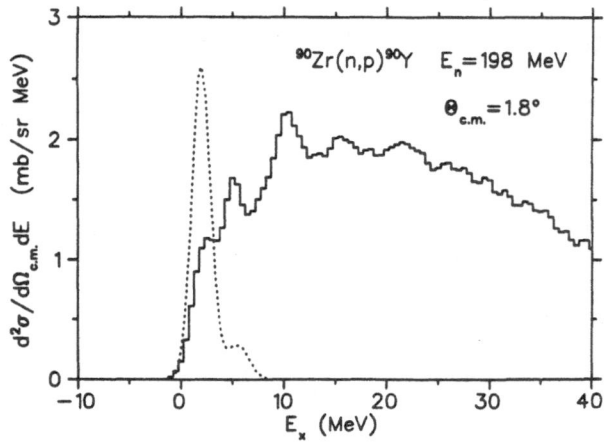

Fig. 10. The histogram represents the data of Yen *et al.* (Ref. 14) for the (n, p) reaction on ^{90}Zr at 1.8°. Shown dotted for comparison is the cross section predicted for B^+_{GT} by the calculation of Bloom *et al.* (Ref. 38) on the basis of their model A (reprinted from Ref. 14).

for the ^{90}Zr(n, p) data can be resolved, conclusions based on the apparent success of theoretical predictions of the (p, n) data must be viewed with caution.

The possible contribution of B^+_{GT} to the ^{90}Zr(n, p) data at forward angles is of particular importance. The RPA calculations of the full nuclear response reported in Refs. 34–37 do not include the effects of 2p-2h correlations in the ^{90}Zr$_{g.s.}$ and therefore give $S^+_{GT}=0$. Bloom *et al.*[38] have calculated B^+_{GT} on the basis of a shell model including g.s. correlations involving the excitation of 2 protons from the $2p$ shell to the $1g_{9/2}$ shell. Figure 10 illustrates the results of this calculation when single nucleon transfer data are used to fix the corresponding $1g_{9/2}$ occupation probability (model A of Ref. 38). The ^{90}Zr(n, p) data at 1.8° are also shown and are clearly inconsistent with this prediction.

We have noted earlier the sensitivity of B^+_{GT} to the influence of space symmetry (SU$_4$) in light nuclei and RPA correlations in ^{54}Fe. Detailed calculations by Towner[39] for the $N=50$ isotones suggest that one can expect very substantial quenching of the strength predicted with a shell model with a restricted set of basis states. In particular core polarization and higher-order effects are estimated to reduce B^+_{GT} at low E_x for ^{92}Mo by a factor of ~9 compared to the result expected if the wave function of the ^{92}Mo$_{g.s.}$ involved simply 2 protons in $1g_{9/2}$ coupled to an $A=90$ closed shell. It seems plausible that these effects could also account for the substantial overestimate of B^+_{GT} at $E_x < 5$ MeV shown in Fig. 10.

CONCLUSIONS

The (n, p) data presented and discussed above are a representative sample of the initial experimental program undertaken at TRIUMF. A number of simple conclusions can already be drawn.

The calibration of the (n, p) reaction as a probe of the Gamow-Teller strength B_{GT}^+ has been achieved for the three targets best suited to this purpose. Features of the design of the TRIUMF (n, p) facility lead to a precision which in these cases exceeds that achieved in analogous (p, n) measurements.

A comparison of the cross sections per unit strength $\hat{\sigma}_{GT}^+$ and $\hat{\sigma}_{GT}^-$ deduced for analogous transitions on ^{13}C reveals a substantial discrepancy. There is at present no adequate explanation of this result.

Detailed theoretical calculations appear to be remarkably successful in accounting in detail for the nuclear response of ^{90}Zr to the (p, n) reaction. The same procedures, however, are not nearly so successful in predicting the ^{90}Zr(n, p) data. A comprehensive understanding of the isovector nuclear response is an important objective which will certainly require further experiments and refinement of the theoretical approach.

Several examples have been given of the sensitivity of the measured distributions of B_{GT}^+ to the detailed nature of the nuclear wave functions. This situation may be contrasted with that for B_{GT}^- where all nuclei with $N > Z$ exhibit a reasonably compact Gamow-Teller resonance located in the vicinity of the isobaric analogue state with a total strength S_{GT}^- given by $3(N - Z)$ to within a factor of 2. In light nuclei the sensitivity of B_{GT}^+ is connected with the influence of spatial symmetry (SU$_4$) on the nuclear wave functions. It is interesting to note that in a very early assessment of limited experimental data on B_{GT} for $A \sim 120$ Fujita and Ikeda[40] proposed a model invoking a "persistent supermultiplet structure" broken in perturbation theory by the spin-orbit interaction. More recent theoretical analyses of this problem in medium and heavy nuclei (such as in Refs. 32, 39 and 41) do not make this connection with SU$_4$ symmetry but arrive at similar conclusions regarding the substantial quenching of B_{GT}^+ strength.

There are two important reasons for an interest in measurements of B_{GT}^+ that extend beyond the field of nuclear structure. One, concerning the rates of electron capture in the collapse of a massive star, has been mentioned in connection with the ^{54}Fe(n, p) data. The other involves critical tests of the nuclear structure calculations used in the analysis of double β decay.[42-45] In general those models which are successful in accounting for the measured suppression of the 2ν transition probabilities also predict strong quenching of B_{GT}^+ for the daughter nucleus and for other nuclei in the same mass region. Measurements of $L=0$ strength in the (n, p) reaction are being used to test the adequacy of such predictions and hence provide additional constraints on the models. ^{48}Ti and ^{76}Se are the first two targets being investigated for this purpose.

A great many people have contributed to the development of the TRIUMF charge exchange facility and to its early use. Experimental collaborators appear as authors in Refs. 10, 13, 14, 15, and 17. We would also like to acknowledge important contributions in the early stages of develpment by D. Clark, G. Clark, D. George, F. Marsiglio, A. Otter, P. Reeve, and V. Verma. P. Machule, R. Churchman, and W. Felske have provided excellent ongoing technical assistance. Theoretical support from N. Auerbach, S. Bloom, A. Brown, J. Engel, G. Fuller, A. Gal, M. Macfarlane, J. Millener, R. Smith, I. Towner, and J. Wambach has been appreciated. We thank Ada Strathdee for her prompt and careful preparation of this manuscript.

References

1. J.D. Anderson and C. Wong, Phys. Rev. Lett. **7**, 250 (1961).

2. D.H. Wilkinson, ed., *Isospin in Nuclear Physics* (North-Holland, Amsterdam, 1969).

3. R.R. Doering, A. Galonsky, D.M. Patterson, and G.F. Bertsch, Phys. Rev. Lett. **35**, 1691 (1975).

4. C.D. Goodman, C.A. Goulding, M.B. Greenfield, J. Rapaport, D.E. Bainum, C.C. Foster, W.G. Love, and F. Petrovich, Phys. Rev. Lett. **44**, 1755 (1980).

5. B.D. Anderson, J.N. Knudson, P.C. Tandy, J.W. Watson, R. Madey, and C.C. Foster, Phys. Rev. Lett. **45**, 699 (1980).

6. C. Gaarde, J. Rapaport, T.N. Taddeucci, C.D. Goodman, C.C. Foster, D.E. Bainum, C.A. Goulding, M.B. Greenfield, D.J. Horen, and E. Sugarbaker, Nucl. Phys. **A369**, 258 (1981).

7. H. Sakai, T.A. Carey, J.B. McClelland, T.N. Taddeucci, R.C. Byrd, C.D. Goodman, D. Krofcheck, L.J. Rybarcyk, E. Sugarbaker, A.J. Wagner, and J. Rapaport, Phys. Rev. C **35**, 344 (1987).

8. D.F. Measday and J.N. Palmieri, Phys. Rev. **161**, 1071 (1967).

9. F.P. Brady and G.A. Needham, in *The (p, n) reaction and the nucleon-nucleon force*, eds. C.D. Goodman, S.M. Austin, S.D. Bloom, J. Rapaport, and G.R. Satchler (Plenum, New York, 1980), p. 357;
   F.P. Brady, G.A. Needham, J.L. Ullmann, C.M. Castaneda, T.D. Ford, N.S.P. King, J.L. Romero, M.L. Webb, V.R. Brown, and C.H. Poppe, J. Phys. G **10**, 363 (1984);
   F.P. Brady, Can. J. Phys. **65**, 578 (1987).

10. M.C. Vetterli, O. Häusser, W.P. Alford, D. Frekers, R. Helmer, R. Henderson, K. Hicks, K.P. Jackson, R.G. Jeppesen, C.A. Miller, M.A. Moinester, K. Raywood, and S. Yen, Phys. Rev. Lett. **59**, 439 (1987).

11. R. Helmer, Can. J. Phys. **65**, 588 (1987).

12. S. Yen, Can. J. Phys. **65**, 595 (1987).

13. K.P. Jackson, A. Celler, W.P. Alford, K. Raywood, R. Abegg, R.E. Azuma, C.K. Campbell, S. El-Kateb, D. Frekers, P.W. Green, O. Häusser, R.L. Helmer, R.S. Henderson, K.H. Hicks, R. Jeppesen, P. Lewis, C.A. Miller, A. Moalem, M.A. Moinester, R.B. Schubank, G.G. Shute, B.M. Spicer, M.C. Vetterli, A.I. Yavin, and S. Yen, Phys. Lett. **B201**, 25 (1988).

14. S. Yen, B.M. Spicer, M.A. Moinester, K. Raywood, R. Abegg, W.P. Alford, A. Celler, T.E. Drake, D. Frekers, O. Häusser, R.L. Helmer, R.S. Henderson, K.H. Hicks, K.P. Jackson, R. Jeppesen, J.D. King, N.S.P. King, K. Lin, S. Long, C.A. Miller, V.C. Officer, R. Schubank, G.G. Shute, M.C. Vetterli, and A.I. Yavin, Phys. Lett. **B206**, 597 (1988).

15. W.P. Alford, R.L. Helmer, R. Abegg, A. Celler, O. Häusser, K.H. Hicks, K.P. Jackson, C.A. Miller, S. Yen, R.E. Azuma, D. Frekers, R.S. Henderson, H. Baer, and C.D. Zafiratos, Phys. Lett. **B179**, 20 (1986).

16. R.G. Jeppesen, these proceedings.

17. R.S. Henderson, W.P. Alford, D. Frekers, O. Häusser, R.L. Helmer, K.H. Hicks, K.P. Jackson, C.A. Miller, M.C. Vetterli, and S. Yen, Nucl. Instrum. Methods **A257**, 97 (1987).

18. J. Rapaport, private communication.

19. E.P. Wigner, Phys. Rev. **56**, 519 (1939).

20. D.R. Inglis, Phys. Rev. **87**, 915 (1952).

21. D. Kurath, Phys. Rev. **101**, 216 (1956); Phys. Rev. **106**, 975 (1957).

22. S. Cohen and D. Kurath, Nucl. Phys. **73**, 1 (1965).

23. J.C. Bergstrom, U. Deutschmann, and R. Neuhausen, Nucl. Phys. **A327**, 439 (1979).

24. J.P. Perroud, A. Perrenoud, J.C. Alder, B. Gabioud, C. Joseph, J.F. Loude, N. Morel, M.T. Tran, E. Winkelmann, H. von Fellenberg, G. Strassner, P. Truöl, W. Dahme, H. Panke, and D. Renker, Nucl. Phys. **A453**, 542 (1986).

25. T.N. Taddeucci, C.A. Goulding, T.A. Carey, R.C. Byrd, C.D. Goodman, C. Gaarde, J. Larsen, D. Horen, J. Rapaport and E. Sugarbaker, Nucl. Phys. **A469**, 125 (1987).

26. R.A. Arndt and L.D. Soper, Scattering analysis interactive dial-in (SAID) program (June 1987) (unpublished).

27. F. Ajzenberg-Selove, Nucl. Phys. **A413**, 1 (1984).

28. R. Schaeffer and J. Raynal, Program DWBA70 (unpublished); J.R. Comfort, extended version DW81 (unpublished).

29. S.D. Bloom and G.M. Fuller, Nucl. Phys. **A440**, 511 (1985).

30. N. Auerbach and A. Klein, Phys. Rev. C **30**, 1032 (1984); N. Auerbach, A. Klein, and W.G. Love, in *Antinucleon and Nucleon-Nucleus Interactions*, ed. G.E. Walker, C.D. Goodman, and C. Olmer (Plenum, New York, 1985), p. 323.

31. J. Rapaport, T. Taddeucci, T.P. Welch, C. Gaarde, J. Larsen, D.J. Horen, E. Sugarbaker, P. Koncz, C.C. Foster, C.D. Goodman, C.A. Goulding, and T. Masterson, Nucl. Phys. **A410**, 371 (1983).

32. N. Auerbach, Phys. Rev. C **36**, 2694 (1987).

33. R.D. Smith and J. Wambach, Phys. Rev. C **38**, 100 (1988).

34. R.D. Smith and J. Wambach, Phys. Rev. C **36**, 2704 (1987) and private communication.

35. F. Osterfeld, D. Cha, and J. Speth, Phys. Rev. C **31**, 372 (1985).

36. A. Klein, W.G. Love, and N. Auerbach, Phys. Rev. C **31**, 710 (1985);
A. Klein, W.G. Love, M.A. Franey, and N. Auerbach, in *Antinucleon and Nucleon-Nucleus Interactions*, ed. G.E. Walker, C.D. Goodman, and C. Olmer (Plenum, New York, 1985), p. 351.

37. M. Yabe, Phys. Rev. C **36**, 858 (1987).

38. S.D. Bloom, G.J. Mathews, and J.A. Becker, Can. J. Phys. **65**, 684 (1987).

39. I.S. Towner, Nucl. Phys. **A444**, 402 (1985).

40. J-I. Fujita and K. Ikeda, Nucl. Phys. **67**, 145 (1965).

41. D. Cha, Phys. Rev. C **27**, 2269 (1983).

42. B.A. Brown, in *Nuclear Shell Models*, ed. M. Vallieres and B.H. Wildenthal (World Scientific, Singapore, 1985), p. 42.

43. K. Grotz and H.V. Klapdor, Nucl. Phys. **A460**, 395 (1986).

44. T. Tomoda and A. Faessler, Phys. Lett. **B199**, 475 (1987).

45. J. Engel, P. Vogel, and M.R. Zirnbauer, Phys. Rev. C **37**, 731 (1988).

(p,n) MEASUREMENTS ABOVE 200 MeV AND THE ISOVECTOR EFFECTIVE

NN INTERACTION

R.G. Jeppesen

Simon Fraser University
Burnaby Mountain
Burnaby, BC, Canada V5A 1S6

INTRODUCTION

The proportionality between forward angle (p,n) cross sections and β-decay matrix elements[1] has most often been used to study distributions of β-decay strength in nuclei. However, under this same approximation the (p,n) cross sections can give information on the isovector parts of the effective NN interaction. This is perhaps best seen in figure 1 which shows the 0° ^{14}C(p,n) cross section as a function of bombarding beam energy. The cross section for the Fermi transition (0^+ state at 2.31 MeV), which is driven by the isovector non-spinflip part of the interaction (V_τ), falls drastically relative to the Gamow-Teller (GT) transition (1^+ state at 3.95 MeV), which is driven by the isovector spinflip part ($V_{\sigma\tau}$).

Under the distorted wave impulse approximation (DWIA), and assuming that only the central parts of the effective interaction are important, the ratio of these cross sections is[1]

$$\frac{\sigma_{GT}(0°)}{\sigma_F(0°)} = \frac{k_f^{GT}}{k_f^F} \frac{B(GT)}{B(F)} \frac{N_{\sigma\tau}}{N_\tau} \frac{|J_{\sigma\tau}|^2}{|J_\tau|^2} \qquad (1)$$

where $B(F) = N - Z$ and B(GT) are the β-decay matrix elements for the states, k_f^F (k_f^{GT}) is the final momentum of the outgoing neutron in the c.m. frame for the Fermi (GT) state, N_τ ($N_{\sigma\tau}$) is the distortion factor[1] for the Fermi (GT) excitation, and J_τ ($J_{\sigma\tau}$) is the $q = 0$ amplitude of the non-spinflip (spinflip) part of the isovector effective NN interaction. The β-decay matrix elements have been measured and the distortion factors can be calculated using the DWIA, so the ratio $|J_{\sigma\tau}|^2/|J_\tau|^2$ can be determined. It is an advantage both experimentally and theoretically to look at the ratio of the cross sections rather than each cross section individually. Experimentally many of the uncertainties in determining an absolute cross section cancel when looking at the ratio of two states in the same nucleus. Theoretically, there are differences of up to 20% in N_τ and $N_{\sigma\tau}$ when different phenomological optical potentials are used[2], but the ratio $N_\tau/N_{\sigma\tau}$ is stable at the few percent level. Taddeucci, *et al*, have analyzed the ^{14}C(p,n) data in this manner and

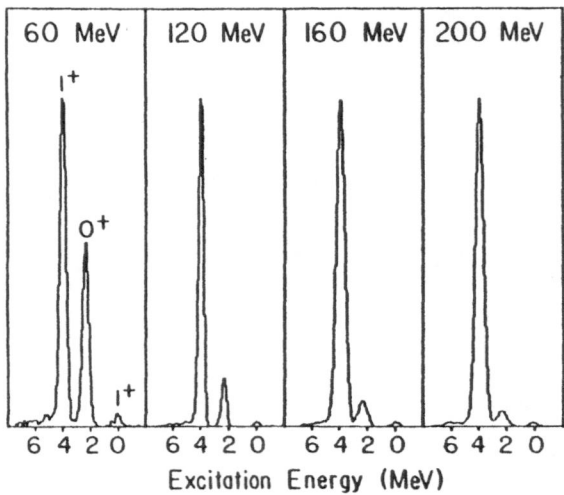

Figure 1. Cross section spectra for the $0°$ $^{14}C(p,n)$ reaction at IUCF energies. The spectra have been normalized to give the 1^+ GT state at 3.95 MeV constant peak height. (From ref. 2)

find the the following linear relationship:[2]

$$\sqrt{\frac{N_{\sigma\tau}}{N_\tau}}\,\frac{|J_{\sigma\tau}|}{|J_\tau|} = \frac{E_p}{55.0 \pm 0.4\ MeV} \qquad\qquad 60 \le E_p \le 200\ \text{MeV} \qquad (2)$$

where E_p is the incident energy of the proton. The ratio $N_{\sigma\tau}/N_\tau$ is roughly constant at 1.2 ± 0.1 over this energy range[3] so the ratio $|J_{\sigma\tau}|/|J_\tau|$ extracted is also linear. This ratio is also found[2] to be independent of nucleus (A).

The ratio $|J_{\sigma\tau}|/|J_\tau|$ can also be predicted from the impulse approximation (IA) using the free NN effective interaction of Franey and Love[4] (see figure 2). This interaction is fit to the NN phase shifts which are in turn fit to measured NN observables. The interaction explicitly includes exchange and the IA prediction in figure 2 uses a knock-on exchange approximation[4] for A=14. The IA prediction is not linear, especially beyond 180 MeV, and falls significantly below the data at 160 MeV and 200 MeV. Experiments beyond the 200 MeV upper limit of IUCF could tell if $|J_{\sigma\tau}|/|J_\tau|$ rolls over as predicted by the IA or not.

EXPERIMENTAL FACILITIES

A number of new facilities for doing high-resolution (p,n) experiments at energies above 200 MeV have come on line in the past few years. I will talk about the long flight path neutron time-of-flight (TOF) facility at LAMPF–WNR and the CHARGEX facility at TRIUMF. More recent results are presented to this conference by John Ullmann from a revised version of the LAMPF–WNR facility[5] and by John McClelland from the new NTOF facility at LAMPF.[6]

LAMPF–WNR. This facility uses the same method as was used for the IUCF studies, neutron time-of-flight. However, the flight path had to be much longer since the energy

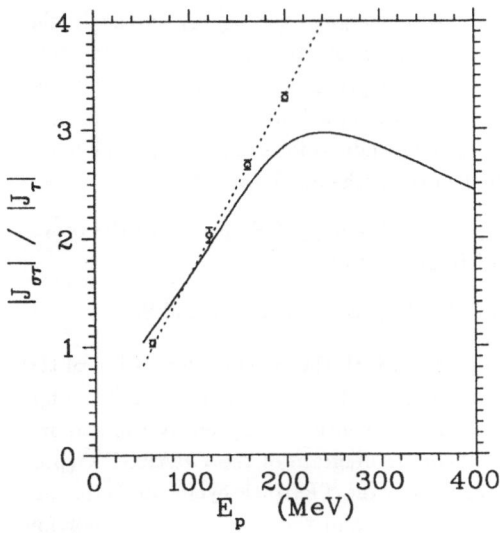

Figure 2. The ratio $|J_{\sigma\tau}|/|J_\tau|$ extracted from $0°$ $^{14}C(p,n)$ measurements at IUCF. The dotted line is a linear fit to the data (see eqn (2)). The solid line is the IA prediction using the Franey-Love interaction (ref. 4).

resolution as a function of timing resolution is given by

$$|dE_n| = (\gamma_n\beta_n)^3 \, m_nc^2 \, \frac{c}{l} \, |dt_n| \tag{3}$$

where l is the flight path length. At 200 MeV the factor $(\gamma_n\beta_n)^3$ is 0.3; at 800 MeV it increases by more than a factor of ten to 3.8. The timing resolution is essentially fixed by the size of the detector necessary for reasonable count rates and is about the same for IUCF detectors and the ones used at LAMPF, so the only thing that can be done is to increase the path length. The length of the flight path for the facility was 213 meters, limited by the edge of the mesa upon which LAMPF is built.

The LAMPF–WNR measurements where made at the Weapons Neutron Research (WNR) area on Line D of the Los Alamos Meson Physics Facility (LAMPF).[7,8] For zero degree measurements the targets were placed just upstream of a 30° bend magnet that swept the proton beam into a well-shielded beam dump in the Blue Room (target 2). The neutrons produced at 0° by the (p,n) reaction were detected at the end of a well collimated 213 meter flight path using a mean-timed 25.4cm x 51.8cm x 7.62cm scintillator detector. This detector had an intrinsic resolution of 375ps. A mean-time signal, position signal, and pulse height were returned to the counting room using fiber optics. The start signal for the time-of-flight came from a capacitive pickoff just upstream of the (p,n) target. Wraparound was suppressed by chopping the beam before acceleration so that the micropulses were separated by $3 - 4\mu$sec. Measurements off zero degrees were made by placing the (p,n) target in the 30° bend magnet along the path of the proton beam, and looking at neutrons down the same flight path as the zero degree measurements.

The resolution obtained with this system was 2.7 MeV at 800 MeV incident energy. This is somewhat larger than what is expected from equation (3) and a timing resolution of 375ps. Moreover, measurements at 318 MeV had no better resolution. The reason for this is that the timing is relative to the centroid of the beam pulse (with the capacitive pickoff) and a proton anywhere in the beam pulse can produce a neutron, so the length of the beam pulse enters into the timing uncertainty. The beam pulses from the LAMPF accelerator start out very compact, but as they drift from the end of the accelerator to the (p,n) target (250m at 800 MeV, longer for lower energies) the energy spread in the beam (≤ 2 MeV) causes the pulse to lengthen. In addition, the energy of the incident proton is translated to the outgoing neutron, so the lengthening continues on the 213m flight path to the detector. This turned out to be the reason for the degraded resolution of this system. New TOF facilities[5,6] at LAMPF use longitudinal focusing to make the beam pulse compact at the (p,n) target rather than the exit of the accelerator.[9]

Absolute cross sections were determined by comparing our TOF spectra to the previously published zero degree ^{nat}Pb(p, n) data of Bonner, et al.[10]

For more information on the LAMPF–WNR facility, see references 7 and 8.

TRIUMF. The TRIUMF measurements were made with the (p,n) configuration of the CHARGEX facility[11]. In this mode the neutrons from the (p,n) reaction scatter from hydrogen in a scintillator to produce protons that are momentum analyzed by the existing Medium Resolution Spectrometer (MRS). A schematic diagram of the CHARGEX (p,n) configuration is shown in figure 3. A proton beam from the TRIUMF cyclotron (typically 200-400 namps) impinges on the (p,n) target of interest and the non-interacting portion of the beam is swept away to a shielded beam dump by a dipole magnet immediately downstream of the target. Neutrons produced in the target are converted to protons in a scintillator 90 cm from the target and the protons are momentum analyzed using the quadrupole-dipole (QD) MRS spectrometer. Protons that scatter from the primary (p,n) target and which would be bent into the scintillator by the sweeping magnet are removed with a copper block. In addition, a veto scintillator just upstream of the conversion scintillator guarantees that the incident particle is a neutron. The standard MRS instrumentation of a front-end wire chamber (FEC0) before the quad, two pairs of vertical drift chambers (VDC) at the focal plane and ten segmented trigger scintillator paddles (FP$_{0-9}$) just above the VDCs, is augmented by a second FEC (FECM) and another scintillator (S1) above the focal plane. The second FEC allows improved traceback to the conversion scintillator to ensure that the proton did come from the scintillator. The additional scintillator above the focal plane is used in the trigger to reduce room backgrounds. The trigger usually consists of signals in the conversion scintillator, one of the trigger paddles, the additional scintillator, and the lack of a signal in the veto scintillator.

The major aberrations to the energy resolution are the vertical position on the conversion scintillator, the recoil kinematic correction for the conversion process, and the energy lost in the scintillator after the conversion. Traceback to the conversion scintillator using the positions in the two FECs allows the first two aberrations to be corrected. The use of an active converter (the scintillator outputs a signal proportional to the energy lost by the proton) allows the third aberration to be corrected. This active converter is important since it allows one to use a thick converter for high efficiency while maintaining good energy resolution. The 2 cm thick conversion scintillator used has a maximum energy loss of about 8 MeV at 200 MeV. The disadvantage of the active converter is that the neutrons can also interact with the carbon in the scintillator and produce protons. The Q-value of this reaction is -12.79 MeV so it does not usually interfere with low excitation studies, and it can be corrected for now that the ^{12}C(n, p) reaction has been measured at TRIUMF.

The resolution that can be obtained with this setup is determined by four factors: the intrinsic resolution of the MRS (≤ 100 keV), the resolution of the energy deposition signal from the conversion scintillator (about 300 keV), the thickness of the (p,n) target

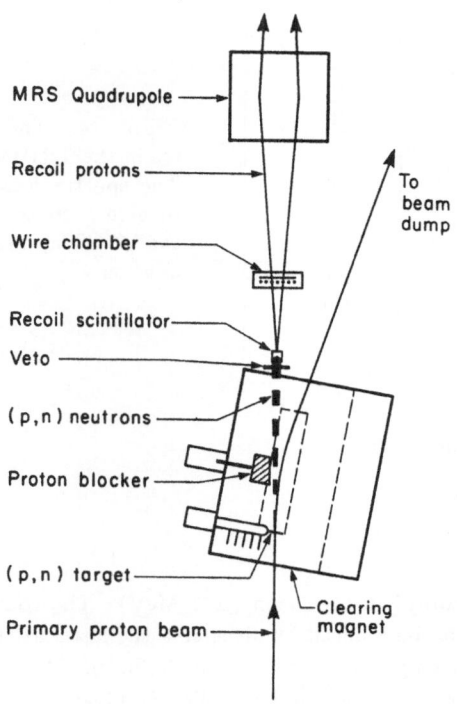

MRS Quadrupole

Recoil protons

To beam dump

Wire chamber

Recoil scintillator

Veto

(p,n) neutrons

Proton blocker

(p,n) target

Clearing magnet

Primary proton beam

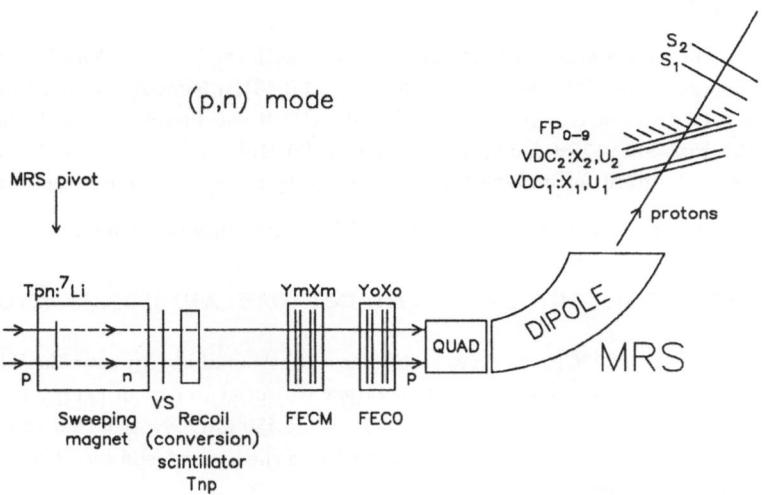

(p,n) mode

MRS pivot

Tpn:^{7}Li

YmXm   YoXo

S_2
S_1

FP_{0-9}
VDC_2:X_2,U_2
VDC_1:X_1,U_1

protons

p   n

QUAD

DIPOLE

MRS

VS

Sweeping magnet

Recoil (conversion) scintillator Tnp

FECM   FECO

P

Figure 3. Schematic view of the (p,n) configuration of the CHARGEX facility, top view and side view. The top view only shows one of the FECs. (from ref. 11)

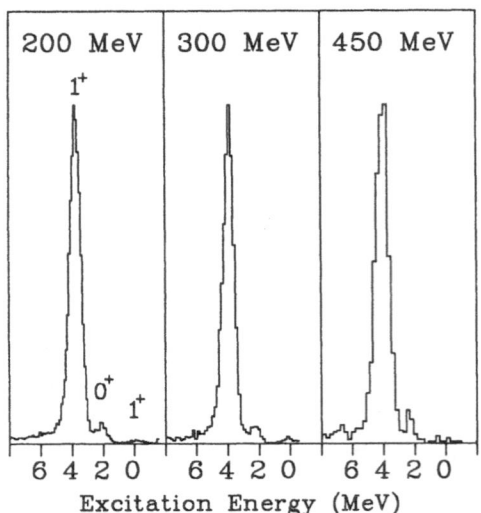

Figure 4. The $0°$ $^{14}C(p,n)$ spectra in the TRIUMF energy range[14]. The spectra have been normalized to give a constant peak height for the 1^+ GT state at 3.95 MeV of excitation.

(variable), and the energy spread in the beam ($\simeq 1$ MeV). The contribution from the energy spread in the beam can be reduced by dispersing the beam at the target and then using a strip target to select only a small portion of the beam. With this technique, an overall resolution of $\simeq 800$ keV can be obtained. The disadvantage is that one can not determine an absolute cross section; a separate run with achromatic beam is needed to normalize the data.

Finally, absolute cross sections have been obtained with the (p,n) CHARGEX setup by measuring an angular distribution for $^7Li(p,n)^7Be(gs+0.43)$, integrating it, and comparing the resulting total cross section to radiochemical data[12]. It was found that the $0°$ lab cross section for this reaction is 35 ± 2 mb/sr from 200 to 400 MeV[13]. This reaction is now used as a comparison standard for making further absolute (p,n) cross section measurements.

For further information on the TRIUMF CHARGEX facility, see reference 11.

EXPERIMENTAL RESULTS AND THE IMPULSE APPROXIMATION

Figure 4 shows the $0°$ $^{14}C(p,n)$ cross section as a function of energy over the TRIUMF energy range[14]. A dispersed beam on a strip target was used to get an energy resolution of 780 KeV at 200 MeV incident energy and 980 keV at 450 MeV. Notice that the ratio of Fermi cross section to GT cross section no longer follows the linear trend seen from 60 MeV to 200 MeV, but is instead roughly constant.

The resolution obtained with the original LAMPF–WNR setup was not sufficient to cleanly separate the Fermi and GT states in $^{14}C(p,n)$. Instead, measurements where made on $^{13}C(p,n)$ and $^{15}N(p,n)$. Both these nuclei have a mixed Fermi and GT ground state and a pure GT state at higher excitation (3.51 MeV in ^{13}N and 6.19 MeV in ^{15}O). The resolution of ~ 2.7 MeV was adequate to resolve these states (see figure 5). The extraction

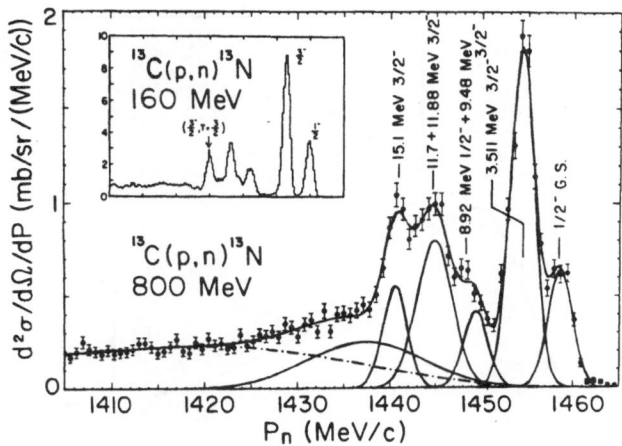

Figure 5. The 0° ^{13}C(p,n) spectrum at 800 MeV. The peak at 3.51 MeV is a pure GT state and the ground state is mixed Fermi and GT. Peak fitting was guided by the 160 MeV data (inset, from ref. 16) and by data at 1°, 3°, 4°, and 6°. (from ref 8)

of the ratio $|J_{\sigma\tau}|^2/|J_\tau|^2$ again relies on the ratio of cross sections for the two states,

$$\frac{|J_{\sigma\tau}|^2}{|J_\tau|^2} = \frac{N_\tau}{N_{\sigma\tau-}}\, \frac{B_m(F)}{B_{GT}(GT)}\, \frac{\sigma_{GT}}{\sigma_m\left[1 - \frac{\sigma_{GT}}{\sigma_m}\frac{B_m(GT)}{B_{GT}(GT)}\right]} \tag{4}$$

where the subscript m refers to the mixed Fermi and GT ground state and the subscript GT refers to the pure GT excited state. The term in square brackets removes the GT contribution of the mixed transition by comparing it to the excited state, so the quantity $\sigma_m[\ldots]$ in the denominator is just the Fermi cross section. This method of extracting the Fermi cross section can introduce large uncertainties in $|J_{\sigma\tau}|^2/|J_\tau|^2$ if the Fermi part of the mixed cross section is small.

The matrix elements for the ground state transitions in ^{13}C(p,n) and ^{15}N(p,n) are known from β-decay, but not the excited state matrix elements. Therefore we must turn to previous (p,n) results to find the B(GT) matrix elements for these transitions. Unfortunately, there are two separate analyses of the same data set that give radically different B(GT)'s for these transitions. The analysis of Watson, et al,[15] assumes that the constant of proportionality between 0° cross section and B(GT) can be given by the DWIA. Using measured cross sections and DWIA calculations they find a $B_{GT}(GT)$ of 1.38 for ^{13}C(p,n). The analysis of Taddeucci, et al,[2] only compares transitions within a single nucleus, and finds $B_{GT}(GT) = 0.82 \pm 0.05$. This subsequently means that the cross section per unit B(GT) is about 50% higher for ^{13}C(p,n) than for neighboring nuclei. Measurements of ^{13}C(n,p) do not show this large enhancement of unit cross section[17]. Our 800 MeV ^{13}C(p,n) data were analyzed using both methods. Using the cross section per unit B(GT) from ^{12}C(p,n)^{12}N(gs) whose B(GT) is known, we find $B_{GT}(GT) = 1.01 \pm 0.10$.

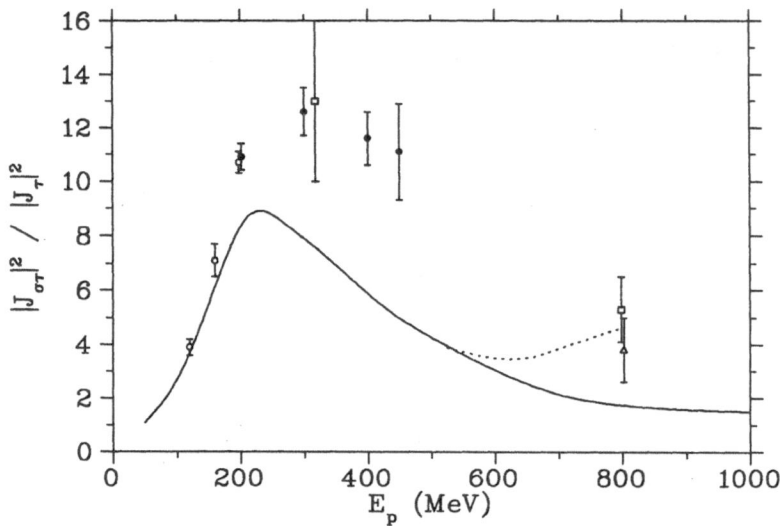

Figure 6. Comparison of the ratio $|J_{\sigma\tau}|^2/|J_\tau|^2$ extracted from (p,n) measurements to impulse approximation (IA) predictions. The open circles are $^{14}C(p,n)$ data from IUCF (ref. 2), the solid circles are $^{14}C(p,n)$ data from TRIUMF (ref. 14), the squares are $^{13}C(p,n)$ data from LAMPF–WNR, and the triangle is $^{15}N(p,n)$ data from LAMPF–WNR. IA predictions for the Franey-Love interaction (solid line) and an interaction fit to the Saclay S670 phase shifts (dotted line) are shown.

(This is equivalent to using a DWIA calculation since the DWIA would predict that the cross section per B(GT) is about the same for ^{12}C and ^{13}C.) Comparing the cross section for the pure GT state with the cross section for another pure GT state at 15.1 MeV ($T = \frac{3}{2}$) we find $B_{GT}(GT) = 0.79 \pm 0.08$. The cross section for the state at 15.1 MeV was determined by the peak fitting shown in figure 5. This fit was aided by the $^{13}C(p,n)$ angular distributions and similar values of $B_{GT}(GT)$ were found for both the 800 MeV and 318 MeV data (only 0° data). As a result of these findings, we have used β-decay matrix elements from the analysis of Taddeucci, *et al.*

The values of the ratio $|J_{\sigma\tau}|^2/|J_\tau|^2$ extracted from the (p,n) measurements at LAMPF–WNR[18] and TRIUMF[14] are shown in figure 6 along with the earlier data from IUCF[2]. The ratio of distortion factors, $N_{\sigma\tau}/N_\tau$, was calculated using the DWIA and phenomenological optical potentials and was found to be roughly constant at 1.2 ± 0.1 in the energy range 100 to 400 MeV, then rising to 1.5 at 800 MeV. The data from IUCF and TRIUMF at 200 MeV are in excellent agreement with each other and the $^{13}C(p,n)$ data at 318 MeV from LAMPF–WNR are in good agreement with TRIUMF results at 300 MeV, though the LAMPF–WNR data have much larger uncertainties.

The solid line is the IA prediction using the Franey-Love[4] interaction with exchange evaluated at A=14. The data above 200 MeV show an even greater deviation from the IA than that seen from the IUCF data. At 800 MeV, it is possible to get better agreement by choosing a different effective interaction, in this case an interaction fit[8] to the Saclay S670 phase shift solution[19]. The main difference between the S670 phase shift solution and the

SP84 solution (which the Franey-Love interaction was fit to) is that the S670 solution is in much better agreement with the magnitude of the nucleon-nucleon NP charge exchange cross section at 800 MeV. The S670 solution is only valid in the energy range 550–800 MeV.

The effects of non-central parts of the interaction are not sufficient to explain the differences between the IA and the values of $|J_{\sigma\tau}|^2/|J_\tau|^2$ extracted from (p,n) data. Full DWIA calculations show that the Fermi cross section in $^{14}C(p,n)$ can decrease by 15% at 400 MeV due mainly to the tensor part of the interaction. The GT transition is largely unaffected. This makes the value of $|J_{\sigma\tau}|^2/|J_\tau|^2$ extracted from (p,n) cross sections a little high, but not enough to explain the differences between the IA and the data. At 800 MeV, the cross section of the GT part of the mixed transition increases by 50%, again because of the tensor part of the interaction. The Fermi part of the mixed transition and the pure GT transition are not affected, so the effect would be to overestimate the Fermi cross section using equation (4) and the ratio extracted from the $^{13}C(p,n)$ data will be underestimated.

DENSITY DEPENDENT INTERACTIONS

A possible explanation for the differences seen between $|J_{\sigma\tau}|^2/|J_\tau|^2$ extracted from the (p,n) data and the IA is that the interaction is modified in the medium. This was suggested many years ago and has recently been looked at in more detail by Love, Nakayama, and Franey[20] for nonrelativistic (Schrodinger) dynamics and by Horowitz[21] for relativistic (Dirac) dynamics. These calculations use a medium-modified interaction, G, given by the Bethe-Goldstone equation[22] (notation here is non-relativistic)

$$G(\omega) = V + V \frac{Q}{\omega - H_1 + i\epsilon} G(\omega) \tag{5}$$

where V is the NN potential, Q is a Pauli projection operator which prohibits scattering to occupied states, H_1 is the single particle Hamiltonian containing both the kinetic energy and a single particle potential, and ω is the starting energy of the two interacting nucleons. This can be compared to the free NN t-matrix which satisfies a Lippmann-Schwinger equation,

$$t(\omega) = V + V \frac{1}{\omega - H_o + i\epsilon} t(\omega) \tag{6}$$

where H_o is now only the kinetic energy of the nucleon. The density effects are then from the Pauli projection operator and the binding of the struck nucleon in the single particle potential.

The G-matrix as given in equation (5) depends on both the radial separation of the two nucleons and the density of the medium. This form is too complicated to be of direct use in distorted wave (DW) calculations so one usually solves for a density-dependent effective interaction which reproduces most of the features of the G-matrix at a given density. These density-dependent effective interactions are then used with a local density approximation (LDA) to do a full DW calculation. Figure 7 shows the central part of the G-matrix at $q = 0$ as a function of density.[23] Note that all but the long range $G_{\sigma\tau}$ part of the interaction are affected. The ratio $|J_{\sigma\tau}|^2/|J_\tau|^2$ using a G-matrix (HM86) derived from the latest Bonn potential[20] is plotted at zero density (free interaction) and at one-third nuclear density in Figure 8a. Differences between the $\rho = 0$ density dependent interaction and the free Franey-Love interaction come about from differences in fitting the NN observables. The Bonn potential fits the NN phase shifts up to about 350 MeV.

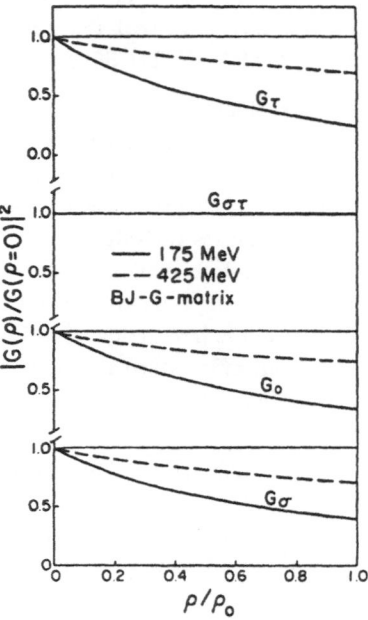

Figure 7. Density dependence of the central $q = 0$ part of the NN effective interaction for a G-matrix based on the HEA one-boson exchange potential of the Bonn group. The solid lines are for 175 MeV and the dashed lines are for 425 MeV. Note that the $\sigma\tau$ part does not show any density dependence. (From ref Lo87a).

The relativistic formalism used by Horowitz[21] has the same general features as the non-relativistic, with the addition of a modification of the Dirac spinors in the presence of the strong scalar and vector potentials in the nuclear medium. The density-dependent G-matrix derived by Horowitz from the latest relativistic Bonn potential, however, only includes the effects of Pauli blocking. As with the nonrelativistic calculations, it is found that the long range spinflip part of the interaction ($G_{\sigma\tau}$) is largely unaffected, but the non-spinflip part (G_τ) is strongly density dependent. The ratio $|J_{\sigma\tau}|^2/|J_\tau|^2$ using the relativistic G-matrix evaluated at zero, half, and full nuclear density is shown in Figure 8b.

The relativistic and non-relativistic calculations are very similar and comparison to the $^{14}C(p,n)$ data indicates that the G-matrix for a density of one-third to one-half nuclear density is appropriate. Horowitz performed an eikonal approximation calculation and found that in the presence of distortions the average density for the $0p_{\frac{1}{2}}$ and $0p_{\frac{3}{2}}$ orbitals is indeed $0.3 - 0.4\rho_o$.

Love, et al,[20] have performed full distorted wave (DW) calculations using their density dependent interaction and a local density approximation. These calculations are shown in figure 9. Instead of comparing to the ratio extracted from the (p,n) data, they compare their calculations to the absolute cross sections and the ratio of these cross sections. This allows them to include non-central parts of the interaction and to compare absolute cross sections to check the prediction of the GT cross section being unaffected while the Fermi cross section is lower than predicted by the free DWIA. As can be seen in figure 9, the agreement is very good over the region that the Bonn potential fits the NN phase shifts (up to 350 MeV). The agreement with absolute cross sections is improved if phenomological optical potentials are used instead of the folding model potentials using the density dependent interaction (see open circles in figure 9).

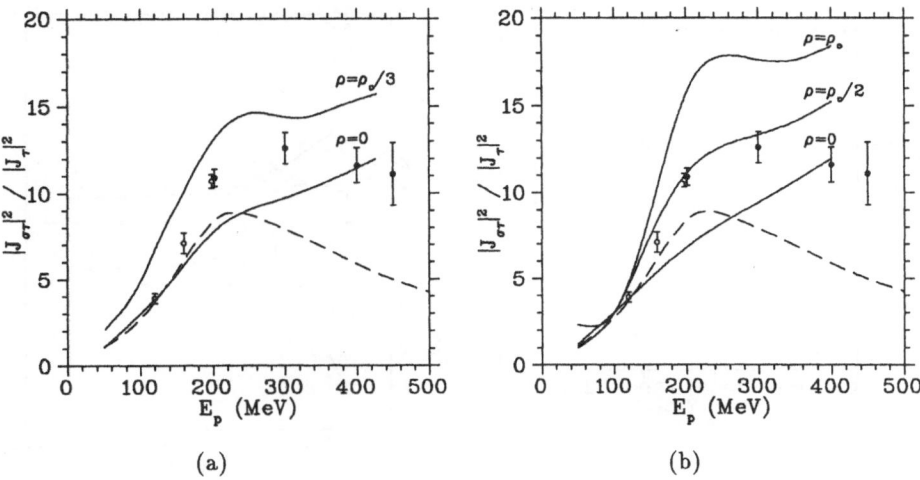

Figure 8. The ratio $|J_{\sigma\tau}|^2/|J_\tau|^2$ for non-relativistic (a) and relativistic (b) density dependent interactions. The solid curves are calculations for different nuclear matter densities (ρ_o is full nuclear matter density). The dashed curve is the Franey-Love free effective interaction.

Absolute cross section measurements have been made recently for the $^{14}C(p,n)$ reaction at TRIUMF energies and are currently being analyzed[24].

CONCLUSIONS

The extractions of $|J_{\sigma\tau}|^2/|J_\tau|^2$ from (p,n) data above 200 MeV show an even greater deviation from the predicted impulse approximation result than that found at 200 MeV. In the range 100–400 MeV this deviation seems to be explained by use of density dependent interactions which have little effect on the $J_{\sigma\tau}$ part of the the interaction but reduce the magnitude of the shorter range J_τ part. At 800 MeV the data seem to be explained by an alternate choice of phase shift solution. However, this alternate solution does not show a tendency to predict larger ratios at lower energies, but rather agrees with the Franey-Love interaction at its lower limit of 550 MeV.

Though the density dependent interactions seem to be able to explain the data up to 400 MeV, it is a little surprising that density dependence can have a large effect at such high energies. The effect is expected to be smaller as the energy increases, though this may be negated by the fact that we are looking at a zero momentum transfer process. It would be interesting to see if this effect continues to be important at higher energies – this will require NN potentials valid beyond 400 MeV.

Comparisons of the density dependent calculations to absolute cross sections on $^{14}C(p,n)$ show that the effect is on J_τ not $J_{\sigma\tau}$. It seems important to continue measurements on this nucleus at higher energies, especially given the uncertainty in the nuclear structure for $^{13}C(p,n)$ and $^{15}N(p,n)$. The data of Ullmann[5] presented at this conference show that this is possible.

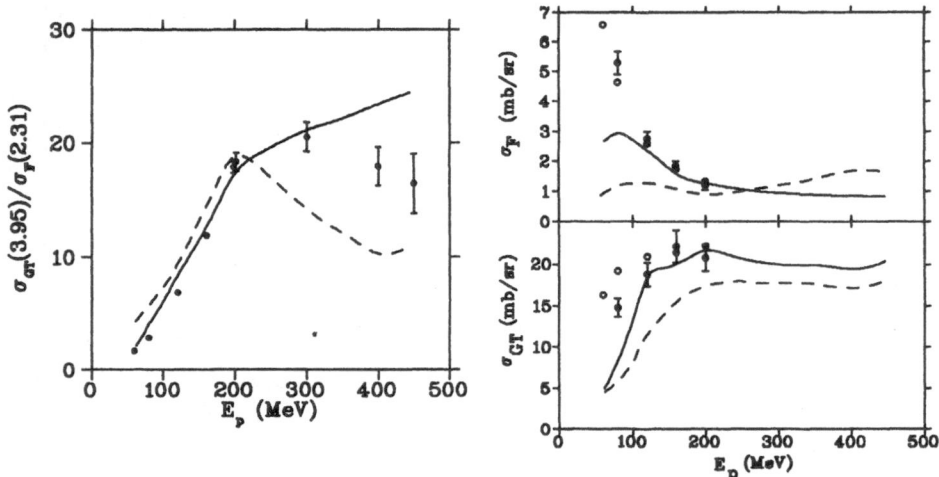

Figure 9. Comparison of the full distorted wave calculations of Love, *et al*, with measured ^{14}C(p, n) cross sections. On the left the ratio of the GT to Fermi cross section is compared to calculations using the HM86 density dependent interaction (solid line) and the free SP84 (Franey-Love) interaction (dashed line). On the right, these calculations are compared to the absolute cross sections for these states (top – Fermi, bottom – GT). The open circles are calculations that use the HM86 interaction and phenomological optical potentials. (From ref. 20).

The measurements at LAMPF–WNR where carried out by a collaboration from Los Alamos National Laboratory, University of Colorado, EG&G, and Indiana Cyclotron Facility (see ref. 18). The measurements from TRIUMF were carried out by members of the CHARGEX collaboration, see references 14 and 17.

REFERENCES

[1] C.D. Goodman, C.A. Goulding, M.B. Greenfield, J. Rapaport, D.E. Bainum, C.C. Foster, W.G. Love, and F. Petrovich, Phys. Rev. Lett. **44**, 1755 (1980).

[2] T.N. Taddeucci, C.A. Goulding, T.A. Carey, R.C. Byrd, C.D. Goodman, C. Gaarde, J. Larsen, D. Horen, J. Rapaport, and E. Sugarbaker, Nucl. Phys. **A469**, 125 (1987).

[3] T.N. Taddeucci, J. Rapaport, D.E. Bainum, C.D. Goodman, C.C. Foster, C. Gaarde, J. Larsen, C.A. Goulding, D.J. Horen, T. Masterson, and E. Sugarbaker, Phys. Rev. **C25**, 1094 (1981).

[4] M.A. Franey and W.G. Love, Phys. Rev. **C31**, 488 (1985). W.G. Love and M.A. Franey, Phys. Rev. **C24**, 1073 (1981).

[5] J.L. Ullmann, contribution to these proceedings.

[6] J.B. McClelland, contribution to these proceedings; J.B. McClelland, "Proceedings of the Workshop on Isovector Excitations in Nuclei", edited by W.P. Alford and K.P. Jackson, Can. Journ. Phys. **65**, 633 (1987).

[7] D.A. Lind, "Proceedings of the Workshop on Isovector Excitations in Nuclei", edited by W.P. Alford and K.P. Jackson, Can. Journ. Phys. **65**, 637 (1987).

[8] R.G. Jeppesen, PhD Thesis, University of Colorado, unpublished (1986).

[9] D.A. Lind, D.L. Prout, W.R. Smythe, C.D. Zafiratos, J.B. McClelland, N.S.P. King, University of Colorado Nuclear Physics Laboratory Progress Report, October 1, 1987, NPL 1028, pg. 110.

[10] B.E. Bonner, J.E. Simmons, C.R. Newsom, P.J. Riley, G. Glass, J.C. Hiebert, Mahavir Jain, and L.C. Northcliffe, Phys. Rev. C**18**, 1418 (1978).

[11] R. Helmer, "Proceedings of the Workshop on Isovector Excitations in Nuclei", edited by W.P. Alford and K.P. Jackson, Can. Journ. Phys. **65**, 588 (1987).

[12] J. D'Auria, M. Dombsky, L. Moritz, T. Ruth, G. Sheffer, T.E. Ward, C.C. Foster, J.W. Watson, B.D. Anderson, and J. Rapaport, Phys. Rev. C**30**, 1999 (1984).

[13] W.P. Alford, J.W. Watson, C.D. Zafiratos, and A. Celler, TRIUMF experiment 379, private communication.

[14] W.P. Alford, R.L. Helmer, R. Abegg, A. Celler, O. Häusser, K. Hicks, K.P. Jackson, C.A. Miller, S. Yen, R.E. Azuma, D. Frekers, R.S. Henderson, H. Baer, and C.D. Zafiratos, Phys. Lett. **179**B, 20 (1986).

[15] J.W. Watson, W. Pairsuwan, B.D. Anderson, A.R. Baldwin, B.S. Flanders, R. Madey, R.J. McCarthy, B.A. Brown, B.H. Wildenthal, and C.C. Foster, Phys. Rev. Lett. **55**, 1369 (1985).

[16] C.D. Goodman, R.C. Byrd, I.J. Van Heerden, T.A. Carey, D.J. Horen, J.S. Larsen, C. Gaarde, J. Rapaport, T.P. Welch, E. Sugarbaker, and T.N. Taddeucci, Phys. Rev. Lett. **54**, 877 (1985).

[17] K.P. Jackson, A. Celler, W.P. Alford, K. Raywood, R. Abegg, R.E. Azuma, C.K. Campbell, S. El-Kateb, D. Frekers, P.W. Green, O. Häusser, R.L. Helmer, R.S. Henderson, K.H. Hicks, R. Jeppesen, P. Lewis, C.A. Miller, A. Moalem, M.A. Moinester, R.B. Schubank, G.G. Shute, B.M. Spicer, M.C. Vetterli, A.I. Yavin, and S. Yen, Phys. Lett. **201**B, 25 (1988).

[18] N.S.P. King, P.W. Lisowski, G.L. Morgan, P.N. Craig, R.G. Jeppesen, D.A. Lind, J.R. Shepard, J.L. Ullmann, C.D. Zafiratos, C.D. Goodman, and C.A. Goulding, Phys. Lett. **175**B, 279 (1986).

[19] R.A. Arndt, phase shift program SAID, unpublished.

[20] W.G. Love, K. Nakayama, and M.A. Franey, Phys. Rev. Lett. **59**, 1401 (1987).

[21] C.J. Horowitz, Phys. Lett. **196**B, 285 (1987).

[22] W.G. Love and Amir Klein, Proc. Sixth Int. Symp. Polar. Phenom. in Nucl. Phys., Osaka, 1985; J. Phys. Soc. Jpn. **55**, 78 (1985).

[23] W.G. Love, Amir Klein, M.A. Franey, and K. Nakayama, "Proceedings of the Workshop on Isovector Excitations in Nuclei", edited by W.P. Alford and K.P. Jackson, Can. Journ. Phys. **65**, 536 (1987).

[24] W.P. Alford and R.L Helmer, TRIUMF experiment 437, private communication.

SPIN-ISOSPIN EXCITATIONS IN NUCLEI AND (p,n) REACTIONS

S.N. Ershov, F.A. Gareev, and N.I. Pyatov

Joint Institute for Nuclear Research, Dubna, USSR

S.A. Fayans

I.V. Kurchatov Institute for Atomic Energy, Moscow, USSR

ABSTRACT

The review is given of theoretical studies of spin-isospin excitations of nuclei in (p,n) reactions at intermediate energies. The microscopic approach used combines the distorted-wave impulse approximation (DWIA) for the reaction mechanism with the nuclear structure description within the theory of finite Fermi-systems (TFFS). The model was applied to the calculation of angular distributions for bound and resonance states, inclusive energy spectra of neutrons and spin observables for some closed shell nuclei.

INTRODUCTION

In recent years charge-exchange nuclear reactions with hadrons and light ions have become an important tool for investigating both the nuclear structure and the reaction mechanisms. In particular, much progress has been achieved with the intermediate energy particles ($\gtrsim 100$ MeV/nucleon) when the direct quasi-elastic processes dominate in the cross sections for the low energy excitations ($E_x \lesssim \epsilon_F$) and small momentum transfers ($q \lesssim p_F$). A new class of spin-isospin resonances was discovered in forward angle inclusive neutron spectra in (p,n) reactions, in particular, the Gamow-Teller resonance (GTR). By comparison with shell-model calculations the quenching of low-energy spin-isospin transitions was established [1-9]. These studies stimulated the discussions of the effects arising from multi-particle excitations as well as the role of both the meson exchange currents and baryon resonances in this phenomenon (see, e.g. refs. [10-14]). The new important information concerning the effective interactions between quasiparticles in nuclei and the nucleon-nucleus interactions have been obtained for the spin-isospin channel.

Recently, much attention has been paid to the reactions with polarized beams in which the analyzing power and polarization transfer coefficients have been measured [15-18]. These quantities are very sensitive to details of the reaction mechanism and the nuclear excitation structure. In principle, one may hope to perform the multipole decomposition of continuum spectra. On the other hand, with the known structure of nuclear excitations one can study individual components of the nucleon-nucleus effective interactions.

In this talk we discuss recent developments in theoretical and experimental studies of (p,n) reactions at intermediate energies.

FORMALISM

To deduce the structure information from the (p,n) data, especially in the continuum, the microscopic models [19-21] have been developed. The basic assumptions of these models are the following.

i) For the low excitation energy of target nuclei ($E_x < \epsilon_F \leq E_p/2$) one-step processes dominate in the reaction and the cross sections are calculated in the distorted wave impulse approximation (DWIA) [22].

ii) The t-matrix NN-interaction [23] deduced from phase-shift data at the corresponding energies is used as an effective nucleon-nucleus interaction.

iii) The nuclear excitations are described in microscopic approaches using the RPA or TFFS. Usually, one includes contributions from all the particle-hole (ph) transtitions with the orbital momentum transfers $0 \leq L \leq 5$ and spin transfers S = 0 and S = 1 (it is the ph excitations with the total spin and parity $J^{\pi} = 0^+, 1^+, 1^-, ..., 4^+, 4^-$).

The distorted waves used in calculations are generated by means of phenomenological optical potential. The most popular are parametrizations [24,25] obtained from elastic proton-nucleus scattering data.

The models of (p,n) reactions differ in details concerning mainly the structure calculations. In papers [19] the transition densities are calculated in RPA using the effective $\pi + \rho$ meson exchange interaction supplemented by a local Landau-Migdal interaction in the spin-isospin channel. The self-consistent Hartree-Fock method with simplified Skyrme forces was used in ref. [20]. In papers [21] the structure calculations were performed in the TFFS with the effective spin-isospin interactions that included the local Landau-Migdal repulsion with g' = 1.1 (G'_0 = 330 MeV·fm^2) and the finite-range one-pion exchange attraction amplitude renormalized in matter by a factor of $e_q^2[\sigma\tau]$ = 0.64 (for the details see ref. [26]).

In the approximations which we have used the (p,n) cross section can be expressed in terms of the nuclear polarization operator for the following external charge exchange field

$$V^{\circ}(\vec{r}) = \int d\vec{r}' \chi_n^{(-)} (\vec{r}, \vec{p}_2) t(\vec{r}, \vec{r}') \chi_p^{(+)}(\vec{r}', \vec{p}_1) , \qquad (1)$$

where $\chi_{p,n}$ are the distorted wave functions, $\vec{p}_1 (\vec{p}_2)$ is the incident (outgoing) proton (neutron) momentum, and t is the effective free NN-interaction (the spin-isospin indices and time-dependent factor exp(iωt) are omitted for simplicity). For the effective field V, which arises in the nucleus as a response to the external field V°, we have the equation

$$V(\vec{r}, \omega) = e_q V^{\circ}(\vec{r}) + \int d\vec{r}_1 d\vec{r}_2 \mathcal{F}(\vec{r}, \vec{r}_1) A(\vec{r}_1, \vec{r}_2; \omega) V(\vec{r}_2; \omega) \equiv e_q V^{\circ} + (\mathcal{F}AV) , \qquad (2)$$

where $\mathcal{F}$ is the effective interaction between quasiparticles, $A(\vec{r}_1\vec{r}_2;\omega)$ is the 1p1h propagator and e_q is the local charge of nucleon quasiparticle with respect to the field V°. The local charge $e_q[\sigma\tau]$ will appear in any low-energy spin-isospin transition matrix element [26], thus renormalizing the shell-model sum rules which should be multiplied by $e_q^2[\sigma\tau]$. This reflects the difference between nucleons and quasiparticles in the TFFS; the factor $e_q^2[\sigma\tau]$ measures the 1p1h-fraction of the Ikeda sum rule 3(N-Z) for the Gamow-Teller field $V^{\circ} \sim \sigma\tau$ (which is model-independent in a pure nucleonic sector), the rest being associated with multipair and other modes, in particular, with non-nucleonic degrees of freedom.

By means of the Landau renormalization procedure the polarization operator can be written in the form (see, e.g., refs.[26,27]).

$$P(\omega) = (e_q V \,^\circ A V) + (V \,^\circ B \frac{1}{a} e_q V \,^\circ). \tag{3}$$

Here, the first term represents the contribution of the ph excitation branch and it is calculated in the TFFS, the second one is due to multipair excitations and is believed to be small at low ω because its imaginary part contains the small parameter $(\omega/2\epsilon_F)^2$.

Now, the differential cross section in the excitation energy interval from ω to $\omega+d\omega$ can be given as $(h=c=1)$

$$\frac{d^2\sigma}{d\Omega d\omega} = - \frac{E_1 E_2 p_2}{(2\pi)^2 p_1 \pi} \, \text{Im} P(\omega). \tag{4}$$

Here, $E_1(E_2)$ is the reduced projectile-target energy in the initial (final) state, E_i and p_i being calculated with relativistic kinematics, and $\text{Im}P(\omega)$ is, in terms of the exact eigenstates $|s>$ of the final nucleus, given by

$$\text{Im} P(\omega) = - \pi \sum_s |<s|V^\circ|0>|^2 \, \delta(\omega - \omega_s). \tag{5}$$

The low-energy nuclear response is given by the strength function

$$S(\omega) = - \frac{1}{\pi} \, \text{Im}(e_q V_0 A V) . \tag{6}$$

The bound and resonance states were separated in $S(\omega)$ calculated for the simple charge-exchange multipole fields[26]. The transition densities for each state with the excitation energy $\omega = \omega_R$ were obtained,

$$\rho_{tr}(\vec{r}; \omega_R) = N \cdot \text{Im} \int d\vec{r}' A(\vec{r}, \vec{r}'; \omega_R) V(\vec{r}'; \omega_R) \tag{7}$$

the normalization constant N being determined from the transition matrix element

$$M^2 = (e_q V^\circ \rho_{tr})^2 \equiv \int_{\Delta\omega} S(\omega) d\omega. \tag{8}$$

For the bound states M^2 were calculated as a residue of $S(\omega)$ at $\omega = \omega_R$ while in the continuum integration over ω in eq. (8) was performed in the vicinity of the resonance state. The contributions from nonresonance excitations in a given energy interval were included in the transition densities, thus allowing one to exhaust the complete multipole strength (sum rules).

For the structure calculations of natural parity excitations in the isovector rr' -channel we use an effective density-dependent interactions satisfying the consistency condition between the isovector mean field potential and the isovector density. No renormalization of the t-matrix in this channel was introduced since the revelant local charge of quasiparticles $e_q[r]$ at small 4-momentum transfer (q, ω) equals unity due to conservation laws[27]. In the spin-isospin channel, the local charge $e_q[\sigma r]$ according to the TFFS can be approximated by a constant with an accuracy to $(q/2p_F)^2$ and $(\omega/2\epsilon_F)^2$.

We should like to emphasize that in our approach the strength functions and transition densities are calculated with the complete 1p1h-basis. This is achieved by using the propagator $A(\vec{r}, \vec{r}'; \omega)$ in the coordinate representation [26] .

171

The excitation of the bound states with the known structure provides
a good test of the (p,n) reaction mechanism. For that purpose we calculated
the angular distributions at E_p = 134 and 160 MeV for the ^{48}Ca(p,n)^{48}Sc reaction with excitation of the $(\pi f_{7/2}, \nu f_{7/2}^{-1})$ ph-band of levels with J^{π} =
= $0^+, 1^+, \ldots, 7^+$. For illustration the angular distributions for some states
are shown in fig. 1. We have used the TFFS transition densities and the optical potential parameters were taken from ref.[25]. The t-matrix interaction[23]
parametrized at E_p = 140 MeV was used in calculations. In the direct channel both the central and tensor components were taken into account, while in
the exchange one only the central force was included in the pseudopotential
approximation[28]. For more details see ref.[29]. It is seen from fig.1 that
for the 0^+ state the experimental angular distributions are well described
without any fitting parameters (i.e. with $e_q[\tau]$ = 1). The angular distributions for the 1^+ state are well reproduced with $e_q[\sigma\tau]$ = 0.8 that corresponds
to 36% quenching of the GT transition strength. For the 3^+ state a good fit
is obtained with $e_q[\sigma\tau]$ = 0.8, while for the 5^+ state the value $e_q[\sigma\tau]$ = 1.0
seems to be more preferable. However, it should be noted that for these sta-
tes the contribution from the exchange tensor force may be important. According to our estimate with the pure ph transition density for the 5^+ state
a 20-30% increase at the maximum of the cross section is expected. Due to
both this reason and the absence of a reliable normalization of the tensor
force we cannot draw a definite conclusion about the quenching of the $\sigma\tau$-
strength for the high spin states. The theory predicts small excitation
cross sections for the $2^+, 4^+$ and 6^+ states of the multiplet considered,
which agrees with the experimental data[9].

The most reliable estimate of the quenching factor (or the local charge
$e_q[\sigma\tau]$) can be obtained from comparing the calculated inclusive energy
spectra of neutrons at small angles with the experimental ones. Usually, one
obtains a discretized excitation spectrum of the daughter nucleus and for
each state the angular distribution is obtained. Then the smooth inclusive
neutron energy spectrum for a given scattering angle is calculated by folding the differential cross sections with the Breit-Wigner functions. This
procedure phenomenologically simulates the spreading of the resonances due
to coupling of simple 1p1h excitations with the complex np-nh ones. In ref.[19]
the asymmetric Breit-Wigner functions with the energy-dependent widths were
used that allowed one to remove the essential part of the GT-strength to
the higher excitation energies (E_x > 20 MeV) so that in the low-energy region
the quenching needed did not exceed 15%; then the inclusion of the coupling
of 1p-1h states with the Δ-isobar - nucleon hole states ($E_x \simeq$ 300 MeV) resulted in a too strong quenching. In ref.[20] the constant width Γ = 2 MeV
was assumed for all the states and the conclusion was drawn that the GT-
transitions in the region $0 \lesssim E_x \lesssim 20$ MeV should be quenched by about 35%.
In refs.[21,29,30] we have used the symmetric Breit-Wigner functions with the
energy-dependent widths which were chosen so as to reproduce the structures
observed in (p,n) spectra for $E_x \lesssim 20$ MeV. Representative examples of the
small-angle (p,n) spectra for ^{48}Ca, ^{90}Zr and ^{208}Pb are shown in fig.2. For
^{208}Pb the optical potential from ref.[24] was used. Calculations for ^{48}Ca
and ^{90}Zr were performed with the optical potential from ref.[25]. Note that
the change of the optical potential for ^{90}Zr as compared with refs.[21,30]
results in approximately 25% reduction of the cross sections. From our
analysis of the inclusive forward angle neutron spectra as well as angular
distributions for individual bound states it has been concluded that the
local charge $e_q[\sigma\tau] \simeq 0.8$ is needed to describe the low-energy $\sigma\tau$-transi-
tions, i.e. the ph $\sigma\tau$-branch is quenched by $\simeq$ 36%. The partial contributions to the cross section from transitions of different multipolarities are
also shown in fig.2. It is seen that the dominating contribution at $\theta = 0^\circ$
comes from GT transitions, the background from L > 0 transitions being small.
But the latter increases quickly with θ. At θ = 4.5° in ^{90}Zr one can clearly
see the maximum in the spectrum in the region $20 \lesssim -Q_{pn} \lesssim 30$ MeV, which is

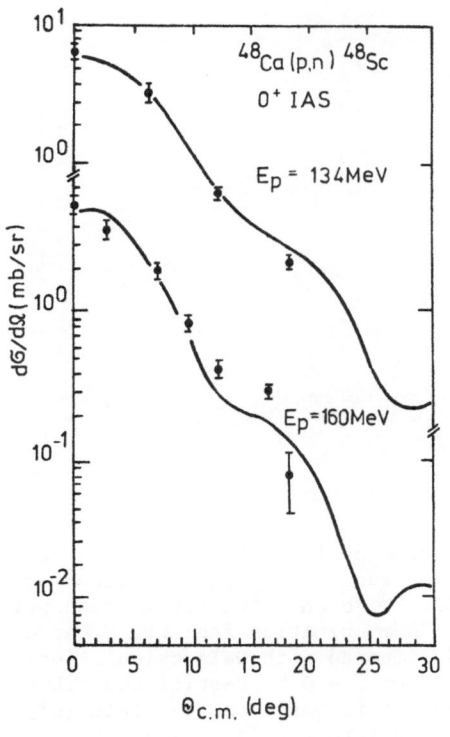

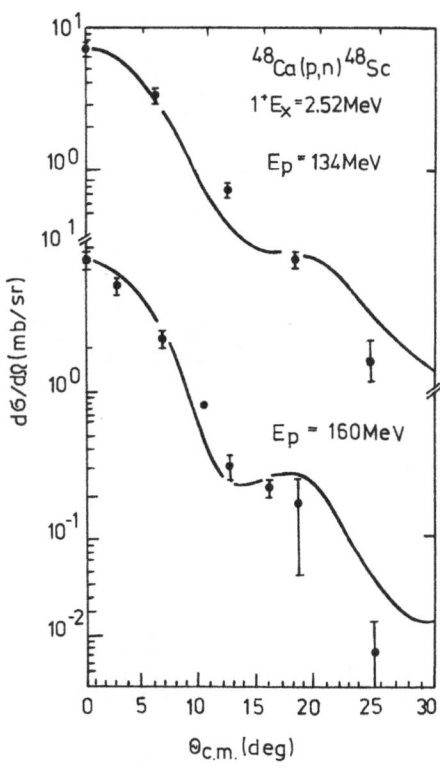

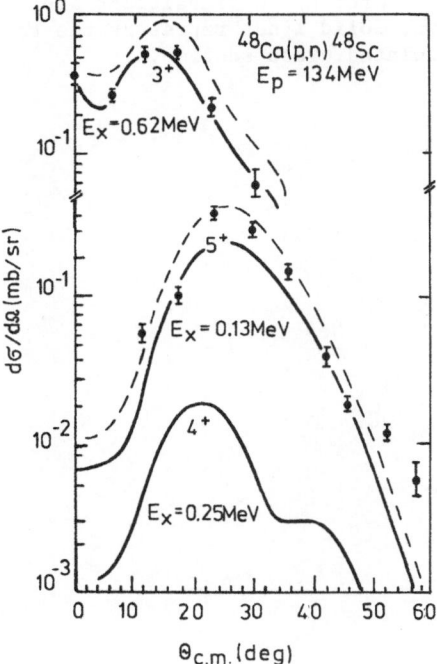

Fig. 1. The experimental data
of Ref.[9] for some states of
the $(\pi f_{7/2}, \nu f_{7/2}^{-1})$ ph-band in ^{48}Sc
are compared with DWIA calculations
performed for $e_q[\tau] = 1.0$ (for 0^+
and 4^+ levels), $e_q[\sigma\tau] = 0.8$ (solid
curves) and $e_q[\sigma\tau] = 1.0$ (dashed
curves).

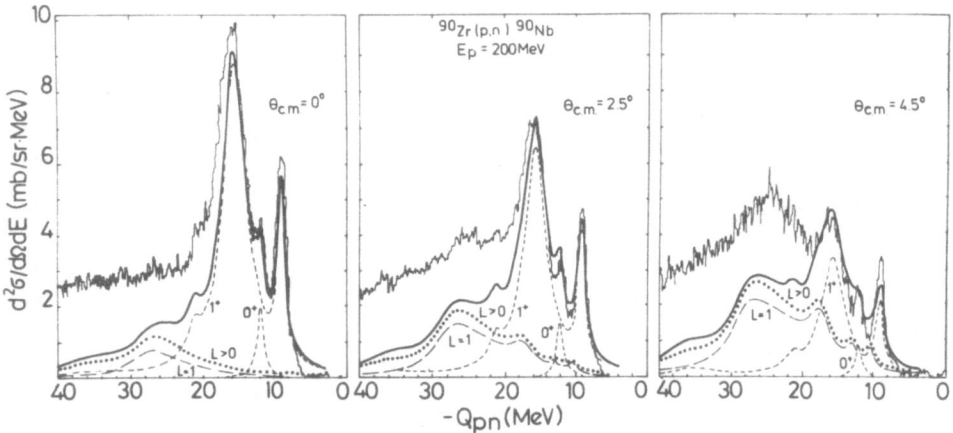

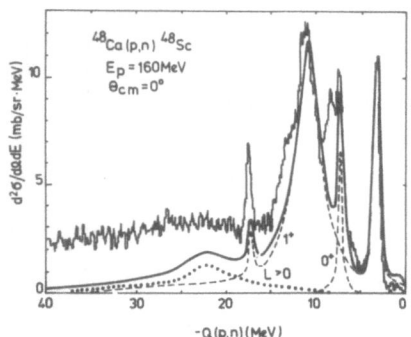

Fig.2. Experimental energy spectra of neutrons from (p,n) reactions on ^{90}Zr, ^{208}Pb and ^{48}Ca (ref.[4]) and private communication from Dr. C.Gaarde) are compared with DWIA calculations for $e_q[\sigma\tau] = 0.8$. Partial contributions of isobaric analog state (0^+), GT transitions (1^+), spin-dipole transitions ($L = 1$) and the background from $L > 0$ transitions are shown. Solid lines represent the total calculated cross section.

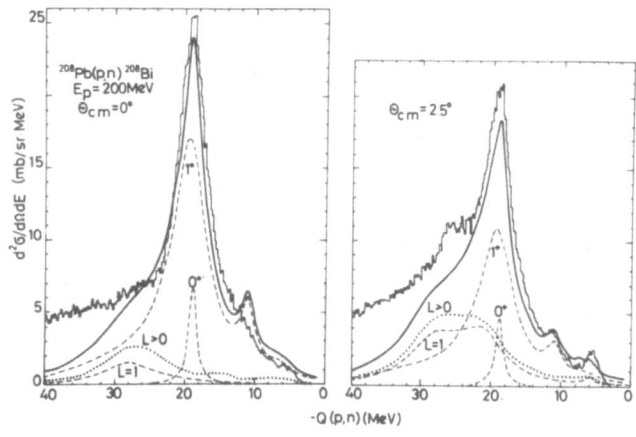

associated with the theoretically predicted spin-dipole resonance. It is formed by the excitations with $J^{\pi} = 0^-, 1^-$ and 2^-.

The conclusion drawn from our calculations is that the one-step mechanism provides a good description of the forward angle (p,n) spectra at $E_p > 100$ MeV in the excitation energy region $0 \leq E_x \leq 20$ MeV. Above GTR the theoretical cross sections are systematically lower than the experimental ones. However, with $e_q [\sigma\tau] = 1$ the calculated energy integrated cross sections in the whole region $0 \leq -Q_{pn} \leq 40$ MeV shown in fig.2 are in good agreement with the corresponding experimental data, i.e. if we replace the "excess" of the cross section from the low-energy region by the region above GTR, DWIA appears to be valid up to $-Q_{pn} \approx 40$ MeV. Such a replacement of the transition strength occurs when 1p1h excitations are coupled with the multipair ones [10].

SPIN OBSERVABLES

Important information on the distribution of the strength of various transitions can be obtained from experimental studies with polarized protons. In the case of transverse polarization the spin observables are related by[31]

$$P_f = (P_y(\theta) + p_i D_{NN}(\theta))(1 + p_i A_y(\theta))^{-1}, \tag{9}$$

where p_i (p_f) is the initial (final) nucleon polarization, $P_y(\theta)$ is the polarization function, $A_y(\theta)$ is the analyzing power and $D_{NN}(\theta)$ is the transverse spin-transfer coefficient. At $\theta = 0^\circ$, $A_y = P_y = 0$, and eq. (9) becomes $p_f = p_i D_{NN}(0^\circ)$. According to the plane-wave estimates[32] the coefficients $D_{NN}(0^\circ)$ take specific values for different L and S transfers. Thus, for transitions to bound 0^+ states $D_{NN}(0^\circ) = 1$, while for 0^- excitations $D_{NN}(0^\circ) = -1$ and for the other unnatural parity transitions (including GTR) $D_{NN}(0^\circ) \leq -1/3$. Our DWIA calculations for bound and resonance states agree well with these estimates.

The coefficient D_{NN} is related to the transverse spin-flip probability S by $D_{NN} = 1 - 2S$. By means of the latter quantity the total unpolarized cross section σ is represented as a sum of the spin-flip cross section σS and the non-spin-flip cross section $\sigma(1 - S)$, so that

$$\sigma D_{NN} = \sigma(1 - S) - \sigma S. \tag{10}$$

The measured data for σS, $\sigma(1 - S)$ and D_{NN} are usually sorted into bins of 1-2 MeV width and averaged to reduce statistical scatter.

Recent experiments[15-18] have provided evidences of the dominance of the spin-flip transitions in the spectra at forward angles. Using the model derived in ref.[21] we have calculated the energy distributions of σS, $\sigma(1-S)$ and D_{NN} for ^{90}Zr(p,n) and ^{48}Ca(p,n) reactions. The optical potential was taken from ref.[25]. As illustrative examples, fig.3 shows the calculated with $e_q [\sigma\tau] = 0.8$ spin-flip and non-spin-flip cross sections together with transverse polarization transfer coefficient $D_{NN}(0^\circ)$ for ^{90}Zr(p,n) reaction and the measured values at $E_p = 160, 200$ MeV. In both σS and $\sigma(1 - S)$ spectra the contribution of GT transitions dominates in the region $-Q_{pn} \leq 22$ MeV. The agreement between the calculated and observed distributions in this part of the excitation spectrum support the suggested quenching of the low-energy spin-isospin transitions. The cross sections above GTR are featureless at forward angle.

Theoretical distributions of $D_{NN}(0^\circ)$ are in good agreement with the measured ones. In the low-energy part of spectra the contribution from the first 1^+ state ($-Q_{pn} \approx 9$ MeV) dominates that determines the value of D_{NN}

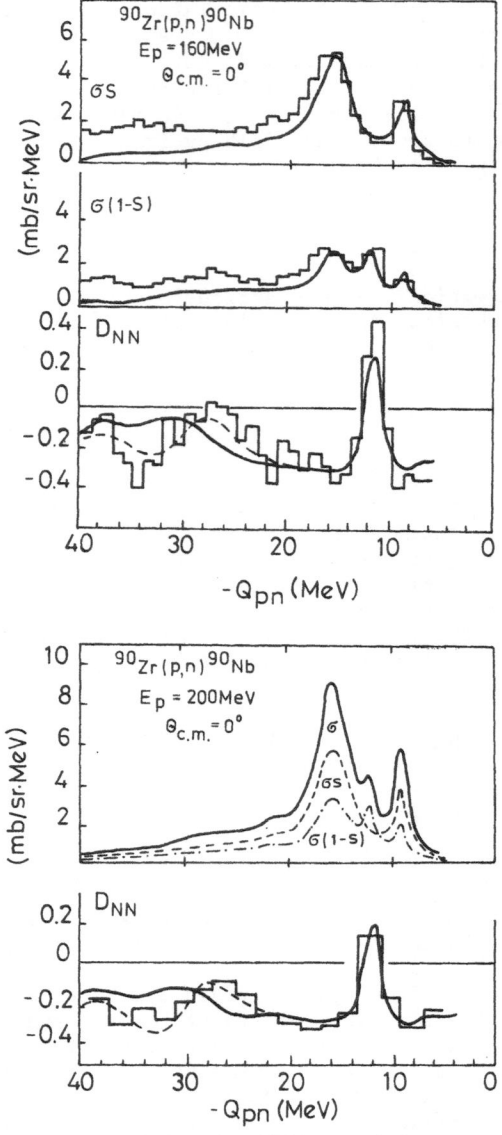

Fig.3. Calculated double differential cross sections and spectra of $D_{NN}(0^{\circ})$ for $^{90}Zr(p,n)$ at E_p = 160 MeV (top) and 200 MeV (bottom) are compared with experimental data of Refs. [15-17] . See text.

≃ -0.3. The same value is found for the whole GTR region ($13 \lesssim -Q_{pn} \lesssim 20$ MeV). In the vicinity of the isobaric analog state ($-Q_{pn} \approx 12$ MeV) the calculated value $D_{NN} \approx 0.3$ is less than unity predicted for the isolated state. This decrease is due to the influence of 1^+ states in the neighborhood of the IAS. In the region of $-Q_{pn} = 20$-30 MeV the distribution of D_{NN} is governed by the competition of contributions from strongly fragmented 2^- transitions, 0^- resonance in the vicinity of $-Q_{pn} = 27$ MeV ($D_{NN} \approx -1$), 3^+ resonance at $-Q_{pn} = 26$ MeV ($D_{NN} \approx -0.3$) and from the "tail" of GTR, and the contributions from the fragmented 1^- transitions for which $D_{NN} > 0$. Among the latter, however, the largest contribution to D_{NN} comes only from the spin-dipole 1^- resonance predicted in the vicinity of $-Q_{pn} \approx 28$ MeV ($D_{NN} \approx 0.3$). Due to the energy proximity of 0^-, 3^+ and 1^- resonances with the comparable cross sections, the resulting $D_{NN} < 0$ in this region. The measured distributions of D_{NN} in the region $-Q_{pn} > 22$ MeV are much better described by the model if one shifts the 0^- resonance at $-Q_{pn} = 32$ MeV (see dashed lines in fig.3). This indicates the high sensitivity of distributions of D_{NN} to the energy localization of different resonances for which the integral cross section is ≥ 1 mb/sr. Note that the above-mentioned 5 MeV shift of the 0^- resonance (it corresponds to the case when the one-pion exchange is neglected in calculations) does not change any noticeably the cross section spectra. The measured distributions of D_{NN} (0^0) do not show convincing signature for the spin-dipole 1^- resonance at $-Q_{pn} \approx 28$ MeV. According to our calculations $D_{NN} < 0$ in this region because of underlying GT strength.

In the region of $-Q_{pn} > 30$ MeV the predicted values of $D_{NN}(0^0)$ are small that is connected with the contributions from 2^+ and 3^- transitions. The spin-quadrupole resonance is predicted to appear in the vicinity of $-Q_{pn} \approx$ ≃ 36 MeV.

The energy distributions of $D_{NN}(\theta)$ are noticeably changing with θ (Fig.4). Note also specific changes with θ associated with the shift of the 0^- resonance (dashed lines in fig.4). All this emphasizes the importance of

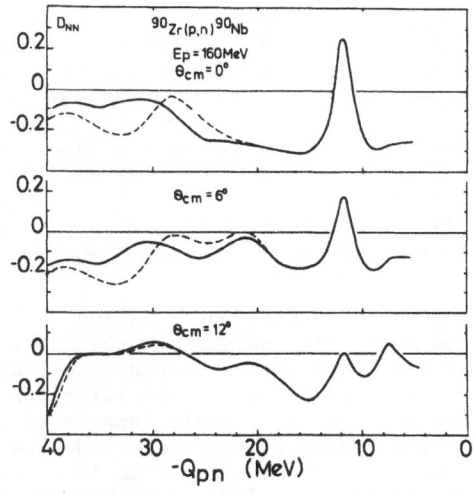

Fig.4. Distributions $D_{NN}(\theta)$ calculated for the reaction ^{90}Zr(p,n)^{90}Nb at E_p =160 MeV. The dashed lines show the distributions if the 0^- resonance is shifted towards $-Q_{pn}$=32 MeV.

measurements of D_{NN} for $\theta > 0^\circ$ and arouses hope to extract new information about the spin-multipole resonances in the continuum from such distributions. Analogous conclusions have been drawn [29] from the analysis of the reaction ^{48}Ca(p,n)^{48}Sc.

The calculated and experimental (taken from refs.[9,33]) cross sections for some incident energies and angles integrated over selected Q_{pn}-value intervals are given in table 1. It is seen that the predicted ratio of the energy integrated cross sections $\sigma(1-S)/\sigma S$ is in good agreement with experimental data at $\theta = 0^\circ$. Note that this ratio increases noticeably with θ. It would be interesting to verify this tendency experimentally.

Table 1. Energy-integrated cross sections for different intervals - Q_{pn} calculated with $e_q[\sigma r] = 0.8$. Partial contributions from the 1^+ (L = 0) transitions $\sigma(1^+)$, the total cross section σ and the background from the L > 0 transitions $\sigma(L > 0)$ are given. The ratio $\sigma(1-S)/\sigma S$ of the non-spin-flip cross sections to the spin-flip one are also given

$-\Delta Q$ (MeV)	$\sigma(1^+)$ (mb/sr)	$\sigma(L>0)$ (mb/sr)	σ (mb/sr) calc.	σ (mb/sr) exp.	$\sigma(1-S)/\sigma S$ calc.	$\sigma(1-S)/\sigma S$ exp.
\multicolumn{7}{c}{^{90}Zr(p,n)^{90}Nb, E_p = 160 MeV, $\theta = 0^\circ$}						
0-10	10.1	1.0	11.5	10.3	0.58	0.50
0-20	48.4	4.5	57.2	62	0.64	0.67
0.40	56.4	20.4	81.9	125	0.66	0.71
\multicolumn{7}{c}{$\theta = 6^\circ$}						
0-10	3.0	2.3	5.4		0.73	
0-20	14.7	16.7	32.8		0.79	
0-40	18.7	48.5	69.0		0.82	
\multicolumn{7}{c}{^{48}Ca(p,n)^{48}Sc, E = 135 MeV, $\theta = 0^\circ$}						
0-5	9.0	0.7	9.9	10	0.50	
0-15	50.1	3.2	59.0	72	0.63	
0-40	60.4	17.5	84.4	136	0.64	
\multicolumn{7}{c}{$\theta = 6^\circ$}						
0-5	4.3	2.0	6.4		0.58	
0-15	24.2	11.0	38.2		0.70	
0-40	30.7	38.2	72.4		0.72	

The measurements of the analyzing power A_y and polarization function P_y can be useful for the multipole analysis in the continuum. Our calculations show that for the GTR at forward angles both A_y and P_y are quite small compared with those for other resonances, especially for L = 1. It occurred that the contribution of the GTR is strongly suppressed in the energy distributions of $P_y - A_y$, as it is shown in fig.5. For the low-energy part of spectra $P_y \approx A_y$. Above the GTR the large maximum is associated with the predicted spin-dipole 1^-(S=1) resonance ($-Q_{pn} \approx 28$ MeV for ^{90}Zr and ≈ 23 MeV for ^{48}Ca) for which $A_y \ll P_y$. For the 0^- resonance $A_y(\theta) = -P_y(\theta)$, , so that it does not contribute to the $P_y + A_y$ distributions. The main strength of 2^- transitions is predicted to appear between the GTR and 1^-(S=1) resonances.

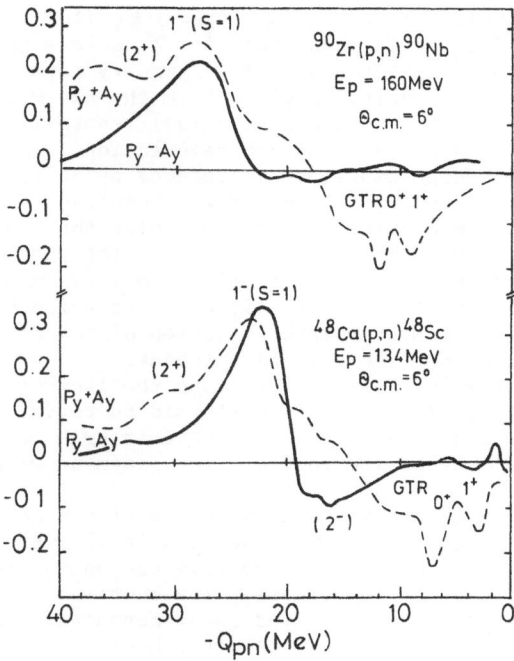

Fig.5.Calculated at $\theta_{c.m.} = 6°$ distributions $P_y \pm A_y$ for $^{90}Zr(p,n)$ and $^{48}Ca(p,n)$. The localization of different resonances giving the main contribution is indicated.

In the high energy tail the noticeable contribution comes from the spin-quadrupole 2^+ transitions.

At present this picture is considered as qualitative because we have neglected the spin-orbital force and tensor exchange in our calculations.

CONCLUSION

Our analysis shows the validity of the DWIA and microscopic TFFS structure calculations for the description of the (p,n) data at intermediate proton energies. Not only the angular distributions for bound states and inclusive neutron spectra at small angles up to excitation energies of $E_x \lesssim 20$ MeV but also more detailed information from polarization measurements are reasonably well reproduced in our model. The dominating contribution to the spectra at $\theta \approx 0°$ comes from the GT-transitions and it turns out necessary to reduce the total low-energy GT-strength by $\approx 1/3$ that corresponds to the TFFS quasiparticle local charge $e_q[\sigma\tau] \approx 0.8$. At $\theta = 0°$ the spin-flip transitions dominate in the whole region $0 \leq -Q_{pn} \leq 40$ MeV, the ratio of the non-spin-flip cross section to the spin-flip one being approximately 2/3, but this ratio is predicted to increase with θ.

According to the TFFS the quenching of the 1p-1h spin-isospin transitions should be universal in nature. We expect therefore that all the local irreducible $\sigma\tau$-vertices are renormalized in the same way. In particular, in nuclear β-decay the local charge $e_q[\sigma\tau]$ can be included in the effective axial-vector weak-coupling constant $G_A = e_q[\sigma\tau] \cdot g_A$ (for the free nucleon $g_A \approx 1.25$ in units of g_v). It means that for a nucleon quasiparticle in nuclear matter we "come back" to the SU(4)-symmetry ratio $G_A/G_V = 1.0$. This is in agreement with the results obtained from the analysis of the β-decay data[34]. We suppose that the same charge $e_q[\sigma\tau]$ renormalizes the πNN coupling constant in nuclei which was done in our calculations. As a result, in our model the one-pion exchange amplitude is reduced by 36%. Due to this reduction and having in mind that the Landau-Migdal constant $g' \approx 1.1$ is larger than it was estimated in earlier studies, we think that there is no chance to observe precritical phenomena associated with the pion condensation. From that point of view the negative results of $(\vec{p}, \vec{p}')$ experiments[35] concerning the collective enchancement in the longitudinal spin-isospin channel is not surprising. This means that the interpretation of the EMC effect in terms of excess pions[36] does not seems to be relevant.

The local charge $e_q[\sigma\tau]$ also enters into the isovector part of the M1 transition operator. The analysis performed in ref.[37] for the M1 resonances in ^{48}Ca, ^{90}Zr and ^{208}Pb has shown that at $e_q[\sigma\tau] = 0.8$ the calculated B(M1) values and (p,p') cross sections are in agreement with the known experimental data.

In the low-energy excitation region the local charge $e_q[\sigma\tau]$ is assumed to incorporate the effects from mixing between 1p-1h states and high-lying np-nh states as well as the contribution from the non-nucleonic degrees of freedom (mesons and baryon resonances). The coupling of 1p-1h excitations with multipair ones leads to the spreading of resonances and fragmentation of the transition strength. This has recently been confirmed by the calculations of the ^{90}Zr(p,n) spectra at $E_p = 200$ MeV taking account of the 2p-2h excitations[38]. The contribution of two-step processes to the reaction cross section was qualitatively estimated in ref.[39] and it turned out to be small in the region of $-Q_{pn} \leq 26$ MeV at $\theta = 0°$. This confirms the reliability of our estimate of $e_q[\sigma\tau] \approx 0.8$.

The contribution of the GT-transitions to the nonresonance part of the continuum above the GTR is still obscure. Here, the essential role can be played also by the ground state correlations not included in the usual RPA or TFFS calculations. It has been mentioned in ref.[40] that in nuclei with closed shells and N = Z these correlations can lead to the appearance of an additional GT-strength in the region of $E_x \approx \epsilon_F$. Perhaps, this problem as well as the spin-multipole decomposition of the continuum would be solved by the measurements of a complete set of spin observables at different angles.

In conclusion, the authors express their gratitude to Dr. J.Bang for the collaboration and to Drs. C.Gaarde and R.Madey for providing us with the experimental data.

REFERENCES

1. D.E. Bainum et al., Phys.Rev.Lett. 44: 1751 (1980).
2. C. D. Goodman et al., Phys.Rev.Lett. 44: 1755 (1980).
3. D. J. Horen et al., Phys.Lett. 95B: 27 (1980).
4. C.Gaarde et al., Nucl. Phys. A369: 258 (1981).
5. C. Gaarde, Nucl.Phys. A396: 127c (1983).
6. C. Gaarde, Physica Scripta V5: 55 (1983).
7. C. Gaarde et al., in: Spin Excitations in Nuclei, ed. F.Petrovich et al. (Plenum, N.Y., 1984), p.65.
8. C. D. Goodman, S. D. Bloom, in: Spin Excitations in Nuclei, ed. F. Petrovich et al., (Plenum, N.Y., 1984), p.143.
9. B. D. Anderson et al., Phys.Rev. C31: 1147, 1161 (1985).
10. G. F. Bertsch, I. Hamamoto, Phys.Rev. C26: 1323 (1982); K. Takayanagi et al., Nucl.Phys. A444: 436 (1985); K. Muto et al., Phys.Lett. 165B: 25 (1985); S. Drozdz et al., Nucl.Phys. A451: 11 (1986).

11. I. S. Towner, F. C. Khanna, Nucl.Phys. A399: 334 (1983).
12. A. Bohr, B. R. Mottelson, Phys.Lett. 100B: 10 (1981).
13. T. Izumoto, Nucl.Phys. A395: 189 (1983).
14. A. Arima et al.,Phys.Lett. 122B: 126 (1983).
15. T. N. Taddeucci et al., Phys.Rev. C33: 746 (1986).
16. T. N. Taddeucci, Suppl.J.Phys.Soc.Japan 55: 156 (1986).
17. T. N. Taddeucci, Can.J.Phys. 65: 557 (1987).
18. R. Madey et al., in: Weak and Electromagnetic Interactions in Nuclei,
    ed. H. V. Klapdor, (Springer, 1986), p.280; B. D. Anderson et al.,
    Phys.Rev. C34: 422 (1986).
19. F. Osterfeld, A. Schulte, Phys.Lett. 138B: 23 (1984); F. Osterfeld
    et al., Phys.Rev. C31: 372 (1985).
20. A. Klein et al., Phys.Rev. C31: 710 (1985).
21. F. A. Gareev et al., Yad.Fiz. 39: 1401 (1984); ibid 44: 1435 (1986);
    J. Bang et al., Nucl.Phys. 440A: 445 (1985); Phys.Scr. 34: 541 (1986).
22. A. K.Kerman et al., Ann. Phys. 8: 551 (1959); G. Bertsch, O. Scholten,
    Phys.Rev. C25: 804 (1982).
23. W. G. Love, M. A. Franey, Phys.Rev. C24: 1073 (1981); ibid C31: 488
    (1985).
24. A. Nadasen et al., Phys.Rev. C23: 1023 (1981).
25. C. M. Crawley et al., Phys.Rev. C26: 87 (1982).
    P. Schwandt et al., Phys.Rev. C26: 55 (1982).
26. N. I. Pyatov, S. A. Fayans, Sov.J.Part.Nucl. 14: 401 (1983).
27. A. B. Migdal, Theory of Finite Fermi-Systems and Properties of Atomic
    Nuclei (Interscience, N.Y., 1967).
28. W. G. Love, Nucl.Phys. A312: 160 (1978).
29. F. A. Gareev et al., JINR preprint P4-87-540, Dubna, 1987; Yad.Fiz.
    48 (1988).
30. S. N. Ershov, F. A. Gareev, N. I. Pyatov, S. A. Fayans, in: Weak and
    Electromagnetic Interactions in Nuclei, ed. H.V.Klapdor (Springer,
    1986), p.287.
31. G. G. Ohlsen, Rep.Prog.Phys. 35: 717 (1972).
32. J. M. Moss, Phys.Rev. C26: 727 (1982).
33. A. Klein, W. G. Love, Phys.Rev. C33: 1920 (1986).
34. D. H. Wilkinson, Phys.Rev. C7: 930 (1973);
    B. A. Brown, At Data Nucl. Data Table 33: 347 (1985);
    I. S. Towner, Nucl.Phys. 444A: 402 (1985).
    G. D. Alkhazov et al., Nucl.Phys. 438: 482 (1985).
35. L. B. Rees et al., Phys.Rev. C34: 627 (1986).
36. C. H. Llewelyn Smith, Phys.Lett. 128B: 107 (1983).
    M. Ericson, A. W. Thomas, Phys.Lett. 128B: 112 (1983);
    A. I. Titov, Yad.Fiz. 40: 76 (1984);
    E. E. Saperstein, M. J. Schmatikov, Pisma JETP 41: 44 (1985).
37. I. N. Borzov et al., Yad.Fiz. 40: 1151 (1984); Yad.Fiz. 42: 625 (1985).
38. J. Wambach et al. Preprint P/87/3/43, University of Illinois, 1987.
39. H. Esbensen, G. F. Bertsch, Phys.Rev. C32: 553 (1985).
    R. D. Smith, J.Wambach, Phys.Rev. C36: 2704 (1987).
40. S. Adachi et al., Nucl.Phys. A438: 1 (1985).
    B. Desplanques, S. Noguera, Phys.Lett. B173: 23 (1986).

ISOVECTOR SPIN OBSERVABLES

IN NUCLEAR CHARGE-EXCHANGE REACTIONS AT LAMPF

J. B. McClelland

Los Alamos National Laboratory
Los Alamos, NM 87545

INTRODUCTION

LAMPF has undertaken a major development program to upgrade facilities for nuclear charge-exchange studies at intermediate energies.[1] The major components of this upgrade are a medium-resolution spectrometer and neutron time-of-flight system for good resolution ($\Delta E < 1$ MeV) charge-exchange programs in (n,p) and (p,n) respectively. Major emphasis is placed on polarization phenomena using polarized beams and analyzing the polarization of the outgoing particle.

BRIEF PHYSICS OVERVIEW

It would seem that this meeting in Telluride is most appropriate to discuss the new charge-exchange facilities at LAMPF since much of the motivation and early direction came in part from a meeting here in 1982. It was at that conference "Spin Excitation in Nuclei," in which both new theoretical ideas and experimental techniques were brought to focus on the issues of the spin-isospin response of the nucleus. As an outgrowth of those discussions a series of experiments were planned at both LAMPF and IUCF aimed at the new physics. As results materialized and a better understanding of the problem developed, it became clear that new facilities for charge-exchange physics were required and that LAMPF was in an excellent position to make a significant contribution to this area of physics.

As other speakers at this conference will address many of the important physics issues in great detail, I will briefly touch on those topics in which we expect LAMPF will play a major role.

Nucleon charge-exchange reactions provide a sensitive tool to to study the isovector response of the nucleus, particularly with respect to spin degrees of freedom. In contrast to proton inelastic scattering, which is primarily sensitive to the isoscalar pieces of the interaction, charge-exchange reactions will sample only the isovector parts of the force, giving rise to isovector discrete transitions and giant resonances. In addition, this reaction can be used to study the isovector effective interaction in the nucleus. In some cases the effective amplitudes come directly from zero-degree cross-section measurements, while more detailed studies of their momentum dependence as well as form factors and convection currents will require polarization transfer measurements to be made. The upgraded facilities, together with a new high-intensity optically pumped polarized ion source, existing beamline polarimeters, and precession systems, all put LAMPF in a unique position to make these types of measurements. In addition, LAMPF spans an energy range (200-800 MeV) where energy- and density-dependent effects are expected to change dramatically. As such, one can "tune" the probe for sensitivity to different aspects (i.e. spin, medium modifications, absorbtion) of the general investigation of the nuclear response. This is a range that extends (p, n) measurements at the Indiana University Cyclotron Facility and both the (p, n) and (n, p) measurements at the new TRIUMF facility. At these higher energies the impulse approximation is expected to be valid even for reactions leading to large energy losses in the target, such as quasi-free nucleon ($\Delta E_x \sim 100$ MeV) scattering and delta ($\Delta E_x \sim 300$ MeV) production. It is also a range of energies where the effective interaction is changing dramatically, offering an additional variable for testing reaction mechanisms and investigating nuclear structure.

A direct connection can be made between polarization transfer (PT) observables and the amplitudes of the effective NN scattering matrix given by

$$M(q) = A + B\sigma_{1\hat{n}}\sigma_{2\hat{n}} + C(\sigma_{1\hat{n}} + \sigma_{2\hat{n}}) + E\sigma_{1\hat{q}}\sigma_{2\hat{q}} + F\sigma_{1\hat{p}}\sigma_{2\hat{p}} \ , \tag{1}$$

where 1(2) denotes the target (projectile) nucleon and the unit vectors $(\hat{n}, \hat{q}, \hat{p})$ are in the $\mathbf{K} \times \mathbf{K}'$, $\mathbf{K} - \mathbf{K}'$, and $q \times n$ directions respectively, with $\mathbf{K}(\mathbf{K}')$ the relative momentum in the NN system before (after) collision. For unnatural parity transitions, it has been shown[2,3] that in the static limit

$$I_0 = (C^2 + B^2 + F^2)X_T^2 + E^2X_L^2 \ , \tag{2.1}$$

$$I_0 D_{nn} = (C^2 + B^2 - F^2)X_T^2 - E^2X_L^2 \ , \tag{2.2}$$

$$I_0 D_{pp} = (C^2 - B^2 + B^2)X_T^2 - E^2X_L^2 \ , \tag{2.3}$$

$$I_0 D_{qq} = (C^2 - B^2 - F^2)X_T^2 + E^2X_L^2 \ , \tag{2.4}$$

$$I_0 D_{n0} = I_0 D_{0n} = 2X_T^2 Re(BC*) \ , \quad \text{and} \tag{2.5}$$

$$I_0 D_{qp} = -I_0 D_{pq} = 2X_T^2 Im(BC*) \ , \tag{2.6}$$

where $X_L^2(X_T^2)$ is the spin-longitudinal (transverse) form factor and the D_{ij}'s are the PT observables. Therefore a complete set of PT observables can be used to map the individual spin-dependent parts of the effective $N - N$ interaction for a state of known nuclear structure. Likewise, this procedure may be inverted to extract the spin-transverse and -longitudinal response functions for a known interaction. Both of these approaches have been successfully taken in inelastic proton scattering at LAMPF's High Resolution Spectrometer (HRS). [4,5] Quasi-free PT measurements provide general nuclear response functions and may be sensitive to the form of the $N - N$ interaction in the Dirac formulation.[6] The spin-longitudinal response function represents new nuclear-structure information not directly available from (e, e') or (π, π'). Observables such as $\sigma(P - A)$ and $\sigma(D_{ls} + D_{sl})$ have been shown to be more sensitive to convection (j) and composite ($j \times \sigma$) currents than unpolarized cross sections alone.[7]

The elastic "Q" data from HRS[8] was the driving force in the development of a relativistic scattering theory based on the Dirac equation. The success of relativistic theory in reproducing the data where nonrelativistic theory has failed is notable. Spin observables have presented the clearest test cases for these theories, but thus far have been restricted to the dominantly isoscalar channels in (p, p').[9] Here the biggest effect is a sensitive cancellation of large scalar and vector potentials predicted by the relativistic theory. In charge-exchange reactions these large potentials are absent, and the largest effect comes from the pion part of the interaction. Most observables are sensitive to how the pion is treated in these theories. Formal ambiguities exist in the construction of the relativistic scattering amplitude that will only be removed by spin-dependent measurements.

Previous speakers have pointed out the selectivity of spin-flip measurements ($S_{nn} = (1 - D_{nn})/2$) to $\Delta S = 1$ strength in medium-energy reactions. It has been shown[10] that a fairly model-independent form of D_{nn} is given by

$$D_{nn} \sim \frac{-1}{1 + (V_t/V_\ell)^2} \cdot \qquad (3)$$

The form of V_ℓ and V_t is determined predominantly by π- and ρ-exchange, producing a large value of the ratio at $q \sim 0.7$ fm^{-1}. This forces D_{nn} to zero, thus drastically reducing the sensitivity of D_{nn} to $\Delta S = 1$ strength. At 160 MeV this momentum transfer corresponds to $E_x \sim 50$ MeV. This is an important concern for experiments searching for missing Gamow-Teller (GT) strength in the continuum. The higher energies available at LAMPF allow one to recover this sensitivity to much higher excitation where much of the missing GT strength is hypothesized to be, either due to ground-state correlations or delta admixtures to the ground-state wave function.

THE CHARGE-EXCHANGE FACILITIES

Figure 1 shows the upgraded experimental areas at LAMPF. The MRS will reside in the rightmost indicated location in the (p, p') mode of operation. This will presumably be the development phase of the MRS when it will be optimized similar to the HRS. After this initial commissioning phase, the MRS will be moved to the adjacent area at the left where a neutron beam produced in the $0°$ ^{7}Li(p, n) reaction will provide the source for the (n, p) reaction analyzed by the MRS. The MRS is shielded from this primary beam by 6 feet of steel and 7 feet of concrete. This is the existing targeting area used by the $n - p$ experiments using a thick liquid-deuterium target. The lithium production target thickness is chosen such that the overall resolution of the system is better than 1 MeV. The (n, p) target thickness is similarly chosen and may be up to 3×30 cm^2 and multiply-targeted similar to the TRIUMF (n, p) system.[11]

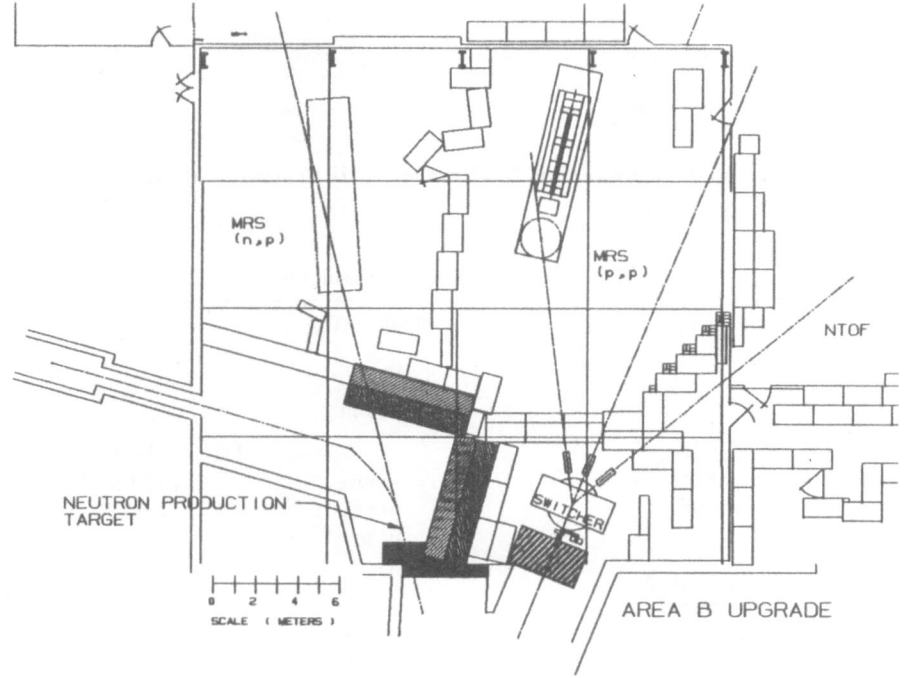

Fig. 1. Upgraded experimental areas at LAMPF.

Figure 2 shows the MRS QD(-D) design chosen in order to keep the net bend angle ~18° while still providing enough dispersion for an energy resolution of 0.5 MeV at 800-MeV incident energy over a restricted acceptance. Table 1 gives the design specification of the MRS. These were dictated by (n,p), NN, and proton-light nucleus requirements all of which the MRS is intended to serve. The small net bend angle is based on the requirement that all three outgoing polarizations at the target produce substantial transverse components at the focal-plane polarimeter for outgoing energies from 200 to 800 MeV. Detailed ion-optical calculations have been carried out on this system and a final design exists to meet all specifications. Construction of the MRS is in progress with an anticipated completion date of mid-1989. The first (n,p) measurements might start in late 1989.

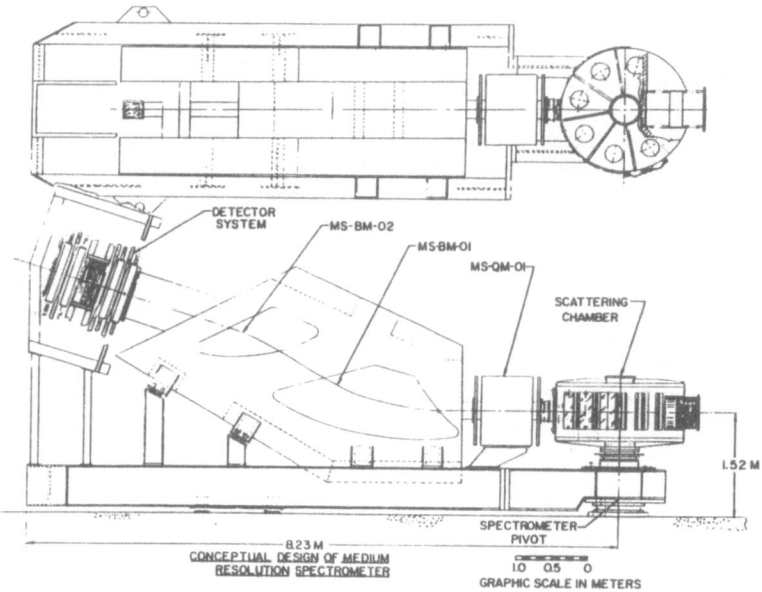

Fig. 2. Medium Resolution Spectrometer (MRS) schematic design.

Table 1. MRS Design Requirements.

Maximum Momentum:	2000 MeV/c	
Angular Range:	0-160°	
Overall Length:	8.5 m	
FULL ACCEPTANCE		For 800 MeV Protons
Momentum Acceptance:	+20%	+150 MeV
Momentum Resolution:	0.4%	5 MeV
Solid Angle:	10 msr	
REDUCED ACCEPTANCE		
Momentum Acceptance:	±1.5%	±20 MeV
Momentum Resolution:	0.04%	0.5 MeV
Solid Angle:	10 msr	
Angular Resolution:	±2 mrad	

Fig. 3 shows the neutron time-of-flight (NTOF) beam-swinger cave and magnet transport system. The beam swinger consists of five large magnets. Due to the rigidity of the high-energy proton beam, these magnets are on movable carriages to accommodate production angles from -5° to +52°. This provides maximum momentum transfers greater than 3 fm^{-1} at all incident energies. The cave is heavily shielded with steel and concrete so as to target up to $1\mu A$ of primary proton beam, consistent with the higher intensities expected from the new polarized ion source. A 4.5-tesla-meter superconducting solenoid resides just outside the main cave. Together with a horizontal-field dipole magnet, this system provides general neutron spin-precession capabilities. A high-energy spin precessor in the primary proton beamline provides three orthogonal proton-beam spin directions at the (p, n) target at all energies and swinger angles.

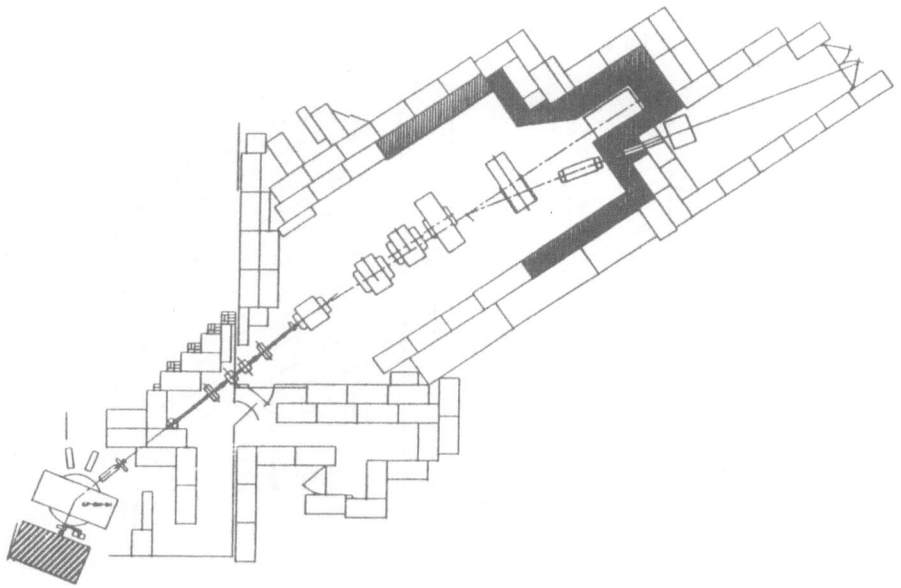

Fig. 3. Neutron Time-of-Flight (NTOF) beam-swinger cave.

A single 600-meter flight path extends from the beam swinger to the east. End stations providing power and data communications will be located at 170, 300, 420, and 550 meters. The detector trailer will move along this flight path to realize the required neutron energy resolution, communicating with the aquisition trailer over a computer network. Excavation is complete on the 600-meter neutron flight path. Power distribution and data communication systems are being installed. The complete flight path should be available by Spring of 1988 in time for start of LAMPF's production running.

The detectors are shown in Fig. 4. Three planes of neutron counters comprise the neutron polarimeter. Each plane is a stainless steel tank with optical isolators and is filled with BICRON BC-5175 mineral-oil-based liquid scintillator. The isolators effectively produce ten cells 10 × 10 × 105 cm^3 per plane viewed by phototubes on both sides. Time-difference information gives the position within a cell while the cell number provides the other coordinate. The mean time of the two ends provides neutron timing for time of flight. The polarimeter is based on the ^{1}H$(n, n)^1$H reaction in the liquid scintillator for the analyzer plane. The two rear catcher planes are used to reconstruct the polar and azimuthal scattering angles in the analyzer plane so as to deduce the two neutron transverse-polarization components. Charged-particle planes between the analyzer and catcher planes separate elastic from charge-exchange scattering in the analyzer. Both types of scattering are believed to be useful for polarimetry at high energy. The free hydrogen scattering is identified by pulse-height versus scattering-angle correlations and absolute velocity measurements between planes.

During July 1987 a development was undertaken to establish a technique to refocus the beam in time on the neutron detector. The time spreading of the LAMPF micropulse due to its energy spread limits the ultimate resolutions obtainable at NTOF. A solution to this problem was proposed to utilized unused RF acceleration cavities in the LINAC at energies less than 800 MeV as time-focusing elements. High-resolution beam-timing monitors and neutron time-of-flight spectra were used to tune a single cavity to focus a 500-MeV beam.

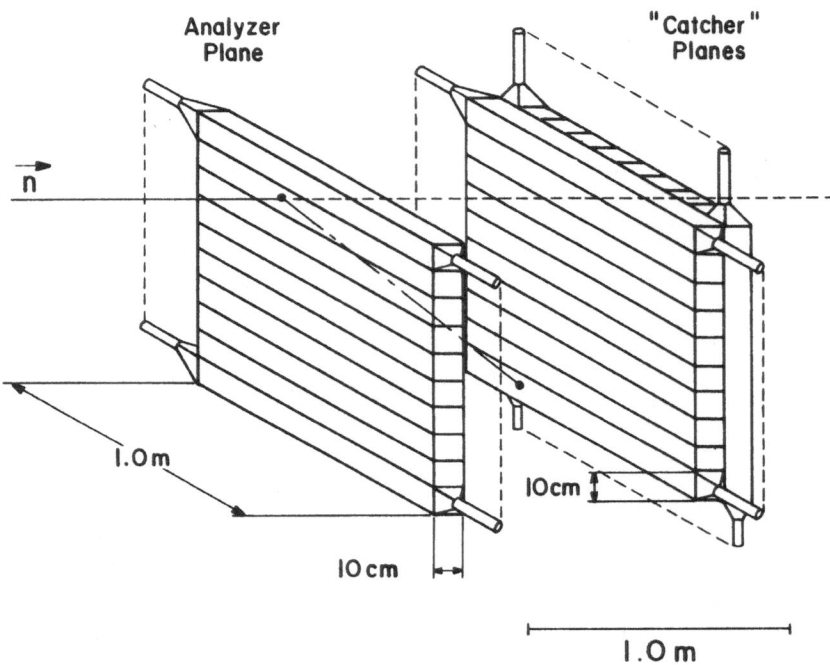

Fig. 4. Schematic of NTOF detector array.

This rebunching technique compressed the time structure of the beam from a nominal 1.5 ns to 300 ps, consistent with calculated values. The effect of the rebuncher on a $^{13}C(p,n)^{13}N$ time-of-flight spectrum is shown in Fig. 5. A special control system was needed to use these cavities as rebunchers as well as still serving as acceleration cavities for the full-energy H^+ beam. The prototype module is planned to be improved and extended to other cavities for other beam energies. This technique has also been applied to 113- and 600-MeV beams for (p,n) measurements.

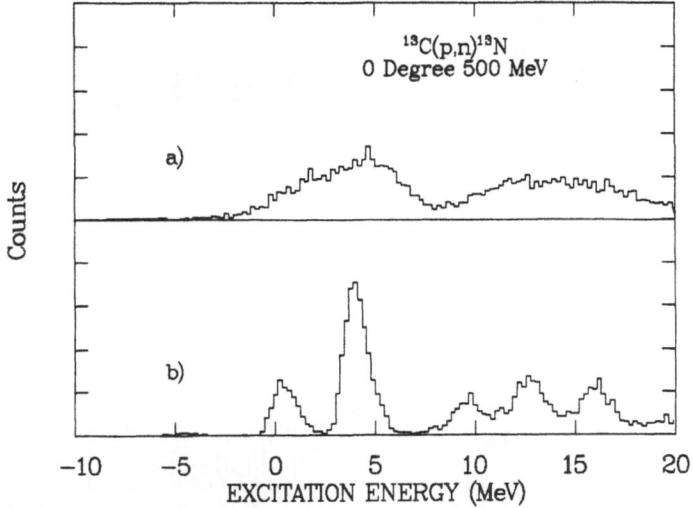

Fig. 5. $^{13}C(p,n)^{13}N$ time-of-flight spectrum at 500 MeV a) without LINAC rebunching and b) with rebunching.

Commissioning of the NTOF beamline, detector, and acquisition systems was started in November 1987. A temporary 85-meter flight path was constructed with a neutron beam stop that also housed the neutron detection system. A 500-MeV, 100-ns pulse-selected polarized beam was used, employing the newly developed high-energy rebunching system to focus the micropulse timing at the detectors. This allowed for better than 600-ps overall timing resolution, corresponding to 3-MeV energy resolution at this short flight path. This is consistent with the intrinsic detector timing of 280 ps, front-to-back flight time in the detectors of 440 ps, a beam-timing contribution of 300 ps, and target energy loss contributions. Cross-section and analyzing power data were taken on CD_2, $^{6,7}Li$, $^{12,13}C$, and $^{14,15}N$ at zero degrees. Angular distributions of $^7Li(p,n)$ and $^{15}N(p,n)$ cross sections and analyzing powers were taken to establish detector efficiencies and perform the first physics measurements with the system. Further development is needed to implement the polarimeter software and analysis structure along the lines of the Focal Plane Polarimeter at the HRS[12] in anticipation of polarization-transfer measurements in 1989 using the new polarized ion source.

Figure 6 is a zero-degree spectrum of $^{15}N(p,n)^{15}O$ taken during this development. The ground state ($\frac{1}{2}^-$) and first excited state ($\frac{3}{2}^-$) seen in the spectrum are separated by 6.2 MeV and are well resolved. This spectrum has had software cuts applied so as to reject a large peak at 366 nsec corresponding to gamma rays from the decay of neutral pions produced in the target from the following micropulse and for events at energies apparently greater than the ground state corresponding to low-energy neutrons produced in preceeding pulses; so called "wrap around." At this energy, flight path, and pulse separation the first frame of wrap around occurs at 250 MeV. Both of these background event types are easily eliminated when one runs the neutron detector in a coincident mode. As such, an absolute time-of-flight measurement is made between planes, and particles with higher or lower velocity than the neutrons of interest may be rejected.

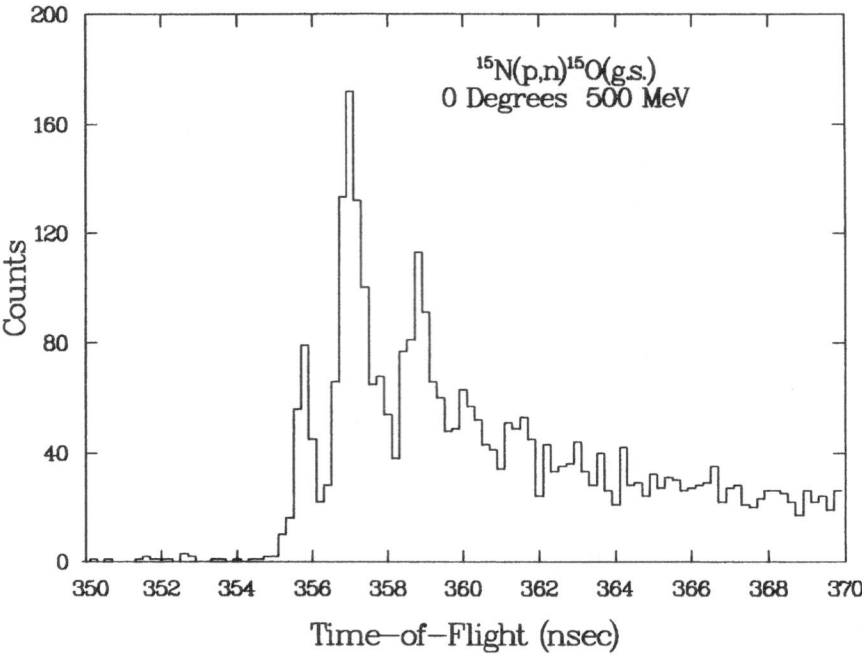

Fig. 6.   Neutron time-of-flight spectrum for the $^{15}N(p,n)^{15}O(g.s.)$ reaction at 500 MeV taken during commissioning phase of NTOF.

Figure 7 is the $^{7}Li(p,n)^{7}Be(g.s. + 0.43$ MeV) differential cross section as measured in this development and at other energies. This reaction can be used to calculate the detector efficiency by comparison to an activation technique.[13] It is notable that the cross section for this reaction is almost independent of incident energy from 80 to 500 MeV as a function of momentum transfer.

Figure 8a and 8b are the preliminary results for cross section and analyzing power in the $^{15}N(p,n)^{15}O(g.s.)$ reaction measured during this commissioning period. Final statistical accuracy per point should be 3-4 percent. This reaction was being considered as a possible calibration reaction for the neutron polarimeter using the symmetry that the induced polarization is equal to the analyzing power for the ground-state mirror transition to a high degree of accuracy. The observed small figure of merit, however, precludes this possibility. It will probably be necessary to use the $^{14}C(p,n)^{14}N$ IAS transition which guarantees $D_{nn} = 1$ to calibrate the neutron polarimeter as has been done at lower energies. These calibrations are very time consuming, and alternate methods need to be developed.

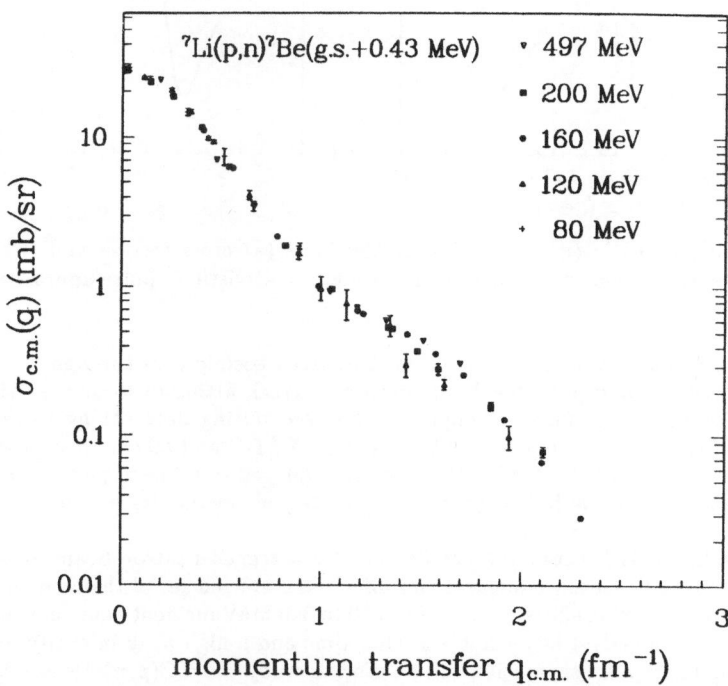

Fig. 7.   The $^7\text{Li}(p,n)^7\text{Be}(g.s. + 0.43\ \text{MeV})$ reaction differential cross section from 80 to 500 MeV.

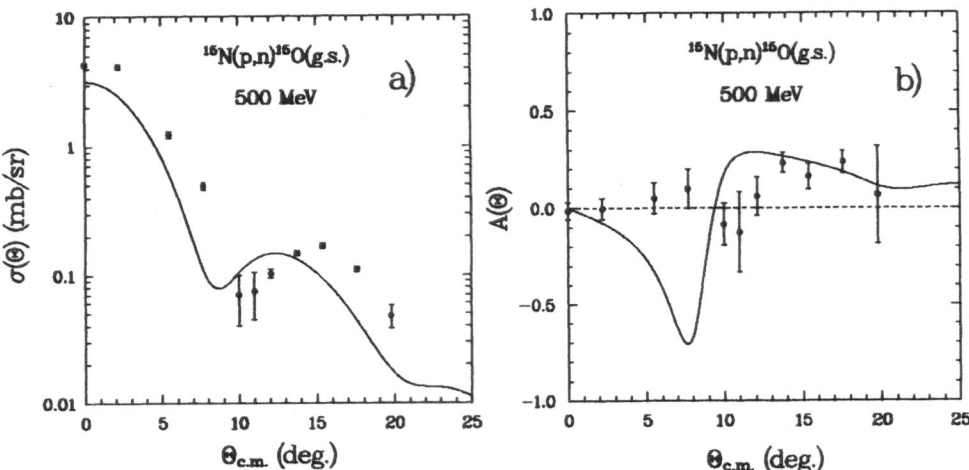

Fig. 8.   The ^{15}N$(p,n)^{15}$O(g.s.) reaction at 500 MeV (a) cross section and (b) analyzing power (Preliminary data). Calculations are relativistic impulse approximation.

Calibration routines have been developed to track cosmic rays through the system to both establish timing and pulse-height calibrations. Track fitting of these rays yields a position resolution of 4 cm. This technique can be used during data taking to monitor the performance of the system comprised of 80 channels of TDC and 80 channels of ADC information. A new LeCroy FERA CAMAC system is employed to perform pedestal subtraction and sparse data scans, while buffering multiple events per macropulse into its memory.

This year's LAMPF running schedule calls for a tagged-neutron-beam experiment to be run in the spring to measure integral and differential efficiencies of the neutron detection array.[8] This should provide efficiency data from 50 to 800 MeV incident neutron energies. The 600 meter flight path should be available at that time and a high-peak-intensity unpolarized beam with long micropulse separation will be used to study the ^{14}C$(p,n)^{14}$N at 500, 650 and 800 MeV. This will serve to measure ratios of interaction strengths for the isovector spin-dependent and spin-independent parts of the effective interaction at these energies. Other speakers at this conference will adress the impact of these measurements and the apparent discrepancy of the existing data with T- and G-matrix calculations. These experiments require high resolution to resolve the two states of interest in this transition separated by 1.6 MeV, as well as background-free spectra due to the small cross sections involved. They should provide a stringent test of the capabilities of the new facility. Seven experiments to use NTOF have already been approved by the Program Advisory Committee. The physics goals of these experiments range from calibrations of the system to effective interaction studies to reaction theory and nuclear response. Those requiring high-intensity polarized beams must wait until the new polarized ion source becomes available in 1989.

ACKNOWLEDGMENTS

This project represents the hard work of many individuals both within the Los Alamos National Laboratory and outside experimental teams. In particular, I would like to acknowledge the contributions of J. Amann, R. Byrd, T. Carey, D. Clark, J. Davis, R. Haight, J. Hurd, R. Johnson, N. King, P. Lisowski, G. Morgan, M. Murray, L. Rybarcyk, T. Taddeucci, N. Tanaka, and J. Ullman (Los Alamos); E. Sugarbaker and D. Marchlenski (OSU); C. Goodman and W. Huang (IUCF); D. Lind, D. Prout, W. Symthe, and C. Zafiratos (U. Col); S. Baker (EG&G); P. Alford (UWO); M. Barlett and D. Ciscowski (UT); and E. Gulmez (UCLA), who participated in the development and commissioning phases of the NTOF project.

REFERENCES

1. J. B. McClelland, et al., "Development Plan for the Nucleon Physics Laboratory Facility at LAMPF," Los Alamos National Laboratory report LA-10278-MS (February 1986).
2. J. M. Moss, *Phys. Rev. C* **26**, 727 (1982).
3. E. Bleszynski, M. Bleszynski, and C. A. Whitten, Jr., *Phys. Rev. C* **26**, 2063 (1982).
4. J. B. McClelland, et al., *Phys. Rev. Lett.* **52**, 98 (1984).
5. T. A. Carey, et al., *Phys. Rev. Lett.* **53**, 144 (1984); L. B. Rees, et al., *Phys. Rev. C* **34**, 627 (1986).
6. C. Horowitz and M. Iqabal, "Relativistic Effects on Spin-Observables in Quasielastic Proton Scattering," *Phys. Rev. C* **33**, 2059 (1986).
7. W. G. Love and Amir Klein, "Nuclear Currents in Inelastic Scattering: Relativistic and Nonrelativistic Approaches," Proc. of the LAMPF Workshop on Dirac Approaches to Nuclear Physics, Los Alamos National Laboratory document LA-10438-C (May 1985); D. A. Sparrow, et al., *Phys. Rev. Lett.* **54**, 2207 (1985).
8. A. Rahbar, et al., *Phys. Rev. Lett.* **47**, 1811 (1981).
9. B. C. Clark, et al., *Phys. Rev. C* **30**, 314 (1984).
10. W. G. Love and Amir Klein, "Non-Relativistic Effective Interactions and Polarization Phenomena," Proc. 6th Int. Symp. Polar. Phenon. in Nucl. Phys., Osaka, 1985, *J. Phys. Soc. Jpn.* **55** (1986); W. G. Love et. al., *Can. J. Phys.* **65**, 536 (1987).
11. R. Helmer, *Can. J. Phys.* **65**, 588 (1987).
12. J. B. McClelland, et al., "A Polarimeter for Analyzing Nuclear States in Proton-Nucleus Reactions Between 200 and 800 MeV," LA-UR-84-1671.
13. J. D'auria, et al., "Activation Measurements of the $^7Li(p,n)^7Be$ Reaction from 60 to 480 MeV," Phys. Rev. **30**, 1999(1984).
14. T. N. Taddeucci, et al., "A Neutron Polarimeter for (p,n) Measurements at Intermediate Energies," Nucl. Inst. and Meth. **A241**, 448 (1985).

NEW NEUTRON TIME-OF-FLIGHT (NTOF)*

FACILITIES AT THE BROOKHAVEN 200-MEV LINAC

T.E. Ward, J. Alessi, J. Brennan, P. Grand,
R. Lankshear, C.L. Snead, N. Tsoupas, and M. Zucker

Neutral Particle Beam Division
Department of Nuclear Energy
Brookhaven National Laboratory
Upton, New York 11973

ABSTRACT

The installation of a new beam chopper and radio-frequency quadrupole (RFQ) preinjector (750 keV) at the Brookhaven National Laboratory (BNL) 200-MeV Linac will enable single micropulse selection (pulse width <1 ns) with periods ranging from 400 ns to 10 μs. The standard micropulse intensity is 1.2×10^9 p/μ pulse with dc-average beam currents of 50 nA-1 μA routinely available. The NTOF facilities consists of 30-100 meter flight paths at angles of 0, 12, 30, 45, 90, and 135°. Lower energies of 93, 117, 139, 161, and 181 MeV are also available as well as polarized beams at much reduced intensities. The present paper describes the new facilities, and the capabilities of future improvements and upgrades, for use in the BNL intermediate energy (p,n) experimental program.

INTRODUCTION

A new high current 750 keV RFQ is being installed at the BNL 200-MeV Linac as a replacement for one of the Cockcroft-Walton preinjectors.[1-3] A double chopper capable of single micropulse selection at frequencies of 1 Hz up to 2.5 MHz has also been constructed and sucessfully tested with the RFQ. The new double-chopper RFQ system is detailed in the present paper as well as the NTOF capabilities that exist at the 200-MeV Linac complex. Future improvements and upgrades that will be described include a new high intensity polarized ion source, a 400 meter zero degree beam line, a debuncher for reducing the beam spread, and a 0-30° beam swinger on the zero degree line. Experimentally, the facilities will be capable of examining the E_n = 1-200 MeV spectral range without pulsed beam wrap-around and with dc-averaged currents of 50 nA or more.

The Double Chopper RFQ System

The RFQ Linac input is from a 35 keV H⁻ magnetron ion source modified to produce an axially-symetric beam matched to the RFQ. The RFQ parameters are summarized in Table 1. Operational tests have routinely produced

---

*Research carried out under the auspices of the U.S. Department of Energy under Contract No. DE-AC02-76CH00016.

TABLE 1

RFQ Parameters:

Ion	H^+H^-
Input Energy	35 keV
Output Energy	753 keV
Current Limit	$\geq$ 100 mA
Operating Frequency	201.25 MHz
Peak Cavity Power	100 kW
Stored Energy	0.5 Joules
Duty Factor	0.007
Structure	4-vane, ringed
Vane Length	1.62 m

outputs of 50 mA from the RFQ, whereas the current limit is rated at $\geq$100 mA. The RFQ emittance at 50 mA was optimized for transport to the 200-MeV Linac.

Figure 1 shows an illustration of the double chopper RFQ system. A fast beam chopper (Chop I) located between the ion source and the RFQ can variably bunch-structure the beam with frequencies of 2.5 MHz or less. The first chopper is a slow wave electrostatic deflection device that rejects the beam at a small aperture located before the RFQ. The second chopper (Chop II) is located after the RFQ and is phase locked to the RFQ. Chop II is a fixed frequency sine wave chopper that selects single microbunches (440 ps width) of the 200 MHz RFQ Linac. The duty factor of the double chopper is adjustable from one bunch every 400 ns to one bunch per 450 μs macropulse. The macropulse frequency is 5 Hz. The standard 200 MHz micropulse intensity with 50 mA averaged current is 1.2×10^9 protons/ μ-pulse. The dc averaged beam current with 10 μs repetition rate and 5 Hz macrostructure is 50 nA, a value comparable with a 10 μsec 200 MeV pulsed beam from a cyclotron such as the Indiana University Cyclotron.

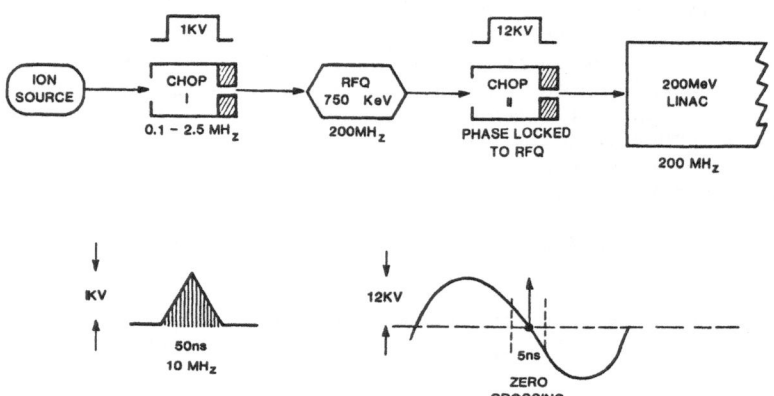

Figure 1.   Double Chopper RFQ Showing Single Micropulse Selection

NTOF Facilities and Facilities Performance

The 200-MeV Linac complex consists of NTOF facilities at the Radiation Effects Facility (REF) and Neutral Beam Test Facility (NBTF). The REF has 30-100 m flight paths at 12, 30, 45, 90, and 135° with approximately 30 m of earth shielding (12" tubes through shielding). The zero degree line located at the NBTF consists of a 10 ft diameter underground tunnel with a present length of 100 meters, a beam dump sweep magnet and a collimation wall (10 ft thick).

The Linac can accelerate proton beams of 92.6, 116.5, 139.0, 160.5, 181.0, and 200.3 MeV with an energy spread of about 140 keV at 200 MeV.[3] The momentum spread in the beam ($\Delta p/p = 7 \times 10^{-4}$), coupled with the 125 m beam transport into the REF and NBTF, widens the width of the micropulse to about 1 ns from the intrinsic width of 440 ps which results from the Linac acceleration. The overall NTOF energy resolution (ΔE_n) that results from a combination of the micropulse width ($\approx$1 ns), detector timing resolution (0.8 ns), and the detector width (see Ref. 4 for details) is shown in Figure 2 for various flight-paths and neutron energies. The detector width contributes 58-82% to the uncertainty in the 200-10 MeV neutron energy range respectively.[4-6] The overall resolution is somewhat less than that of other intermediate energy (p,n) facilities but uniquely provides high dc averaged beam currents with repetition rates of 400 ns-10 μs or greater. These low frequency micropulse modes, with 100 kHz and 100 m path lengths, allow the NTOF spectral range of E_n = 1-200 MeV to be acquired without troublesome wrap-around backgrounds. This capability would allow experimenters to look high into the continuum where missing Gamow-Teller (GT) strength may be found.

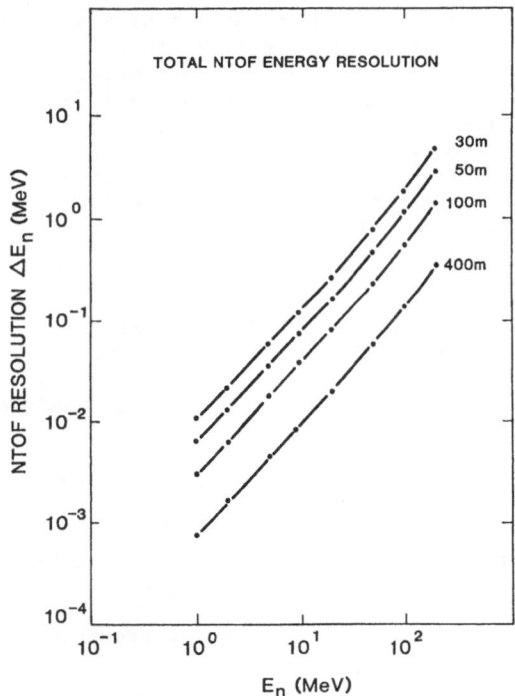

Figure 2. NTOF Energy Resolution as a Function of Neutron Energy and Flight Path

Future Improvements and Upgrades

A new polarized H⁻ source is planned to be installed with the double chopper RFQ system after the commissioning and first-year operation of the NTOF facilities. This source will operate at much reduced intensities compared with unpolarized beams but otherwise comparable in intensity to other NTOF facilities at IUCF or LAMPF. Both polarized and unpolarized dc averaged beam intensities can effectively be doubled by increasing the frequency of the macrostructure from the present 5 Hz to the Linac design limit of 10 Hz.

Two improvements to the beam resolution will effectively reduce the 1.4 MeV resolution at 200 MeV (10 µs rep-rate and 100 m path) to less than 350 keV as illustrated in Figure 2. The first is the installation of a debuncher in the transport beam line that will reduce the momentum spread without loss of intensity from the present 950 ns to 700 ns resulting in a increased resolution of about 90 keV. The second improvement is the planned extension of the 100 m zero degree line in the NBTF to 400 m, which results in an overall reduction of 960 keV to the energy uncertainty.

A preliminary design study for a zero degree beam swinger to be installed at the NBTF is currently underway.[7] The beam swinger design calls for a dynamic range of 0-30°, an angular region of much interest in the investigation of GT (p,n) reactions. A one year construction schedule would preclude the use of such a device in the first or second year. The design of a swinger facility is currently being investigated.

Summary

The installation of a new double-chopper RFQ preinjector at the BNL 200-MeV Linac complex will allow high current NTOF studies at intermediate energies. The NTOF facilities consist of 30-100 meter flight paths at angles of 0, 12, 30, 45, 90, and 135°. Lower energies of 93, 117, 139, 161, and 181 MeV are also available. Future plans call for a 400 m flight path at the zero degree line, a polarized H⁻ source, and a 0-30° beam swinger to be installed on the zero degree line. This facility uniquely provides single micropulse selection (width <1 ns) with periods of 400 ns to 10 µs or more thereby effectively ensuring no wrap-around background in the E_n = 1-200 MeV spectral range (10 µs period, 100 m flight path).

References

1. R. Gough, et. al., "Design of an RFQ for BNL/FNAL," 1986 Linac Conference, Abstract #WE 3-8, to be published.

2. J.G. Alessi, et. al., Proc. IEEE Part. Accel. Conference 1, 276 (1987), and 1, 304 (1987).

3. G.W. Wheeler, K. Batchelor, R. Chasman, P. Grand, and J. Sheehan, PLACBD 9, 1 (1979).

4. T.N. Taddeucci, Proc. AIP Conf. #124, pg. 394 (1985), J. Rapaport, R.W. Finlay, S. Grimes, and F. Deitrich, editors (AIP, New York 1985).

5. C.D. Goodman, J. Rapaport, D. Bainum, M. Greenfield, and C. Goulding, IEEE NS-25, 577 (1978).

6. C.D. Goodman, J. Rapaport, D. Bainum, and C. Brient, NIM 151, 125 (1978).

7. N. Tsoupas (private communication, Feb. 1988).

THE ^{208}Pb$(\vec{p},\vec{n})^{208}$Bi REACTION AT 135 MeV

J. W. Watson,[a] Marco R. Plumley,[a] P. J. Pella,[b] B. D. Anderson,[a] A. R. Baldwin,[a] R. Madey,[a] and C. C. Foster[c]

a) Kent State University, Kent, OH, U.S.A. 44242
b) Gettysburg College, Gettysburg, PA, U.S.A. 17325
c) Indiana Univ. Cyclotron Facility
   Bloomington, IN, U.S.A. 47405

INTRODUCTION

For nucleon inelastic scattering, i.e. (p,p′), (p,n), (n,p) or (n,n′) reactions, spin modes of nuclear excitation play an important role at medium-energies. Because of the special character of purely isovector transitions, we are interested in measuring spin-observables for (p,n) reactions at medium energies. These measurements include analyzing powers, $A_y(\theta)$, reaction polarizations, $P(\theta)$, and transverse polarization-transfer coefficients, $D_{nn}(\theta)$, as well as differential cross sections, $\sigma(\theta)$. These four observables are sensitive to different aspects of the reaction: $\sigma(\theta)$ is sensitive to ΔL; $A_y(\theta)$ is sensitive to whether the transition involves $\not{j} = \ell \pm \frac{1}{2} \to \not{j} = \ell \pm \frac{1}{2}$ or $\not{j} = \ell \pm \frac{1}{2} \to \not{j} = \ell \mp \frac{1}{2}$; $D_{nn}(\theta)$ is sensitive to ΔJ and $\Delta \pi$; $P - A_y$ is sensitive to non-localities. Previously, we reported measurements of $A_y(\theta)$[1,2] and $D_{nn}(\theta)$[3,4] for the ^{40,48}Ca(p,n)^{40,48}Sc reactions at 135 MeV. In this paper, we report on measurements of $D_{nn}(\theta)$ for the ^{208}Pb$(\vec{p},\vec{n})^{208}$Bi reaction at 135 MeV.

THE NEUTRON POLARIMETER

We developed a high-efficiency neutron polarimeter that utilizes the analyzing power of n-p scattering from hydrogen nuclei in BC-517L mineral-oil scintillator; BC-517L has an H:C ratio of 2:1. The physical layout of this polarimeter is shown in Fig. 1; the operation and performance are described in Ref. 5. We obtained calibration data from the ^{12,14}C$(\vec{p},\vec{n})^{12,14}$N reactions at 65, 100, and 135 MeV. In Figs. 2 and 3, we

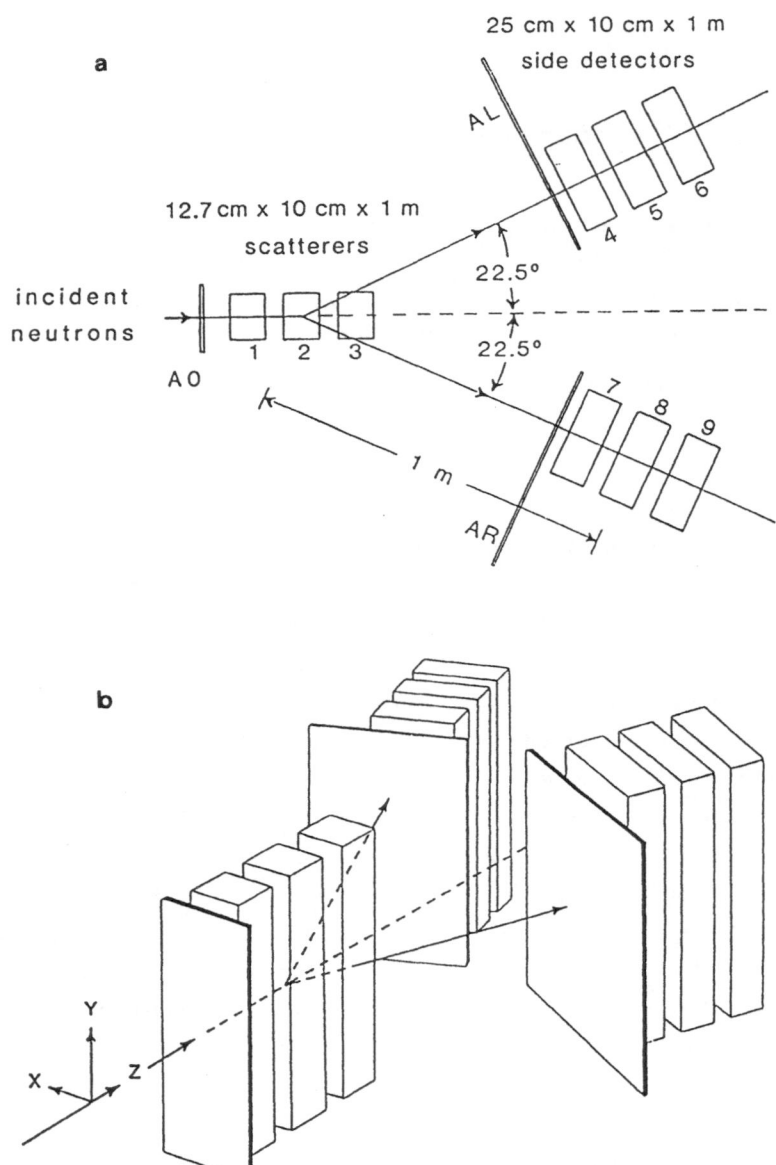

25 cm x 10 cm x 1 m
side detectors

a

AL

22.5°

12.7 cm x 10 cm x 1 m
scatterers

incident
neutrons

22.5°

A0

1 m

AR

4 5 6

7 8 9

1 2 3

b

Y

X Z

Figure 1. The configuration of the neutron polarimeter (a) as seen from above and (b) in perspective. AO, AL and AR are anti-coincidence detectors.

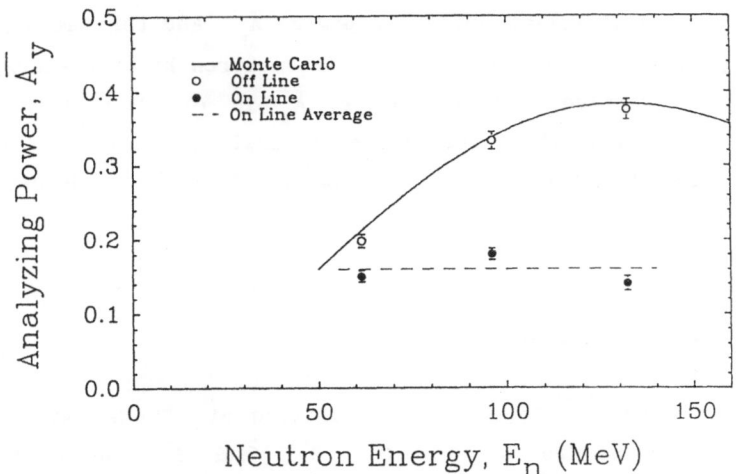

Figure 2. The analyzing power of $\overline{A}_y$ of the neutron polarimeter with mineral-oil scatterers as a function of the incident neutron kinetic energy E_n. Shown are on-line results, off-line results for the final software cuts, and results from a Monte-Carlo simulation assuming n-p scattering alone.

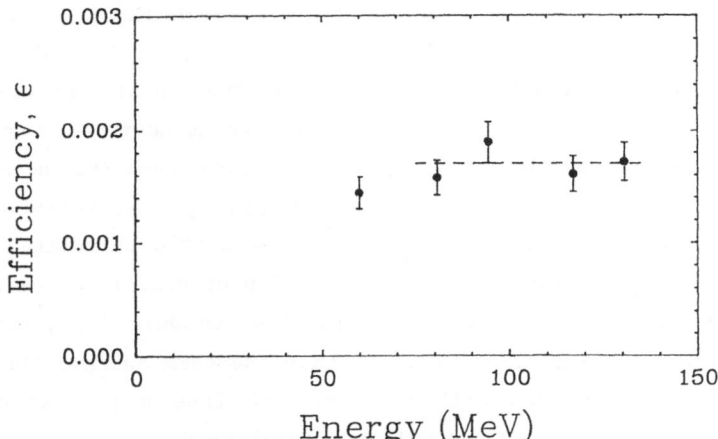

Figure 3. The efficiency of the polarimeter with (BC-517L) mineral-oil scatterers as a function of the incident neutron kinetic energy E_n. These efficiencies were determined from the $^{12}C(p,n)^{12}N$ (g.s.) and $^{14}C(p,n)^{14}N$(3.95 MeV) reactions with the final choice of software cuts. The weighted average of the four highest energy points = 0.17%.

show data for the effective analyzing power, $\overline{A_y}$, and for the efficiency, ε, of the polarimeter as a function of the neutron kinetic energy E_n. At 130 MeV, the effective analyzing power, $\overline{A_y}$ is +0.375 ± 0.014 and the efficiency ε is 0.17%; thus, the figure-of-merit, $\eta = (\overline{A_y})^2 \cdot \varepsilon$, is nearly 50% higher than for our earlier polarimeter design[6] which used NE-102 plastic scintillators.

THE $^{208}Pb(\vec{p},\vec{n})^{208}Bi$ REACTION AT 135 MeV

We studied the $^{208}Pb(\vec{p},\vec{n})^{208}Bi$ reaction at 135 MeV with this new polarimeter. For angles of 0°, 3°, 6°, and 9°, we measured the polarization-transfer coefficient $D_{nn}(\theta)$; the spin-flip probability $S = (1 - D_{nn}(\theta))/2$. Figure 4 shows S spectra for the $^{208}Pb(\vec{p},\vec{n})^{208}Bi$ reaction at 0° along with data from the $^{48}Ca(\vec{p},\vec{n})^{48}Ca$ and $^{40}Ca(\vec{p},\vec{n})^{40}Ca$ reactions.[3] The Gamow-Teller Giant Resonances (GTGR), 0^+ isobraric-analog states, 1^- (non-spin-flip) components of the dipole resonance, and other known 1^+, 2^+, and 2^- states are indicated. For Q-values larger than 20 MeV (E_n < 115 MeV), the isovector spin-response is essentially identical for all three nuclei, suggesting a common mechanism for production of the continuum at large Q-values. The solid and dashed lines in Fig. 4 are spin-flip probabilities for free p-n scattering calculated from phase shifts and from the Bonn potential, respectively, with the computer code SAID.[8] These free p-n spin-flip probabilities were obtained from the observable D_t as $S = (1 - D_t/2)$. D_t, the polarization transfer from the projectile to the target, is the free p-n observable corresponding to $D_{nn}(\theta)$ for a (p,n) reaction. These calculations assume that the (p,n) reaction can be described as p-n scattering from stationary, bound neutrons. For a series of binding energies, two-body kinematics were used to determine the final-state laboratory neutron kinetic energy, E_n, the momentum transfer q, and the final state p-n c.m. energy; D_t was then calculated for these latter two kinematic variables. The spin-flip probabilities calculated in this simple model agree remarkably well with the data for Q-values < -30 MeV; for neutron energies as small as half the beam energy, the continuum clearly carries the spin-flip signature of free p-n scattering. This simple model is equally successful in describing data for Q < -30 MeV for the $^{208}Pb(\vec{p},\vec{n})^{208}Bi$ reaction at 3°, 6°, and 9°, and $(\vec{p},\vec{n})$ data for other targets at other energies reported by other groups.[8] Similar calculations do not seem to do a good as job of fitting medium-energy $(\vec{p},\vec{p}')$ data,[9] apparently[10] due to the presence of strong, collective, isoscaler, $\Delta S = 0$

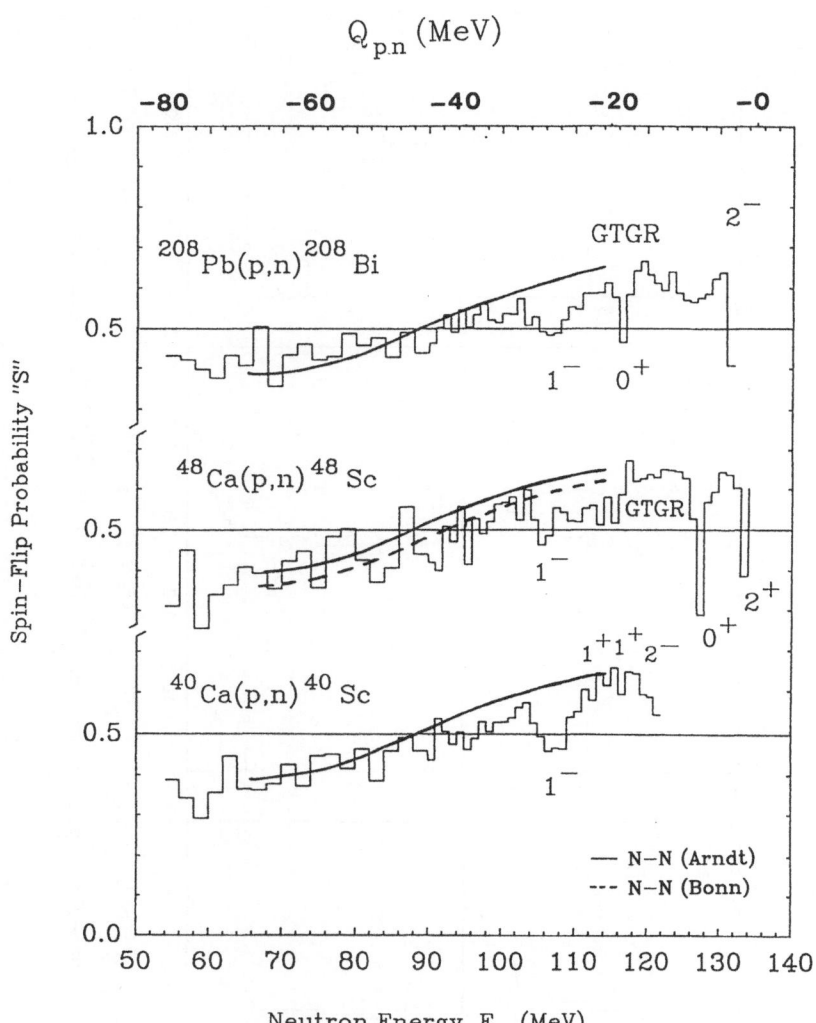

$Q_{p.n}$ (MeV)

Figure 4. Spin-Flip probabilities for the $^{208}Pb(\vec{p},\vec{n})^{208}Bi$, $^{48}Ca(\vec{p},\vec{n})^{48}Ca$ and $^{40}Ca(\vec{p},\vec{n})^{40}Ca$ reactions at 135 MeV and 0°.The solid and dashed lines are for free p-n scattering calculated from phase shifts, and from the Bonn potential, respectively.

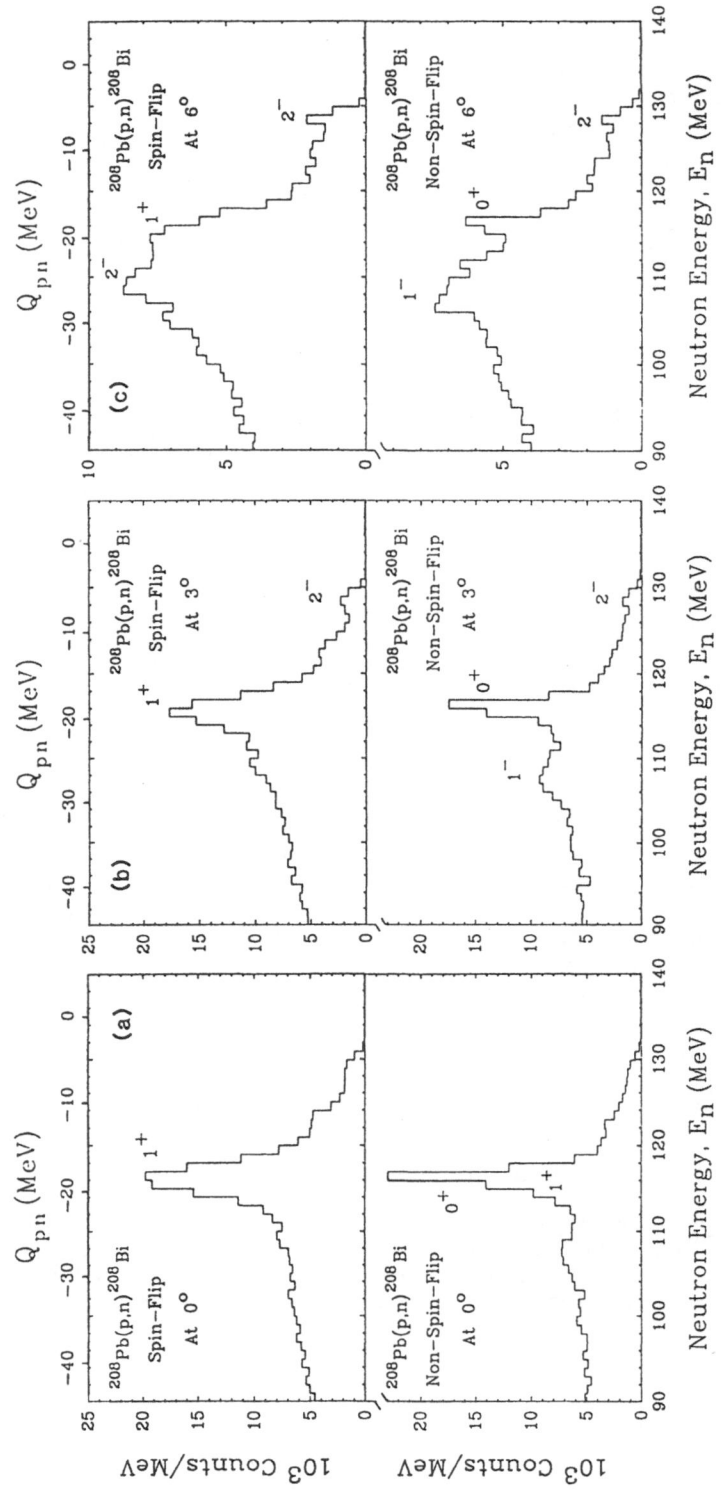

Figure 5. Spin-flip and non-spin-flip spectra for the 135 MeV ^{208}Pb$(\vec{p},\vec{n})^{208}$Bi reaction at (a) 0°, (b) 3°, and (c) 6°.

resonances which deplete non-spin-flip strength in the continuum.

Figure 5 shows spin-flip spectra $[S \times \sigma(\theta)]$ and non-spin-flip spectra $[(1 - S) \times \sigma(\theta)]$ for the $^{208}Pb(\vec{p},\vec{n})^{208}Bi$ reaction at $0°$, $3°$, and $6°$. In the non-spin-slip spectra, the 0^+ IAS and the 1^- natural-parity component of the spin-dipole resonance are separated clearly from the GTGR and the unnatural parity components of the spin-dipole resonance which dominate the spin-flip spectra. This ability of spin-flip measurements to identify and separate the various components of the spin-dipole resonance was noted also for the $^{40}Ca(\vec{p},\vec{n})^{40}Ca$ reaction.[3,4]

REFERENCES

1) T Chittrakarn, B. D. Anderson, A. R. Baldwin, C. Lebo, R. Madey, J. W. Watson and C. C. Foster, Phys. Rev. C. 34, 80 (1986).

2) B. D. Anderson, T. Chittrakarn, A. R. Baldwin, A. Fazeley, C. Lebo, J. W. Watson, and C. C. Foster, Phys. Rev. C 34, 422 (1986).

3) J. W. Watson, P. J. Pella, B. D. Anderson, A. R. Baldwin, T. Chittrakarn, B. S. Flanders, R. Madey, C. C. Foster and I. J. van Heerden, Phys. Lett. B181, 47 (1986).

4) J. W. Watson, B. D. Anderson and R. Madey, Can. J. Phys. 65, 566 (1987).

5) J. W. Watson, Marco R. Plumley, P. J. Pella, B. D. Anderson, A. R. Baldwin, and R. Madey, Submitted to Nucl. Instrum. and Meth.

6) J. W. Watson, B. D. Anderson, A. R. Baldwin, T. Chittrakarn, B. S. Flanders, R. Madey, P. J. Pella, C. C. Foster and I. J. van Heerden, in Nucleon and Anti-nucleon Scattering from Nuclei, ed by G. E. Walker, C. D. Goodman and C. Olmer (Plenum, New York, 1985) p.371.

7) R. A. Arndt, computer code SAID (unpublished).

8) T. N. Taddeucci, Can. J. Phys. 65, 557 (1987).

9) C. Glashausser, K. Jones, F. T. Baker, L. Bimbot, H. Esbensen, R. W. Fergerson, A. Green, S. Nanda, and R. D. Smith, Phys. Rev. Lett. 58, 2404 (1987).

10) P. M. Boucher, B. Castel, Y. Okuhara, I. P. Johnstone, J. Wambach, and T. Suzuki, Phys. Rev. C37, 906 (1988).

SPIN-ISOSPIN MODES IN QUASI-FREE REGION

M. Ichimura and K. Kawahigashi

Institute of Physics, University of Tokyo
Komaba, Tokyo 153, Japan

T. S. Jorgensen and C. Gaarde

Niels Bohr Institute, University of Copenhagen
DK-2100, Copenhagen, Denmark

INTRODUCTION

Put rather high energy ω (30~120MeV) and high momentum q (1.0~2.5 fm^{-1}) together with spin and isospin into nucleus and see how it responds. This is one of the main subjects of this meeting. Inelastic and charge exchange scatterings of nuclear probes are very promising tools for this purpose. If a probe interacts a nucleon inside the nucleus almost freely, one expects the broad bump with the peak at $\omega = \dot{q}^2/2m$ (m being the nucleon mass). This bump region is called the quasi-free region.

Measurement of spin observables of nuclear probes is expected to be very useful to decompose the various modes such as spin-longitudinal and spin-transverse modes, etc.

At this conference 1982, M. Ericson[1] predicted very interesting contrasts between the above two modes in isovector channel. Namely, as to the isovector spin-longitudinal mode the peak shifts downward (softening) and the strength is enhanced (especially large at lower energy side)[2], while as to the isovector spin-transverse mode the peak shifts upwards (hardening) and the strength is quenched (especially large at lower energy side)[3].

However, as was repeatedly discussed in this meeting, the experimental results of the complete polarization transfer measurements of (p,p') at LAMPF[4,5] and the tensor analysing power of $(\vec{d},2p)$ at SATURNE[6] apparently contradict the prediction. For instance, the LAMPF results show that the ratio of the isovector spin-longitudinal response function R_L to the isovector spin-transverse response function R_T at q = $1.75fm^{-1}$ is close to or even slightly smaller than unity in contrast to the theoretical prediction[1] that it is much larger than unity (even larger than 10 at lower energy).

On the other hand the softening of the energy spectra are found in the (^{3}He,t) and (d,2p) at the quasi-free region measured at SATURNE[7,6] which may support the prediction (see Fig. 5).

The original version of the theoretical analysis[1,2] was based on the Fermi gas model, the ring approximation (simple RPA) and the plane wave impulse approximation (PWIA). Stimulated by the LAMPF results, various improvements of theoretical analysis have been performed both in nuclear structure and reaction sides. The main concerns are how to cope with the finiteness of the nucleus and how to cope with the distortions on the incident and outgoing nuclear probes.

Here we discuss them separately, and report our theoretical developments and numerical analysis on both subjects.

NUCLEAR RESPONSE FUNCTIONS

First let us consider how to calculate the nuclear response functions of the isovector spin-longitudinal and -transverse modes. What technique of many body theory can we use? The ring approximation where the exchange term in RPA is approximated by the contact interaction is practically used. What effective particle-hole interaction should we use? Most usually used one is the one-pion- plus one-rho-exchange plus g'-term (π + ρ + g' model)[2,3]. We will comment them later.

But most serious concern we want to discuss is how to take account of the effect of finiteness beyond the Fermi gas model. It has been pointed out[8,9] that the effects show up as the density dependence and the mixing of the longitudinal and the transverse modes.

People applied a) the Fermi gas model with the local density approximation[10,11] b) the semi-infinite slab model (SIS)[10,12] c) the bound state approximation for the single particle states (such as the use of harmonic oscillator wave function)[8,13] d) the semi-classical approach[8,14] and e) the continuum RPA of Shlomo-Bertsch[15] where continuum of the single particle states are rigorously taken into accout[16].

Advancing one step more, here we report our calculation by the continuum RPA with the orthogonality condition[17].

a) Formalism

Definitions and formalism are briefly given first. We consider the external fields of the isovector spin-longitudinal and -transverse modes,

$$O_L(\mathbf{q},\mathbf{r}) = \tau(\sigma \cdot \mathbf{q})\exp(i\mathbf{q}\cdot\mathbf{r})/\sqrt{2}, \tag{1}$$

$$O_T(\mathbf{q},\mathbf{r}) = \tau(\sigma \times \mathbf{q})\exp(i\mathbf{q}\cdot\mathbf{r})/2. \tag{2}$$

The response functions for them are given by[18]

$$R_{\alpha\beta}(\mathbf{q}\ \mathbf{q}',\omega) =$$

$$- (1/\pi)\operatorname{Im}\iint O_\alpha^\dagger(\mathbf{q},\mathbf{r})\Pi(\mathbf{r},\mathbf{r}';\omega)O_\beta(\mathbf{q},\mathbf{r})d\mathbf{r}d\mathbf{r}' \tag{3}$$
$$(\alpha,\beta = L \text{ or } T)$$

In the ring approximation the polarization propagator $\Pi(\mathbf{r},\mathbf{r}';\omega)$ obeys the integral equation

$$\Pi(\mathbf{r},\mathbf{r}';\omega) = \Pi^{(0)}(\mathbf{r},\mathbf{r}';\omega)$$

$$+ \int\int \Pi^{(0)}(\mathbf{r},\mathbf{r}_1;\omega)V_{ph}(\mathbf{r}_1,\mathbf{r}_2)\Pi(\mathbf{r}_2,\mathbf{r}';\omega)d\mathbf{r}_1 d\mathbf{r}_2 \qquad (4)$$

with the particle hole interaction V_{ph}. The free polarization propagator $\Pi^{(0)}(\mathbf{r},\mathbf{r}';\omega)$ is given by

$$\Pi^{(0)}(\mathbf{r},\mathbf{r}';\omega) =$$

$$\sum_{p,h} [\frac{\phi_h^\dagger(\mathbf{r})\phi_p(\mathbf{r})\phi_h(\mathbf{r}')\phi_p^\dagger(\mathbf{r}')}{\omega - (\varepsilon_p - \varepsilon_h) + i\eta} + \frac{\phi_h^\dagger(\mathbf{r}')\phi_p(\mathbf{r}')\phi_h(\mathbf{r})\phi_p^\dagger(\mathbf{r})}{-\omega - (\varepsilon_p - \varepsilon_h) + i\eta}] \qquad (5)$$

where sums p and h run over single particle and hole states, respectively. Shlomo and Bertsch[15] rewrite it in the form

$$\Pi^{(0)}(\mathbf{r},\mathbf{r}';\omega) =$$

$$\sum_h [\phi_h^\dagger(\mathbf{r})g(\mathbf{r},\mathbf{r}';\omega+\varepsilon_h)\phi_h(\mathbf{r}') + \phi_h^\dagger(\mathbf{r}')g(\mathbf{r}',\mathbf{r};-\omega+\varepsilon_h)\phi_h(\mathbf{r})] \qquad (6)$$

in terms of the single particle Green's function

$$g(\mathbf{r},\mathbf{r}';E) = \langle\mathbf{r}|(E - H_{sp} + i\eta)^{-1}|\mathbf{r}'\rangle. \qquad (7)$$

This method does not work if the particle and hole wave functions are not orthogonal to each other, for instance,[19] if one uses an optical potential in H_{sp}. Izumoto and Mori[20] gave the expression of the particle Green's function in which the particle states are orthogonal to the hole states,

$$g_p(\mathbf{r},\mathbf{r}';E) = \sum_{p(\neq h)} \frac{\phi_p(\mathbf{r})\phi_p^\dagger(\mathbf{r}')}{E - \varepsilon_p + i\eta} = g - g\Gamma(\Gamma g\Gamma)^{-1}\Gamma g. \qquad (8)$$

with $\Gamma = \sum_h |\phi_h\rangle\langle\phi_h|$. Izumoto[17] utilized g_p instead of g in Eq.(6). We adopt this method and call it the continuum RPA with the orthogonality condition.

The response functions of the isovector spin-longitudinal and -transverse modes conventionally called[18] are the diagonal element of Eq. (3) as

$$R_\alpha(\mathbf{q},\omega) = R_{\alpha\alpha}(\mathbf{q},\mathbf{q};\omega), \qquad (\alpha = L, T). \qquad (9)$$

The $(\pi + \rho + g')$ model is used for the p-h interaction. The delta-hole (Δ-h) configurations are included in a standard way[3] and the universality is assumed for g'. The woods-Saxon potential is used for the single particle potential, the depth of which is determined by the separation energy of the proton and neutron, respectively. The transferred angular momenta are taken into account up to 11, the dependence on which will be discussed in the next section.

b) Numerical results

Fig. 1 shows the energy spectra of the response functions, R_L and R_T, at $q = 1.75\text{fm}^{-1}$ in the ring approximation with $g' = 0.6$. The response functions of the spin-longitudinal and -transverse modes without nuclear correlation, $R_{L,0}$ and $R_{T,0}$, are in principle different to each other due to the single particle spin-orbit force, but practically the difference is negligibly small. So we do not distinct them and denote simply by R_0.

Many sharp resonances can be seen when the particles and holes obey the same real Woods-Saxon potential(full lines). This is because the spreading width is completely neglected. To smear out these structures we introduced the imaginary potential for the particle states, the depth of which is taken to be $W = 5.0$ MeV (dashed lines). In this case we must use the continuum RPA with the orthogonality condition. One sees that this prescription well average out the sharp resonances.

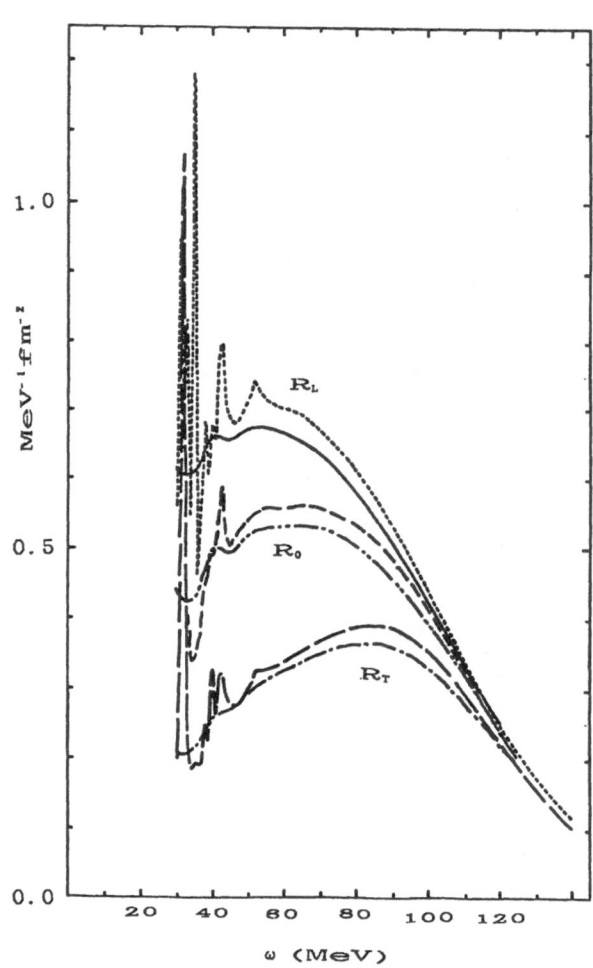

Fig. 1

Response functions in the continuum RPA with the orthogonality condition.

The results with and without the imaginary potential W for the particle states are compared

The full, dash-dot-dot and dash-dot lines are those with $W = 5.0$MeV for R_L, R_T and R_0, respectively.

The other lines are those without W.

In Fig 2, the results for $q = 1.75\text{fm}^{-1}$ of the Fermi gas model with the Fermi momentum $k_F = 1.2\text{fm}^{-1}$ are compared with the continuum RPA results with $W = 5.0$ MeV. One sees that Fermi gas calculation extremely overestimates the enhancement of R_L at the softening region. This is one of the main reasons that the huge enhancement was predicted originally. Correct treatment for the finiteness of the nucleus is definitely necessary.

In Fig. 3 we compare our continuum RPA results with those of the Alberico et al.'s bound state approximation[8] (harmonic oscillator wave function) together with the approximate solution of RPA equation of Toki and Weise.[21] The calculation are carried out at $q = 330\text{MeV/c}$ $(= 1.67\text{fm}^{-1})$ with $g' = 0.7$. The bound state approximation works semi-quantitatively. However, to get a continuum spectrum one must introduce the width parameter. Shigehara et al.[16] warned strong dependence on it.

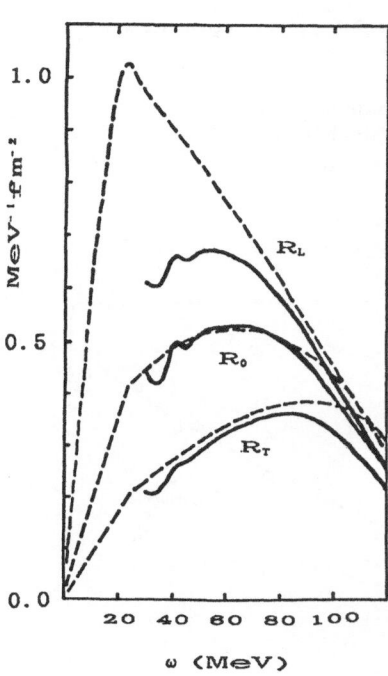

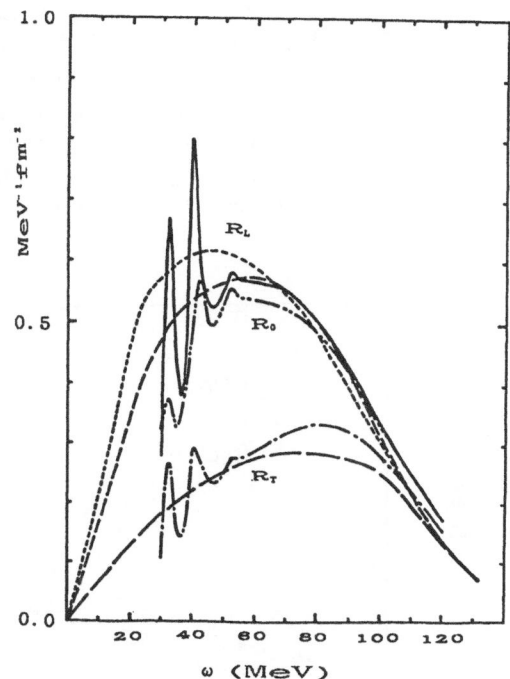

Fig. 2 Response functions in the Fermi gas model with $k_F = 1.2\text{fm}^{-1}$ (dashed lines) and in the continuum RPA with the orthogonality condition (the full lines).

Fig. 3 Response functions in the bound state approximation (ref.8) and in the continuum RPA(the lines with rapid oscillation).

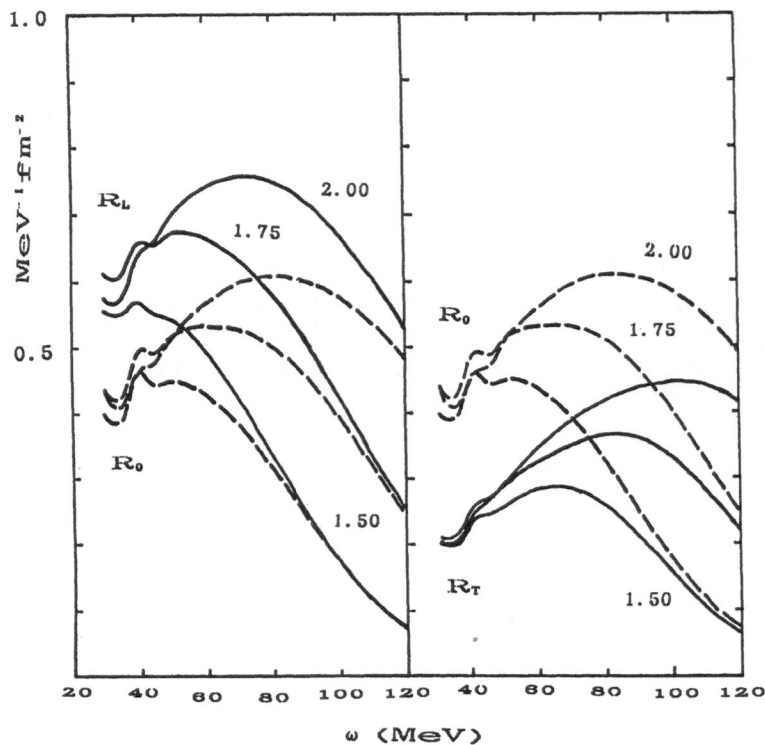

Fig. 4   The energy spectra of the response functions
R_L, R_T (full lines) and R_0 (dashed lines)
for the momenta q = 1.50, 1.75 and 2.00 fm^{-1}
attached to the lines.

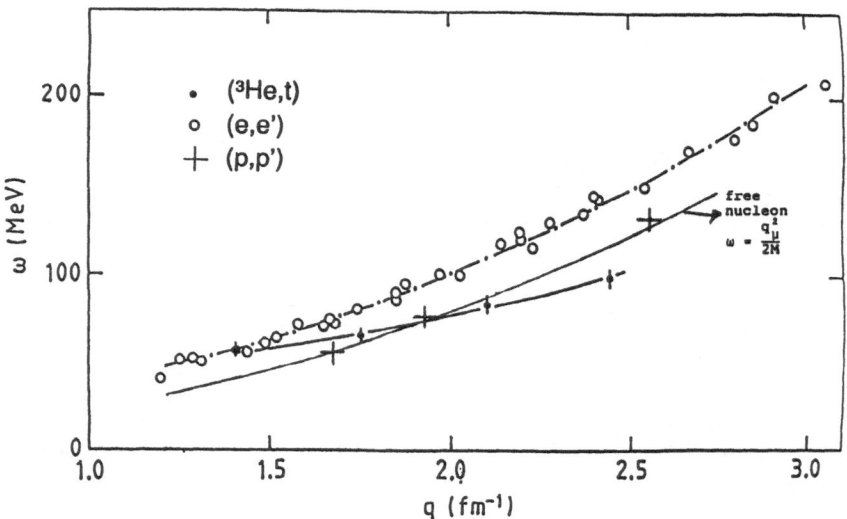

Fig. 5   The peak lines of (e,e'), (^{3}He,t) and
the quasi-free scattering on q-ω plane.

Fig. 4 shows how the energy spectra change as the momentum q changes. Comparison of the peaks of (^{3}He,t), (e,e') and (p,p') with the quasi-free peak is shown in Fig. 5.[7,22] (Note that (p,p') does not show the softening.) We found that the peak of the calculated R_0 runs on the quasi-free line $\omega = q^2/(2m)$, that of R_T runs close to the (e,e') line and that of R_L does about 10 MeV below the quasi-free line.

DWIA ANALYSIS

As we discussed, we cannot compare the calculated response functions directly with the experimental data. Before that we must take into account the distortion of the incident and outgoing nuclear probes. Here we discuss the distortion effects in terms of DWIA.

a) Formalism

First one must prepare the nucleon-nucleon t-matrix. Following Bugg-Wilkin's convention,[23] the t-matrix in the isovector channel is written as

$$t = \alpha - i\gamma[(\sigma_1 \cdot \hat{\mathbf{n}}) + (\sigma_2 \cdot \hat{\mathbf{n}})] +$$

$$\beta(\sigma_1 \cdot \hat{\mathbf{n}})(\sigma_2 \cdot \hat{\mathbf{n}}) + \delta(\sigma_1 \cdot \hat{\mathbf{q}})(\sigma_2 \cdot \hat{\mathbf{q}}) + \varepsilon(\sigma_1 \cdot \hat{\mathbf{p}})(\sigma_2 \cdot \hat{\mathbf{p}}) \qquad (10)$$

where $\alpha, \cdots, \varepsilon$ are the function of q^2 and the incident energy E. The term of α does not contribute to the spin modes in DWIA. We also neglect the γ-term in the following, keeping in mind that the LAMPF data are summarized in the form of the special linear combinations of D_{ij}, the polarization transfer observables, which pick up the δ- and ε-term contribution selectively in PWIA.

Furthermore if

$$\beta = \varepsilon \ (\equiv \eta), \qquad (11)$$

the t-matrix is written as

$$t(\mathbf{q},E) \approx \eta(\sigma_1 \times \hat{\mathbf{q}})(\sigma_2 \times \hat{\mathbf{q}}) + \delta(\sigma_1 \cdot \hat{\mathbf{q}})(\sigma_2 \cdot \hat{\mathbf{q}}) \qquad (12)$$

which depends only on q and E. Then we can have the energy dependent local t-matrix in the r-representation

$$t(\mathbf{r},\mathbf{r}') = \int t(\mathbf{q},E)\exp[i\mathbf{q}\cdot(\mathbf{r} - \mathbf{r}')]d\mathbf{q}/(2\pi)^3 \qquad (13)$$

Note that the t-matrix is in principle the four point function and we need the condition (11) to get the velocity independent local interaction (13).

Using the t-matrix (13) and the polarization propagator (4), the inclusive (p,p') cross section is given by[17,25]

$$\frac{d^2\sigma}{d\Omega d\omega} = K \cdot \text{Im} \int S^\dagger(\mathbf{r}) \Pi(\mathbf{r},\mathbf{r}';\omega) S(\mathbf{r}') d\mathbf{r} d\mathbf{r}' \qquad (14)$$

where K is the kinematical factor and

$$S(\mathbf{r}) = \int \chi^{(-)*}(\mathbf{k}_f,\mathbf{r}') t(\mathbf{r},\mathbf{r}') \chi^{(+)}(\mathbf{k}_i,\mathbf{r}') d\mathbf{r}' \qquad (15)$$

with the incident and outgoing distorted waves, $\chi^{(+)}(\mathbf{k}_i,\mathbf{r})$ and $\chi^{(-)}(\mathbf{k}_f,\mathbf{r})$, respectively.

Comparing Eqs. (12) and (13) with Eqs. (1) and (2), one can write $S(\mathbf{r})$ by the sum of the fields $O_\alpha(\mathbf{q},\mathbf{r})$ as

$$S(\mathbf{r}) = \sum_{\alpha=L,T} \int \tilde{S}(\mathbf{q}) O_\alpha(\mathbf{q},\mathbf{r}) d\mathbf{q}/(2\pi)^3. \qquad (16)$$

Thus one can connect the cross section with the response function. In PWIA, $\tilde{S}(\mathbf{q}) \propto \delta(\mathbf{q} - (\mathbf{k}_i - \mathbf{k}_f))$ and hence the cross section is proportional to the response function.

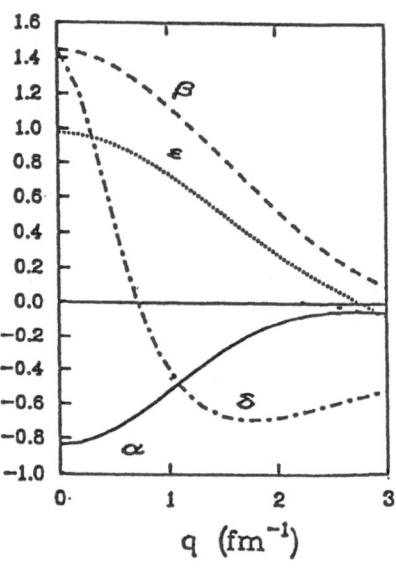

Fig. 6

Real part of the t-matrix in the isovector channel given by ref. 24.

In reality the condition (11) does not hold well (see[24] Fig. 6), but in the following we present the results under the condition with the averaged η as

$$|\eta|^2 = (|\beta|^2 + |\varepsilon|^2)/2. \qquad (17)$$

The formula for other reactions such as (^{3}He,t) and (d,2p) can be written straightforwardly by introducing the form factors.

b) Numerical results

We made DWIA calculation for the inclusive (p,p') cross section with the incident energy E = 500MeV, by the t-matrix

$$t_T = \eta(\sigma_1 \times \hat{\mathbf{q}})(\sigma_2 \times \hat{\mathbf{q}}), \quad \text{and} \quad t_L = \delta(\sigma_1 \cdot \hat{\mathbf{q}})(\sigma_2 \cdot \hat{\mathbf{q}}). \qquad (18)$$

The cross section given by t_T is called the isovector spin-transverse cross section and denoted by σ_T, and that given by t_L is called the isovector spin-longitudinal cross section and denoted by σ_L. The optical potential is taken from the set 4 of the Dirac phenomenological analysis by Clark et al.[26]

Fig. 7 shows the energy spectra of the cross sections at the scattering angle θ_{cm}= 19.2 , 21.7 , 24.2 which roughly corresponds to q = 1.75, 1.96, 2.18fm^{-1}, respectively. θ_{cm}= 19.2 corresponds to θ_{lab}= 18.5 ,the angle of LAMPF experiment. The full lines show the RPA cross section, σ^{RPA}, the results given by use of the polarization propagator π of RPA, and the dashed lines denote the cross section without correlation, σ^{0th}, those given by $\pi^{(0)}$, the polarization propagator without correlation.

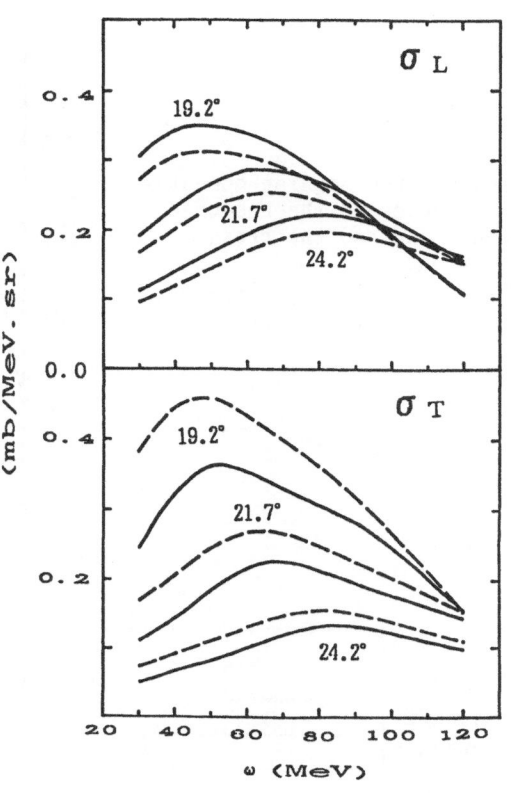

Fig. 7

The cross sections by
the RPA (full lines)
and without correlation
(dashed lines)
for the fixed angles
indicated by the numbers
attached to the lines.

The upper box shows the
spin-longitudinal cross
sections and the lower
box shows the spin-
transverse ones.

One sees the drastic change of the shape of the spectra from the response functions (compare with Fig. 2), especially in the transverse mode. For a fixed angle, all of the peaks of , σ_L^{RPA}, σ_T^{RPA}, σ^{0th} appeared at almost the same energy which is about 10MeV lower than the quasi-free peak $\omega = q^2/(2m)$. This must be essentially due to the distortion. The softening and hardening are now hardly seen though the peak of σ_T is slightly higher than that of σ_L. However, one can still see considerable collectivity in the strength. Qualitatively this is the same trend as was seen in the previous eikonal approximation calculations[10,11].

Fig. 8 shows to what extent each transferred angular momentum J contributes to the response function at $q = 1.75$ fm^{-1} and the cross section of $\theta_{cm} = 19.2$. Here $R_L(J=10)$ and $\sigma_L(J=10)$ are normalized unity and $R_T(J=10)$ and $\sigma_T(10)$ are set 0.768 , the value of $R_T(10)/R_L(10)$. One sees the attenuation due to distortion is larger as J is smaller, hence the dominant multipolarity is shifted to larger J. The dominant J is larger for the transverse mode than for the longitudinal mode. We also noticed that the larger q is, the larger J becomes important.

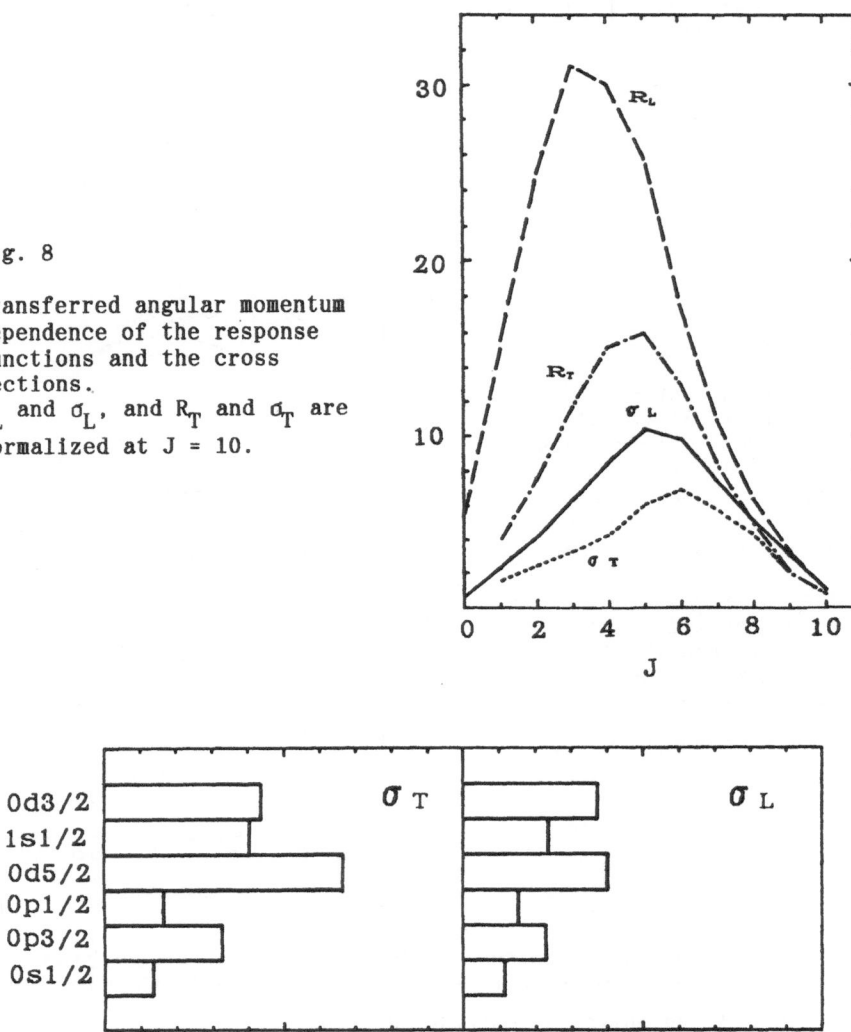

Fig. 8

Transferred angular momentum dependence of the response functions and the cross sections.
R_L and σ_L, and R_T and σ_T are normalized at J = 10.

Fig. 9 Contributions from different hole states to the cross section without correlation

In Fig. 9, the contributions from the different hole states are compared for σ^{0th} at $\theta_{cm} = 19.2$ and $\omega = 60MeV$. Attenuation factors of the hole state h

$$a_\alpha(h) = \sigma_\alpha^{0th}(h)/R_0(h) \tag{19}$$

are given in Table 1, relative to that of $1s_{1/2}$. The deeper hole contributions are attenuated more with the shell gap. Surprisingly, the contributions of the inner levels ($0p_{1/2}$, $0p_{3/2}$, $0s_{1/2}$) still share considerable fraction as is seen in Fig. 9.

Table 1 Relative attenuation factor
of the hole contributions

holes	trans.	long.
0d3/2	0.688	0.858
1s1/2	1.000	1.000
0d5/2	0.701	0.823
0p1/2	0.415	0.584
0p3/2	0.413	0.542
0s1/2	0.238	0.345

COLLECTIVITY RATIO

The LAMPF experiment[5] discussed the collectivity ratio

$$r_1 = \left(\frac{R_L/R_{L,0}}{R_T/R_{T,0}}\right) = \frac{R_L}{R_T} \tag{20}$$

Recall that $R_{L,0}$ and $R_{T,0}$ are well to be identified and both of them have been written as R_0. In the (p,p') scattering, one must take into account the isoscalar contribution. Assuming that the isoscalar response functions are non-collective and approximated to be the same as R_0, Rees et al.[5] introduced the ratio

$$r_{1+0} = \frac{(3.62R_L + R_0)/4.62}{(1.15R_T + R_0)/2.15} \tag{21}$$

Here the experimental data $|\delta|^2/|\delta_0|^2 = 3.62$, and $|\varepsilon|^2/|\varepsilon_0|^2 = 1.15$ for E = 500MeV are used where δ_0 and ε_0 are the isoscalar correspondents of δ and ε, respectively.

Including the distortion effects, the effective collectivity ratio may be written as a generalization of Eq. (20) by

$$\bar{r}_1 = \frac{\sigma_L^{RPA} / \sigma_L^{0th}}{\sigma_T^{RPA} / \sigma_T^{0th}} \tag{22}$$

Following the same idea as Eq. (21), we also introduce the ratio

$$\bar{r}_{1+0} = \frac{(3.62 \, \bar{r}_1 + 1 \,)/4.62}{(1.15 \, \bar{r}_1 + 1 \,)/2.15} \tag{23}$$

to take into account the isoscalar contributions.

In Fig. 10, the collectivity ratios r_1 , $\bar{r}_1$, r_{0+1} and $\bar{r}_{0+1}$ are compared with the 'experimental' results. The ratios are very much reduced by the effects of the finiteness and the distortion but still of course larger than unity due to remaining collectivity. The results are similar but slightly smaller than those of the previous analysises.

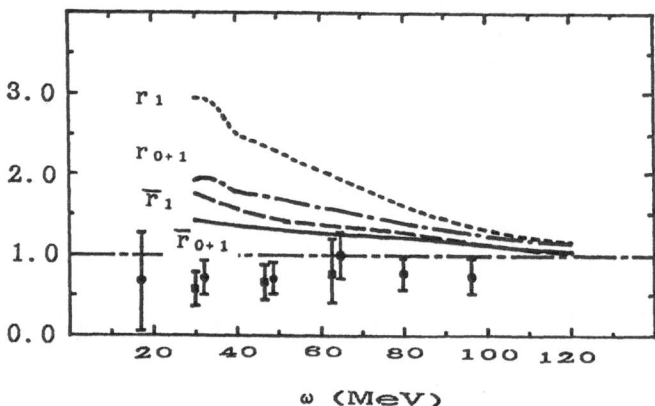

Fig. 10   Collectivity ratios.   Experimental data from ref.5

CONCLUSIONS AND REMARKS

We calculated the isovector spin-longitudinal and -transverse response functions by the continuum RPA with the orthogonality condition. This honest treatment of the finiteness of the nucleus reduces the collectivity comparing to the results of various methods previously used.

We then performed DWIA calculation of the spin-longitudinal and -transverse cross sections of ^{40}Ca(p,p') at E = 500MeV. The shape of the energy spectra are found very much different from those of the corresponding response functions. The position of the peaks of the cross sections are extremely affected by the distortions. It is often said that this reaction is the surface one, but we found that the deeper hole states still contribute in some fractions. The ratio of R_L/R_T reduced more than the previous analysises.

We must keep in mind that very simple model is assumed to obtain the ratio from the experiment. To get a reliable information, we must directly calculate the observed polarization transfers D_{ij} in the framework of DWIA . At moment, however, we calculated only σ_L and σ_T given by t_L and t_T, separately. To make a reliable analysis of spin observables, we have to take off the approximation $\beta = \varepsilon$ and to prepare the response functions of other velocity dependent operators, and to take into account the interference between the different terms of the t-matrix (10). This is a future problem we must challenge.

In the calculation of the response function, we used Woods-Saxon potential as the mean field, but in principle we should use the Hartree-Fock potential consistent to the effective interaction used. The effect of the non-locality of the H-F potential (or of the effective mass) may affects the spectrum.
The importance of the spreading width of the p-h states (especially of the 2p-2h configurations) are often stressed[27].
The choice of g' parameter is of course unsettled question. We used g'= 0.6 in most cases here except for in Fig. 3 where g' = 0.7 was used. We see rather large g' dependence. This problem must be investigated in more detail.
We used the universality of g' for the RPA effective interaction, which does not have any theoretical basis[11] A reasonable interaction must be found.

The ring approximation is also in question. Shigehara et al[28] carried out RPA calculation in which the exchange term is completely treated and reported the ring approximation holds very well in the isovector spin modes, though it does not hold for the Coulomb modes.

As some examples are mentioned above, there are many problems left unsolved but we think our approach is most complete at moment and the development on this line should be extended.

ACKNOWLEDGMENTS

M.I. would like to thank NORDITA very much for the financial support of his stay in LUND, and his thanks are also due to Prof. I. Hamamoto for her kind hospitality and valuable discussions during his stay in LUND.

References

1. M. Ericson, Proc. International Conf. on Spin Excitations in Nuclei, (ed.,F. Petrovich et al., Plenum Press, NY, 1984) p.27
2. W.M. Alberico, M. Ericson and A. Molinari, Phys. Lett. **92B**:153(1980)
3. W.M. Alberico, M. Ericson and A. Molinari, Nucl. Phys. **A379**:429(1982)
4. T.A. Carey et al., Phys. Rev. Lett. **53**:144(1984)
5. L.B. Rees et al., Phys. Rev. **C34**:627(1986)
6. C. Ellegaard et al., Phys. Rev. Lett. **59**:974(1987) and private communication.
7. I. Bergqvist et al., Nucl. Phys. **A469**:648(1987)
8. W.A. Alberico et al., Phys. Rev. **34**:977(1986)
9. W.A. Alberico et al., Phys. Rev. Lett. **183B**:135(1987)
10. H. Esbensen, H. Toki and G.F. Bertsch, Phys. Rev. **C31**:1816(1985)
11. E. Shiino, Y. Saito, M. Ichimura and H. Toki, Phys. Rev. **C34**:1004 (1986)

12. H. Esbensen and G.F. Bertsch, Ann. Phys.(NY) **157**:255(1984)
13. Y. Okuhara et al., Phys. Lett. **186B**:113(1987)
14. U. Stroth et al., Phys. Lett. **156B**:291(1985)
15. S. Shlomo and G.F. Bertsch, Nucl. Phys. **A243**:507(1975)
16. T. Shigehara, K. Shimizu and A. Arima, Nucl. Phys. To be published
17. T. Izumoto, Nucl. Phys. **A395**:189(1983)
18. W.M. Alberico, A. De Pace and A. Molinari, Phys. Rev. **31**:2007(1985)
19. Y. Horikawa, F. Lenz, N.C. Muhkopadhyay, Phys. Rev. **C22**:1680(1980)
    F.A. Brieva and A. Dellafiore, Phys. Rev. **C36**:899(1987)
20. T. Izumoto and A.Mori, Phys. Lett. **82B**:163(1979)
21. H. Toki and W. Weise, Phys. Rev. Lett. **42**:1034(1979),
    Z. Phys. **A292**:389(1979)
22. C. Gaarde, Proc. International Conf. PANIC, 1987
23. D.V. Bugg and C. Wilkin, Phys. Letts. **152B**:37(1985),
    Nucl. Phys. **A467**:575(1987)
24. D.V. Bugg and C. Wilkin, private communication
25. T. Izumoto, M. Ichimura, C.M. Ko and P.J. Siemens, Phys. Lett.
    112B:315(1982)
26. B.C. Clark et al., Phys. Lett. **122B**:211(1983)
27. K. Shimizu, M. Ichimura and A. Arima, Nucl. Phys. **A226**:282(1974)
    G.F. Bertsch and I. Hamamoto, Phys. Rev. **C26**:1323(1982),
    Z.E. Meziani et al., Phys. Rev. Lett. **54**:1233(1985)
    S. Drozdz et al., Phys. Lett. **185B**:287(1987)
28. T. Shigehara, Dissertation, University of Tokyo, 1988

CHARGE EXCHANGE REACTIONS AT SATURNE

Clive Ellegaard

Niels Bohr Institutet
University of Copenhagen
DK-2100 Copenhagen, Denmark

1. The (^{3}He,t) Reaction and Introduction

In this paper we concentrate on a region of the nuclear response that lies higher in excitation energy than the previously discussed nuclear resonance region and quasifree region. Figure 1 shows one of the main results of the Saturne charge exchange experiments from a couple of years ago[1]. The spectra show how the charge exchange reaction (^{3}He,t) strongly excites the Δ resonance, the $\sigma \tau$ excitation of the nucleon.

The spectrum on carbon shows a strong peak at low excitation energy containing the Gamow-Teller resonance in addition to the Δ excitation. The spectrum on the proton shows only the excitation of the Δ^{++}.

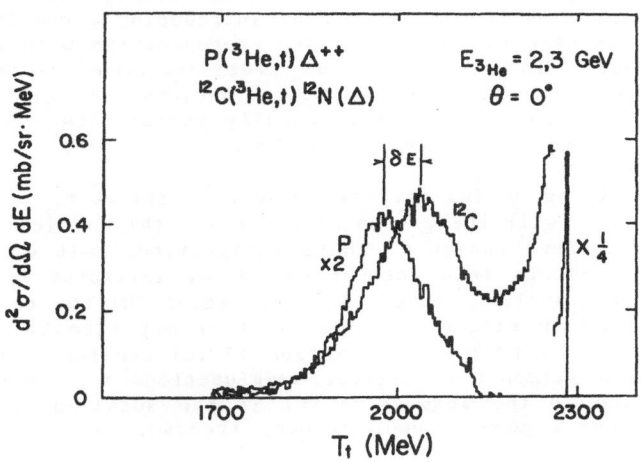

Fig.1. Triton spectra from the (^{3}He,t) reaction at 0^0 and 2.3 GeV, emphasising the Δ-region.

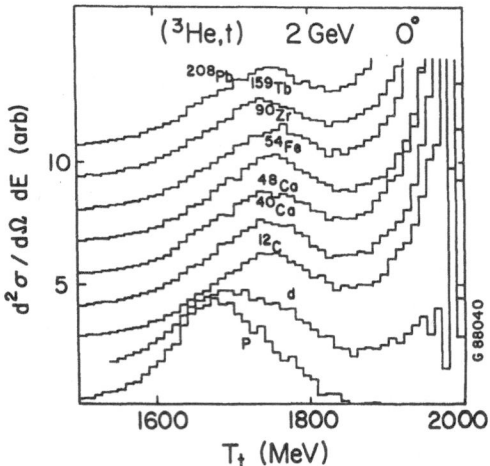

Fig.2. Triton spectra from the (^{3}He,t) reaction at 0^0 and 2 GeV on targets ranging from the proton to ^{208}Pb.

One point of much discussion over the past several years has been the shift down in energy of the Δ excitation in carbon relative to that on the free proton. Fig. 2 emphasises how this is a universal property of all nuclei from ^{12}C to ^{208}Pb. The only exception is the very small nucleus, the deuteron. Strong correlations are expected in the nuclear medium in the spin longitudinal channel from one pion exchange for both particle-hole states and Δ-hole states. These correlations are what at some density and momentum transfer could lead to pion condensation.

Before we can say that we are probing these correlations, there are a number of features of the reaction that need to be examined. These are symbolised in fig. 3.

Fig. 3a gives the basic diagram of the nucleon-nucleon going to nucleon-Δ, showing exchange of a pion or whatever else is necessary (derived empirically or from models, depending on the approach), and with the subsequent decay of the Δ giving it its width. Introducing a complex projectile (fig. 3b) is taken care of in the impulse approximation with a ^{3}He→t form-factor as a factor in the production cross section. A complex target (fig.3c) leads to Fermi motion of the struck nucleon and also to additional distortion (absorption) of the beam. Finally correlations may modify the observed energy of the Δ produced on a nucleus.

The correlations of (d) are the subject of interest. The basic reaction (a) is measured with the proton as a target. (b) and (c) cause modifications to the observed energy of the Δ excitiation, that we call trivial and which must be taken into account before we interpret our results as correlations in the nuclear medium. The formfactor (b) can be taken as the formfactor measured in electron scattering[2] or may also be measured with the proton as a target[3]. The Fermi motion of (c) can for example be taken into account with harmonic oscillator wavefunctions in momentum space. A good approximation to the absorption (c) is the socalled eikonal approximation. This gives a good account of many trends in the data e.g. target

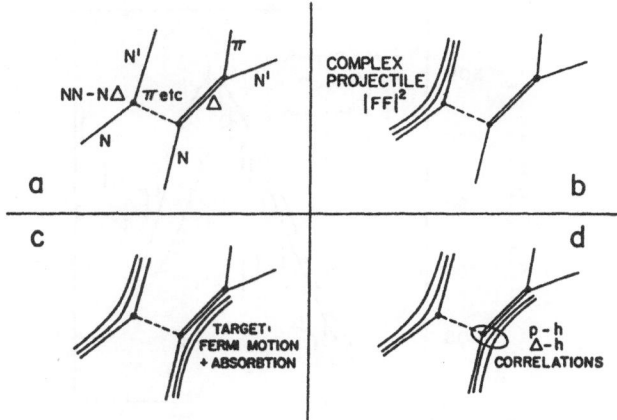

Fig. 3. Diagrams illustrating the ingredients of a charge exchange reaction
        leading to Δ excitation with a complex projectile on a complex
        target (see text).

mass dependence for both heavy ion and (d,^{2}He) reactions (see later).

        One of the first calculations of the expected correlations (d) was
done by M. Ericsson et.al[4]). The results of their calculation for infinite
nuclear matter is shown in fig. 4. The shift in energy of the Δ peak in the
longitudinal channel of over 100 MeV is far beyond what is observed. However, the finiteness of the nucleus is important. There have been a number
of calculations moving from a local Fermi gas model[5]) through a surface
response model[6]) to the most recent calculation by Delorme et al.[7]). The
result of this calculation is shown in fig. 5.

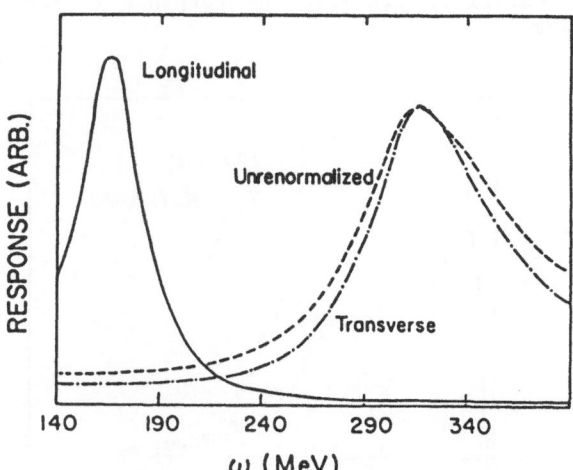

Fig. 4. The calculated response of the nuclear medium for a σ τ probe in
        the longitudinal and transvers channels, according to M. Ericsson
        et al. The figure is drawn from results in ref. 4.

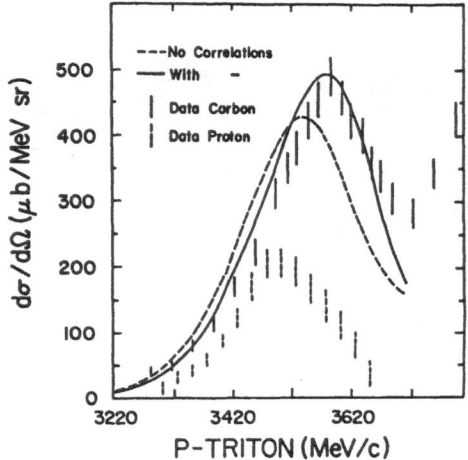

Fig. 5. The response of a finite nucleus calculated by Delorme and Guichon[7])

The contributions to the shift are roughly:
~30 MeV from Fermi motion and (^{3}He,t) formfactor, ~5 MeV from distortion and ~30 MeV from correlations. The calculation reproduces the shape at the data very well.

At the same time there is still the suspicion[8]) that distortion effects may account for a larger part of the shift. So at present there is no consensus as to the interpretation of the (^{3}He,t) data. We therefore produce new data that might shed light on the situation.

First we turn to heavy ion reactions that primarily are absorbed very differently and probe different densities (a decisive factor in many of the calculations), and with the ($\bar{d}$,^{2}He) reaction we study spin observables that allow us in more detail to separate the response in the longitudinal and transverse channels.

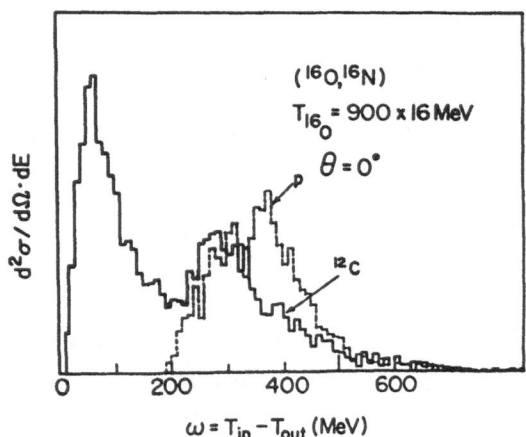

Fig. 6. Spectra of ^{16}N from the (^{16}O,^{16}N) reaction at 0⁰ and 14.4 GeV on proton and ^{12}C targets.

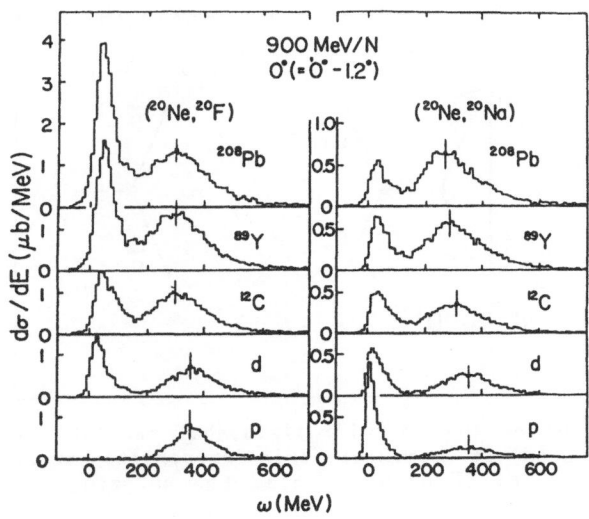

Fig. 7. Mass systematics of the two reactions (^{20}Ne, ^{20}F) and (^{20}Ne, ^{20}Na) at 0^0 and 18 GeV.

2. Heavy Ion Reactions

Heavy ion charge exchange reactions at intermediate energies[9]) are just as simple and selective as the (p,n) and (n,p) reactions. Here just one point will be made: the proton vs. nucleus energy shift of the Δ.

Fig. 6 shows a pair of typical spectra from (^{16}O, ^{16}N) and it shows the same shift between proton and nucleus as (^{3}He,t). Figure 7a shows the mass systematics of this for the also (p,n)-like reaction (^{20}Ne, ^{20}F). There is a fixed shift between the proton and all nuclei from ^{12}C to ^{208}Pb. The deuteron in this case is identical with the proton (apart from the addition of the peak from d → 2n). Fig.7b shows the corresponding mass systematics for the (n,p)-like reaction (^{20}Ne, ^{20}Na). The change from the proton to nuclei is still approximately the same 70 MeV. The deuteron Δ position is again identical with the proton. For the first time we here observe a continuous change in energy from ^{12}C to ^{208}Pb. It has been pointed out by Kisslinger[9]) that the shift is close to being just what is expected from the Coulomb energy difference in going from ^{12}C to ^{208}Pb. Why the (p,n) side has a constant shift and the (n,p) side shows the variation is not clear, but the difference in position is consistent with the Coulomb energy difference. This, on the other hand means that whatever the nuclear potential is doing is remarkably constant.

3. The (d,2He) Reaction

In the (d,2p) reaction[11]) the charge exchange transforms the deuteron in its triplet state to the 2-p system in the singlet S-state. It is the restrictions the spectrometer puts on the relative momenta of the two protons that guarantee that they are in the relative $^1S_0$-state. The d → 2p($^1S_0$) transition is what we allow ourselves to call (d,^{2}He). This reaction is a charge exchange reaction with spin transfer. Typical spectra are shown in figure 9.

225

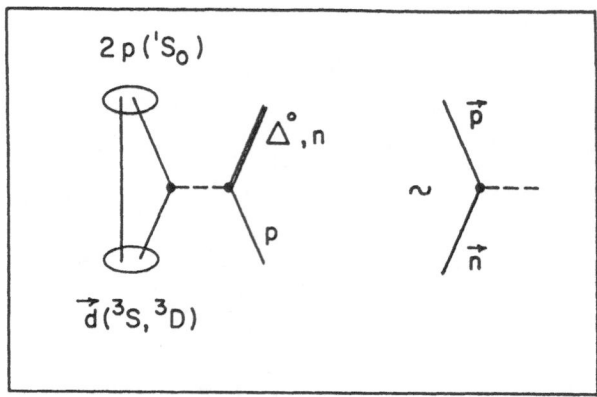

Fig.8. Illustration of the $(d,2p(^1S_0))\equiv(d,^2He)$ reaction. With a polarised deuteron beam it measures spin properties equivalent of those measured in the $(\vec{p},\vec{n})$ where also the outgoing spin must be measured.

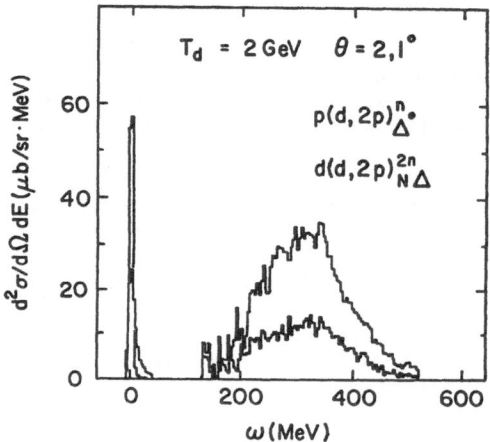

Fig. 9. Spectra of the diprotons from the $(d,2p)$ reaction at 2.1^0 and 2 GeV on the proton and on deuteron.

The spectrum on the proton shows the Δ excitation as in $(^3He,t)$ but now to the Δ^0 state. Here in addition there is the sharp peak at zero excitation energy from the charge exchange to the neutron. Only a small fraction (5-15%) of the total cross section is observed in the spectrograph, and the neutron peak is used to normalize the reaction.

The basic assumption is that at these intermediate energies the impulse approximation is a good approximation. The cross section for the (d,2p) reaction can then be written in terms of the nucleon-nucleon charge exchange cross section and a formfactor for the d → 2p transition. Because of the D-state in the deuteron there are different formfactors for the longitudinal and transvers transitions. In the notation of ref.12 the cross section is written as:

$$\frac{d\sigma}{dt} = \frac{1}{3}[(|\beta|^2+|\epsilon|^2+|\gamma|^2)|S^-(t)|^2+|\delta|^2|s^+(t)|^2]\ast (abs.factor)$$

β and ε are the transvers amplitudes and δ the longitudinal. γ is the spin orbit amplitude.

Here the absorption factor, arising mainly from deuteron breakup, is assumed to be independent of spin and t (see below). The formfactor is calculated from the known deuteron wave function (S and D wave functions in the Paris[13] potential) and the S-state of the 2-p system in a realistic potential (Reid[14] plus Coulomb potential). The absorption factor is included in the normalisation. The independence of t of the absorption factor is illustrated in fig. 9. The peak area of d(d,2p)2n is 0.69 times the peak of p(d,2p)n. In both cases there is one proton that can convert to a neutron. The discrepancy is accounted for by an increase in absorption going from proton target to deuteron target. The expected cross section of the Δ peak is four times that for p → Δ⁰ since both proton and neutron can be excited and an isospin Clebsch-Gordon coefficient gives the n → Δ⁻ a factor of three in cross section. The observed Δ cross section is only about twice the cross section for the proton. The discrepancy is accounted for by the same absorption factor (0.69) in the Δ region. When the reaction is normalised at 2.1⁰ we get the result of figure 10. The calculated curve goes through the data points over more than three decades, documenting the quality of the impulse approximation.

In a similar way we compare the polarization response with that measured in (p,n) → (n,p). The measured quantity is:

$$M = \frac{1}{2} \frac{\sigma(\uparrow\uparrow) + \sigma(\downarrow\downarrow) - 2\sigma(0)}{\sigma(\uparrow\uparrow) + \sigma(\downarrow\downarrow) + \sigma(0)}$$

In terms of the tensor analysing power components $T_{2\mu}$ it may be written:

$$M = P_{20} \frac{1}{2} (T_{20} + \sqrt{6}\ T_{22} \cos 2\varphi)$$

P_{20} (=0.6) is the beam polarization and φ is the angle between the beam polarization axis and the normal to the scattering plane. In the impulse approximation M may be given in terms of the charge exchange spin amplitudes:

$$M \approx P_{20} \frac{1}{\sqrt{2}} \frac{[2(\beta^2+\gamma^2)-\epsilon^2]}{(\beta^2+\gamma^2+\epsilon^2)} \frac{|S^-|^2 - \delta^2|S+|^2}{|S^-|^2 + \delta^2|S+|^2}$$

(under the assumption of cos2φ=1, good for θlab >0⁰).

In figure 11 our measured results for the p(d,2p)n reaction are compared with the results of the nucleon-nucleon work, with the amplitudes of ref.12 extrapolated to 1 GeV. There is perfect agreement out to a momentum transfer of 2 fm⁻¹. After that there is a large deviation. This is where one would expect rescattering in the deuteron to become important.

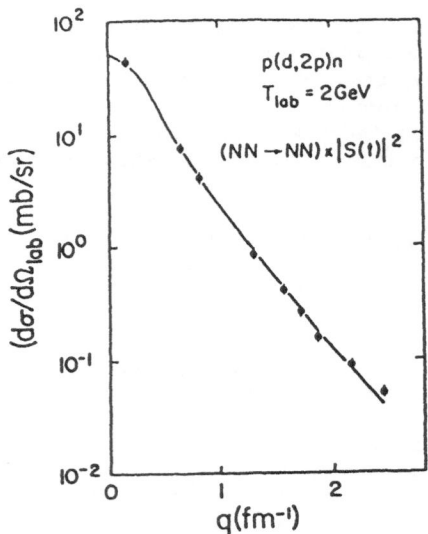

Fig.10. The measured cross section (points) of the p(d,2p)n reaction as a
function of momentum transfer, compared with the cross section
calculated in the impulse approximation (curve). The curve and the
points are normalised at 0.64 fm^{-1} (2.1^0).

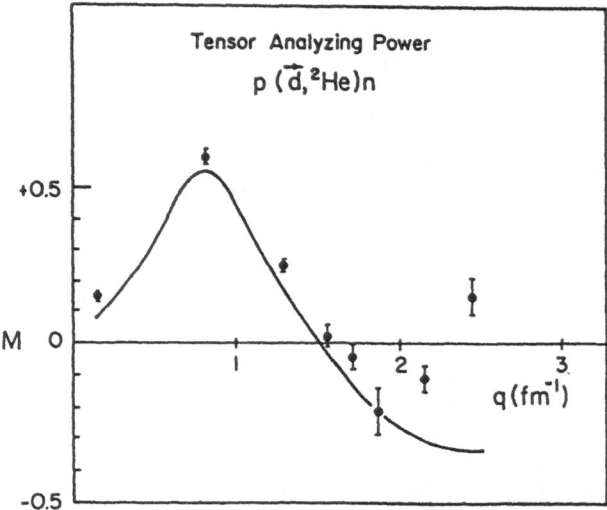

Fig.11. The measured tensor analysing power (points) of the p($\vec{d}$,2p)n reaction as a function of momentum transfer, compared with that calculated from the nucleon-nucleon data (curve).

With the probe tested thus we turn to polarization data for the Δ-
region.   Figure 12 shows the result for 0^0 for the proton and carbon as
targets. The energy position of the points may be identified by looking at
the spectra of figure 9 which has the same energy loss scale (ω). The polarization data for the deuteron target (not shown) are identical with the
proton whereas the carbon data lie significantly higher. On the figure are
given curves for the expectation for a pure longitudinal response, for a
pure transvers response and for mixtures with different ratios of the two.

The proton data in the Δ-region are seen to be close to a ratio of 2:1 in favour of transvers excitation; far from the pure longitudinal response of one-pion exchange. For carbon the ratio is closer to 3:1; even further from expectations. The present expectation is that correlating should build up in the longitudinal channel and enhance this. The contrary is observed.

The calculations refered to in section 1. that reproduce the energy shift assume a predominantly longitudinal driving force. In the light of our polarization data this ought to be changed, with the probable result that the shift arizing from correlations becomes smaller.

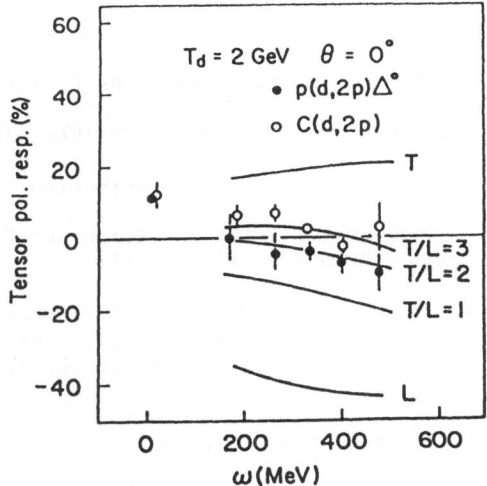

Fig. 12. The tensor analysing power measured at 0⁰ with the $(\vec{d},2p)$ reaction on the proton and on ^{12}C. The curves represent the values predicted for pure longitudinal (L), pure transvers (T) and for the ratios T:L of 1:1, 2:1 and 3:1 respectively.

The problem that the current models do not account for our data is very much related to the same problem in the quasifree region discussed so much at this conference, especially the $(\vec{p},\vec{p}')$. In fact we observe a similar favouring of the transvers channel in the quasifree region with the $(d,^2He)$ reaction as we do in the Δ region.

The results discussed in this paper are obtained in collaboration with M. Bedjidian, D. Contardo, J.Y. Grossiord, A. Guichard, R. Haroutunian and J.R. Pizzi from IPN, Lyon, D. Bachelier, J.L. Boyard, T. Hennino and M. Roy-Stephan from IPN, Orsay, P. Radvanyi and J. Tinsley from LN Saturne, C. Goodman from Indiana, I. Bergqvist, A. Brockstedt and P. Ekstrom from Lund and C. Gaarde, T.S. Jørgensen, J.S. Larsen and B. Million from the Niels Bohr Institute. This work has been supported in part by the Danish Natural Science Research Council.

References

1. I. Bergqvist, A. Brockstedt, L. Carlén, L.P. Ekström, B. Jakobsson, C. Ellegaard, C. Gaarde, J.S. Larsen, C. Goodman, M. Bedjidian, D. Contardo, J.Y. Grossiord, A. Guichard, R. Haroutunian, J.R. Pizzi, D. Bachelier, J.L. Boyard, T. Hennino, J.C. Jourdain, M. Roy-Stephan, M. Boivin and P. Radvanyi, Nucl. Phys. A469 (1987) 648.

2. J.M. Cavedon, B. Frois, D. Goutte, M. Huet, Ph. Leconte, J. Martino, X.H. Phan, S.K. Platchow, S.E. Williamson, W. Breglin, I. Sick, P de Witt-Huberts, L.S. Cardman and C.N. Papanicolas, Phys. Rev. Lett. 49 (1982) 986.

3. C. Ellegaard, C. Gaarde, J.S. Larsen, V. Dmitriev, O. Sushkov, C. Goodman, I. Bergqvist, A. Brockstedt, L. Carlén, P. Ekström, M. Bedjidian, D. Contardo, J.Y. Boyard, T. Hennino, M. Roy-Stephan, M. Boivin and P. Radvanyi, Phys. Lett. 154B (1985) 110.

4. G. Chanfray and M. Ericsson, Phys. Lett. 141B (1984) 163.

5. V.F. Dmitriev and T. Suzuki, Nucl. Phys. A438 (1985) 697.

6. H. Espensen and T.S.H. Lee, Phys. Rev. C32(1985) 1966.

7. J. Delorme and B. Guichon, private communication (1988).

8. T. Udagawa and F. Osterfeld, in "Giant Resonance Excitations in Heavy-Ion Reactions", Padova, Italy (1987).

9. D. Bachelier, J.L. Boyard, T. Hennino, J.C. Jourdain, M.Roy-Stephan, C. Contardo, J.Y. Grossiord, A. Guichard, J.R. Pizzi, P. Radvanyi, J. Tinsley, C. Ellegaard, C. Gaarde and J.S. Larsen. Phys. Lett. 172B, 23 (1986).

10. L.S. Kisslinger, Private communication.

11. C. Ellegaard, C. Gaarde, T.S. Jørgensen, J.S. Larsen, C. Goodman, I. Bergqvist, A. Brockstedt, P. Ekström, M. Bedjidian, D. Contardo, J.Y. Grossiord, A. Guichard, D. Bachelier, J.L. Boyard, T. Hennino, J.C. Jourdain, M. Roy-Stephan, P. Radvanyi and J. Tinsley, Phys. Rev. Lett. 59 (1987).

12. D.V. Bugg and C. Wilkin, Phys. Lett. 152B (1985) 37; C. Wilkin and D.V. Bugg, Phys. Lett. 154B (1985) 243; D.V. Bugg and C. Wilkin, Nucl. Phys. A467 (1987) 575

13. M. Lacombe, B. Loiseau, R. Vinh Mau, J. Côté, P. Pirés, R. de Tourreil, Phys. Lett. 101B (1981) 139.

14. R.V. Reid, Jr. Ann. Phys. (N.Y.) 50 (1968) 411.

SPIN-ISOSPIN RESPONSE OF A NUCLEON

AT HIGH EXCITATION ENERGY*

V. F. Dmitriev

Institute of Nuclear Physics
Novosibirsk

Charge-exchange reactions have been used recently to study excitation of internal degrees of freedom of a nucleon inside nuclei. The simplest example is the excitation of a delta-isobar in nuclei seen in the reactions (^{3}He,t) [1,2] and (d,2p) [3]. An important feature of the latter reaction is that it presents the possibility of using polarized deuterons to study the spin structure of the reaction amplitude.

It was shown in [4] that the delta-isobar excitation in the (^{3}He,t) reaction on a proton is well described in the one pion exchange model. This model reproduces well both the absolute value of the cross-section and the shape of the tritium spectrum. The same is true for the ^{2}He spectrum in the (d,^{2}He) reaction (see Fig. 1).

If the reaction takes place on a nucleus, a virtual pion can propagate inside the nuclear medium, changing the position and the width of the delta [5]. These effects are, however, very sensitive to the distortion of the incoming and outgoing waves. It is therefore difficult to estimate the degree of pion renormalization. In addition, all the estimates were carried out assuming the dominant contribution of pion exchange. This hypothesis must be the subject of a study as well.

Therefore, in this paper, we discuss how the (d,^{2}He) reaction with polarized deuterons can be used to separate the contribution of mechanisms other than pion exchange that may contribute to isobar production.

THE TENSOR ANALYZING POWER OF THE (d,^{2}He) REACTION IN OPE APPROXIMATION

The experiment [3] involved two precisely measured protons in the $^1S_0$ state. According to the authors, the admixture of higher angular momentum states was only a few percent. With this accuracy we shall consider below the reaction as the transformation of a particle with spin 1 and positive parity into a particle with spin 0 and positive parity. A particle with spin 1 can be described by a polarization vector

---

Note: The editors have tried to improve the English of this paper. We apologize for the errors we have introduced.

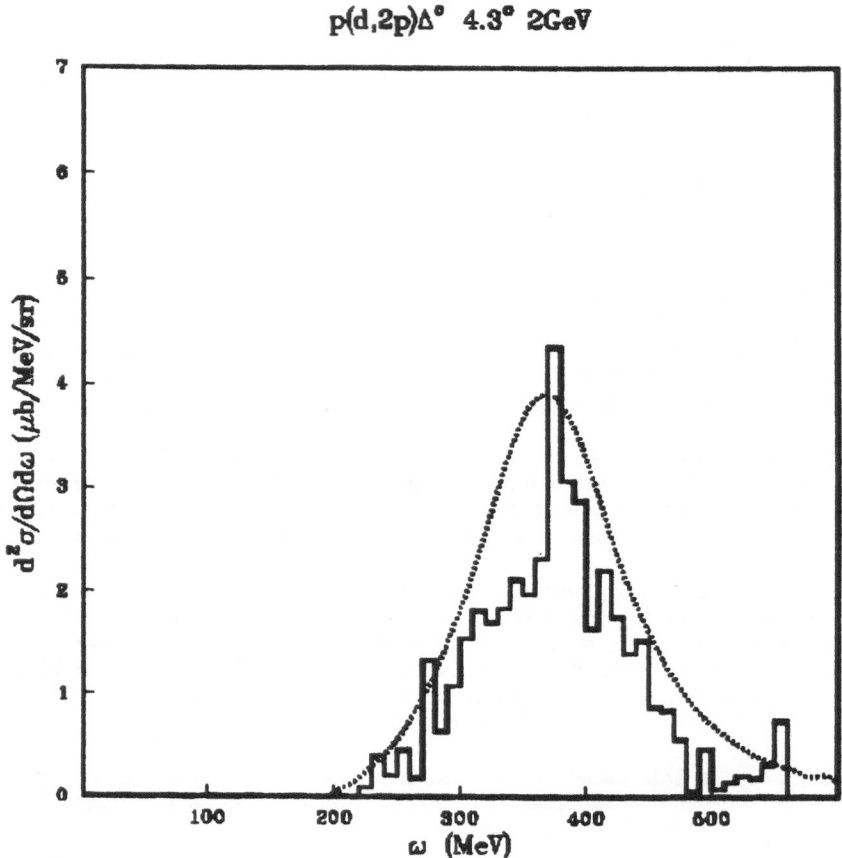

Fig. 1. The ^{2}He spectrum in the Δ-isobar region. *Calculated curve*: OPE.

with the following components:

$$\mathcal{E}_+^\mu = (0, -\tfrac{1}{\sqrt{2}}, \tfrac{i}{\sqrt{2}}, 0)$$
$$\mathcal{E}_-^\mu = (0, \tfrac{1}{\sqrt{2}}, \tfrac{i}{\sqrt{2}}, 0) \tag{1}$$
$$\mathcal{E}_0^\mu = (\tfrac{P}{M}, 0, 0, \tfrac{E}{M}) \;.$$

The polarization vector satisfies the condition $p_\mu \mathcal{E}_\lambda^\mu(p) = 0$. In eq.(1) M is the deuteron mass and p_μ is the deuteron momentum.

The reaction amplitude in one pion exchange approximation may be written

$$T_\lambda = -\sqrt{\tfrac{2}{3}} F(q^2) \frac{f_\pi}{m_\pi} (\mathcal{E}_\lambda^\mu(p) \cdot q_\mu) \frac{1}{q^2 - m_\pi^2} \Gamma_{\pi N \Delta} \tag{2}$$

where $F(q^2)$ is the transition form factor and $\Gamma_{\pi N \Delta}$ is the $\pi N \Delta$-vertex. Due to this factorization, the tensor analyzing power does not depend on the structure of the $\pi N \Delta$-vertex. From the definitions of T_{20} and T_{22} we may write

$$T_{20} = \frac{1}{\sqrt{2}} \cdot \frac{\overline{T_+ T_+^*} + \overline{T_- T_-^*} - 2\,\overline{T_0 T_0^*}}{\sum_\lambda \overline{|T_\lambda|^2}} \;. \tag{3}$$

After a simple calculation we obtain

$$T_{20} = -\sqrt{2} \cdot \left(1 + \frac{3}{2} \frac{q_\perp^2}{q^2(1 - \frac{q^2}{4M^2})} \right) \;; \tag{4}$$

$$T_{22} = \frac{\sqrt{3}}{2} \cdot \frac{q_\perp^2}{q^2 \cdot (1 - \frac{q^2}{4M^2})} \;. \tag{5}$$

For

$$A_y = \tfrac{1}{2} T_{20} + \sqrt{\tfrac{3}{2}}\, T_{22} \cos 2\phi$$

we find

$$A_y = -\frac{1}{\sqrt{2}} + \frac{3}{2\sqrt{2}} (1 - \cos 2\phi) \frac{q_\perp^2}{-q^2(1 - \frac{q^2}{4M^2})} \;; \tag{6}$$

where ϕ is the angle between the spin quantization axes and the normal to the scattering plane.

The polarization response (6) does not depend on the target and must be the same for a proton and for carbon. In Figs. 2 and 3 the measured response for these targets is shown taking into account the polarization of the deuteron beam, $\rho_{20} = 0.61$ [6]. One can see the difference between the responses for a carbon and proton target. Note that for the proton the response is approximately one half the prediction of the OPE model.

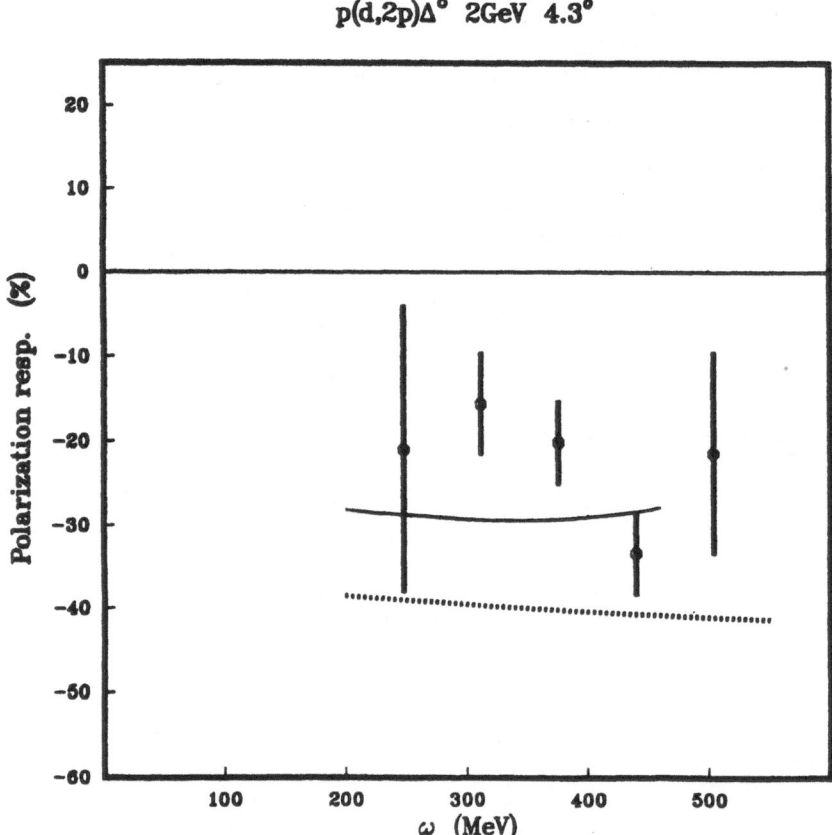

Fig. 2. Polarization response in reaction $p(d, {}^2\mathrm{He})$ at scattering angle $\theta = 4.3°$, $\cos 2\phi = 0.909$. *Solid line*: $\pi + \rho$ model; *dotted line*: OPE.

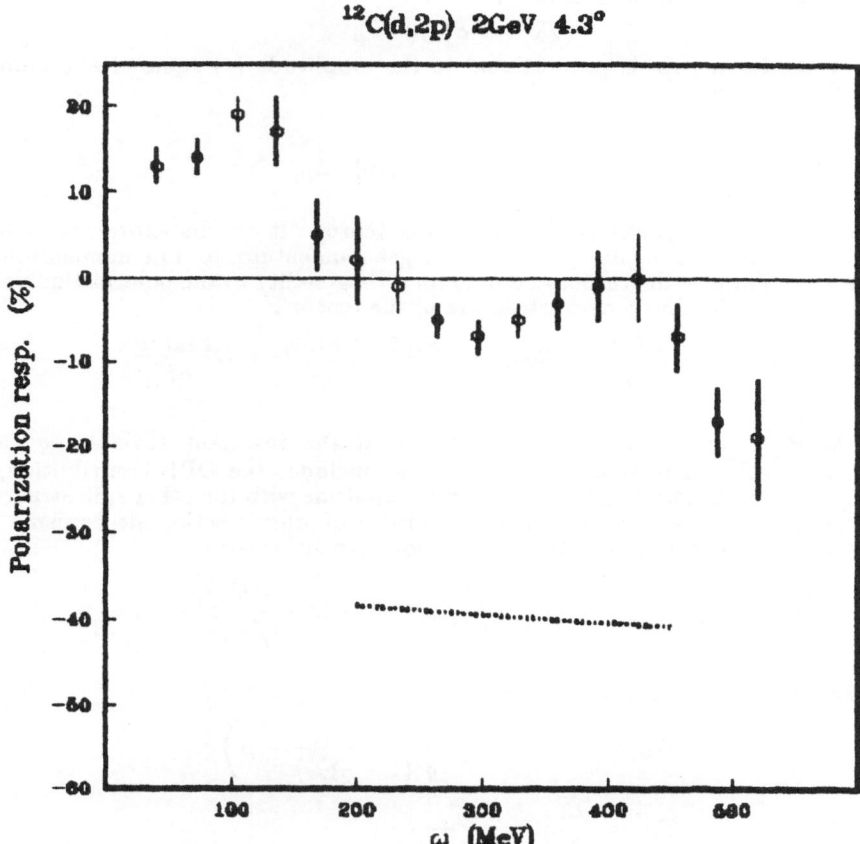

Fig. 3. Polarization response in reaction $^{12}C(d,\ ^2He)$ at scattering angle $\theta = 4.3°$, $\cos 2\phi = 0.909$.

The large difference between the proton and carbon responses indicates the importance of distortion effects. In this talk we shall not discuss these effects but shall try to analyze the elementary amplitude of the reaction on a proton.

The deviation of the measured proton response from the OPE predictions indicates the presence of transverse spin components in the amplitude which are absent in the OPE contribution. To estimate the contribution of these components let us discuss the general structure of the polarization response for the spin transition $1^+ \rightarrow 0^+$.

The amplitude of the reaction is given by

$$T_\lambda = -\mathcal{E}_\lambda^\mu(p) \cdot \Gamma_\mu \ . \tag{7}$$

The spin density matrix is proportional to the amplitude (7) squared and summed over all final states.

$$\overline{T_\lambda \cdot T_{\lambda 1}^*} = \mathcal{E}_\lambda^\mu(p)\mathcal{E}_{\lambda 1}^\nu(p) \cdot \Lambda_{\mu\nu} \ . \tag{8}$$

The tensor $\Lambda_{\mu\nu}$ is a symmetrical second rank tensor. It can be expressed in terms of two independent momenta, q_μ and the target momentum, p. The momentum, p_μ, does not contribute to the response due to its orthogonality to the polarization vector, i.e., $p_\mu \cdot \mathcal{E}_\lambda^\mu(p) = 0$. The general structure of the tensor is

$$\Lambda_{\mu\nu} = -A\frac{q_\mu q_\nu}{q^2} - Bg_{\mu\nu} + C\frac{p_{1\mu}q_\nu + p_{1\nu}q_\mu}{M\sqrt{-q^2}} + D\frac{p_{1\mu}p_{1\nu}}{M^2} \ . \tag{9}$$

where A, B, C and D are scalar functions of the invariant variables q^2 and $S = (p_\mu + p_{1\mu})^2$. The structure function, A, includes the OPE contribution; the function C describes interference of the OPE amplitude with the other spin structures. The functions B and D correspond to contributions of other reaction mechanisms. The cross-section is proportional to the trace of spin density matrix

$$\sigma \sim (\frac{p_\mu p_\nu}{M^2} - g_{\mu\nu})\Lambda_{\mu\nu} = A(1 - \frac{q^2}{4M^2}) + 3B - C\frac{w - E\frac{q^2}{4M^2}}{\sqrt{-q^2}} + D\frac{\vec{p}^2}{M^2} \ . \tag{10}$$

For the tensor analyzing powers we obtain

$$T_{20} = -\sqrt{2}\left(1 + \tfrac{3}{2}\alpha\frac{q_\perp^2}{q^2(1 - \frac{q^2}{4M^2})} - \beta\right) \ ; \tag{11}$$

$$T_{22} = \frac{\sqrt{3}}{2}\alpha\frac{q_\perp^2}{q^2(1 - \frac{q^2}{4M^2})} \ ; \tag{12}$$

where $\alpha = A \cdot (1 - q^2/4M^2)/\sigma$ and $\beta = 3B/\sigma$ are the relative contributions of the corresponding structure functions to the cross-section. For pure OPE, $\alpha = 1$ and $\beta = 0$. For the polarization response we have:

$$p = \rho_{20}\left(-\frac{1}{\sqrt{2}} + \frac{3}{2\sqrt{2}}(1 - \cos 2\phi)\alpha \cdot \frac{q_\perp^2}{-q^2(1 - \frac{q^2}{4M^2})} + \frac{1}{\sqrt{2}}\beta\right) \ . \tag{13}$$

From (13) one can estimate the contribution of the non-OPE part of the cross-section. From fig. 2 we see that for the Δ-isobar region $p \approx -0.2$. Taking into account the value of $< \cos 2\phi > = 0.9 = 4.3°$, we obtain $\beta = 0.5$. Therefore, the results of the measurements [3] yield contributions of non-OPE structures of the same order as the OPE.

THE MODEL OF $\pi + \rho$ EXCHANGE

As a possible generalization of the OPE model let us discuss the contribution of two pion exchange. Since the main contribution in the two pion channel comes from the rho-resonance, we shall restrict ourselves to rho-meson exchange. Similar to what has been done for the OPE amplitude (2), it is convenient to introduce a relativistic vertex for ρd ^{2}He-interaction. The only combination satisfying all conditions of invariance and yielding, at small deuteron momentum, the matrix element of $\vec{\sigma} \times \vec{q}$ calculated with wave functions of the d and ^{2}He ($\vec{\sigma}$ is the nucleon spin matrix) is

$$V_\lambda = \sqrt{2}\frac{f_\rho}{m_\rho}e_{\mu\nu\gamma\sigma}\mathcal{E}^\nu_\lambda(p)q_\gamma\frac{p_\sigma}{M}F(q^2) \tag{14}$$

where M is the deuteron mass and $F(q^2)$ the same transition form factor as in (2). The Δ-isobar created in the laboratory system is nonrelativistic, therefore for the vertices $\pi N\Delta$ and $\rho N\Delta$ we can use nonrelativistic expressions. Let $p_{1\mu} = (m, 0)$ be the momentum of target, then the $\rho N\Delta$ vertex can be written as follows:

$$\Gamma^\mu_{\rho N\Delta} = \sqrt{\frac{1}{3}}\frac{f_{\rho N\Delta}}{m_\rho}e_{\mu\nu\gamma\sigma}S_\nu q_\gamma\frac{p_{1\sigma}}{m} \tag{15}$$

Since the only nonzero component of $p_{1\sigma}$ is p_{10}, $\Gamma^\mu_{\rho N\Delta}$ eventually coincides with its nonrelativistic expression. In (15) S_ν is the usual matrix of the $N\Delta$-transition. For the $\pi N\Delta$ vertex we obtain $\Gamma_{\pi N\Delta} = \sqrt{\frac{1}{3}}\frac{f_{\pi N\Delta}}{m_\pi}(\vec{S} \cdot \vec{q})$.

Combining (14), (15) and (2) and calculating the product of two antisymmetric tensors we find

$$T_\lambda = -\sqrt{\frac{2}{3}}\,F(q^2)(\mathcal{E}^\mu_\lambda \cdot q_\mu) \cdot \left[\frac{f_\pi f_{\pi N\Delta}}{m^2_\pi}\frac{(\vec{S} \cdot \vec{q})}{q^2 - m^2_\pi} - \right.$$
$$\left. \frac{f_\rho f_{\rho N\Delta}}{m^2_\rho}\frac{\frac{E}{M}(\vec{S} \cdot \vec{q}) - \frac{W}{M}(\vec{S} \cdot \vec{p})}{q^2 - m^2_\rho}\right] + (\mathcal{E}^\mu_\lambda \cdot \tilde{S}_\mu)\frac{q^2(E + E')}{2M} \tag{16}$$
$$\cdot \frac{f_\rho f_{\rho N\Delta}}{m^2_\rho}\frac{1}{q^2 - m^2_\rho}\sqrt{\frac{2}{3}}F(q^2) \; ;$$

here p' is the momentum of the ^{2}He and the vector $\tilde{S}_\mu$ is

$$\{\tilde{S}_0, \vec{\tilde{S}}\} = \left\{\frac{(\vec{p} + \vec{p}') \cdot \vec{S}}{E + E'}, \vec{S}\right\} \; .$$

From eq.(16) we see that, as in the nonrelativistic case, the rho-meson contribution partially cancels the π-meson contribution and creates another spin structure in the reaction amplitude.

The results for the polarization response, in this model, are shown as a solid line in Fig. 2 and Fig. 4. From these figures we see that the response does not change appreciably. The contribution of the additional spin structure decreases with

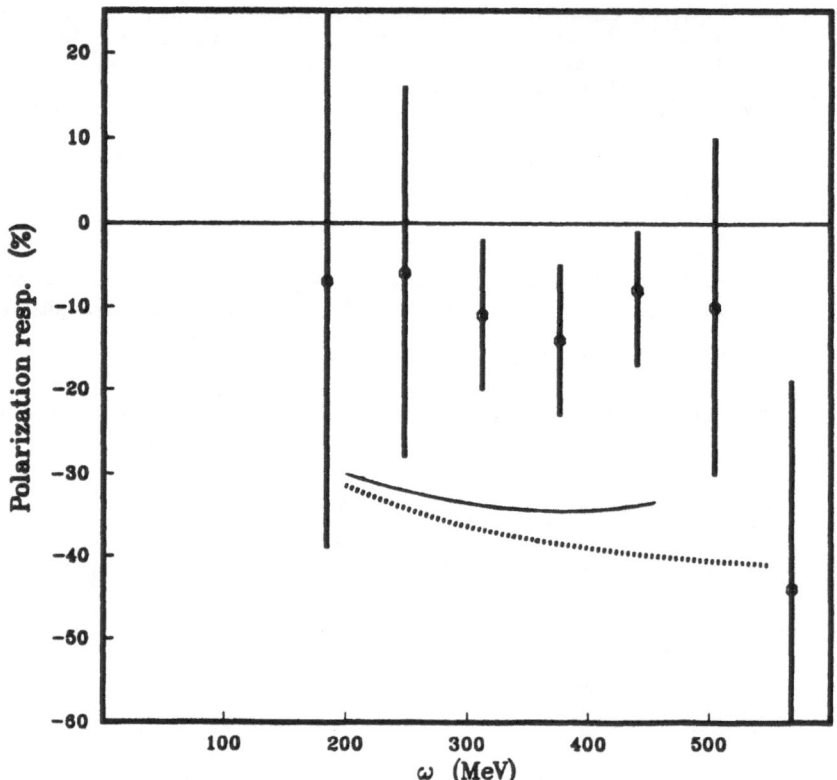

p(d,2p)Δ° 2GeV 2.14°

Fig. 4. Polarization response in reaction $p(d, {}^2\text{He})$ at full scattering angle $\theta = 2.14°$, $\cos 2\phi = 0.68$. *Solid line:* $\pi + \rho$ model; *dotted line:* OPE.

decreasing scattering angle. The response becomes similar to that predicated by the OPE model. The experimental polarization response decreases at a small angle.

Strictly speaking, the model of rho-exchange discussed above is unsatisfactory from the theoretical point of view. The contribution of rho-exchange in (16) rises with the energy of the deuteron. Such a behavior is common for exchange by any particle with nonzero spin. Thus, the Born approximation must be modified in order to satisfy the unitarity condition. At high energy the standard modification is reggeziation which changes the dependence on S. Analysis allows one to find the dominant trajectory.

The existing data (7) for the reaction leading to delta production indicates the cross-section decreases with a 1/S dependence at least up to 20 GeV. This indicates the dominance of pion exchange since the rho-trajectory yields a decrease of the cross-section dropping off more slowly than 1/S.

Summarizing the discussion we can say that the simple picture of meson exchange does not describe the polarization data for delta-isobar production. The strong target dependence of the response may indicate the importance of absorptive effects.

I would like to acknowledge C. Gaarde and T. S. Jorgensen for presentation of the data and helpful discussions. This work was started during my visit to NBI and I would like to express my gratitude to the Institute for its warm hospitality.

REFERENCES

1. V. G. Ableev, G. G. Vorob'ev, S. M. Eliseev et al., JETP Letters 40 (1984) 763.

2. D. Contardo et al., Phys. Lett. 168B (1986) 331.

3. C. Ellegaard et al., Phys. Rev. Lett. 59 (1987) 974.

4. C. Gaarde, V. F. Dmitriev, O. P. Sushkov, Nucl. Phys. A459 (1986) 503.

5. V. F. Dmitriev, Preprint INP 86-118, Novosibirsk.

6. C. Gaarde, private communication.

7. V. Flaminio et al., CERN-HERA 84-01 (1984).

MOMENTUM DISTRIBUTIONS OF FRAGMENTS OF D, ^{3}He AND ^{4}He NUCLEI AND T_{20} IN 0° DEUTERON BREAK-UP

V.G. Ableev, B. Naumann, L. Naumann, A.A. Nomofilov,
L. Penchev, N.M. Piskunov, V.I. Sharov, I.M. Sitnik,
E.A. Strokovsky, L.N. Strunov, and S.A. Zaporozhets

Joint Institute for Nuclear Research, Dubna, USSR

A.P. Kobushkin

ITP, Ukr. Academy of Science, Kiev, USSR

Kh. Dimitrov

CLANP, Bulg. Academy of Science, Sofia, Bulgaria

L. Vizireva

HCTI, Sofia, Bulgaria

B. Kuhn and W. Neubert

ZfK, Rossendord, GDR

To investigate in detail the fragmentation of relativistic light nuclei as well as their internal structures at small distances, we have performed a series of measurements of the deuteron fragmentation on hydrogen [1], ^{3}He and ^{4}He nuclei on carbon [1] and also tensor polarized deuterons on carbon [2].

0° invariant cross sections of the ^{12}C(^{4}He,t), ^{12}C(^{4}He,p), ^{12}C(^{3}He,d), ^{12}C(^{3}He,p) and p(d,p) reactions are shown (Fig.1) depending on the relativistic internal motion momentum K of the nuclear fragment. The variable K naturally arises when the fragmentation process is described by analogy with the parton model: before the interaction the projectile nucleus (pr) dissociates virtually into a fragment-participant (p) and a detected fragment-spectator (s). At a zero emission angle the variable K is related to the detected fragment momentum as follows [3]:

$$K^2 = \lambda (M^2_{eff} , m^2_s , m^2_p) / 4M^2_{eff},$$

$$\lambda (a, b, c) = a^2 + b^2 + c^2 - 2ab - 2ac - 2bc ,$$

$$M^2_{eff} = (m^2_s (1 - \alpha) + m^2_p)/(\alpha (1 - \alpha)) ,$$

where $\alpha = (E_s + P_s)/(E_{pr} + P_{pr})$. In the infinite momentum frame, α is a fraction of the projectile momentum carried away by the fragment spectator. At large energies α coincides with the Feynman variable x.

The shapes of the K-spectra of the fragments are very similar throughout almost the whole measured region of K for the complementary $^{12}C(^4He,t)$ and $^{12}C(^4He,p)$ reactions as well as for the $^{12}C(^3He,d)$ and $^{12}C(^3He,p)$ reactions. At K<250 MeV/c there is a good agreement between the data and impulse approximation calculations using the ^4He, ^3He and d wave functions for the Reid-soft-core NN potential as well as the Paris NN potential (for the deuteron wave function). But at large K the theory underestimates the data. A maximum discrepancy occurs at K ≈ 350 MeV/c in all cases. It reaches a factor of about 2 in the deuteron fragmentation and about 4 in the ^3He or ^4He fragmentations.

In Fig.2 we present the K-dependence of the nucleon momentum distribution in the deuteron (NMDD) obtained from our data on (d,p) fragmentation the d(e,e'p)n data [4] and from the analysis [5] of the d(e,e') data. One can observe a good consistency between the NMDDs, extracted from various processes, and their disagreement with calculations using the Paris wave function of the deuteron. Using our measured NMDD, the energy dependence of the elastic backward dp scattering can be quantitatively described over the whole investigated range of energies[6] in the framework of a simple one-nucleon-exchange model.

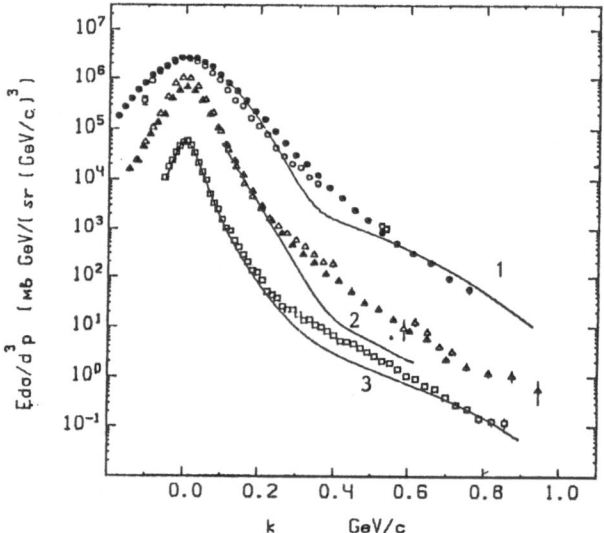

Fig.1. Invariant cross sections of the fragmentation reactions as a function of the internal momentum K: $\circ$ - $^{12}C(^4He,p)*100$ at $P_{^4He}$ = 18.1 GeV/c ;
$\bullet$ - $^{12}C(^4He,t)*100$, $\blacktriangle$ - $^{12}C(^3He,d)*10$,
$\triangle$ - $^{12}C(^3He,p)*10$ at $P_{^4He,^3He}$ = 10.8 GeV/c ;
$\square$ - p(d,p) at P_d = 9.1 GeV/c . The curves correspond to: 1,2 - impulse approximation with wave functions for the Reid-soft-core NN potential, 3 - impulse approximation with the Paris wave function.

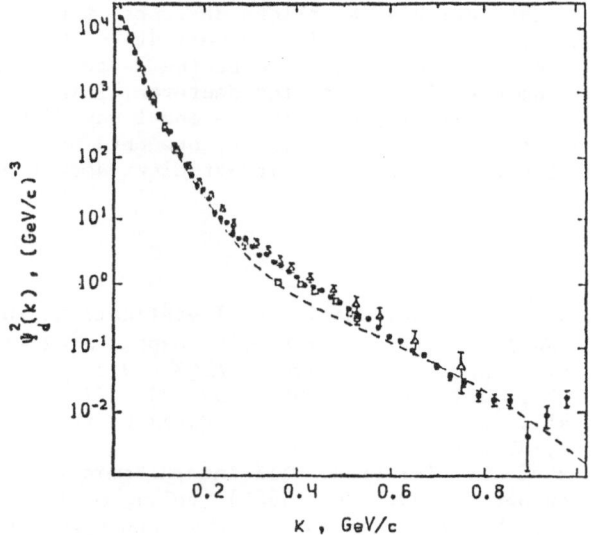

Fig. 2. Nucleon momentum distribution in the deuteron extracted
from the reactions: ● - p(d,p), Δ- d(e,e'), □ - d(e,e'p)n.
The dashed line presents the Paris wave function versus K.

Figure 3 presents our new data on tensor analyzing power T_{20} of the $^{12}C(d,p)$ zero angle fragmentation at 9.1 GeV/c. As for the differential cross sections, the calculation with the Paris deuteron wave function (solid line) does not yield a good description of the data at q > 200 MeV/c.

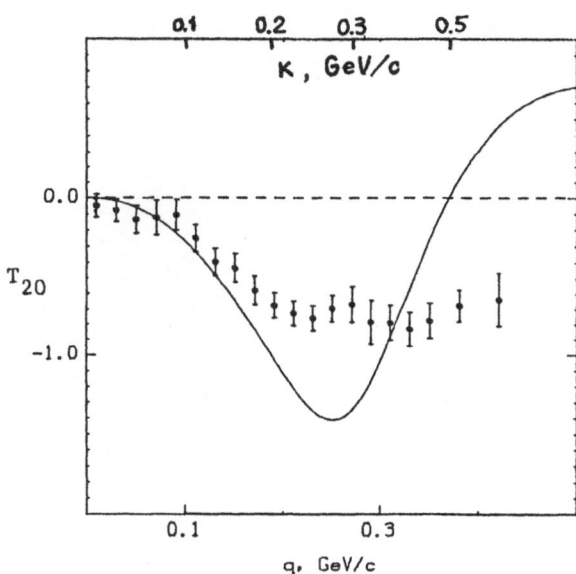

Fig. 3. Tensor analyzing power T_{20} of the (d,p) fragmentation
as a function of the fragment momentum q taken in the
deuteron rest frame.

Thus, the use of the variable K allows us to see that in the fragmentation reactions, deuteron disintegration as well as in the elastic backward dp scattering just the same characteristic of the deuteron structure, namely the nucleon momentum distribution in the deuteron, is measured. Comparing the present data and calculation with the model wave functions of light nuclei based on the realistic NN potentials, one obtains a strong indication that such wave functions contain an insufficient amount of their high momentum component.

References

1. V.G. Ableev et al., in: Proc. Meeting on Investigations in the Field of Relativistic Nuclear Physics, D2-82-568, Dubna, JINR (1982) p.130.
   V.G. Ableev et al., Nucl.Phys., (1983), A393, 491; A411(541E).
   V.G. Ableev et al., JETP Lett., (1983), 37, 233; (1987), 45, 596.
   V.G. Ableev et al., in: Proc. Fewbody and Quark-Hadronic Systems, D4-87-692, Dubna, JINR (1987), p.140.
2. S.A. Zaporozhets et al., in: Proc. VIII Intern.Seminar on High-Energy Physics, D1, 2-86-668, Dubna, JINR (1986), vol.1, p.341.
   B. Naumann et al., in: Proc. VII Intern. Symp. on High Energy Spin Phys., Protvino, (1986), vol.2, p.111.
3. V.B. Berestetsky and M.V. Terentyev, Yad.Fiz. (1976), 24, p.1044.
4. S. Turck-Chieze et al., Phys.Lett., (1984), B142, p.145.
5. P. Bosted et al., Phys.Rev.Lett., (1982), 49, p.1380.
6. A.P. Kobushkin, J.Phys.G: Nucl.Phys., (1986), 12, p.487.

RATIO OF GAMOW-TELLER TO FERMI STRENGTH OBSERVED IN ^{13,14}C(p,n) AT 492 AND 590 MeV

J.L. Ullmann, J. Rapaport, P.W. Lisowski, R.C. Byrd, T. Carey, T.N. Taddeucci, J. McClelland, L. Rybarcyk, R.C. Haight, N.S.P. King, and G.L. Morgan

Los Alamos National Laboratory
Los Alamos, NM 87545

D.A. Lind, R. Smythe, C.D. Zafiratos, and D. Prout

University of Colorado
Boulder, CO 80309

E. Sugarbaker

The Ohio State University
Colombus, OH 43212

W.P. Alford

TRIUMF
Vancouver, B.C., Canada V6T 2A3

W.G. Love

University of Georgia
Athens, GA 55455

INTRODUCTION

It has been recognized for a number of years that certain spin-isospin components of the nucleon-nucleus effective interaction can be inferred from (p,n) reactions to states of known nuclear structure. For L = 0, S = 0 and L = 0, S = 1 transitions, the 0-degree (p,n) cross section can be related respectively to Fermi and Gamow-Teller beta decay matrix elements[1]. If these transitions occur in the same nucleus, the ratio of isovector spin-flip to non-spin-flip effective interactions can be measured without regard for absolute normalization. The best reaction to measure this is ^{14}C(p,n) which goes by a pure Gamow-Teller transition to the 1$^+$ state at 3.95 MeV in ^{14}N, and Fermi transition to the 2.31 MeV 0$^+$ state. This work extends the ratio measurements made at lower energies (ref. 1, 2, 3) to 492 and 590 MeV.

We also report on the ^{13}C(p,n) reaction which goes by a pure GT transition to the 3.51 MeV 3/2$^-$ state in ^{13}N, but by a mixed Fermi plus Gamow-Teller transition to the 1/2$^-$ ground state.

EXPERIMENTAL METHOD

The measurements were made on the 240 m zero-degree flight path of the WNR Target-Two facility at LAMPF. Neutrons were detected in a 25 cm × 50 cm × 7.5 cm thick plastic scintillator coupled at both ends to a RCA C31024 phototube, as was used in previous measurements. (ref. 4). The intrinsic time resolution of this system is about 300 ps FWHM. For beam energies less than 800 MeV, the last sections of the LAMPF linac are turned off and normally act as a passive drift section. Over this drift space, the normal beam energy spread ($\Delta E/E \approx 0.1\%$) slews into a time spread that is much larger than the intrinsic resolution of the detector. This width seriously limited previous measurements at energies below 800 MeV. The present measurements were made by tuning one of the unused linac cavities to adjust the phase space of the beam to produce a time focus at the detector position. (See ref. 5.) With this technique, beam pulse widths of 300 ps were observed at the target position. The effect of this bunching scheme is shown in Figure 1.

The ^{14}C target was made from amorphous carbon enriched to 89% in ^{14}C and encased in a 0.005 cm thick nickel cell. The ^{14}C thickness was 170 mg/cm^2. This is the same target as was used in the TRIUMF measurements. (ref. 3) Figure 2 shows the time spectrum for $^{14}C(p,n)$ at 492 MeV. A spectrum at 590 MeV was not obtained because of experimental difficulties. The dominant feature of the spectrum is the 1^+ Gamow-Teller state at 3.95 MeV excitation. The IAS at 2.31 MeV appears as a small shoulder on the larger GT peak. Neutron yields were obtained by fitting an asymmetric Gaussian line shape to the data. The peak-shape parameters were obtained from isolated states in ^{13}C. Special care was taken to insure that the yields obtained for the IAS were not sensitive to reasonable small changes in the fitting parameters or procedure.

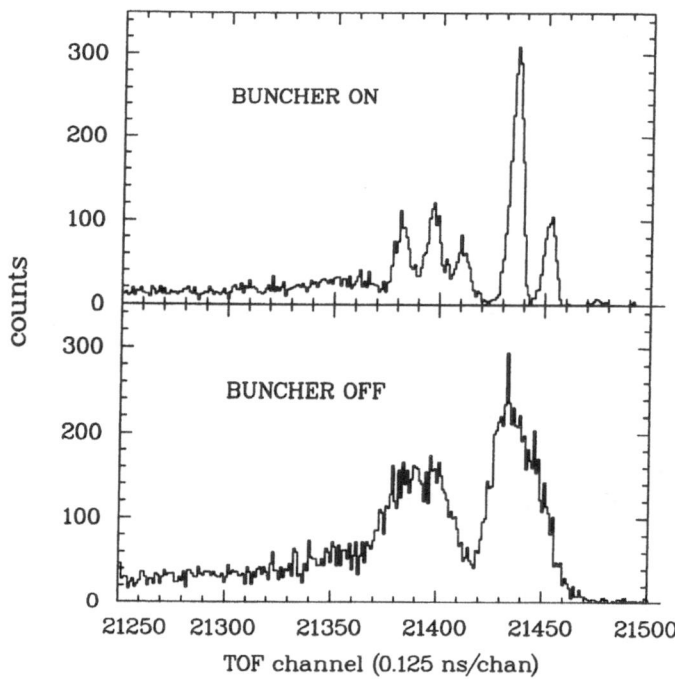

Fig. 1.   Buncher on vs. buncher off for $^{13}C(p,n)$ at 492 MeV.

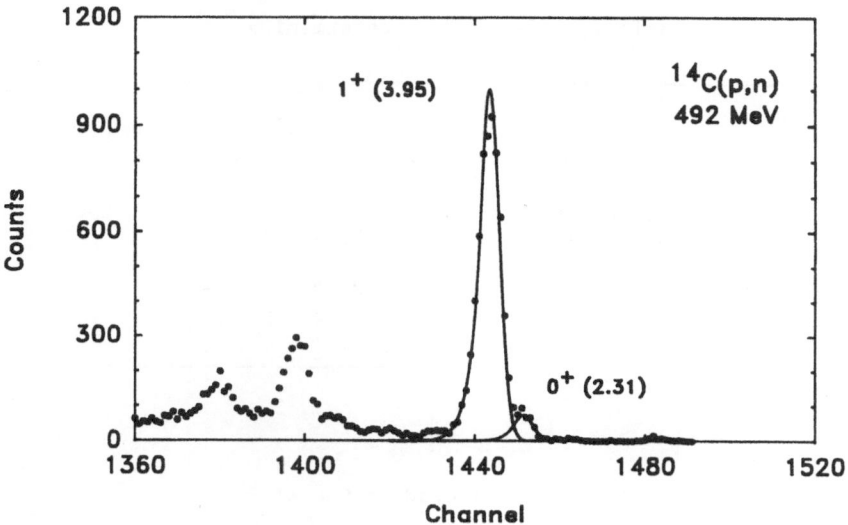

Fig. 2.   Zero-degree time spectrum of ^{14}C(p,n) at 492 MeV (125 ps/chan). The curve fitted to the 1^+ GT state and the 0^+ IAS is shown.

The ^{13}C target was a 209 mg/cm^2 self-supporting foil consisting of ^{13}C enriched to 99% mixed with 15% binder enriched to 90%. Figure 3 shows the spectra obtained at both energies. The $1/2^-$ gs and $3/2^-$ state at 3.51 MeV are well resolved. Neutron yields were determined by simply summing counts.

RESULTS

Taddeucci (ref. 1) has given an expression for the cross section of a given type α (α = GT or F):

$$\sigma(q,\omega) = \hat{\sigma}(E,A)\ F_\alpha(q,\omega)\ B(\alpha) \tag{1}$$

where $\hat{\sigma}$ is the "unit cross section" and depends on mass and bombarding energy, B is the beta-decay transition strength that contains the nuclear structure information, and F_α is a factor that describes the shape of the angular distribution [$F_\alpha(q = 0, \omega = 0) = 1$].

The unit cross section is the factor of primary interest, and may be written approximately as

$$\hat{\sigma}_{GT} = K(E_p,0)N^D_{\sigma\tau}\ |J_{\sigma\tau}|^2$$

$$\hat{\sigma}_F = K(E_p,0)N^D_{\tau}|J_{\tau}|^2 \tag{2}$$

where K is a kinematic factor, N^D is the distortion factor, and $J_{\sigma\tau}$ and J_{τ} are the volume integrals of the spin-flip ($V_{\sigma\tau}$) or non-spin-flip (V_{τ}) isovector central interaction. We note that

$$\hat{\sigma}_{GT} = \sigma_{GT}(q = 0, \omega = 0)/B(GT) \quad \text{and} \quad \hat{\sigma}_F = \sigma_F(q = 0, \omega = 0)/B(F). \tag{3}$$

247

Table 1. B's used in determining $\hat{\sigma}$
(from ref. 1)

			B(GT)	B(F)
$^{14}C(p,n)^{13}N$ 2.31 MeV	(0+)		-	2.0
3.95	(1+)		2.76	-
$^{13}C(p,n)^{13}N$ gs	(1/2-)		0.20	1.0
3.51	(3/2-)		0.83	-

Table 2. Measured Values of R^2

^{14}C	492 MeV	9.4 ± 0.7
^{13}C	492 MeV	13.8 ± 2.3
	590 MeV	11.4 ± 0.7

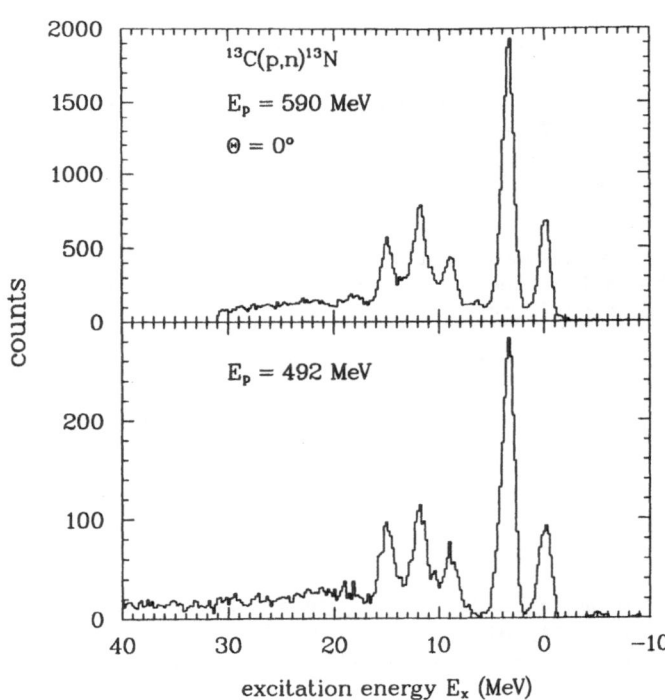

Fig. 3. Zero-degree energy spectra for $^{13}C(p,n)$ at 590 and 492 MeV.

The ratio of these unit cross sections is defined as R^2,

$$R^2 = \hat{\sigma}_{GT}/\hat{\sigma}_F \qquad (4)$$

which, in the factorized approximation, may be written as

$$R^2 = \frac{|J_{\sigma\tau}|^2}{|J_\tau|^2} \frac{N^D_{\sigma\tau}}{N^D_\tau} . \qquad (5)$$

We elect to compare values of R^2 to calculated values. R^2 can be determined directly from the data, while eq. (5) involves an approximation. We prefer to lump the uncertainty in calculating the distortion factors and other reaction details into the theoretical calculations and report numbers more closely related to experiment.

For the pure transitions in $^{14}C(p,n)$, R^2 can be obtained directly from eq. 3. However, for the mixed transitions in ^{13}C, we need to use the incoherent sum of the cross sections, and obtain

$$R^2 = \frac{B_1(F)}{(\sigma_1/\sigma_2)B_2(GT) - B_1(GT)} \qquad (6)$$

The values of the beta decay transition strengths we used are those suggested by Taddeucci (ref. 1), and are obtained from β decay for all transitions except $^{13}C(p,n)^{13}N(3.51)$. The value of the latter was determined by a comparison of (p,n) data to the ground state or 15.1 MeV transitions. The values of the B's are shown in Table 1. Table 2 shows the values of R^2 that we extracted from the data.

The measured values of R^2 are compared to calculations for $^{14}C(p,n)$ in figure 4. The data points at 200 MeV and below are from ref. 1 and 2, and those between 200 and 450 from ref. 3. The calculations are described in detail in a recent paper (ref. 7). Briefly, the calculations are non-relativistic DWIA and include direct and exchange terms explicitly with central, spin-orbit, and tensor parts for each interaction. Results using two nucleon-nucleon interactions are shown. The first used a free t-matrix interaction based on Arndt's SP84 phase shifts[8]. The second is an unpublished[9] density-dependent G-matrix interaction (HM86) based on the Bonn potential[10]. The G-matrix calculation was extended to 500 MeV using the 425 MeV g-matrix. This is not, of course, well justified, but used for comparison.

The density-dependent G-matrix interaction was used to include the relatively strong medium effects on the short range processes that contribute to V_τ. The results reproduce the data well up to 300 MeV, but the trend of the data above that seems to follow more the trend of the t-matrix calculations. We note, however, that the Bonn potential used in the G-matrix calculation was only fit to the N-N data up to 300 MeV, and that above 300 MeV even the zero-density G-matrix calculations differ significantly from the t-matrix calculations[6]. A similar good reproduction of the lower energy ratios was obtained by Horowitz[11] in a relativistic calculation for the ratio $|J_{\sigma\tau}/J_\tau|^2$ that included only Pauli blocking as a medium effect. Cross section calculations were not presented, however, so a direct comparison to our data cannot be made.

Also shown in Figure 4 are our R^2 values for ^{13}C. We observe that the ratio is somewhat different in value from that measured for ^{14}C, but follows a similar trend with energy.

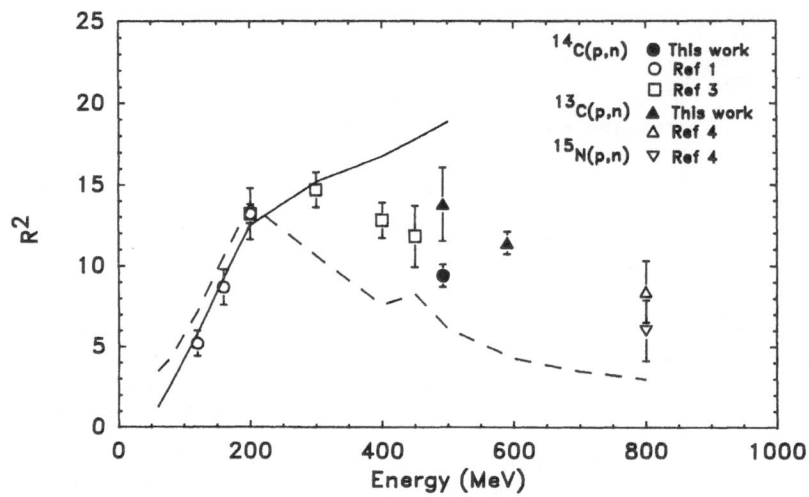

Fig. 4.   Values for the ratio $R^2 = \hat{\sigma}_{GT}/\hat{\sigma}_F$ measured in the $^{14}C(p,n)$
reaction from 100 to 600 MeV. Data from $^{13}C(p,n)$ at 492 and
590 MeV, and $^{13}C(p,n)$ and $^{15}N(p,n)$ at 800 MeV are also shown.
The solid line is the result of a G-matrix calculation using
the Bonn potential, and the dashed line is a calculation
using a t-matrix interaction.

REFERENCES

1.   T.N. Taddeucci, et al., Nucl. Phys. A469, 125 (1987).
2.   J. Rapaport, et al. Phys. Rev. C 24, 335 (1981).
3.   W.P. Alford, et al., Phys. Lett. B179, 20, 1986.
4.   N.S.P. King, et al., Phys. Lett. B175, 279 (1986), R.G. Jeppesen,
        "Observation of Gamow-Teller and Fermi Strength in light nuclei
        using the 800 MeV (p,n) reaction." PhD Dissertation, University
        of Colorado, 1986 (unpublished).
5.   R. Smythe, et al., to be published.
6.   W.G. Love, K. Nakayama, and M.A. Franey, Phys. Rev. Lett. 59, 1401
        (1987).
7.   M.A. Franey and W.G. Love, Phys. Rev. C 31, 488 (1985).
8.   R.A. Arndt, et al., Phys. Rev. D 28, 97, (1983).
9.   K. Nakayama and W.G. Love, unpublished.
10.   R. Machleidt, K. Holinda, and C. Elster, Phys. Rep. 149, 1 (1987).
11.   C. J. Horowitz, to be published.

SPIN-ISOSPIN NUCLEAR RESPONSES WITH HADRONIC PROBES

W. M. Alberico

Dipartimento di Fisica Teorica
Università di Torino
and Istituto Nazionale di Fisica Nucleare
Torino, Italy

INTRODUCTION

Since many years, it has been emphasized the interest of the spin-isospin nuclear response in the quasi-elastic peak region, due to the possible onset of collective effects. The latter were found to be quite sizable within an RPA treatment of infinite nuclear matter [1]. In this framework (and in the so-called *ring* approximation) the spin-longitudinal (transverse) response is proportional to the imaginary part of the following particle-hole (ph) polarization propagator

$$R_{L(T)}(q,\omega) \propto -\,\mathrm{Im}\,\frac{\Pi^0(q,\omega)}{1 - V_{L(T)}(q,\omega)\Pi^0(q,\omega)}, \tag{1}$$

Π^0 being the sum of the nucleon-hole and Δ-hole free propagators.

The spin-isospin longitudinal and transverse excitation operators are

$$O_L = \vec{\sigma} \cdot \hat{q}\tau_\lambda e^{i\vec{q}\cdot\vec{r}} \tag{2a}$$

$$\text{and} \qquad O_T = \vec{\sigma} \times \hat{q}\tau_\lambda e^{i\vec{q}\cdot\vec{r}}, \tag{2b}$$

respectively. They are entirely decoupled in nuclear matter, due to momentum conservation. Thus the corresponding responses are ruled by the longitudinal

$$V_L(q,\omega) = V_\pi(q,\omega) + \Gamma_\pi^2(q_\mu^2)\frac{f_\pi^2}{\mu_\pi^2}(\vec{\sigma}_1\cdot\vec{\sigma}_2)(\vec{\tau}_1\cdot\vec{\tau}_2)g' \tag{3a}$$

and transverse ph interaction

$$V_T(q,\omega) = V_\rho(q,\omega) + \Gamma_\pi^2(q_\mu^2)\frac{f_\pi^2}{\mu_\pi^2}(\vec{\sigma}_1\cdot\vec{\sigma}_2)(\vec{\tau}_1\cdot\vec{\tau}_2)g' \tag{3b}$$

as implied by equation (1). In the above V_π and V_ρ are the direct particle-hole matrix elements of the one-pion and one-rho exchange potentials, g' is the usual Landau-Migdal parameter, which accounts for short range correlations between nucleons as well as for the exchange matrix elements of V_π and V_ρ.

When V_L and V_T are displayed versus q, there appears, at momenta of the order of $2 \div 3\mu_\pi$, a marked contrast between the two, which bears direct consequences on the collective responses (1). In particular a quite large (at low frequencies) ratio R_L/R_T is predicted. The measurements of polarized (p, p') scattering performed in Los Alamos [2], have shown instead a ratio close to one or even below unity.

On the one side, this outcome could point to the absence of collectivity in the spin-isospin responses at intermediate momentum transfers. On the other, at least in the transverse channel, the available data from deep inelastic electron scattering[3] seem to support some collective quenching and hardening of R_T.

Additional experimental information on the $\sigma\tau$ responses in the quasi-free region has been recently provided by the charge-exchange $(^3He, t)$ reactions measured at Saturne[4]. The observed softening of the peak at large momentum transfers, where the longitudinal channel dominates, could be ascribed as well to pionic collective effects.

These rather controversial outcomes led us to consider the spin-isospin responses in finite nuclei, with the twofold purpose of investigating the effects, on the collective behaviour of the responses, of both the nuclear finite size and the surface character of a hadron induced excitation.

In the following I will first introduce the RPA $\sigma\tau$ volume responses of a finite nucleus, in which case any nucleon in the nucleus can be directly excited by the external probe. Then I will consider surface responses, where the direct excitation can only affect the external nucleons: the surface responses are more specifically suited to describe the interaction of a nucleus with a hadronic probe.

FINITE NUCLEI: VOLUME RESPONSES

In finite systems, due to the loss of translational invariance, one can define three different ph polarization propagators, according to the nature of the external vertices, O_L and O_T: Π_{LL}, Π_{TT} and Π_{LT}. Correspondingly, a set of three coupled integral RPA equations for these propagators replaces the two uncoupled algebraic equations of nuclear matter (for Π_{LL} and Π_{TT})[5].

Indeed, at variance with nuclear matter, the spin-transverse and spin-longitudinal couplings are no-longer orthogonal already at the level of the free ph propagator. This implies that both V_L and V_T enter into the RPA equations for Π_{LL}, Π_{TT} and Π_{LT}, obviously with different weights.

By performing the usual multipole expansion in the angular momentum basis, the three above equations reduce to a *unique* set of RPA integral equations for the different multipolarities (which embody the dynamical part of the propagators):

$$[\hat{\Pi}_J^{RPA}(q, q'; \omega)]_{ll'} = [\hat{\Pi}_J^0(q, q'; \omega)]_{ll'} +$$
$$+ \frac{1}{(2\pi)^3} \int_0^\infty dk\, k^2 \sum_{l_1 l_2} [\hat{\Pi}_J^0(q, k; \omega)]_{ll_1} [U_J(k, \omega)]_{l_1 l_2} [\hat{\Pi}_J^{RPA}(k, q'; \omega)]_{l_2 l'}, \quad (4)$$

where

$$[U_J(k, \omega)]_{l_1 l_2} = a_{Jl_1} V_L(k, \omega) a_{Jl_2} + V_T(k, \omega)(\delta_{l_1 l_2} - a_{Jl_1} a_{Jl_2}), \quad (5)$$

the a_{Jl} being proportional to Clebsch-Gordan coefficients.

We describe the nucleus within an Harmonic Oscillator shell model with the residual ph-interaction (3). The equations (4) have been solved [6] with an approximate method, originally proposed by Toki and Weise[7], which exploits the quasi-diagonality of the propagators in momentum space, a property well satisfied for medium-heavy nuclei and in the quasi-elastic peak region.

The responses one obtains with these RPA polarization propagators *do* include the effects of a non-uniform nuclear density as well as the mixing of the $\vec{\sigma} \cdot \hat{q}$ and $\vec{\sigma} \times \hat{q}$ modes; the importance of the latter is confined at small energies and depends on q, reflecting the momentum behaviour of V_L and V_T.

However one should keep in mind that these polarization propagators allow the external excitation, brought in by the operators (2), to penetrate everywhere inside the nucleus, as it happens, e.g., when an electromagnetic probe is employed. In this connection we will call *volume* responses the ones derived from (4); the transverse response can be fairly tested against the inelastic electron scattering data, as it is shown in Fig.1.

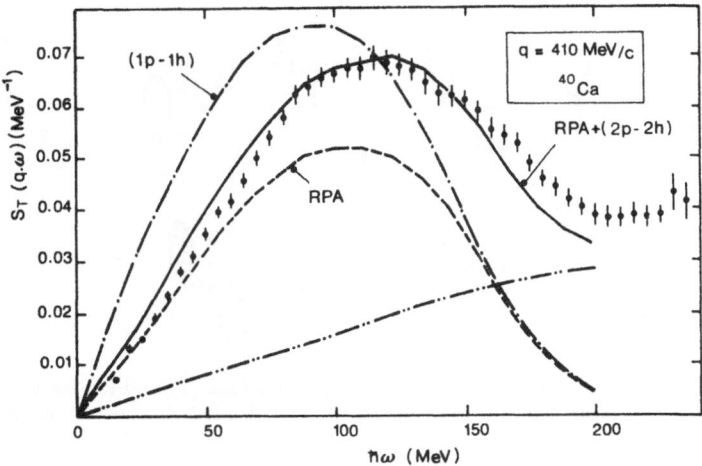

Fig.1 The transverse structure function of ^{40}Ca measured in (e, e') inelastic scattering as a function of $\hbar\omega$, at $q = 410$ MeV/c. The experimental points are taken from ref.[3]. The double-dot-dashed line is the 2p-2h contribution of ref.[8].

FINITE NUCLEI: SURFACE RESPONSES

As anticipated in the introduction, the most recent measurements of the spin-longitudinal and transverse responses have been obtained with hadronic probes. For them the nucleus is a strongly absorptive medium and the scattering is confined to the surface region.

Accordingly, we have considered the $\sigma\tau$ *surface* responses, which we obtain by modifying the external vertices of the polarization propagators as follows:

$$O_{L,T} \rightarrow O_{L,T}^{surf} = F(r)O_{L,T}. \qquad (6)$$

Before sticking to the determination of $F(r)$, which will be pursued according to Glauber's theory, we notice that the vertex (6) entails the introduction of two new propagators, one with a volume and a surface vertex and the other with two surface vertices. Then, instead of eq.(4), we get the following two chain integral equations:

$$[\hat{\Pi}_J^{RPA,ss}(q, q'; \omega)]_{ll'} = [\hat{\Pi}_J^{0,ss}(q, q'; \omega)]_{ll'} +$$
$$+ \frac{1}{(2\pi)^3} \sum_{l_1 l_2} \int_0^\infty dk \, k^2 [\hat{\Pi}_J^{0,s}(q, k; \omega)]_{ll_1} [U_J(k)]_{l_1 l_2} [\hat{\Pi}_J^{RPA,s}(k, q'; \omega)]_{l_2 l'}, \qquad (7)$$

$$[\hat{\Pi}_J^{RPA,s}(q, q'; \omega)]_{ll'} = [\hat{\Pi}_J^{0,s}(q, q'; \omega)]_{ll'} +$$
$$+ \frac{1}{(2\pi)^3} \sum_{l_1 l_2} \int_0^\infty dk \, k^2 [\hat{\Pi}_J^0(q, k; \omega)]_{ll_1} [U_J(k)]_{l_1 l_2} [\hat{\Pi}_J^{RPA,s}(k, q'; \omega)]_{l_2 l'}. \qquad (8)$$

They are diagrammatically illustrated in Fig.2, where a black vertex represents the vertex (6). It is clear from this figure that the excitation is bound to be produced in the outer region of the nucleus, but it can then propagate to the interior through the residual interaction.

The equations (8) can be solved with the same approximate method utilized for the RPA volume propagators in (4); one obtains the following expressions for the transverse,

$$R_T^{surf}(q, \omega) = R_T^{0,surf}(q, \omega) - \frac{1}{16\pi^2} \text{Im} \sum_{J=1}^\infty \left\{ \frac{(2J+1)[\hat{\Pi}_J^{(1),ss}(q, \omega)]_{J,J}}{1 - \frac{\gamma\bar{q}^2}{(2\pi)^3} V_T(\bar{q}, \omega)\hat{\Pi}_J^0(\bar{q}, \omega)} + \right.$$

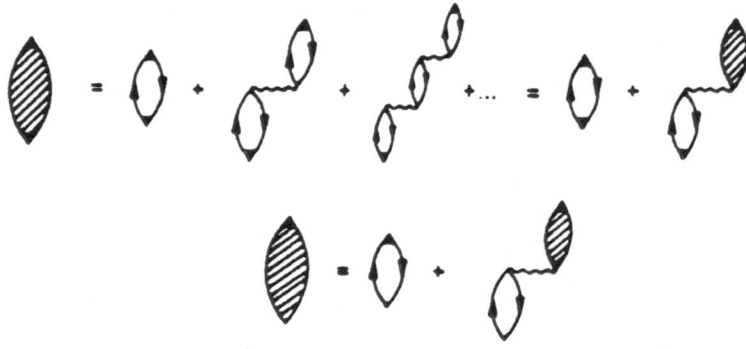

Fig.2 Diagrams representing the surface RPA equations (7) and (8) for the dynamical propagator; the p-h propagator includes both N-h and Δ-h excitations.

$$+ \frac{(J+1)[\hat{\Pi}_J^{(1),ss}(q,\omega)]_{J-1,J-1} + \sqrt{J(J+1)}[\hat{\Pi}_J^{(1),ss}(q,\omega)]_{J+1,J-1}}{1 - \frac{\gamma \bar{q}^2}{(2\pi)^3} V_T(\bar{q},\omega) \hat{\Pi}_{J-1}^0(\bar{q},\omega) + \frac{J}{2J+1} \mathcal{F}_{J+1}(\bar{q},\omega)} +$$

$$+ \left. \frac{J[\hat{\Pi}_J^{(1),ss}(q,\omega)]_{J+1,J+1} + \sqrt{J(J+1)}[\hat{\Pi}_J^{(1),ss}(q,\omega)]_{J-1,J+1}}{1 - \frac{\gamma \bar{q}^2}{(2\pi)^3} V_T(\bar{q},\omega) \hat{\Pi}_{J+1}^0(\bar{q},\omega) + \frac{J+1}{2J+1} \mathcal{F}_{J-1}(\bar{q},\omega)} \right\} \qquad (9)$$

and for the longitudinal response function

$$R_L^{surf}(q,\omega) = R_L^{0,surf}(q,\omega) -$$

$$- \frac{1}{8\pi^2} \operatorname{Im} \sum_{J=0}^{\infty} \left\{ \frac{J[\hat{\Pi}_J^{(1),ss}(q,\omega)]_{J-1,J-1} - \sqrt{J(J+1)}[\hat{\Pi}_J^{(1),ss}(q,\omega)]_{J+1,J-1}}{1 - \frac{\gamma \bar{q}^2}{(2\pi)^3} V_L(\bar{q},\omega) \hat{\Pi}_{J-1}^0(\bar{q},\omega) + \frac{J+1}{2J+1} \mathcal{G}_{J+1}(\bar{q},\omega)} + \right.$$

$$+ \left. \frac{(J+1)[\hat{\Pi}_J^{(1),ss}(q,\omega)]_{J+1,J+1} - \sqrt{J(J+1)}[\hat{\Pi}_J^{(1),ss}(q,\omega)]_{J-1,J+1}}{1 - \frac{\gamma \bar{q}^2}{(2\pi)^3} V_L(\bar{q},\omega) \hat{\Pi}_{J+1}^0(\bar{q},\omega) + \frac{J}{2J+1} \mathcal{G}_{J-1}(\bar{q},\omega)} \right\} \qquad (10)$$

In the above formulas $\bar{q}$ is a suitably chosen average momentum[7] (usually close to q), $\gamma \approx \pi/R$ (R = rms radius of the nucleus) and the $[\hat{\Pi}_J^{(1),ss}]_{J,J'}$ are the first order surface propagators. Moreover

$$\mathcal{F}_{J-1}(\bar{q},\omega) = \frac{\gamma \bar{q}^2}{(2\pi)^3} \left[V_T(\bar{q},\omega) - V_L(\bar{q},\omega) \right] \frac{\hat{\Pi}_{J+1}^0(\bar{q},\omega) - \hat{\Pi}_{J-1}^0(\bar{q},\omega)}{1 - \frac{\gamma \bar{q}^2}{(2\pi)^3} V_L(\bar{q},\omega) \hat{\Pi}_{J-1}^0(\bar{q},\omega)}, \qquad (11)$$

$$\mathcal{F}_{J+1}(\bar{q},\omega) = \frac{\gamma \bar{q}^2}{(2\pi)^3} \left[V_T(\bar{q},\omega) - V_L(\bar{q},\omega) \right] \frac{\hat{\Pi}_{J-1}^0(\bar{q},\omega) - \hat{\Pi}_{J+1}^0(\bar{q},\omega)}{1 - \frac{\gamma \bar{q}^2}{(2\pi)^3} V_L(\bar{q},\omega) \hat{\Pi}_{J+1}^0(\bar{q},\omega)} \qquad (12)$$

and

$$\mathcal{G}_{J+1}(\bar{q},\omega) = \mathcal{F}_{J+1}(\bar{q},\omega; V_L \leftrightarrow V_T), \qquad (13)$$

$$\mathcal{G}_{J-1}(\bar{q},\omega) = \mathcal{F}_{J-1}(\bar{q},\omega; V_L \leftrightarrow V_T) \qquad (14)$$

In order to determine $F(r)$ we resort to Glauber's theory[9]. Within this framework, the effective number of nucleons taking part in the (one-step) reaction is given by the expression

$$N_{eff} = \frac{\sigma^{(1)}}{\sigma_{tot}} = \frac{1}{\sigma_{tot}} \int_0^\infty db \, 2\pi b \, \chi(b) e^{-\chi(b)}, \qquad (15)$$

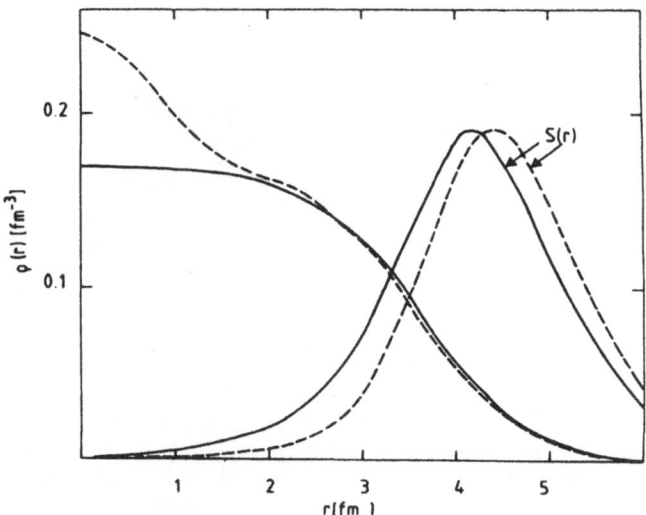

Fig.3 The function $S(r)$, integrand of the rhs of eq.(17), is drawn with an arbitrary scale for $\sigma_{tot} = 40$ mb (solid line) and $\sigma_{tot} = 55$ mb (dashed line). The scale on the left refers to the two curves displaying the empirical (solid line) and H.O. (dashed line) density of Ca^{40}.

where $\chi(b)$ is the phase-shift function

$$\chi(b) = \sigma_{tot} \int_{-\infty}^{\infty} dz\, \rho \left(r = \sqrt{z^2 + b^2} \right), \tag{16}$$

ρ being the nuclear density and σ_{tot} the total *probe-nucleon* cross section.

On the other hand, for a spherical nucleus and at momenta large enough to neglect the Pauli correlations, N_{eff} can also be expressed in terms of the sum rule:

$$N_{eff} = \int_0^{\infty} d\omega\, R^{surf}(q, \omega) = 4\pi \int_0^{\infty} dr\, r^2 |F(r)|^2 \rho(r). \tag{17}$$

By comparing (17) and (15) one can thus fix the form of $F(r)$. In Fig.3 the integrand of the rhs of eq.(17) is reported for two values of σ_{tot}, namely 40 mb and 55 mb, together with the density $\rho(r)$ of Ca^{40}: the excitation is clearly peaked in the surface region, at $\rho = \bar{\rho} = 0.28\rho_0$ and $\rho = \bar{\rho} = 0.20\rho_0$, respectively ($\rho_0$ is the central nuclear density).

There remain to be seen whether in such dilute media the p-h interaction is still able to set up collective effects, either locally or by spreading inside the system, and how the corresponding surface responses compare with the experiment. We remind that our surface RPA formalism for finite nuclei accounts, notwithstanding the above outlined approximations, for the mixing between the $\vec{\sigma} \cdot \hat{q}$ - $\vec{\sigma} \times \hat{q}$ couplings, for the non-uniformity of the nuclear density (which involves the coupling of different angular momenta in the multipole expansion of $R_{L,T}$) and finally for the surface absorption of the external probe. As a matter of fact, *all these factors work against collectivity*, but not necessarily wash it completely out.

COMPARISON WITH THE EXPERIMENT

We can now test our surface spin-isospin responses towards the experimental data from (p, p') scattering and $(^3He, t)$ charge exchange reactions.

The first measurement of the separated spin longitudinal and transverse response functions has been performed in Los Alamos a few years ago[2] using polarized protons of 500

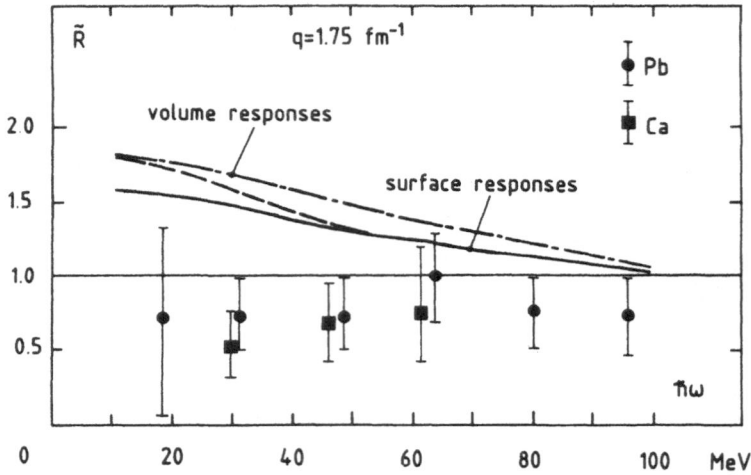

Fig.4 The ratio $\tilde{\mathcal{R}}$ at $q = 1.75 \text{fm}^{-1}$ as a function of $\hbar\omega$. The dot-dashed line is the prediction of the RPA theory of the volume responses; the surface RPA prediction is given by the solid and dashed lines, with and without the $\vec{\sigma} \cdot \hat{q} - \vec{\sigma} \times \hat{q}$ mixing, respectively; $g' = 0.7$.

MeV on Ca^{40} and Pb^{208} targets. This experiment cannot separate the isovector contribution ($\tau = 1$) from the isoscalar one ($\tau = 0$); thus it actually measures the combination

$$\tilde{\mathcal{R}} = \frac{2.15}{4.62} \frac{3.62 R_L^{\tau=1}(q,\omega) + R_L^{\tau=0}(q,\omega)}{1.15 R_T^{\tau=1}(q,\omega) + R_T^{\tau=0}(q,\omega)}. \tag{18}$$

Both the numerator and the denominator are obtained as ratios of the corresponding quantities in H^2 and Ca^{40} (or Pb^{208}).

To compare with the experiment we have evaluated the ratio (18) utilizing our surface RPA responses (with $\sigma_{tot} = 40$ mb) in the isovector channel and the independent particle surface responses in the isoscalar one; the results are displayed in Fig.4, at $q = 1.75 \text{fm}^{-1}$.

With respect to the old predictions of nuclear matter the ratio R_L/R_T appears to be considerably reduced: in particular the surface character of the process helps in bringing $\tilde{\mathcal{R}}$ down towards unity, as it can be inferred from the comparison with the corresponding ratio between the *volume* responses, which is also shown in the figure. The effect of the mixing between the two spin modes is quite sizable, since at this momentum transfer the rather large transverse p-h interaction strongly affects R_L.

With respect to the experimental points there remain discrepancies on the low energy side. However it is worth pointing out that the measured $\tilde{\mathcal{R}}$ lies even below unity: this could be an indication of the presence of some collective effects in the isoscalar channel. Indeed, treating the $R^{\tau=0}$ as independent particle responses seems to be a suitable approximation, since the central $\tau = 0$ p-h force is known to be rather weak. But from sum rule considerations there is some evidence[10] for a collective character of the isoscalar responses (and of opposite nature than the isovector ones), due to the exchange matrix elements of the π- and ρ-tensor interactions. The future (p, n) experiments with polarized beams, presented in this Conference[11], will certainly help in clarifying this long-standing question.

Turning now to the analysis of the charge exchange $(^3He, t)$ reaction cross sections measured at Saturne[4], a few points should be kept in mind: (i) the ^3He projectiles undergo an even more peripheral scattering than the protons; indeed the *effective* σ_{tot} to be used in the determination of $F(r)$ is $\sigma_{tot} = 55$mb (this value, and not three times σ_{NN}, is required to reproduce the total experimental cross sections within Glauber's theory). (ii) The charge

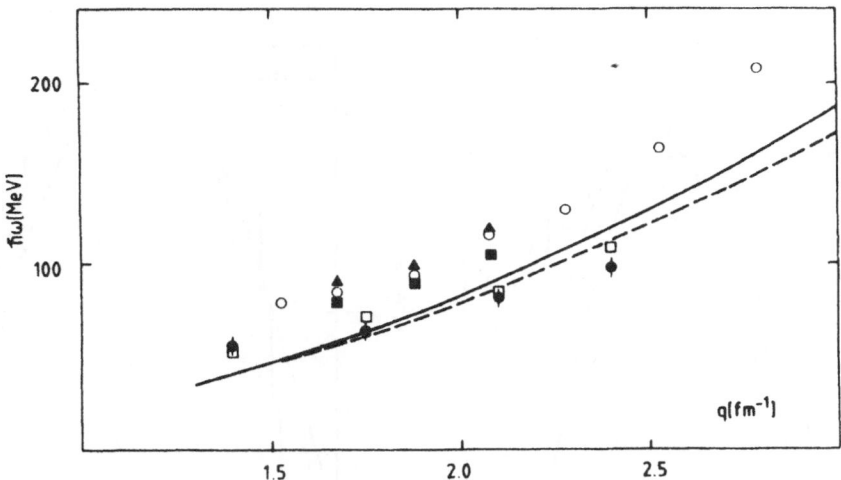

Fig.5 Peak position of the QEP in the reaction $C^{12}(^3He,t)$ at 2 GeV[4] (black dots) and $Ca^{40}(e,e')$[3] (circles); theoretical predictions from ref.|6] for the volume p-h RPA (black squares) and the p-h RPA + 2p-2h (triangles) isovector transverse responses are also displayed; the empty squares are the surface RPA predictions for the $(^3He,t)$ reaction; solid and dashed lines represent the peak position of the non-relativistic and relativistic Fermi gas responses, respectively.

exchange reactions are pure isovector processes, but no observation of polarization transfers were made in the Saturne experiment; thus Bergqvist et al. essentially measured a *mixture* of the $\vec{\sigma} \cdot \hat{q}$ and $\vec{\sigma} \times \hat{q}$ nuclear responses (with the ^3He bombarding energy of 2 GeV the non-spin-flip NN amplitudes are practically negligible).

In the single step approximation, the cross section for the $(^3He,t)$ reaction may be written as follows

$$\frac{d^2\sigma}{d\Omega\, d\omega} = N_{eff}\left\{(|\beta|^2 + |\epsilon|^2)R_T(q,\omega) + |\delta|^2 R_L(q,\omega)\right\}FF^2, \tag{19}$$

FF being the $(^3He,t)$ form factor, β and ϵ the charge exchange spin-transverse NN amplitudes and δ the spin-longitudinal one.

Before considering the cross sections in detail, it is interesting to look at the position of their peaks (ω_M) for different momentum transfers (actually the measurements are performed at fixed scattering angle, but the corresponding q-value does not vary substantially over the energy range spanned by the quasi-elastic peak): ω_M is displayed in Fig.5 as a function of q, together with the curves representing the peak positions in a free Fermi gas (both non-relativistic and relativistic). While at small momenta the experimental points exhibit a hardening with respect to the Fermi gas, at larger q they display a progressively substantial softening.

This outcome could be nicely interpreted in terms of collective effects providing that at small q the transverse response (which is quenched and hardened by the RPA correlations) be dominant, whereas at large q the longitudinal response (enhanced and softened) be the major component in the rhs of (19). This seems to be the case at the incident energy of about 700 MeV/nucleon of the Saturne experiment; indeed the ratio $|\delta|^2/(|\beta|^2 + |\epsilon|^2)$ is almost linearly increasing from 0.4 at $q = 1.4\mathrm{fm}^{-1}$ to 1.5 at $q = 2.4\mathrm{fm}^{-1}$. Thus, although R_L and R_T cannot be separated, one can envisage a natural explanation of the above mentioned behaviour of ω_M, by ascribing the low momentum hardening to the dominant transverse component and the high momentum softening to the longitudinal one.

We have evaluated the cross section (19) by utilizing our surface RPA responses with the vertex function $F(r)$ corresponding to $\sigma_{tot} = 55$ mb; in spite of the low density at which

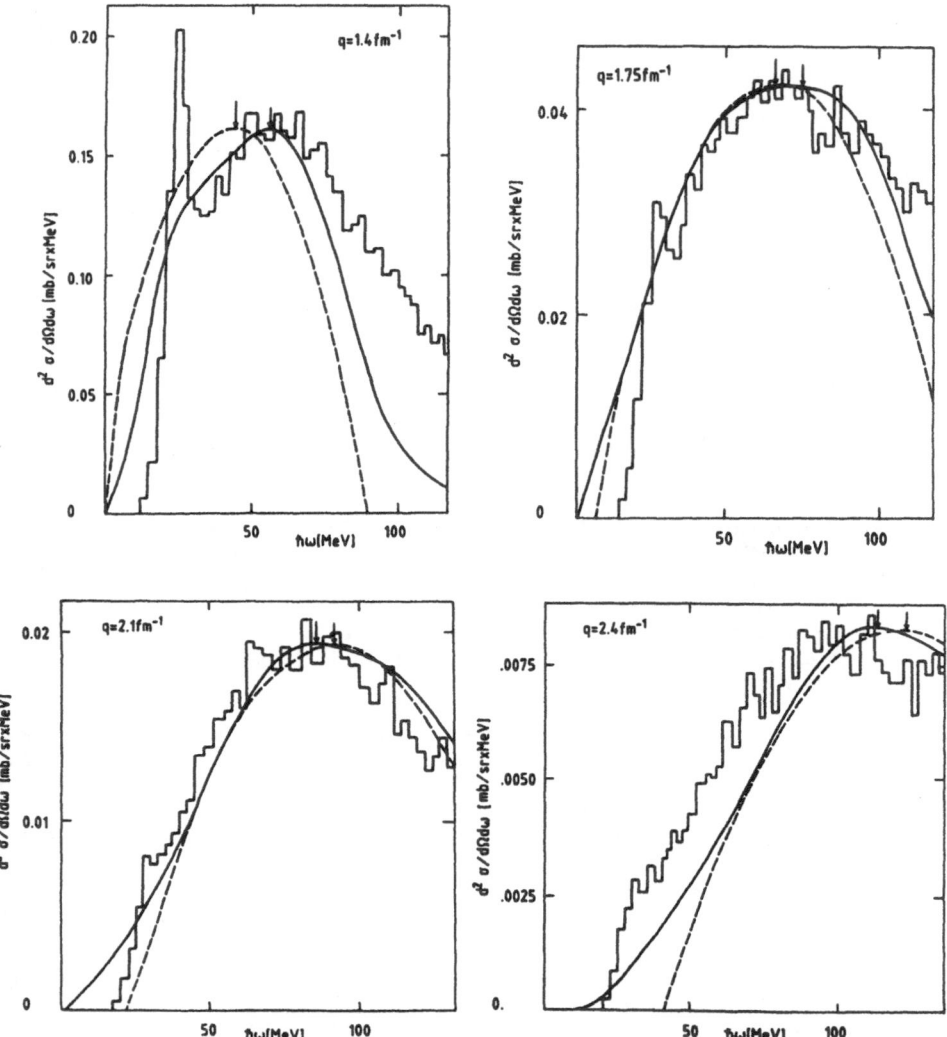

Fig.6 Experimental and surface RPA (solid line) cross sections for the $(^3He, t)$ reaction as a function of ω; $g' = 0.6$. The cross section for a non-relativistic Fermi gas (with $k_F = 0.79$) is also shown for comparison. Arrows indicate the peak position.

the excitation takes place, we still find some collective effects in both channels. In particular, at $q = 2.4^{-1}$ and with $g' = 0.7$ (the value employed to confront with electron scattering data), we obtain a softening of the peak, with respect to the Fermi gas, of about 8 MeV, to be compared with the experimental value of ~ 18 MeV. However, a smaller value of g' might be more suitable here, since at such a low density the assumption of universality is no longer valid and the Δ is likely to experience a weaker short range repulsion[12]. With an effective value $g' = 0.6$, for example, we obtain a downward shift in the cross section of ~ 11 MeV.

In Fig.6a→d we compare our surface RPA cross sections with the available experimental data, for q ranging from 1.4 to 2.4 fm^{-1}. The agreement between theory and experiment is fairly good, but for the highest momentum, where the calculated softening does not fully account for the observed one. In order to cure this failure one should probably utilize a better treatment of the distortion[13], rather than the Glauber's approach employed here. Also, one should not forget the influence of relativistic effects: indeed, as it appears in Fig.5, already at the level of the free Fermi gas, the use of relativistic kinematics provides a substantial softening of the peak (with respect to the non-relativistic one) at the higher momenta.

In conclusion we believe that the observed features of the spin-isospin nuclear responses are not incompatible with a collective interpretation of the processes. In particular it seems natural and appealing to ascribe the rather spectacular effects measured in the charge exchange $(^3He, t)$ reactions, at such low densities and short wavelengths, to the long range nature of the pion, whose role in nuclear structure deserves further investigations.

Acknowledgements This contribution is based on several works done in collaboration with A. De Pace, M. Ericson, Mikkel B. Johnson and A. Molinari, to whom I wish to express my gratitude.

REFERENCES

1. W.M. Alberico, M. Ericson and A. Molinari, Nucl. Phys. **A379** (1982) 429
2. T.A. Carey et al., Phys. Rev. Lett. **53** (1984) 144; L.B. Rees et al., Phys. Rev. **C34** (1986) 627
3. Z.E. Meziani et al., Phys. Rev. Lett. **54** (1985) 1233
4. I. Bergqvist *et al.*, Nucl. Phys. **A469** (1987) 648
5. W.M. Alberico, A. De Pace and A. Molinari, Phys. Rev. **C31** (1985) 2007
6. W.M. Alberico, A. Molinari, A. De Pace, M. Ericson and M.B. Johnson, Phys. Rev. **C34** (1986) 977
7. H. Toki and W. Weise, Phys. Rev. Lett. **42** (1979) 42; Z. Phys. **A292** (1979) 389
8. W.M. Alberico, M. Ericson and A. Molinari, Ann. Phys. (N.Y.) **154** (1984) 356
9. R.J. Glauber, in: *Lectures in theoretical Physics*, vol.1, eds. W. Brittin *et al.* (Interscience, New York, 1959)
10. G. Orlandini, M. Traini and M. Ericson, Phys. Lett. **B179** (1986) 201
11. J. McClelland, these Proceedings
12. A. Hosaka and H. Toki, Progr. Theor. Phys. **76** (1986) 1306
13. M. Ichimura *et al.*, these Proceedings.

WIGNER-KIRKWOOD EXPANSION OF THE QUASI-ELASTIC NUCLEAR

RESPONSES AND APPLICATION TO SPIN-ISOSPIN RESPONSES

Guy CHANFRAY

Institut de Physique Nucléaire (and IN2P3), Université Lyon-1
43, Bd du 11 Novembre 1918 - 69622 Villeurbanne Cedex (France)

We derive a semi-classical Wigner-Kirkwood expansion ($\hbar$ expansion) of the linear response functions. We find that the semi-classical results compare very well to the quantum mechanical calculations. We apply our formalism to the spin-isospin responses and show that surface-peaked $\hbar^2$ corrections considerably decrease the ratio longitudinal/transverse as obtained through the Los Alamos $(\vec{p}, \vec{p}\,')$ experiment.

I. INTRODUCTION

The aim of this paper is twofold. First we present a semi-classical method, going beyond the Thomas-Fermi (TF) approximation, for the calculation of the quasi-elastic nuclear response functions at intermediate momentum transfer ($q \sim$ 2-3 m_π) and second we apply the formalism to the problem of spin-isospin responses which has been one of the central topics of this conference.

In recent years, semi-classical methods have proven to be very convenient and accurate for describing gross nuclear properties where the influence of shell effects has been averaged out. In particular, it has been found that the Thomas-Fermi single-particle response compares extremely well to its quantum mechanical analog[1]. Moreover, realistic TF-RPA calculations give a good description of structure functions measured in electron scattering experiments[2-4].

We will calculate the correction to the T-F response by performing an $\hbar$ (or Wigner-Kirkwood) expansion. At the level of the mean-field (i.e. single particle) response, we will find that the surface-peaked $\hbar^2$ corrections to the TF (i.e. $\hbar^0$) response are, as expected, very small. In addition, we will show that the semi-classical RPA response compares also very well to an exact model calculation which is half quantal in the sense that, for the mean-field response, the semi-classical TF response is used as the input of the integral (Bethe-Salpeter) equation[5]. Again, the TF result is very good and the $\hbar^2$ corrections, although small, improve the agreement.

The main effect of the surface-peaked $\hbar^2$ corrections to the RPA response is, in general, to reduce the collective reshaping predicted by TF theory. This effect remains small as far as we are concerned with volume response, as probed in electron scattering, but may become important for surface responses probed by hadrons. We have applied our formalism[6], to the calculation of the ratio R_L/R_T between the longitudinal and transverse surface spin-isospin response functions measured through a $(\vec{p}, \vec{p}\,')$ experiment performed in Los Alamos[7]. Due to the presence of the one-pion exchange potential, TF theory[3] predicts a ratio R_L/R_T much larger than the measured one which is close to one. We will show that the $\hbar^2$ corrections considerably improve the agreement with data.

II. OUTLINE OF THE SEMI-CLASSICAL METHOD

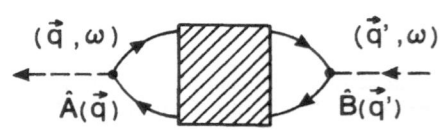

$(\vec{q}, \omega)$ $(\vec{q}\,', \omega)$

$\hat{A}(\vec{q})$ $\hat{B}(\vec{q}\,')$

Figure 1 - The polarization propagator $\pi_{AB}(\vec{q}, \vec{q}\,'; \omega)$

The response of the nucleus to any external excitation can be obtained from the polarization propagator $\pi_{AB}(\vec{q}, \vec{q}\,'; \omega)$ depicted schematically on fig. 1. Here the "end-point" operators $\hat{A}(\vec{q})$ and $\hat{B}(\vec{q}\,')$ may involve the identity operator or spin and isospin operators such as $\sigma_\alpha \tau_\beta$ and $\sigma_\alpha \tau_\beta$. The polarization propagators considered as matrix elements in momentum space of one-body operators and are expressible in terms of their Wigner-transforms[8] (WT) $[\pi_{AB}]_W(\vec{R}, \vec{K}; \omega)$ according to :

$$\pi_{AB}\left(\vec{K} + \frac{\vec{k}}{2}, \vec{K} - \frac{\vec{k}}{2}; \omega\right) = \int d^3R\ e^{i\vec{k}\cdot\vec{R}/\hbar}\ [\pi_{AB}](\vec{R}, \vec{K}; \omega) \tag{II.1}$$

The semi-classical method consists in an expansion in power of $\hbar$ of the WT $[\pi_{AB}]_W$. In practice we will limit ourselves to order h^2. The response to an external weakly interacting probe, transferring to the nucleus a momentum-energy $(\vec{q}; \omega)$ and interacting with the nucleus through the one-body operator $\hat{A}(\vec{q})$ is :

$$R_A(\vec{q}; \omega) = \sum_n \left| \langle n| \sum_{i=1}^{A} \hat{A}(\vec{q})\ e^{i\vec{q}\cdot\vec{x}} | 0\rangle \right|^2 \delta(E_n - \omega) \tag{II.2}$$

where $|0\rangle$ is the ground state and $|n\rangle$ is an excited state with energy E_n. This response is obtained as :

$$R_A(\vec{q}; \omega) = -\frac{1}{\pi} \operatorname{Im} \left\{ \int d^3R\ [\pi_{AA}]_W(\vec{R}, \vec{q}; \omega) \right\} \tag{II.3}$$

In the following we will evaluate semi-classically the response (II.3) at the mean-field level and at the RPA level. For simplicity we will present the formalism at zero temperature but the actual calculations are performed at finite temperature[5] in order to avoid tedious manipulations involving distributions. The zero temperature case is recovered at the end as a limit. In practice the T = 1 MeV result is equal to the T = 0 result with an accuracy better, and most of the time far better, than one per cent.

III. THE PARTICLE-HOLE (OR MEAN-FIELD) RESPONSE FUNCTION

We consider a spherical nucleus with N = Z where the nucleons move in a mean-field schematized by a central local single particle V(R) but all what we will say can be easily extended to a non local mean-field potential. The pure particle-hole response is independent of the spin-isospin structure of the excitation operators and can be written in terms of WT of one-body operators[5]:

$$R_o(\vec{q}; \omega) = \int d^3R\ \frac{d^3p}{(2\pi\hbar)^3}\ \frac{dt}{2\pi}\ e^{i\omega t} \left\{ [\theta(\varepsilon_F - \hat{H})e^{i\hat{H}t}]_W(\vec{R}, \vec{p}) \right.$$
$$\left. \times\ [\theta(\hat{H} - \varepsilon_F)e^{-i\hat{H}t}]_W(\vec{R}, \vec{p} + \vec{q}) \right\} \tag{III.1}$$

where $\hat{H}$ is the single-particle Hamiltonian and ϵ_F is the Fermi energy. The one-body operators appearing in Eq.(III.1) can be expanded in powers of $\hbar^2$ [8]. We thus write the response as :

$$R_0(\vec{q};\omega) = -\frac{1}{\pi} \, Im \left\{ \int d^3R \left(\alpha_0(R,q;\omega) + \hbar^2 \, \alpha_2(R,q;\omega) \right) \right\} \qquad (III.2)$$

The familar Thomas-Fermi result[1] is the first term (i.e. $\hbar^0$ term) of the expansion. In the momentum range of interest the surface-peaked $\hbar^2$ correction, which depends on the derivatives of V(R), gives only a few per cent corrections. The exact expression of α_0 et α_2 may be found in ref. 5 and 6. We show on **fig. 2** the semi-classical responses together with the exact quantum mechanical result[9] (^{40}Ca, Woods-Saxon potential, q = 2.15 Fm^{-1}). We see that the semi-classical results are very close to the quantum one and that $\hbar^2$ corrections, which are positive (negative) at low (high) energy, are very small. To be complete we have to signal a minor pathology, not shown on the figures, which is inherent to all $\hbar$ corrections of the Wigner-Kirkwood type. At very high energy (where the TF response vanishes) the $\hbar^2$ response may become negative or even divergent in such a way that the S_1 sum rule remains fulfilled[5]. Such a problem could be solved by use of partial resummation methods.

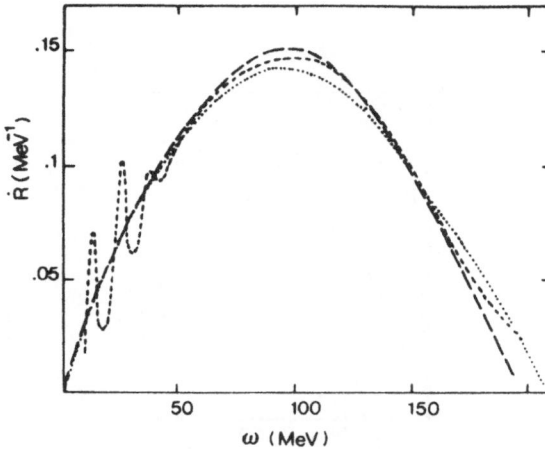

Figure 2 - *Mean field response for ^{40}Ca at momentum transfer q = 2.15 Fm^{-1} versus transferred energy ω. The dotted line corresponds to the Thomas-Fermi calculation and the long-dashed line refers to the $(\hbar^0 + \hbar^2)$ calculation. Also shown (short-dashed line) is the result of the exact quantum mechanical calculation[9].*

IV. THE RPA RESPONSE (CENTRAL FORCE)

In order to show how the semi-classical method works, we consider a very simple case. We calculate the response for can external operator $\hat{A} = \mathbb{1}$ in the RPA ring approximation.Thus, we only need the central component $V_c(\vec{k})$ of the residual interaction. The RPA polarization propagators obeys an integral equation :

$$\pi_c(\vec{q},\vec{q}';\omega) = \pi_0(\vec{q},\vec{q}';\omega) + \int \frac{d^3k}{(2\pi)^3} \, \pi_0(\vec{q},\vec{k};\omega) \, V_c(\vec{k}) \, \pi_c(\vec{k},\vec{q}';\omega) \qquad (IV.1)$$

Taking the WT of Eq. (IV.1), we get an equivalent equation, which is solved order by order up to $\hbar^2$. For doing that, we essentially need the WT of mean-field polarization propagator i.e $[\pi_0]_W = \alpha_0 + \hbar^2 \, \alpha_2$. Its imaginary part has been obtained in the previous section and its real part is obtainable from a dispersion relation. The RPA response is thus written as[5] :

$$R_c(q;\omega) = -\frac{1}{\pi} \, Im \left\{ \int d^3R \left(\tilde{\alpha}_0 \, (R,q;\omega) + \hbar^2 \sum_{i=1}^{3} \tilde{\alpha}_2^{(i)} \, (R,q;\omega) \right) \right\} \qquad (IV.2)$$

The renormalized TF contribution is formally identical to the nuclear matter RPA response :

$$\tilde{\alpha}_0 \ (R,q;\omega) = \alpha_0(R,q;\omega) \ / \ [1 - V_c(q) \ \alpha_0(R,q;\omega)] \qquad (IV.3)$$

We have divided the $\hbar^2$ term in three distinct pieces. The first one :

$$\tilde{\alpha}_2^{(1)} \ (R,q;\omega) = \alpha_2 \ (R,q;\omega) \ / \ [1 - V_c(q) \ \alpha_0(R,q;\omega)]^2 \qquad (IV.4)$$

corresponds to the renormalization of the $\hbar^2$ correction to the mean-field response. The expressions of $\tilde{\alpha}_2^{(2,3)}$, given in ref. 5, depend on derivatives of α_0 with respect to q and R . Here, we simply notice that $\tilde{\alpha}^{(2)}$ is a term wich is proportionnal to V_c or V_c^2, and $\tilde{\alpha}_2^{(3)}$, involving derivatives, with respect to the momentum q, of the residual interaction $V_c(q)$ describes finite range effects and thus disappears in case of a contact force. We have compared the exact RPA response [10], solution of Eq (IV.1) with the semi-classical one, taking in both cases the TF expression of π_0 as an input. On **fig. 3** we show the results of the calculations (^{40}Ca, q = 427.3 MeV) for an attractive contact force V_0 = -118 MeV.Fm^3. We see that the semi-classical approximation is excellent and the small $\hbar^2$ correction brings the semi-classical response practically equal to the exact one. **Fig. 4** corresponds to the result

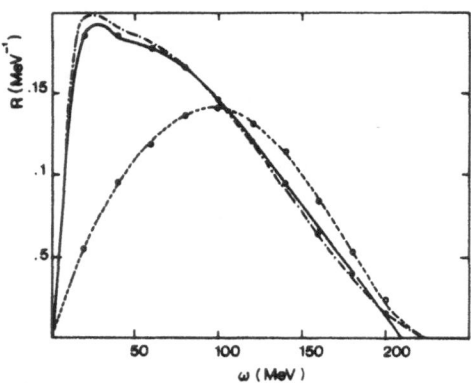 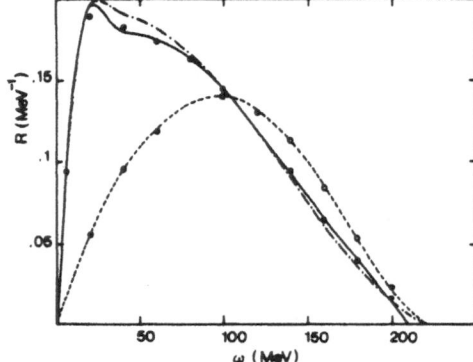

Figure 3 - Mean field and RPA responses for ^{40}Ca at momentum transfer q = 427.3 MeV versus transferred energy ω. The dashed line is the Thomas-Fermi mean-field result. The open circles represent the same calculation but obtained by J. Delorme after expansion in partial waves. The dot-dashed line is the Thomas-Fermi RPA response and the full line refers to the $(\hbar^0+\hbar^2)$ RPA response, ignoring $\hbar^2$ corrections coming from the mean-field response. These curves have to be compared with the exact RPA result (full circle) of J. Delorme[10] where the Thomas-Fermi mean-field propagator (expanded in partial waves) is used as an input. The residual interaction is an attractive contact force V_0 = - 118 MeV-Fm^3.

Figure 4 - The same as **fig. 3** but for a finite range force (see text).

with an interaction of the same strength but with a momentum dependence $\exp(-q^2 r_o^2/4)$ with $r_o = 0.5$ Fm. If we increase the range of the interaction ($r_o = 1$ Fm) the agreement remains excellent apart for the very low energy region where the slight resonance which is present in the exact calculation is overestimated in the semi-classical approach.

V. THE SPIN-ISOSPIN RESPONSES AND INTERPRETATION OF THE LOS ALAMOS $(\vec{p}, \vec{p}')$ EXPERIMENT

We are now interested in the responses to an external excitation mediated by either the longitudinal spin-isospin operator

$$L(\vec{q}) = \vec{\sigma}.\hat{q}\ \tau^{\alpha} \qquad (V.1a)$$

or a component of the transverse spin-isospin operator :

$$T_{\mu}(\vec{q}) = (\vec{\sigma} \times \hat{q})_{\mu}\ \tau^{\alpha} \qquad (V.1b)$$

The residual (p-h) interaction V_L (V_T) consists of pion (rho) exchange potential plus the short range g' interaction (see ref. 6 for details). The effect of antisymmetrization is assumed to be incorporated in the Landau-Migdal parameter g'. This procedure has been justified by Stroth et al. at the TF level i.e antisymmetrization yields a simple redefinition of g'. The longitudinal RPA polarization propagator is obtained as a solution of coupled integral equations which write in a schematic way[6] :

$$\pi_{LL} = \pi^o_{LL} + \pi^o_{LL}\ V_{LL}\ \pi_{LL} + \Sigma_{\mu}\pi^o_{LT_{\mu}}\ V_T\ \pi_{T_{\mu}L} \qquad (V.2a)$$

$$\pi_{T_{\mu}L} = \pi^o_{T_{\mu}} + \pi^o_{T_{\mu}L}\ V_L\ \pi_{LL} + \Sigma_{\nu}\pi^o_{T_{\mu\nu}}\ V_T\ \pi_{T_{\nu}L} \qquad (V.2b)$$

As before, we take the WT of these equations and solve order by order up to $\hbar^2$. Apart from the inclusion of an additionnal $\hbar^2$ correction proportionnal to $\nabla^2_R \alpha_o(R,q;\omega)$ the WT $[\pi^o_{LL}]_W(R,q;\omega)$ is essentially identical to $[\pi_o]_W$ once the Δ-hole contribution, specific to the spin-isospin channel, is taken into account[6]. From Eq. (V.2), we see that there is a coupling between longitudinal and transverse channels. The driving term of this L-T coupling term has the form[6] :

$$\pi^o_{T_{\mu}L}(\vec{R},\vec{q};\omega) = -\frac{i\hbar}{q}\ \left(\hat{q} \times \vec{\nabla}_R\ \alpha_o\ (R,q;\omega)\ \right) + O(\hbar^2) \qquad (V.3)$$

This L-T mixing effect occurs only to order $\hbar$ and thus disappears in the TF approximation. The RPA longitudinal spin-isospin response is found to be

$$R_L\ (q;\omega) = -\frac{1}{\pi}\ \text{Im}\ \left\{ \int d^3R\ \left(\tilde{\alpha}_{oL}\ (R,q;\omega) + \hbar^2\ \sum_{i=1}^{4}\ \tilde{\alpha}^{(i)}_{2L}\ (R,q;\omega)\ \right) \right\} \qquad (V.4)$$

Here $\tilde{\alpha}_{oL}$ is the well-known expression given by Eq. (IV.3) once the central interaction V_c is replaced by the longitudinal spin-isospin residual interaction. Due to the presence of the pion exchange potential, it yields to renormalization and enhancement of the response with a magnitude strongly dependent on the value of g' [3,11]. The surface peaked $\hbar^2$ correction receives four contributions. The first three terms $\tilde{\alpha}^{(1, 2, 3)}_{2L}$ have already been discussed in the previous section. The last term $\alpha^{(4)}_{2L}$ is specific to the spin-isospin channel and describes the surface coupling of the longitudinal and transverse channels[6] :

$$\tilde{\alpha}^{(4)}_{2L} = \frac{1}{3\ q^2}\ [\nabla^2_R \alpha_o + 2\ V_T\ (\vec{\nabla}_R\ \alpha_o)^2 \left(\frac{1}{1-V_T\alpha_o}\right)] \left(\frac{1}{1-V_L\alpha_o}\right)^2 \qquad (V.5)$$

This is, in our case of interest, the most important $\hbar^2$ term. The net effect of this correction is to attenuate the collective reshaping predicted by T-F theory.

However, as shown on **fig. 5** (^{40}Ca, q = 350 MeV/c, g' = 0.7), we see that this reduction is not extremely important. The transverse response R_T, is obtained from R_L, by simply exchanging the role of V_L and V_T. The effect of the $\hbar^2$ terms is now reversed i.e the quenching of R_T, predicted by TF theory is reduced.

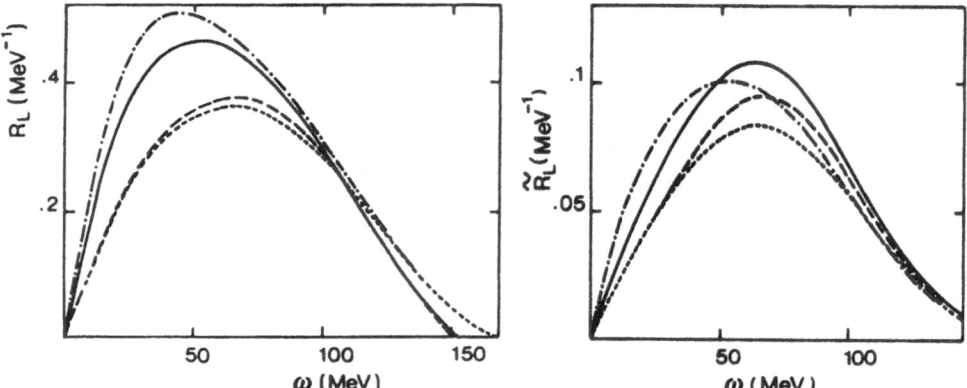

Figure 5 - *Volume longitudinal spin-isospin responses for ^{40}Ca at q = 350 MeV/c. The dotted line (mean-field response) and the dot-dashed line (RPA response) correspond to a Thomas-Fermi calculation. The dashed line(mean-field response) and the full line (RPA response) are obtained after inclusion of the $\hbar^2$ corrections.*

Figure 6 - *The same as* **fig. 5** *but for surface longitudinal responses probed through inelastic scattering of 500 MeV protons.*

Unfortunately, the volume longitudinal response is not known experimentally. The only available information comes from the $(\vec{p}, \vec{p}')$ Los Alamos experiment[7] which essentially measures the surface response. Thus the surface-peaked $\hbar^2$ term will obviously acquire a much more important relative weight leading to a considerable reduction of the collective effects. In addition the isoscalar spin-responses also enter. In this experiment, Carey and collaborators measured in fact the ratio :

$$\tilde{X} = (1/2.3) \left(3.6 \, \tilde{R}_L (\sigma \tau) + \tilde{R}_L (\sigma) \right) / \left(\tilde{R}_T(\sigma \tau) + \tilde{R}_T(\sigma) \right) \qquad (V.6)$$

at momentum q = 350MeV/c. The tilde mean that we are dealing with surface responses and the distorsion effect is taken into account through a position dependent weight factor $C(R)$ [2,6] multiplying the integrand of Eq. (V.4). On **fig. 6**, we show the surface response $\tilde{R}_L$ (^{40}Ca, q = 350 MeV/c, g' = 0.7). We see that there is still a small enhancement effect but the shift of the peak disappears almost completely due to the $\hbar^2$ terms. In Eq. (V.6) we have approximated the isoscalar response $\tilde{R}_{L,T}(\sigma)$ by the mean-field one. The calculated ratio for calcium (lead) is shown of **fig. 7** (**fig. 8**). It is apparent that the TF theory (dashed line) fails and the $\hbar^2$ corrections considerably improve the agreement with data. Similar conclusion have been reached by Alberico[12] et al. and Ichimura[13] whithin quantum mechanical RPA frameworks. The results for lead are very similar to those for calcium reflecting the fact that the surface properties do not depend very much on the mass number. I also would like to mention that improvements in the theoretical description, such as inclusion of 2p-2h contribution or renormalization of the isoscalar responses, are likely to further reduce the calculated ratio $\tilde{X}$.

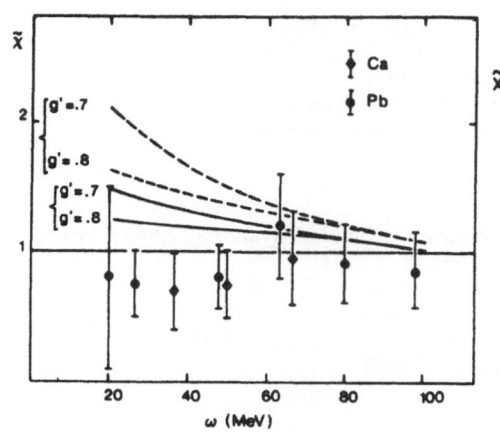

Figure 7 - The ratio $\tilde{X}$ for ^{40}Ca as a function of the transferred energy ω at q = 350 MeV/c. The dashed curves refer to a TF-RPA calculation and the full curves correspond to the RPA result including $\hbar^2$ corrections. The experimental data are from ref. **6**.

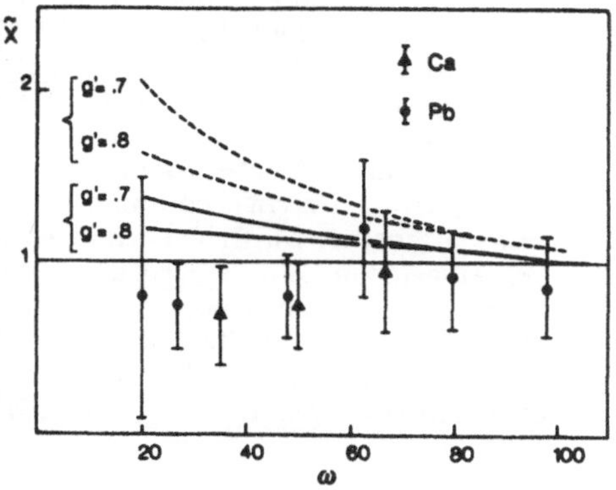

Figure 8 - The same as **fig. 7** but for lead.

VI. CONCLUSION

As a first conclusive remark, I would like to emphasize the power and the accuracy of the semi-classical method. This approach has the merit of flexibility and allows the calculation of a RPA-type response, at intermediate momentum transfer with very small computer time for any nuclei. This theoretical tool is particularly useful for heavy nuclei for which the exact quantum mechanical calculation requires of huge basis of shell model states and becomes extremely time consuming.

The final remark concerns the experimental knowledge about the longitudinal spin-isospin response. From the arguments developped in this paper, it is thus clear that the only existing data (i.e the Los Alamos data) do not provide decisive information concerning the behaviour of this response. The collectivity, if present, is masked by the isoscalar contamination and by the very important surface effects originating from the hadronic nature of the probe. A first interesting possibility would be the use of (p,n) probes in order to eliminate the isoscalar responses. Further experimental information, less affected by surface effects would be charged electroproduction[14]. It has been shown that

$(e,e'\pi^{\pm})$ with pion detected in the direction of the virtual photon yields an exact separation of the longitudinal and transverse spin-isospin responses, through a Rosenbluth plot. In addition, the distorsion of the outgoing pion wave function is less important than the combined distortion of the incident and outgoing

proton wave functions. However, such an $(e,e'\pi^{\pm})$ coincidence experiment is feasible only with the next generation of high duty cycle machines.

REFERENCES

[1] U. Stroth, R. Hasse and P. Schuck, <u>J. de Phys.</u>, C6-:343 (1984)
[2] U. Stroth et al., <u>Phys. Lett</u>, 156-B:291 (1985)
[3] U. Stroth, H. Hasse and P. Schuck, <u>Nucl. Phys.</u>, A-462:45 (1987)

4   W..M. Alberico et al., <u>Nucl. Phys.</u>, A-462:269 (1987)
5   G. Chanfray and P.Schuck, Preprint LYCEN-8767 I.P.N. Lyon
6   G. Chanfray, <u>Nucl. Phys.</u>, A-474:114 (1987)
7   T.A. Carey et al., <u>Phys. Rev. Lett.</u>, 53:144 (1984)
8   P. Ring and P. Schuck, "The nuclear many-body problem",
    Springer-Verlag, Berlin (1980)
9   N. Van Giai, Private communication
10 J. Delorme, Private communication
11 W.M. Alberico, M. Ericson and A. Molinari, <u>Nucl. Phys.</u>, A-379:429 (1982)
12 W. Alberico et al., <u>Phys.Lett.</u>, 183-B:135 (1987)
    W. Alberico, Contribution to this conference
13 M. Ichimura, Contribution to this Conference
14 G. Chanfray and J. Delorme, <u>Phys. Lett.</u>, 129-B:167 (1983)
    G. Chanfray, <u>Nucl. Phys.</u>, A-429:489 (1984)

THE RELATIVISTIC NUCLEAR RESPONSE AND VACUUM POLARIZATION

C. J. Horowitz

Physics Department and Nuclear Theory Center
Indiana University
Bloomington, Indiana 47405

ABSTRACT

The response of nuclear matter to electromagnetic probes is calculated in a relativistic random phase approximation both with and without vacuum polarization. Magnetic moments, the stability of uniform nuclear matter against density fluctuations, elastic magnetic electron scattering from ^{15}N and ^{209}Bi and quasielastic electron scattering from ^{40}Ca are all considered. Vacuum polarization substantially changes results even at modest momentum transfers.

INTRODUCTION

Several new experiments in coming years (for example at CEBAF) will probe nuclear systems at a range of momentum transfers around the nucleon mass $q \approx M$ where relativity is clearly needed. However, q is not yet high enough for perturbative QCD to be directly useful. Therefore, we will describe the nuclear system in terms of an interacting nucleon and meson relativistic quantum field theory.

In this paper we examine relativistic effects at finite momentum transfer q by considering the nuclear response to a variety of probes. There has been much previous work looking for relativistic effects at very low q. Relativistic impulse approximation [12] calculations have found large relativistic effects on proton elastic scattering spin observables. These lead to much better agreement with experiment. However the effect is largest in forward angle diffraction minima where q is small.

There have been many relativistic calculations of magnetic moments (which of course are defined in the q goes to zero limit) [10]. A strong Lorentz scalar field enhances the convection current of a valence nucleon. This greatly changes the Schmidt magnetic moments for closed shell ± one nuclei and is in sharp disagreement with experiment.

However, a valence nucleon will polarize both core and Dirac sea nucleons. Many authors have shown [10,11] that the vector interaction with the valence particle mixes negative energy sea with positive energy core nucleons and leads to a strong reduction in the total convection current. This effect is the *Pauli blocking* of virtual nucleon-antinucleon excitations (vacuum polarization) which are no longer possible in the medium.

If the scalar and vector interactions almost cancel in the binding energy then the enhancement of the current from the strong scalar field will be cancelled by the Pauli blocking of vacuum polarization. Therefore, the isoscalar Schmidt magnetic moments are recovered. Thus, the convection current at q=0 is insensitive to relativistic effects.

However, at finite q there are a number of new relativistic features which may show up in elastic magnetic electron scattering. First, normal particle-hole excitations will now contribute. These have zero phase space at q=0 (in nuclear matter). The transverse vector particle-hole interaction is attractive. Furthermore, this interaction is greatly enhanced as the nucleon effective mass decreases. This particle-hole core polarization will increase convection currents at finite q.

Furthermore, there are other vacuum polarization effects in addition to the correction from Pauli blocking. These may be important because an experiment will be very sensitive to any *changes* in vacuum polarization with increasing nuclear density. Nucleons in both the Dirac and Fermi seas have a reduced effective mass because of the strong scalar potential. Therefore, virtual nucleon-antinucleon excitations may be very different in the medium because the particles now have a different mass. We will examine this change in vacuum polarization and its effects on magnetic electron scattering below.

There is much interest in the longitudinal response for quasielastic electron scattering [1]. Within the simplified model of nonrelativistic quantum mechanics and an impulse approximation for the electron-nucleon current operator a Coulomb sum rule can be derived. This says that the energy integrated longitudinal response at high momentum transfers should go to Z times the square of the proton charge. Experiments [1,2] have suggested that the integrated response is below this prediction (however this is not without some errors and uncertainties).

There have been a large number of relativistic calculations for quasielastic electron scattering [3-8]. These show that within a relativistic model there is no exact Coulomb sum rule [9]. Nevertheless, most relativistic calculations have ignored vacuum polarization and over-estimate experimental results.

The sum rule is calculated by summing over a complete set of intermediate states. In relativistic calculations, the positive energy states alone are not complete. Furthermore, because of the dynamical quantum vacuum, particle number is not conserved. Therefore the sum of the *squares* of the charges, which is what the sum rule tries to measure, is not conserved. A virtual nucleon antinucleon pair will not change the net baryon number or total charge but the pair will change the sum of the squares of the charges. Thus, a pair can contribute to quasielastic scattering and change the sum rule if there is enough momentum transfer to resolve the individual charges.

Vacuum polarization is clearly important at very high momentum transfer q. Here the Coulomb sum rule would measure the square of the nucleon charge at the momentum scale q, $e(q_\mu^2)^2$. This would be different from $e(0)^2$ because the effective charge changes due to vacuum polarization.

In principle, nucleons in the Dirac sea also have an effective mass M^*. Since M^* is significantly smaller than M, vacuum polarization of an M^* sea was found to be very different from the polarization of the free vacuum [13]. In this work, the screening of the bare charge from vacuum polarization was found to increase strongly as the effective mass decreases.

To investigate how this *change* in vacuum screening could effect the Coulomb sum rule, we calculate, in section 6, the longitudinal response of nuclear matter in a relativistic RPA approximation including vacuum polarization. First, section 2 presents the relativistic RPA formalism for the Walecka model. Then, magnetic moments of closed shell ± one nucleon nuclei are reviewed in section 3. The stability of uniform nuclear matter against density fluctuations is briefly discussed in section 4 as a prerequisite for examining magnetic electron scattering in section 5. Finally, conclusions are presented in section 7.

FORMALISM

Our starting point is a mean field model of the ground state of nuclear matter including vacuum fluctuation effects. This is the simplest approximation with the characteristic strong scalar and vector potentials of many relativistic approaches. The Walecka model is a relativistic quantum field theory with nucleon fields interacting with isoscalar scalar (sigma, ϕ) and vector (omega, V_μ) meson fields. The lagrangian is [14]

$$\mathcal{L} = \bar{\psi}[\gamma_\mu(i\partial^\mu - g_v V^\mu) - (M - g_s\phi)]\psi + \frac{1}{2}(\partial^\mu\phi\partial_\mu\phi - m_s^2\phi^2)$$

$$-\frac{1}{4}F_{\mu\nu}F^{\mu\nu} + \frac{1}{2}m_v^2 V_\mu V^\mu + \delta\mathcal{L} \tag{1}$$

Here neutral scalar mesons of mass m_s couple to the nuclear scalar density with strength g_s, and neutral vector mesons of mass m_v couple to the conserved Baryon current with g_v. The vector field tensor is $F_{\mu\nu} = \partial_\mu V_\nu - \partial_\nu V_\mu$, and $\delta\mathcal{L}$ is the counter term lagrangian. The scalar meson provides the intermediate range attraction in the NN interaction, while the vector meson gives the short-range repulsion.

This model has been used for many calculations of nuclear matter and finite nuclei at the mean-field level and in more sophisticated approximations [15]. In the mean-field theory (MFT), the scalar and vector meson field operators are replaced by their ground state expectation values, which are classical fields:

$$\phi \rightarrow < \phi > \equiv \phi_0,$$

$$V_\mu \rightarrow < V_\mu > \equiv \delta_{\mu 0}V_0. \tag{2}$$

This approximation is expected to become better as the density increases. Corrections to the MFT have been discussed by many authors; see the references in [15].

The MFT can be solved exactly and the couplings adjusted to reproduce the saturation density (corresponding to a Fermi momentum of $k_F = 1.30$ fm^{-1}) and binding energy of nuclear matter. Table I contains parameter sets for mean field calculations where vacuum fluctuations are included, RHA (relativistic Hartree approximation), or neglected, MFT. These parameters have been used for finite nucleus calculations, either in the MFT [16] or RHA [17] approximations, where they provide a good description of ground state charge densities, spin-orbit splittings and single particle energies.

The mean scalar field shifts the mass of all of the nucleons from M to M^* (both in the Dirac and Fermi seas), where

$$M^* \equiv M - g_s\phi_0. \tag{3}$$

This change in mass of the Dirac sea produces a vacuum fluctuation contribution to the energy density that is included in the RHA. In either the MFT or the RHA, one

Table I. Model Parameters

Model	g_s^2	g_v^2	m_s(MeV)	m_v(MeV)	M^*/M
MFT	109.626	190.431	520.	783.	0.541
RHA	54.289	102.77	458.	783.	0.730

solves self-consistently for the Dirac spinors and mean scalar fields at each density to arrive at the values of M^* in Table I.

We now examine the linear response of nuclear matter in the MFT or RHA to virtual photons from electron scattering. The starting point is the lowest order polarization insertion $\Pi_{\mu\nu}$ for vector mesons [18]. This describes the coupling of a virtual vector meson or photon of momentum q to a particle-hole or nucleon-antinucleon excitation.

$$\Pi_{\mu\nu}(q) = -ig_v^2 \int \frac{d^4k}{(2\pi)^4} Tr[\gamma_\mu G(k)\gamma_\nu G(k+q)] \tag{4}$$

Here G(k) is the self-consistent RHA baryon propagator which we can write

$$G(k) = G_F(k) + G_D(k). \tag{5}$$

The first part is the usual Feynman propagator G_F,

$$G_F(k) = \frac{[\gamma^\mu k_\mu^* + M^*]}{k_\lambda^{*\,2} - M^{*2} + i\epsilon}, \tag{6}$$

where the mass is given by eq. (3) and the energy has been shifted by the mean vector field,

$$k^{*\mu} = \{k_0 - g_v V_0, \vec{k}\}. \tag{7}$$

[We will not need to know V_0 since the dummy integration variable k_0 can be shifted to $k_0 - g_v V_0$ once the integrals are regularized.] The remaining part

$$G_D(k) = [\gamma_\mu k^{*\mu} + M^*]\frac{i\pi}{E_k^*}\delta(k_0^* - E_k^*)\Theta(k_F - |\vec{k}|), \tag{8}$$

where $E_k^* = \sqrt{\vec{k}^2 + M^{*2}}$ and k_F is the Fermi momentum, changes the positions of the poles in G_F because of the occupied states in the Fermi sea. Using this decomposition of G the polarization is

$$\Pi_{\mu\nu}(q) = \Pi_{\mu\nu}^{RF}(q) + \Pi_{\mu\nu}^{D}(q), \tag{9}$$

where Π^D is the explicitly density dependent part involving at least one power of G_D, and Π^{RF} is the renormalized vacuum polarization which involves only G_F.

$$\Pi_{\mu\nu}^{RF}(q) = \Pi_{\mu\nu}^{F}(q) - CTC \tag{9}$$

$$\Pi_{\mu\nu}^{F}(q) = -ig_v^2 \int \frac{d^4k}{(2\pi)^4} Tr[\gamma_\mu G_F(k)\gamma_\nu G_F(k+q)] \tag{10}$$

Here the subscript R implies renormalized and CTC indicates the renormalization counter term contributions. The renormalization scheme and the computation of $\Pi_{\mu\nu}^{RF}$

are discussed in ref. [13]. There a wave function counter term is subtracted in order that $\Pi_{\mu\nu}^{RF}$ vanish at $q_\mu^2 = 0$ and $M^* = M$. Other choices for renormalization point give similar results. The vacuum polarization is

$$\Pi_{\mu\nu}^{RF}(q) = (q_\mu q_\nu / q_\lambda^2 - g_{\mu\nu})\Pi_v^{RF}(q), \tag{11}$$

which defines Π_v^{RF} and

$$\Pi_v^{RF}(q) = \frac{\lambda_{vac} g_v^2 q_\lambda^2}{2\pi^2} \int\limits_0^1 d\alpha\, \alpha(1-\alpha)ln\{\frac{M^{*2} - \alpha(1-\alpha)q_\lambda^2}{M^2}\}. \tag{12}$$

We assume the isospin degeneracy of the vacuum to be $\lambda_{vac} = 2$. Our results in section 6 for the Coulomb sum rule stem from the M^* and q_λ^2 dependence of Π_v^{RF}.

MAGNETIC MOMENTS

The convection current of a valence particle is enhanced by the strong scalar potential. Therefore, the valence nucleon contribution to the Schmidt magnetic moment of a closed shell $\pm$ one nucleus will be substantially modified. This modification alone is in strong disagreement with experiment.

However, the vector interaction from the valence particle will polarize both the core and the Dirac sea. This can be taken into account with the following effective current operator.

$$J_{eff}^\mu = lim_{q_0 \to 0}\, lim_{q \to 0}\gamma^\mu\,[\alpha_T(q, q_0) + \tau_3]/2 \tag{13a}$$

$$\alpha_T(q, q_0) = \frac{1}{1 - \Pi_t(q, q_0)D(q)} \tag{13b}$$

Note, the polarization is isoscalar because our simple model has uncharged mesons (τ_3 is an isospin operator).

Here the transverse polarization insertion is [18]

$$\Pi_t(q) \equiv \Pi_{11}(q) \tag{14}$$

where the coordinate system has been chosen so that the momentum transfer lies along the three axis. The density dependent part of this has been calculated in ref [4]. (Note, with our conventions $\Pi_t^D(q) = \frac{-g_v^2}{(2\pi)^4}\Pi_T$ where Π_T is given by eq (2.45) of ref [4].) The renormalized vacuum polarization part is given in eqs (11-12). Finally the interaction is

$$D(q) = \frac{1}{q_\mu^2 - m_s^2}. \tag{15}$$

A number of authors have shown that isoscalar magnetic moments of closed shell $\pm$ one nucleon nuclei calculated with J_{eff} (either in a local density approximation [10] or exactly [21]) are very close to the original nonrelativistic Schmidt values. This comes about from a cancellation. First the matrix element of $\vec{\gamma}$ is increased because of the enhanced lower component of the relativistic wavefunction. However, the factor α_T is substantially less then one. Therefore, the product of an enhanced matrix element times a reduced effective current is insensitive to the relativistic dynamics.

It is important to identify the parts of the polarization which are contributing to this reduction in α_T. At $q = 0$ there is no phase space for normal particle-hole excitations (in nuclear matter). Therefore, in the nonrelativistic limit $\alpha_T = 1$ at q=0. Furthermore, because of the choice of renormalization condition in reference [13] the vacuum polarization also vanishes at q=0. However, there is a new piece of the polarization which describes negative energy-hole excitations. This piece of Π_t^D which corrects for the Pauli blocking of vacuum polarization has no nonrelativistic analog and is responsible for the reduction in the current. Therefore, one can not simply ignore all negative energy states for this is inconsistent and gives dramatically incorrect currents and magnetic moments.

STABILITY OF UNIFORM NUCLEAR MATTER

We would like to examine the effective relativistic current at finite momentum transfer with elastic magnetic electron scattering. However, first we consider the stability of uniform nuclear matter against density fluctuations. This is because any instabilities at high density may have precursors which will be visible in electron scattering.

It is easy to solve for the energies of small amplitude density fluctuations of three momentum q. One simply searches for an energy q_0 at which there is a pole in the relativistic RPA meson propagator. For example, for transverse modes one searches for a q_0 at which $\alpha_T(q, q_0) = \infty$. If the energy q_0 of the mode is less than or equal to zero the system will be unstable. This search was done for longitudinal and transverse modes in both the RHA and MFT approximations in reference [13].

The stability of the uniform mean field theory is indicated schematically in fig. 1. At low density both the RHA and MFT are unstable to low q modes. This represents the well known liquid-vapor phase transition of bulk nuclear matter below its saturation density. Interestingly, there is another instability of the MFT (but not the RHA) at high density and medium q. This is do to a very attractive relativistic particle-hole interaction from the exchange of transverse vector mesons. This interaction is a relativistic effect and becomes very large at high density where M^* is small. We will refer to this instability as a "transverse particle-hole condensate".

Finally the RHA (but not MFT) is unstable at very large q because of vacuum polarization [13]. The effective meson-nucleon coupling increases with momentum transfer in the nonasymptotically free Walecka model. This will eventually lead to a pole in the meson propagator at very high q. This is just like the Landau ghost in the RPA photon propagator in QED.

Much further work is needed to deal with these vacuum polarization poles in meson and baryon descriptions of nuclear physics. The poles may represent an upper limit to the momentum transfer domain where it is useful to describe the system in hadron rather than quark or gluon coordinates. We will not consider these very high momentum transfers in the remainder of the paper.

ELASTIC MAGNETIC ELECTRON SCATTERING

We now consider elastic magnetic electron scattering in a relativistic RPA approximation using a local density approximation. We use the effective current operator of eq. (13b) where α_T is taken from nuclear matter results at the appropriate local density in a finite nucleus. We add to this current the contribution of the anomalous moment which we assume is unchanged by the RPA response. This is

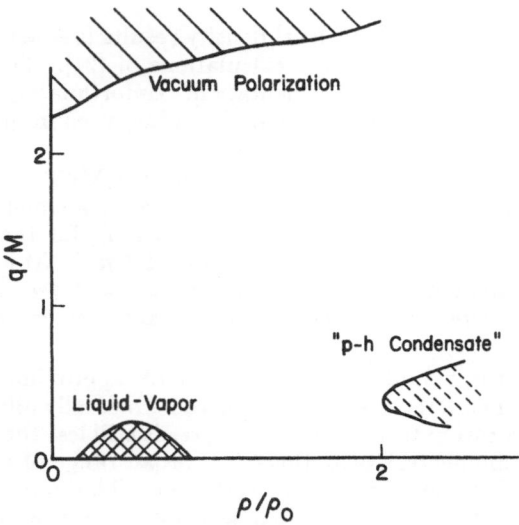

Fig. 1. Sketch of the unstable modes of uniform nuclear matter as a function of the momentum of the mode (vertical scale) in units of the nucleon mass vs. density in units of normal nuclear density. The liquid-vapor phase transition region is unstable in both the MFT (without vacuum) and RHA approximations while the "particle-hole condensate" is unstable only in the MFT and the vacuum polarization region is unstable only in the RHA.

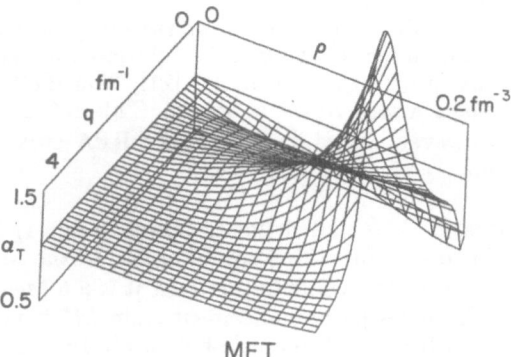

Fig. 2. Random phase approximation enhancement factor α_T eq. (13b) for the MFT as a function of both density ρ and three momentum transfer q.

because the anomalous moment is primarily isovector while our simple model has only isoscalar mesons. The effects of isovector mesons remains to be evaluated.

We note, for ^{15}N our MFT RPA local density results (see below) are qualitatively similar to the exact finite nucleus RPA calculations of [21]. Thus the local density approximation provides a useful first orientation. Unfortunately, there are no finite nucleus calculations for ^{209}Bi or for the RHA including vacuum polarization.

The factor $\alpha_T(q, q_0, \rho)$ is plotted in figure 2 for the MFT as a function of both q and density ρ. The energy transfer q_0 is arbitrarily set to a small value $q_0 = 10$ MeV rather then zero in order to assure the correct q=0 limit for the magnetic moment. The choice of q_0 does not effect the result for q above 1 fm^{-1}. At low q α_T is below one as we saw for the magnetic moments. However above $q = 1$ fm^{-1} α_T is substantially above one. This enhancement is a precursor to the transverse particle-hole condensate.

Figure 3 shows the results for α_T in the RHA approximation which includes vacuum polarization. The RHA response is seen to be radically different then the MFT even at moderate momentum transfers! At low q α_T is still less then one in order to get the correct magnetic moments. Next, there is a broad range of momentum transfers and densities over which the response is nearly one. This represents a cancellation between enhancements from the particle-hole response and a reduction from both vacuum polarization and the Pauli blocking of vacuum polarization.

Note, at zero density the response is no longer unity. Vacuum polarization increases the response slightly by q=4 fm^{-1} (which is the highest momentum plotted). In the limit of very high q this increase will lead to the vacuum polarization pole discussed in the last section.

Finally at high q and high density the response is significantly less than one. This reduction is from vacuum polarization. As the effective nucleon mass decreases virtual pairs are more effective at screening the original charge. Therefore, because of the *change* in vacuum polarization with M^* less of the charge is visible to the electron probe.

The transverse form factor for elastic magnetic electron scattering from ^{209}Bi is shown in fig. 4. The dashed curves are impulse approximation or Hartree calculations assuming the most simple $1h_{9/2}$ wavefunction for a valence particle moving in the inert core potentials of ^{208}Pb. The MFT Hartree curve is already about a factor of two higher then a similar nonrelativistic calculation (not shown) because of the enhancement in the convection current from M^* [21]. The MFT RPA response (upper solid curve) is below the Hartree result at low q. This reflects the reduction in the magnetic moments. However, above q=1 fm^{-1} the RPA curve is above the already enhanced Hartree result.

If vacuum polarization is included (lower curves in fig. 4) there are a number of changes. First the valence wavefunctions are different. In the MFT the $1h_{9/2}$ binding energy is somewhat low at 2 MeV while in the RHA it is 3.5 MeV. Second because M^* is somewhat larger in the RHA the enhancement from M^* is not as big. Finally, the further increase at high q from RPA particle-hole excitations is not seen in the RHA RPA because of vacuum polarization. Therefore, the RHA RPA curve is substantially below the MFT RPA result.

It may be premature to compare these simple calculations to data. As a minimum, one still needs to consider polarizations from isovector mesons and meson exchange currents. Nevertheless, we compare MFT and RHA results with data in figs

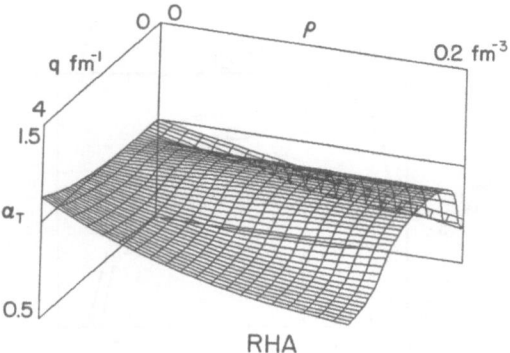

RHA

Fig. 3. Random phase approximation enhancement factor α_T eq. (13b) for the RHA including vacuum polarization as a function of both density ρ and three momentum transfer q.

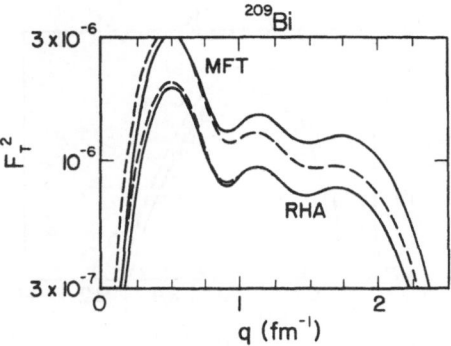

Fig. 4. Transverse elastic magnetic electron scattering response for ^{209}Bi vs. momentum transfer q. The solid curves are RPA results using the α_T factors from figures 2 and 3 while the dashed lines are Hartree calculations. The upper two MFT curves ignore vacuum polarization which is included in the lower two RHA curves.

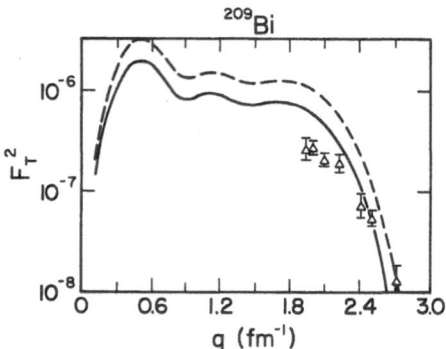

Fig. 5. Transverse elastic magnetic response for ^{209}Bi in a relativistic RPA approximation. The dashed curve is for the MFT while the solid curve is the RHA result.

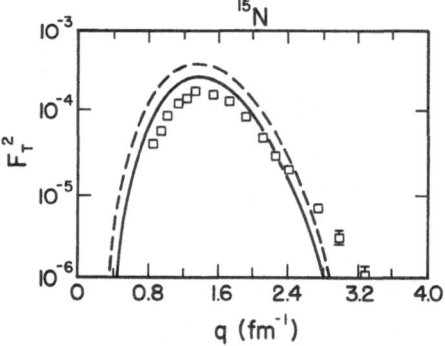

Fig. 6. Transverse elastic magnetic response for ^{15}N in a relativistic RPA approximation. The dashed curve is for the MFT while the solid curve is for the RHA.

5 (^{209}Bi) and 6 (^{15}N). Unless futher corrections produce a substantial cancellation, the strong enhancements predicted in the MFT are not seen in the data.

These enhancements may be an artifact of the neglect of vacuum polarization. When vacuum polarization is included in the RHA the results are closer to nonrelativistic calculations and to data. We conclude that a proper treatment of vacuum polarization may be necessary to calculate convection currents in relativistic models at finite q. This is consistent with earlier results showing the Pauli blocking of vacuum polarization is crucial for the correct magnetic moments.

QUASIELASTIC ELECTRON SCATTERING

In the Hartree approximation the longitudinal response function $S_L(q)$ measured in electron scattering is simply given by

$$S_L(q) = \frac{-ZF_{1p}^2(q)}{g_v^2\pi\rho} Im\Pi_{00}(q),$$ (16)

where Z is the nuclear charge, ρ the density and the proton form factor F_{1p} is parameterized as in ref. [6]. The RPA approximation for the polarization Π_{00}^{RPA} goes beyond the Hartree approximation and sums all of the ring diagrams. This incorporates the polarization of the rest of the medium (including the vacuum). This involves vector and scalar meson mixing which can be expressed through a 2x2 matrix (see [20]). In the Walecka model there are only isoscalar mesons. Therefore the isovector RPA response will be equal to the Hartree one. Thus, vacuum polarization only effects the isoscalar response.

The longitudinal response of the proton charge is half isovector and half isoscalar. Thus the total RPA response function is

$$S_L^{RPA}(q) = \frac{-ZF_{1p}(q)^2}{g_v^2\pi\rho} \left\{ \frac{Im\Pi_{00}^{RPA} + Im\Pi_{00}}{2} \right\}.$$ (17)

We now add to either eq. (16) or eq. (17) the small contribution of the anomalous moments. However, since the anomalous moment is nearly isovector we simply evaluate it in a Hartree approximation.

The longitudinal response function is shown in Figure 7 for nuclear matter at a three-momentum transfer of q=550 MeV and $k_F = 1.30$ fm^{-1} using the RHA parameters in Table I. The RPA result using eq. (17) is substantially below the Hartree calculation based on eq. (16). The integrated strength (or Coulomb sum rule), C(q),

$$C(q) = \int_0^q dq_0 S_L(q, q_0),$$ (18)

is shown in Figure 8 for nuclear matter (although multiplied by Z=20) as a function of three-momentum transfer. The strength is about thirty percent lower in the RPA calculation than for the Hartree response. This is a much larger reduction than the approximately 15 percent found for relativistic RPA calculations omitting vacuum polarization [4,6,7,8].

We emphasize these calculations are for nuclear matter. Although we have not yet performed a finite nucleus calculation including the vacuum, it is possible the

Longitudinal Response

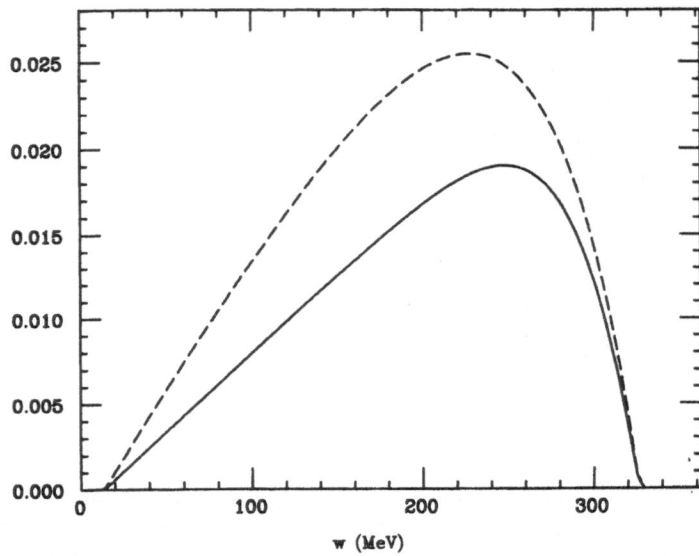

Fig. 7. Longitudinal response function (in MeV^{-1}) vs. excitation energy ($q_0 = \omega$) for infinite nuclear matter at q = 550 MeV and $k_F = 1.30$ fm^{-1} (Z=20) using the RHA parameters in Table I. The dashed curve is the Hartree response of eq. (16) while the RPA result of eq. (17) is indicated by a solid line.

Coulomb Sum

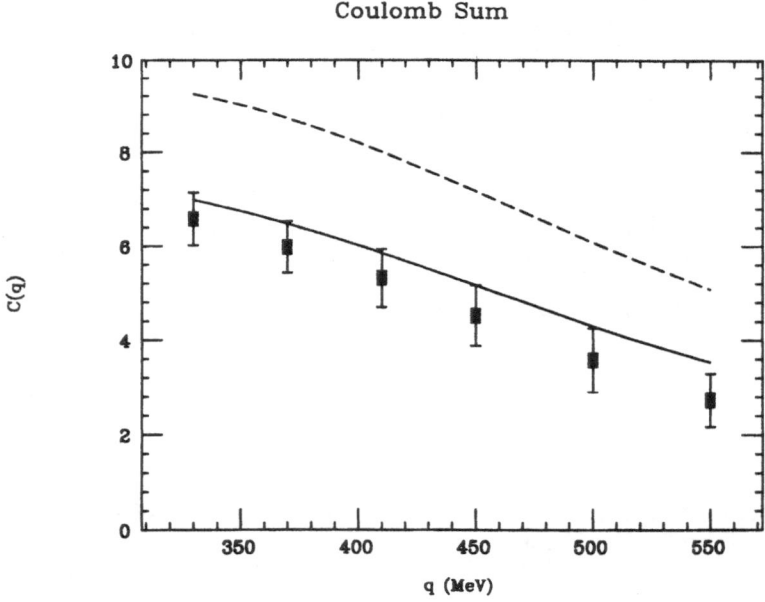

Fig. 8. Coulomb sum C(q) of eq. 18 in Hartree (dashed) or RPA (solid) approximations vs. momentum transfer. The Saclay ^{40}Ca electron scattering results of ref. [1] are also shown.

integrated strengths will be similar. Therefore, we show in Figure 8 the Saclay results for ^{40}Ca from Ref. [1]. These are consistent with our RPA results.

Longitudinal electron scattering is related to the square of the proton's charge at the momentum scale q of the experiment. In our calculation, this charge is being substantially modified by vacuum polarization as follows,

$$e(q_\mu^2) = \frac{1}{1 - \Pi_v^{RF}(q_\mu^2)/(q_\mu^2 - m_v^2)} e(q_\mu^2 = 0). \qquad (19)$$

If the polarization in eq. (12) is expanded for small q this can be approximated (note, this is only qualitative!)

$$\frac{e(Q^2)^2}{e(0)^2} \approx \left\{ 1 - \frac{g_v^2}{m_v^2 + Q^2} \frac{Q^2}{3\pi^2} [\ln\frac{M^*}{M} + \frac{2Q^2}{5M^{*2}}] \right\}^{-2} \cdot \qquad (20)$$

where we have introduced the space-like momentum transfer $Q^2 = -q_\mu^2 > 0$. The polarization in eq. (20) has two terms. The second term is present even if $M^* = M$ and increases the charge. However, the first $ln\frac{M^*}{M}$ term is the new feature of these calculations. It describes how the vacuum polarization changes with density because of the decrease in the effective mass. It is negative $(M^* < M)$ and will dominate for the small to moderate momentum transfers of present experiments. This change in the vacuum polarization with decreasing effective mass is responsible for the large reduction in our RPA response.

Vacuum polarization in our model is isoscalar. Therefore the transverse response, which is dominated by the isovector anomalous moment, is almost unchanged. Indeed, the RPA transverse response is less then five percent below the Hartree result at q=550 MeV (not shown).

The momentum transfer dependence of vacuum polarization is very different from the normal particle-hole response. The latter decreases rapidly for q above $2k_F$. Although relativistic RPA calculations without the full vacuum do find some small decrease in the sum rule it is not from the particle-hole response. The particle-hole response should be very small at momentum transfers of q=550 MeV and above.

The 15 percent decrease found in refs. [4], [6], [7] and [8] is due to the Pauli blocking of vacuum polarization in the medium [19] (which they call 'backflow'). Indeed, the Pauli blocking correction to vacuum polarization is already contained in the density dependent polarizations, Π^D. The reduction in the Coulomb response is thus coming from *changes* in vacuum polarization with density even in these earlier works. It appears that about half of our total reduction is coming from vacuum Pauli blocking and half from the dependence of vacuum polarization on the effective mass (which has been ignored up to now).

Earlier works separated the polarization in to two parts, eq. (9), and then ignored Π_v^{RF} without any justification. Clearly, the separation of the total polarization into density dependent and vacuum parts is arbitrary. This is because the quantum vacuum effects both pieces of the polarization (through Pauli Blocking). Thus, it is inconsistent to keep the density dependent parts of the polarization and simply neglect Π_v^{RF}. A calculation which neglects Π_v^{RF} is not a good approximation to the full calculation even at low momentum transfers.

We find that the full vacuum polarization gives a sizeable decrease in the response for a broad range of momentum transfers. The RPA integrated response is reduced by

24, 30, 34 and 31 percent compared to the Hartree response at momentum transfers of 330, 550, 1000 and 1200 MeV. Vacuum polarization is not expected to be less important at high momentum transfer. However, the reduction in the sum rule finally decreases by q=1200 MeV because of the second term in eq. (20), which increases the charge. Indeed, in the limit of very high q, vacuum polarization will substantially *enhance* the sum rule.

It is important to perform finite nucleus calculations to examine the A dependence of vacuum polarization effects. We speculate that vacuum effects could be significantly smaller in a light nucleus because of the higher average effective mass than in a heavy target. The effective mass increases because of the larger surface (where M^* is large) to volume ratio. A larger M^* will reduce the first term in eq. (20) and could have a sizeable effect because there is significant cancellation between the two terms.

The effect of nucleon substructure on this change in vacuum polarization remains to be investigated. However, the momentum transfers involved are very low. Typical one boson exchange potential form factors for the omega-nucleon vertex will not change our results very much at $q = 550$ MeV. Nevertheless, the effect of substructure at this small a momentum transfer is an open problem.

CONCLUSIONS

We have calculated the response of nuclear matter in a relativistic RPA approximation for the Walecka model. The convection current of a valence proton is enhanced because of its small effective mass. However, in the q=0 limit this enhancement is cancelled by Pauli blocking corrections to vacuum polarization. This effect (often called 'backflow') restores the isoscalar Schmidt magnetic moments.

At finite q, relativistic models without a full treatment of vacuum polarization still predict a sizeable enhancement in convection currents. This is true even with 'backflow' corrections from the density dependent parts of the polarization. This enhancement appears to disagree with data for elastic magnetic electron scattering from ^{15}N and [209]Bi.

Alternatively, if vacuum polarization is included, this enhancement is greatly reduced. Thus it appears to be necessary to consistently treat both the Pauli blocking of vacuum polarization at q=0 and the *change* in vacuum polarization with decreasing effective mass at finite q. If this is done then relativisitic calculations are in good agreement with both isoscalar magnetic moments and elastic magnetic form factors.

The longitudinal quasielastic response function for nuclear matter was also calculated. The RPA response (which includes the vacuum) is very different from earlier calculations where the vacuum was ignored. The polarization of the $M^* = M$ vacuum is not important at these momentum transfers. However, the *change* in vacuum polarization when the nucleon mass is reduced from M to M^* has dramatic effects. The relativistic RPA Coulomb sum rule is reduced by thirty percent from the Hartree result.

Finite nucleus calculations are underway to examine vacuum polarization effects further. Also, there is a need to consider better approximations than the simple RPA. One should look at vacuum polarization for other meson nucleon models (including nonlinear self-couplings and/ or isovector mesons). Finally, it is important to try to observe this polarization (or lack of it) in quark models or QCD.

REFERENCES

1. Z. E. Meziani et al., Phys. Rev. Lett. **52** (1984) 2130.

2. M. Deady et al., Phys. Rev. **C28** (1983) 631.

3. R. Rosenfelder, Ann. of Phys. (N.Y.) **128** (1980) 188.

4. H. Kurasawa and T. Suzuki, Nucl. Phys. **A445** (1985) 685.

5. H. Kurasawa and T. Suzuki, Phys. Lett. **B173** (1986) 377.

6. S. Nishizaki et al., Phys. Lett. **B171** (1986) 1.

7. K. Wehrberger and F. Beck, Phys. Rev. **C35** (1987) 298 and Erratum **C35** (1987) **2337**.

8. K. Wehrberger and F. Beck, Phys. Rev. C in press.

9. G. Do Dang et al., Phys. Rev. **C35** (1987) 1637.

10. J. A. McNeil et al., Phys. Rev. **C34** (1986) 746.

11. R. J. Furnstahl and B. D. Serot, Nucl. Phys. **A468** (1987) 539.

12. J. R. Shepard et al., Phys. Rev. Lett., **50** (1983) 1443; L. Ray and G. Hoffman, Phys. Rev. **C31** (1985) 538; D. P. Murdock and C. J. Horowitz, Phys. Rev. **C35** (1987) 1442.

13. C. J. Horowitz and R. J. Furnstahl, "Vacuum Polarization Effects on Meson Propagators", Indiana University Nuclear Theory Center Preprint IU/NTC 87-6 Nucl. Phys. **A** in press.

14. J. D. Walecka, Ann. of Phys. (N.Y.) **83** (1974) 491.

15. B. D. Serot and J. D. Walecka, Advances in Nucl. Phys. **16**, J. W. Negele and E. Vogt eds. (Plenum, NY 1986).

16. C. J. Horowitz and B. D. Serot, Nucl. Phys. **A368** (1981) 503.

17. C. J. Horowitz and B. D. Serot, Phys. Lett. **140B** (1984) 181.

18. S. A. Chin, Ann. of Phys. (N.Y.) **108** (1977) 301.

19. H. Kurasawa and T. Suzuki, Nucl. Phys. **A454** (1986) 527.

20. C. J. Horowitz, "Vacuum Polarization and the Coulomb Sum Rule", Indiana University Nuclear Theory Center Preprint IU/NTC 88-4, submitted to Phys. Lett. B.

21. B. D. Serot, "Elastic Electron-Nucleus Scattering in a Relativistic Theory of Nuclear Structure", Phys. Lett. **107B** (1981) 263.

DIRAC RPA FOR FINITE NUCLEI

James R. Shepard
Department of Physics
University of Colorado
Boulder, CO 80309

1. Introduction

The Dirac-Hartree model for ground states of doubly closed-shell nuclei has been extensively studied[1,2] . In this paper I wish to discuss a version of the Random Phase Approximation (RPA) for nuclear excited states which has the Dirac-Hartree for finite nuclei as its starting point. This RPA treatment is self-consistent throughout and automatically incorporates a number of physically desireable features including exact conservation of the electromagnetic current and complete separation of spurious and non-spurious levels. I will discuss two distinct computational approaches to this Dirac RPA problem, one of which avoids basis truncations by automatically including both the positive and negative energy continua with correct continuum boundary conditions. A unified approach for extracting the nuclear response for both discrete and continuum excitations will be outlined. The sum rules for this version of the Dirac RPA will also be discussed. Finally directions for future study will be addressed.

2. Review of Dirac-Hartree

In the QHD-1 model of finite nuclei,[2] one begins with a Lagrangian density

$$\mathcal{L} = \mathcal{L}(\psi, \phi, V^\mu) \qquad (2.1)$$

which specifies the behavior of interacting nucleon and isoscalar scalar (σ) and vector (ω) meson fields represented by ψ, ϕ, and V^μ, respectively. The interacting single particle Green function is obtained in Hartree approximation by summing tadpole diagrams[1,2] :

$$
\begin{aligned}
G(x,y) &= -\imath \langle \Psi_0 | T[\hat{\psi}'(x)\hat{\bar{\psi}}'(y)] | \Psi_0 \rangle \\
&\simeq G_H(x,y) = -\imath \langle \Phi_0 | T[\hat{\psi}(x)\hat{\bar{\psi}}(y)] | \Phi_0 \rangle \\
&= G_0(x,y) + \int d^4z \, G_0(x,z)\Sigma_H(z)G_H(z,y)
\end{aligned}
\qquad (2.2)
$$

with

$$\Sigma_H = \Sigma_s + \gamma_\mu \Sigma_v^\mu,$$
$$\Sigma_s = -\imath g_s^2 \, \Delta_0 \, \mathrm{Tr}[G_H] \tag{2.3}$$
$$\Sigma_v^\mu = +\imath g_v^2 \, D_0 \, \mathrm{Tr}[\gamma^\mu G_H]$$

where G, G_0 and G_H are the full, free and Hartree propagators, respectively, Ψ_0 (Φ_0) is the full (Hartree) ground state, and $\hat{\psi}'$ ($\hat{\psi}$) is the full (Hartree) field operator. Also, Σ_H is the Hartree self-energy consisting of scalar (Σ_s) and vector (Σ_v^μ) contributions depending on the scalar and vector coupling constants, g_s and g_v, respectively, and the meson propagators, Δ_0 and D_0. In the simple model used here, the self-energies are evaluated subject to a further approximation which can be understood from the structure of G_H. The Hartree basis consists of eigenfunctions of the following Dirac equation:

$$(\gamma^0 \epsilon_\beta + \imath \vec{\gamma} \cdot \vec{\nabla} - m - \Sigma_s(r) - \gamma^0 \Sigma_v^0(r)) \psi_\beta(\vec{r}) = 0 \tag{2.4}$$

where we have used the fact that, for doubly closed-shell nuclei, $\Sigma_v^\mu \to \Sigma_v^\mu \, \delta^{\mu,0}; i.e.,$ only the time-like component of Σ_v^μ is non-vanishing in the nuclear rest frame. This implies the eigenfunctions have total angular momentum as a good quantum number. We may express G_H in terms of the basis $\{\psi_\beta\}$ via a "spectral" expansion:

$$G_H(\vec{x}, \vec{y}; \omega) = \sum_\beta \psi_\beta(\vec{x}) \bar{\psi}_\beta(\vec{y}) G_\beta(\omega)$$

where

$$G_\beta(\omega) = \left[\frac{\theta(\epsilon_\beta)}{\omega - \epsilon_\beta + \imath\eta} + \frac{\theta(-\epsilon_\beta)}{\omega - \epsilon_\beta - \imath\eta} \right]$$
$$+ 2\pi\imath\delta(\omega - \epsilon_\beta)\theta(F - \beta)\theta(\epsilon_\beta) \tag{2.5}$$
$$= G_{V,\beta}(\omega) + G_{D,\beta}(\omega).$$

In these expressions, β labels a Hartree basis state, F represents the Fermi surface and the structure of the vacuum contribution, $G_{V,\beta}$, ensures positive energy states propagate forward in time while negative energies go backward. The density contribution, $G_{D,\beta}$, represents the correction due to the fact that positive energy levels up to and including $\beta = F$ are occupied and should propagate backward, too. The self-energies of Eq. (2.3)are divergent due to the vacuum contributions entering via $G_{V,\beta}$ and must be renormalized[2,3]. In the present work however, the effect of the vacuum is assumed to be incorporated in the effective masses and coupling constants of the original Lagrangian density. This amounts to making the mean-field approximation[1,2] but, with the reader's indulgence, will be referred to hereafter as the Hartree approximation. We then explicitly treat only the contributions to Σ_H arising from $G_{D,\beta}$. This yields, for example,

$$\Sigma_s(|\vec{x}|) = \imath g_s^2 \int \frac{d\omega}{2\pi} \, d^3 y \, \Delta_0(|\vec{x} - \vec{y}|; \omega' = 0) \, \mathrm{Tr}[G_D(\vec{y}, \vec{y}; \omega)]$$
$$= -g_s^2 \int d^3 y \frac{e^{-m_s|\vec{x} - \vec{y}|}}{4\pi |\vec{x} - \vec{y}|} \sum_{F \geq \beta \geq 0} \psi_\beta(\vec{y}) \bar{\psi}_\beta(\vec{y}) \tag{2.6}$$

where m_s is the σ mass and where the index β runs over positive-energy, occupied levels only. Calculations of this kind, using the QHD-1 meson masses and coupling constants of Ref. 1, have been shown to give a good reproduction of properties of doubly closed-shell nuclei, including binding energies and charge distributions.

In closing this section, we present an alternative representation of the Hartree propagator which appears in "spectral" form in Eq. (2.5). We write

$$
G_H(\vec{x}, \vec{y}; \omega) = \sum_{ljm} \Big[\psi_{u;ljm}(\vec{x}; \omega) \bar{\psi}_{v;ljm}(\vec{y}; \omega) \theta(y - x)
$$
$$
+ \psi_{v;ljm}(\vec{x}; \omega) \bar{\psi}_{u;ljm}(\vec{y}; \omega) \theta(x - y) \Big] \tag{2.7}
$$

where ψ_u and ψ_v each satisfy Eq. (2.4) with $\epsilon_\beta \to \omega$ but where ψ_u (ψ_v) is regular at $r = 0(\infty)$ and where a suitable normalization is obtained by dividing by the Wronskian of the functions associated with the upper components of ψ_u and ψ_v. Note that the linear independence of ψ_u and ψ_v is equivalent to specifying that ω is *not* an eigenvalue of Eq. (2.4). The details of the construction of this propagator are presented in Ref. 4. This "non-spectral" propagator can be thought of as a summation of the spectral form and is entirely equivalent to it.

3. Dirac RPA

There are numerous techniques for deriving the RPA equations for nuclear excitations beginning with the Hartree (or Hartree-Fock) ground[5][6][7][8] state.[6,7,8] In order to be as consistent as possible with the treatment of the Dirac Hartree problem outlined above, I will employ directly the Feynmann rules of field theory.[8]

We first consider some general properties of the fully interacting polarization insertion defined by

$$
\imath \Pi^{ij;kl}(x, y) \equiv \langle \Psi_0 | T[\hat{\psi}'^i(x) \bar{\hat{\psi}}'^j(x) \hat{\psi}'^k(y) \bar{\hat{\psi}}'^l(y)] | \Psi_0 \rangle \tag{3.1}
$$

where i, j, k, l are Dirac indices. This object desribes the propagation of a particle hole excitation in the ground state of the fully interacting system just as G defined in Eq. (2.2) describes the propagation of a single particle (or hole). We may go over to the Lehmann representation[8] ,

$$
\Pi(\vec{x}, \vec{y}; \omega) = \sum_n \Bigg[\frac{\langle \Psi_0 | \hat{\psi}'_S(\vec{x}) \bar{\hat{\psi}}'_S(\vec{x}) | \Psi_n \rangle \langle \Psi_n | \hat{\psi}'_S(\vec{y}) \bar{\hat{\psi}}'_S(\vec{y}) | \Psi_0 \rangle}{\omega - \omega_n + \imath \eta}
$$
$$
- \frac{\langle \Psi_0 | \hat{\psi}'_S(\vec{y}) \bar{\hat{\psi}}'_S(\vec{y}) | \Psi_n \rangle \langle \Psi_n | \hat{\psi}'_S(\vec{x}) \bar{\hat{\psi}}'_S(\vec{x}) | \Psi_0 \rangle}{\omega + \omega_n - \imath \eta} \Bigg] \tag{3.2}
$$

where Ψ_n is the full wavefunction for the n-th excited state whose excitation energy is $\omega_n = E_n - E_0$ and where $\hat{\psi}'_S(\vec{x})$ is the full field operator in the Schroedinger picture. Evidently, Ψ_n must be, at least in part, a 1 particle-1 hole state with respect to the ground state. Other excited states make no contribution to Π.

Eq. (3.2) makes it clear that $\Pi(\vec{x}, \vec{y}; \omega)$ has poles at $\omega = \omega_n$ and that the residue (for $\omega > 0$) is the outer product of the configuration space transition densities:

$$Res \lim_{\omega \to \omega_n} [\Pi(\vec{x}, \vec{y}; \omega)] = \bar{\mathcal{F}}^{(n)}(\vec{x}) \, \mathcal{F}^{(n)}(\vec{y})$$

where

$$\mathcal{F}^{(n)}(\vec{y}) \equiv \langle \Psi_n | \hat{\psi}'_S(\vec{y}) \tilde{\bar{\psi}}'_S(\vec{y}) | \Psi_0 \rangle \tag{3.3}$$

and where

$$\bar{\mathcal{F}} \equiv \gamma^0 \mathcal{F}^\dagger \gamma^0.$$

The transition amplitude for a one body operator $\mathcal{O}$ is given by

$$\langle \mathcal{O}_n \rangle = \int d^3 y \, \mathrm{Tr} \, [\mathcal{O}(\vec{y}) \mathcal{F}^{(n)}(\vec{y})] \tag{3.4}$$

where the trace is over Dirac indices. In a somewhat more general way, linear response theory[8] gives the response to an external probe whose interaction is represented by $\mathcal{O}$ as

$$S_{\mathcal{O}}(\omega) = -\frac{1}{\pi} \mathrm{Im} \int d^3 x \, d^3 y \, \mathrm{Tr} \, [\bar{\mathcal{O}}(\vec{x}) \Pi_{Ret}(\vec{x}, \vec{y}; \omega) \mathcal{O}(\vec{y})] \tag{3.5}$$

where Π_{Ret} is the retarded[8] (or causal) version of Π and where $\bar{\mathcal{O}} \equiv \gamma^0 \mathcal{O}^\dagger \gamma^0$. From Eqs. (3.2) through (3.5), we have

$$S_{\mathcal{O}}(\omega) = \sum_n |\langle \mathcal{O} \rangle_n|^2 \left[\delta(\omega - \omega_n) - \delta(\omega + \omega_n) \right]. \tag{3.6}$$

The standard techniques of field theory may be used to expand Π in terms of Hartree propagators and meson interactions. A diagramatic representation of this expansion appears in Fig. 1. Retaining only the "ring" contributions and summing to all orders yields the following RPA integral equation:

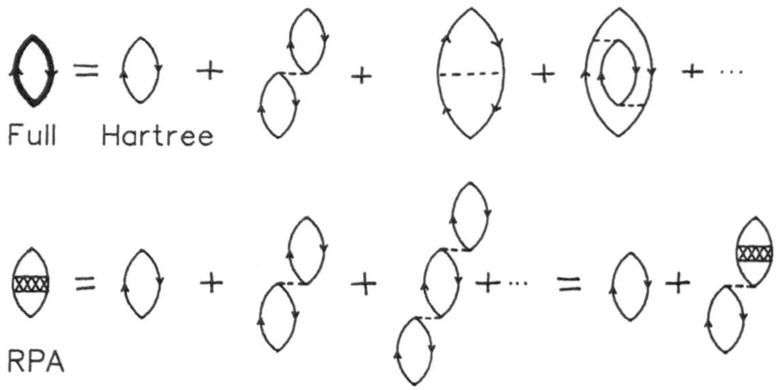

Fig. 1. Diagramatic representation of relations among full, Hartree and RPA polarization insertions. Solid lines are Hartree nucleon propagators while dashed lines are free meson propagators.

$$\imath\Pi^{ij;kl}(\vec{x},\vec{y};\omega) \simeq \imath\Pi_{RPA}^{ij;kl}(\vec{x},\vec{y};\omega) = \imath\Pi_H^{ij;kl}(\vec{x},\vec{y};\omega)$$

$$+ \sum_{\substack{mm'\\nn'}} \int d^3x_1 d^3x_2 \quad \imath\Pi_H^{ij;mm'}(\vec{x},\vec{x}';\omega)$$

$$\times \sum_N (\Gamma_N^a)^{m'm} \imath D_{N;ab}^0(\vec{x}_1 - \vec{x}_2;\omega)(\Gamma_N^b)^{n'n} \imath\Pi_{RPA}^{nn';kl}(\vec{x}_2,\vec{y};\omega)$$

$$(3.7)$$

where N represents the meson type, a and b refer to; $e.g.$, Lorentz indices, and m, m', etc. refer to Dirac indices. Also, Γ_N is a meson-NN vertex and $D_{N;ab}^0$ is the configuration space form of the free meson propagator. Finally, the Hartree polarization insertion is

$$\imath\Pi_H^{ij;kl}(x,y) \equiv \langle\Phi_0| T[\hat{\psi}^i(x)\hat{\bar{\psi}}^j(x)\hat{\psi}^k(y)\hat{\bar{\psi}}^l(y)] |\Phi_0\rangle \qquad (3.8)$$

or

$$\imath\Pi_H^{ij;kl}(\vec{x},\vec{y};\omega) = \int \frac{d\omega'}{2\pi} \; G_H^{il}(\vec{x},\vec{y};\omega+\omega') \; G_H^{kj}(\vec{y},\vec{x};\omega'). \qquad (3.9)$$

It should be emphasized that field theory provides a complete and systematic way of identifying *all* contributions to the full polarization insertion, Π. Therefore we know precisely what physical processes are omitted in making the approximation $\Pi \to \Pi_{RPA}$ and, in principle, how to correct for these omissions. The Π_{RPA} of Eq. (3.9) is divergent due to vacuum contributions. We drop these contributions and retain only the "density-dependent" terms. The justification for this step is that the vacuum effects are again accounted for in the phenomenological masses and coupling constants determined in the Hartree problem alluded to above. However, this is only approximately true since, as discussed by Horowitz at his conference, there are true dynamical effects arising from vacuum fluctuations which should be included in the Dirac RPA. Such effects are ignored in the discussion which follows. The "density-dependent" Hartree polarization propagator is then, schematically,

$$\imath\Pi_H(\omega) \to \int \frac{d\omega'}{2\pi} [G_D(\omega+\omega')G_H(\omega') + G_H(\omega+\omega')G_D(\omega')]$$

which gives

$$\imath\Pi_H^{ij;kl}(\vec{x},\vec{y};\omega) = \sum_{F \geq \beta \geq 0} [\psi_\beta^i(\vec{x})\bar{\psi}_\beta^l(\vec{y})G_H^{kj}(\vec{y},\vec{x};\epsilon_\beta-\omega)$$

$$+ G_H^{il}(\vec{x},\vec{y};\epsilon_\beta+\omega)\psi_\beta^k(\vec{y})\bar{\psi}_\beta^j(\vec{x})]. \qquad (3.10)$$

4. Spectral RPA

We now examine two distinct techniques for solving the integral equation which results from combining Eqs. (3.7) and (3.10) . The first makes use of a spectral expansion of the Hartree propagator (Eq. (2.5)). This yields

$$\Pi_H^{ij;kl}(\vec{x}, \vec{y}; \omega) = \sum_{\substack{F \geq \beta \geq 0 \\ \alpha}} \left[\frac{\psi_\alpha^i(\vec{x}) \bar{\psi}_\beta^j(\vec{x}) \psi_\beta^k(\vec{y}) \bar{\psi}_\alpha^l(\vec{y})}{\omega - \omega_{\alpha\beta} + i\eta_\alpha} - \frac{\psi_\beta^i(\vec{x}) \bar{\psi}_\alpha^j(\vec{x}) \psi_\alpha^k(\vec{y}) \bar{\psi}_\beta^l(\vec{y})}{\omega + \omega_{\alpha\beta} - i\eta_\alpha} \right]$$

(4.1)

where $\eta_\alpha = +(-)\eta$ for $\alpha > (\leq)F$ and $\omega_{\alpha\beta} \equiv \epsilon_\alpha - \epsilon_\beta$. The similarity of the Lehmann representation of the full Π and the spectral form of Π_H motivates us to define the *particle-hole* transitions densities:

$$\mathcal{F}_{\alpha\beta}^{kl}(\vec{y}) \equiv \psi_\beta^k(\vec{y}) \bar{\psi}_\alpha^l(\vec{y})$$

(4.2)

in analogy with the full $\mathcal{F}^{(n)}$ of Eq. (3.3). This allows us to rewrite Eq. (4.1) in matrix form as

$$\Pi_H^{ij;kl}(\vec{x}, \vec{y}; \omega) = \sum_{\substack{F \geq \beta \geq 0 \\ \alpha}} (\bar{\mathcal{F}}_{\alpha\beta}^{ij}(\vec{x}), \quad \bar{\mathcal{F}}_{\beta\alpha}^{ij}(\vec{x}))$$
$$\times \begin{pmatrix} G_{\alpha\beta}(\omega + i\eta_\alpha) & 0 \\ 0 & -G_{\beta\alpha}^*(\omega + i\eta_\alpha) \end{pmatrix} \begin{pmatrix} \mathcal{F}_{\alpha\beta}^{kl}(\vec{y}) \\ \mathcal{F}_{\beta\alpha}^{kl}(\vec{y}) \end{pmatrix}$$

(4.3)

where we have defined

$$G_{\alpha\beta}(\omega) \equiv \frac{1}{\omega - \omega_{\alpha\beta}}.$$

We now assume that the RPA transition density can be written as

$$\mathcal{F}_{RPA} = \sum_{\substack{F \geq \beta \geq 0 \\ \alpha}} \begin{pmatrix} X_{\alpha\beta}^{(n)} \mathcal{F}_{\alpha\beta} \\ Y_{\alpha\beta}^{(n)} \mathcal{F}_{\beta\alpha} \end{pmatrix}$$

(4.4)

where we have dropped the Dirac indices. The RPA integral equation (Eq. (3.7)) can then be recast into the familiar spectral RPA matrix relations:

$$(\omega_n - \omega_{\alpha\beta}) X_{\alpha\beta}^{(n)} - \sum_{\substack{F \geq \beta' \geq 0 \\ \alpha'}} \left[K_{\alpha\beta;\alpha'\beta'}(\omega_n) X_{\alpha\beta}^{(n)} + K_{\alpha\beta;\beta'\alpha'}(\omega_n) Y_{\alpha\beta}^{(n)} \right] = 0$$

$$(\omega_n + \omega_{\alpha\beta}) Y_{\alpha\beta}^{(n)} + \sum_{\substack{F \geq \beta' \geq 0 \\ \alpha'}} \left[K_{\beta\alpha;\alpha'\beta'}(\omega_n) X_{\alpha\beta}^{(n)} + K_{\beta\alpha;\beta'\alpha'}(\omega_n) Y_{\alpha\beta}^{(n)} \right] = 0$$

(4.5)

where

$$K_{\alpha\beta;\alpha'\beta'}(\omega_n) \equiv \sum_{\substack{mm' \\ nn'}} \int d^3 x_1 d^3 x_2 \, \mathcal{F}_{\alpha\beta}^{mm'}(\vec{x}_1)$$
$$\left[\sum_N (\Gamma_N^a)^{m'm} D_{N;ab}(\vec{x}_1 - \vec{x}_2; \omega_n)(\Gamma_N^b)^{n'n} \right] \mathcal{F}_{\alpha'\beta'}^{nn'}(\vec{x}_2)$$

(4.6)

is the RPA kernel whose ω-dependence is typically ignored. This amounts to dropping "retardation effects".

An important property of the Dirac RPA presented here can be derived using the RPA matrix equations. We observe that the uncorrelated Hartree particle-hole current is conserved in the following sense. Define the particle-hole transition current via

$$J_{\alpha\beta}^{\mu} \equiv \int d^3r \; e^{-i\vec{q}\cdot\vec{r}} \operatorname{Tr}\left[J^{\mu}\mathcal{F}_{\alpha\beta}(\vec{r})\right] \qquad (4.7)$$

where, omitting formfactors and isospin labels,

$$J^{\mu} = e\gamma^{\mu} + \frac{i\kappa}{2m}\sigma^{\mu\nu}q_{\nu}$$

is the free nucleon current operator. Since the anomalous moment contribution is divergenceless by construction, we can let $J^{\mu} \to e\gamma^{\mu}$ for a discussion of current conservation. Because the single particle wavefunctions which appear in $\mathcal{F}_{\alpha\beta}$ both satisfy the same Dirac equation (Eq. (2.4)), we have

$$q \cdot J_{\alpha\beta} = 0$$

for $q^{\mu} \equiv (\omega_{\alpha\beta}, \vec{q})$. The RPA transition current is

$$J_{(n)}^{\mu} = \int d^3r \; e^{-i\vec{q}\cdot\vec{r}} \operatorname{Tr}\left[J^{\mu}\mathcal{F}^{(n)}(\vec{r})\right] \qquad (4.8)$$

and, by the RPA equations (Eq. (4.5)), satisfies

$$q \cdot J_{(n)} = 0$$

for $q^{\mu} \equiv (\omega_n, \vec{q})$. Thus, in the present model, the electromagnetic RPA transition current is exactly conserved, ensuring that, among other things, Siegert's theorem is automatically satisfied.

The spectral Dirac RPA equations have been solved by Furnstahl[10] , by Blunden and McCorquodale[11] , and by Furnstahl and Dawson[12] at various levels of approximation. For example, in Refs. 10 and 11, contributions from the negative energy sea (i.e., from $\alpha < 0$ in Eq. (4.5)) were ignored. In all these cases, the continuum (or *continua* in the case of Ref. 12 where negative energy contributions were included) was discretized by "putting the system in a box". This treatment has little effect on the properties of discrete excited states but gives an incorrect picture of the nuclear response in the continuum. Finally, as a necessary consequence of using the spectral method, the discretized Hartree basis is truncated. In Refs. 10 and 11 only positive energy levels with kinetic energies below $\approx$ 40 MeV were retained. In Ref. 12 a much larger basis was used, including positive energy levels up to several hundred MeV as well as all bound negative energy states. Altogether, a total of about 200 basis states was employed. Specific results of the spectral Dirac RPA calculations of Refs. 10 -12 are discussed in in R. Furnstahl's presentation at this conference. They may be briefly summarized by noting that the excitation energies and transition strengths of low-lying T=0 collective levels in, e.g., ^{16}O (negative parity only) are reasonably described. When the Hartree-based RPA outlined above is extended to include π and ρ couplings, the resulting description of T=1 level is found to be poor. Inclusion of exchange (or Fock) terms along with use of a consistent Hartree-Fock basis has been found to greatly improve this situation[11] .

5. Non-spectral RPA

An alternative to the "spectral" formulation is the direct solution of the RPA integral equation, Eq. (3.7). This "non-spectral" approach is implemented by computing Π_H as given in Eq. (3.10) using the non-spectral Hartree propagator, G_H, as given in Eq. (2.7). Beginning with this form of G_H, a partial-wave version of the RPA integral equation may be solved numerically. Specifically, this radial equation is

$$
\imath\Pi^{ij;kl}_{LL'SS'J}(x,y;\omega) = \imath\Pi^{ij;kl}_{H:LL'SS'J}(x,y;\omega) + 2\sum_{L''S''}\sum_{\substack{mm'\\nn'}}\int_0^\infty x_1^2 dx_1 \int_0^\infty x_2^2 dx_2
$$

$$
\times \imath\Pi^{ij;mm'}_{H:LL''SS''J}(x,x_1;\omega)\sum_{N;a}\left(\Gamma^a_{N.}\right)^{m'm}\delta_{S_aS''}
$$

$$
\times \imath f^{(N)}_{L''}(x_1,x_2;\omega)\left(\Gamma_{N;a}\right)^{n'n}\delta_{S_bS''}\,\imath\Pi^{nn';kl}_{L''L'S''S'J}(x_2,y;\omega)
$$

$$
(5.1)
$$

where $\Pi^{ij;kl}_{H:LL'SS'J}(x,y;\omega)$ comes from the partial wave decomposition of Π_H (Eq. (3.10)) and $f^{(N)}_{L''}(x_1,x_2;\omega)$ comes from the Slater expansion of the free meson propagator for meson N.

There are several advantages to the "non-spectral" method. For example, discretization is avoided and the correct continuum boundary conditions are automatically included. This provides a more realistic continuum response and is crucial for treating, e.g., the quasi-elastic response in (e, e') or (p, p'). In addition, the entire Hartree single particle spectrum - including positive and negative energy continua - is automatically incorporated. This means that uncertainties inherent in any basis truncation do not arise. Finally, retardation effects (ω-dependence of $D^0_{N;ab}(\vec{x}_1 - \vec{x}_2; \omega)$ in Eq. (3.7)) can be trivially included. (This dependence, however, is expected to be very weak in the direct-only RPA discussed here.)

Typically the numerical solution to Eq. (5.1) is found either by iteration or by matrix inversion after the integrals are put on a radial grid (see, e.g., Ref. 13 for a similar non-relativistic approach). The non-spectral Dirac RPA method has been applied to finite nuclei by Shepard et.al.[4], Iichi et.al.[15], and Wehrberger and Beck[16]. The first two of these studies were aimed at evaluating RPA-like core-polarization corrections to the isoscalar magnetic response of of closed shell±1 nuclei. Here the iterative and matrix-inversion techniques give essentially identical results. In Ref. 16, the iterative technique was used to find the transverse and longitudinal (e, e') responses in the quasi-elastic region. In this instance, the iterative technique was found to be unreliable in certain kinematical regions. We[17] have recently developed a code for solution via matrix inversion of Eq. (5.1) as it stands for arbitrary complex ω. The calculation is done in configuration space and yields the general result $\Pi^{ij;kl}_{LL'SS'J}(x,y;\omega)$. The flexibility of the code allows us to extract several interesting quantities. We can find the response to any one-body operator at arbitrary ω by solving for Π and taking the traces over the Dirac indices $ij; kl$ as indicated in Eq. (3.5). Alternatively, for a discrete transition, we can evaluate Π at $\omega = \omega_n + i\eta$ (*i.e.*, slightly above the pole of Π in the complex ω-plane), extract the residue, and determine the RPA transition density $\mathcal{F}^{(n)}_{RPA}$, analogous to the full transition density defined in Eq. (3.3). This object can then be used in standard scattering codes for (e, e') and (p, p') to compute observables with RPA transiton densities without resorting to the usual representation in terms of particle-hole pairs $(\alpha\beta)$ and associated $X^{(n)}_{\alpha\beta}$ and $Y^{(n)}_{\alpha\beta}$ coefficients (see Eq. (4.4)). We note that the iterative technique for finding Π fails utterly for discrete transitions.

We are just beginning to perform non-spectral Dirac RPA calculations using the code described above but have already illuminated some interesting physical issues. For example, comparison with the large-basis spectral calculations of Furnstahl and Dawson[12] has allowed assessment of sensitivities to basis truncation in the latter work. Such sensitivities appear, for example, in the description of the "frontflow" phenomenon previously observed in connection with the RPA-like corrections to the magnetic response of closed-shell$\pm$1 nuclei. In contrast to the low-q quenching of this response known as "backflow"[4,15,14], frontflow enhances the response at $q > 1 fm^{-1}$. Both backflow and frontflow are observed in Dirac RPA calculations for inelastic magnetic and transverse electric transitions. Frontflow is an especially collective phenomenon and comparison of spectral and non-spectral results show that positive energy particle states (*i.e.*, those corresponding to the label α in Eq. (4.5)) with up to 10 radial nodes and kinetic energies up to several hundred MeV are required for convergence. Thus the frontflow phenomenon would never appear in the small basis calculations of Refs. 10 or 11.

Non-spectral calculations also provide a benchmark against which to judge the level of convergence of spectral results for the low-lying, collective, isoscalar transitions which the RPA typically describes so well. As will be demonstrated in the paper of Furnstahl at this conference, the large basis calculations of Ref. 12 agree very well with our non-spectral results for, *e.g.*, the lowest 3^- transition in ^{16}O indicating convergence has been achieved. An even more stringent test of such convergence is the description of $1^- T = 0$ levels. As is well known[7], the lack of explicit translational invariance in the general RPA approach implies, in a consistent formulation, the existence of a "spurious" 1^- $T = 0$ level at zero excitation energy which contains nearly all of the isoscalar E1 strength. A numerical realization of this requires (*i.*) consistency between the single particle basis used and the RPA kernel, K, (*ii.*) the completeness of the basis and (*iii.*) the covariance of the NN interaction. As a consequence of the latter requirement, the spurious state will not lie at zero excitation energy if the space-like omega interaction, which is irrelevant in determining the Hartree basis (in the nuclear rest frame), is omitted in the RPA. Because the physical 1^- $T = 0$ states can have essentially no E1 strength, the RPA correlations must greatly alter the qualitative behavior of the transition densities at low q. The longitudinal (e, e') formfactor for the lowest, non-spurious $1^- T = 0$ level in ^{16}O (found at $E_x = 8.467$ MeV in the present Dirac RPA model) is compared with the simple Hartree result in

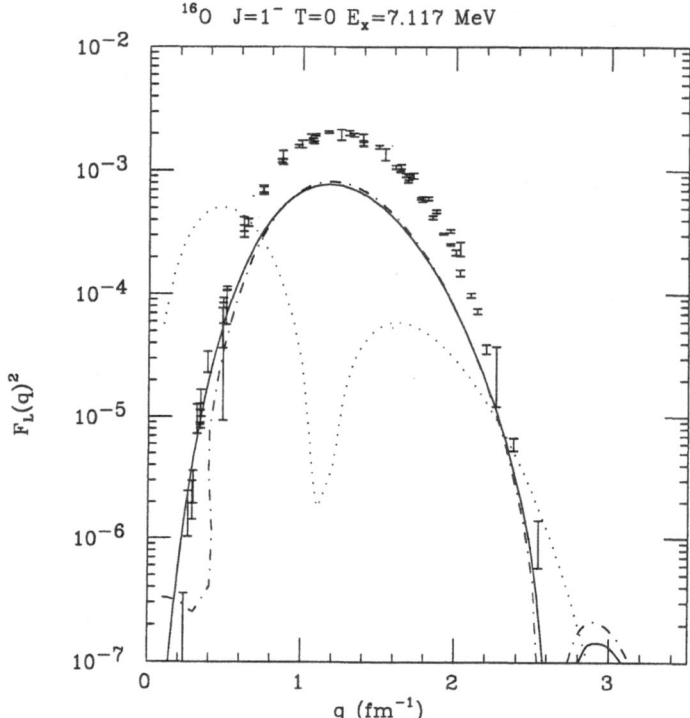

^{16}O J=1$^-$ T=0 E$_x$=7.117 MeV

Fig. 2. The Coulomb form factor for the 7.117 MeV 1$^-$ T=0 level in ^{16}O. The dotted curve is the simple Hartree particle-hole result for a $(p_{1/2})^{-1}(s_{1/2})$ configuration ($E_x = 11.2$ MeV) while the solid line is the RPA result ($E_x = 8.47$ MeV). The dashed dotted line is an RPA result without spacelike ω coupling ($E_x = 10.085$ MeV). Data are from Ref. 18.

Fig. 2. The enormous suppression of the RPA result at low q is an indication that all "spuriosity" has correctly been removed from levels with non-zero excitation energies. (We note that the $B(E1)$ drops from $8.0 \times 10^{-2}e^2 \cdot fm^2$ in the Hartree case to $1.3 \times 10^{-5}e^2 \cdot fm^2$ for the RPA.) Consistent with this finding is the fact that the spurious state is found at an excitation energy of $\approx 0.5\imath$ MeV. The high level of agreement between the spectral[12] and non-spectral calculations for these 1$^-$ T = 0 levels is again strong evidence of convergence in the former.

Non-spectral calculations also facilitate evaluations of local density approximations (LDA) since the full continua are included in both treatments. For example, we find that the overall backflow effect in magnetic transitions, which comes principally form negative energy particle contributions[12,14], is well-reproduced in general by the LDA. Again, this topic is addressed in detail by Furnstahl in his paper at this conference.

One of the shortcomings of the non-spectral RPA method is the difficulty of including exchange (Fock) contributions. The exchange integrals which would appear in the Eq. (5.1) would enormously complicate numerical solutions of that equation.

6. Sum Rules in Dirac RPA

It is straightforward (see, *e.g.*, Ref. 7) to derive sum rules for simple particle-hole (or Hartree) matrix elements of one-body operators. These sum rules depend on the completeness of the particle-hole basis. Furthermore, since the particle-hole and RPA transition densities (Eqs. (4.2) and (4.4)) are related by a unitary transformation, these sum rules carry over directly to the RPA provided the *total* response is accounted for.

We now consider the particle-hole energy weighted sum rule. In a non-relativistic picture, the energy weighted summed response is, using Eq. (3.6),

$$S_{\mathcal{O}} \equiv \int_0^\infty d\omega \, \omega S_{\mathcal{O}}(\omega) = \sum_{\substack{\beta \leq F \\ \alpha}} \omega_{\alpha\beta} |\langle \mathcal{O} \rangle_{\alpha\beta}|^2 \tag{6.1}$$

where the hole index, β, runs over all occupied orbitals. The α index is originally taken to run over all unoccupied states but, because of the antisymmetry of $\omega_{\alpha\beta}$ under interchange of indices, the range of α can be extended to include the range of β. Thus, in this expression, the range of α is unrestricted. Still, the only nonvanishing contributions to $S_{\mathcal{O}}$ arise from terms with $\alpha > F$ which ensures $\omega_{\alpha\beta}$ is positive and that $S_{\mathcal{O}}$ is positive-definite. Assuming, in analogy with Eq. (2.4), that the single particle wavefunctions are eigenfunctions of a single particle Hamiltonian,

$$h \, |\alpha\rangle = \epsilon_\alpha \, |\alpha\rangle$$

we can show

$$S_{\mathcal{O}} = \frac{1}{2} \sum_{\beta \leq F} \langle \beta | \, [\mathcal{O}, [h, \mathcal{O}]] \, |\beta\rangle \tag{6.2}$$

where we have made use of the completeness relation

$$\sum_\alpha |\alpha\rangle \langle \alpha| = 1. \tag{6.3}$$

In the Dirac RPA presented here, the analogue to Eqs. (6.1) and (6.2) is (remembering to construct the response from $\Pi_{H:Ret}$ instead of from the Π_H of Eq. (3.8))

$$S_{\mathcal{O}} = \sum_{\substack{F \geq \beta > 0 \\ \alpha}} \omega_{\alpha\beta} |\langle \mathcal{O} \rangle_{\alpha\beta}|^2 = \frac{1}{2} \sum_{F \geq \beta > 0} \langle \beta | \, [\mathcal{O}, [h, \mathcal{O}]] \, |\beta\rangle \tag{6.4}$$

where now the hole label, β, is restricted to positive energy, occupied levels and the particle label, α, runs over the entire Hartree spectrum, including the negative energy levels. When α refers to a negative energy level, the corresponding term in the energy weighted sum rule (Eq. (6.4)) is *negative*-definite since $\omega_{\alpha\beta} = \epsilon_\alpha - \epsilon_\beta < 0$. Thus the sum rule strength is no longer positive-definite.

A example which illustrates the significance of these differences between Dirac and non-relativistic sum rules is the TRK or photo-absorption sum rule. The relevant one-body operator is the electric-dipole operator:

$$\mathcal{O} \to \frac{e}{2}(1 + \tau_z)\vec{r} \cdot \hat{z}. \tag{6.5}$$

(This expression is valid both relativistically and non-relativistically assuming the operator is to be evaluated between single particle wavefunctions and their Hermetian conjugates, and not, in the relativistic case, between the wavefunctions and their adjoints.) Non-relativistically, assuming the potentials in h to be local and independent of isospin, we have

$$[\mathcal{O}, [h, \mathcal{O}]] = \frac{1}{2}(1 + \tau_z)\frac{1}{m} \tag{6.6}$$

which yields the (lowest order) TRK sum rule

$$\sigma_0 = \frac{2\pi^2\alpha^2}{m}\frac{NZ}{A}. \tag{6.7}$$

Relativistically, as pointed out by Price and Walker[19],

$$[\mathcal{O}, [h, \mathcal{O}]] = 0$$

and the summed photo-absorption strength vanishes! What is the physical interpretation of this result? This question is best answered by recalling that, as discussed in Section 3, the Dirac RPA treated here has the vacuum-vacuum contribution subtracted off. Recalling Eqs. (3.9) and (3.10), this subtraction is represented schematically by

$$\Pi_H = G_H G_H = G_H G_D + G_D G_H + G_V G_V$$
$$\rightarrow G_H G_D + G_D G_H = G_H G_H - G_V G_V.$$

Thus the response in this Dirac RPA should be thought of as the difference between the *full* response of the nucleus and the response of *its* vacuum which is defined by $F \rightarrow 0$ in the Hartree basis. This interpretation is illustrated in Fig. 3 where two distinct contributions are indicated. The first is the usual particle-hole contribution arising from the promotion of nucleons in the positive energy occupied levels to states above the Fermi surface. The second contribution, unique to this version of the Dirac RPA, comes from the promotion of negative energy nucleon into positive energy occupied orbitals. This contribution is accompanied by a minus sign reflecting the fact that this portion of the vacuum response is blocked by the occupied positive energy levels. When the double commutator of an operator with the Dirac Hartree Hamiltonian vanishes, as is the case for the

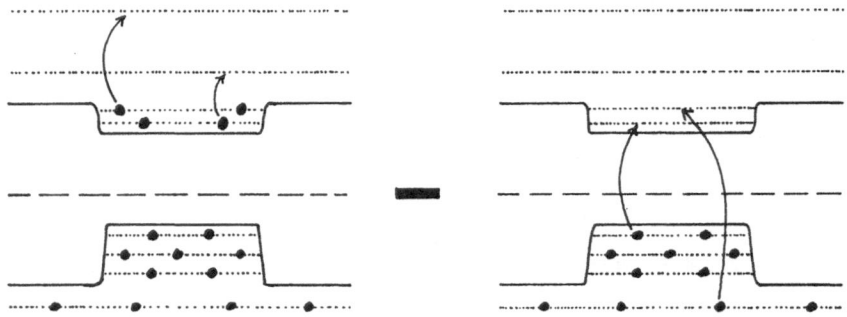

Fig. 3. Schematic representation of nuclear response contained in "density dependent" Hartree polarization insertion. See text.

electric dipole operator, the two contribtions exactly cancel. This accounts for the zero TRK sum rule. As pointed out in Ref. 19, the negative contributions to the sum rule correspond to pair creation and therefore to very high excitation energies. By restricting the range of energy integration in the sum rule (Eq. (6.1)) to exclude pair processes, these negative contributions are dropped and, instead of a summation over all particle states α in Eq. (6.1), only positive energy α are included. Then ,using

$$\sum_{\alpha>0} |\alpha\rangle \langle\alpha| = \Lambda^{(+)}$$

where $\Lambda^{(+)}$ is the positive energy projection operator, the partial sum rule becomes

$$S_{\mathcal{O}}^{(+)} = \frac{1}{2} \sum_{F \geq \beta > 0} \langle\beta| \left(\mathcal{O}\Lambda^{(+)}[h, \mathcal{O}] - [h, \mathcal{O}]\Lambda^{(+)}\mathcal{O} \right) |\beta\rangle \qquad (6.8)$$

and the non-relativistic sum rule (Eq. (6.7)) is (approximately) recovered.[19]

In the RPA, contributions from the ordinary particle-hole excitations are mixed with pair contributions and only the *full* sum rule of Eq. (6.2) is preserved. Thus only the empty zero TRK sum rule is rigorously satisfied. Any "sum rule" for excitations below $N\bar{N}$ threshold will be approximate and model dependent. The awkwardness of the sum rules in this version of the Dirac RPA is related to the fact that we have omitted the vacuum fluctuation contributions to the nuclear response. As indicated in C.Horowitz' contribution to this conference, important physics is ignored in his approximation. Presumably, inclusion of vacuum fluctuation terms in future formulations will lead to more physically satisfying sum rules. It is also clear that an extended Dirac RPA will be required to account for the nuclear response at energies appropriate to $N\bar{N}$ excitations.

7. Summary and Future Directions

In this paper, I have discussed general features of the RPA description of the nuclear response and have presented a specific RPA treatment based on a Dirac Hartree model of the nuclear ground state. This approach has several appealing features including exact conservation of the electromagnetic current and proper treatment of spurious excitations. Two distinct methods of solving the RPA equations were discussed and some of the advantages of one - the non-spectral method - were stressed. Finally, sum rules in this particular version of the Dirac RPA were examined and it was found that no physically meaningful and mathematically rigorous sum rule exists. More useful sum rules may result when dynamics of the nuclear vacuum - not treated here - are explicitly included.

As discussed by R.Furnstahl in a contribution to this conference, comparison of Dirac RPA spectra and (e, e') formfactors with data is encouraging, especially for low-lying collective excitations. However, relatively few such comparisons have yet been made. Still fewer comparisons with non-relativistic RPA calculations have been undertaken. Therefore a prime objective of the near future will be extension of such studies as those reported in the recent review article of Frois and Papanicolas[21] . These authors noted that non-relativistic mean-field calculations of charge densities for the ground states of doubly magic nuclei show substantially more structure in the nuclear interior than is observed experimentally. They went on to demonstrate that a similar excess of structure is observed in the RPA transition densities for the lowest 3^- excitations in these nuclei. As noted at the beginning of the present paper, Dirac mean-field calculations[1,2] give a good description of the ground state densities and in particular show less structure than the non-relativistic results and are therefore in better agreement with data. We should now extend the analysis to include the Dirac RPA transition densities. Very few such comparisons have been made to date. However, Dirac RPA calculations have been done[17] for the lowest 3^- in ^{48}Ca and the resulting charge and current transition densities are compared in Figs. 4 and 5, respectively, with non-relativistic RPA results and with experiment[20]. (These Dirac RPA calculations employ a σ coupling constant which is slightly reduced from the QHD-1 value so as to better reproduce the electron elastic scattering formfactor and 3^- excitation energy for for ^{40}Ca.) The detailed structure of the charge transition density is substantially better described by the Dirac calculation than by its non-relativistic counterpart. A similar situation is observed[17] for ^{40}Ca where the longitudinal formfactor for the lowest 3^- level is *very* accurately reproduced by the Dirac RPA. (See R. Furnstahl's paper, this conference.) The current transition density is less well described and there are sizeable differences between the Dirac and non-relativistic predictions. The origin of these differences is not clear

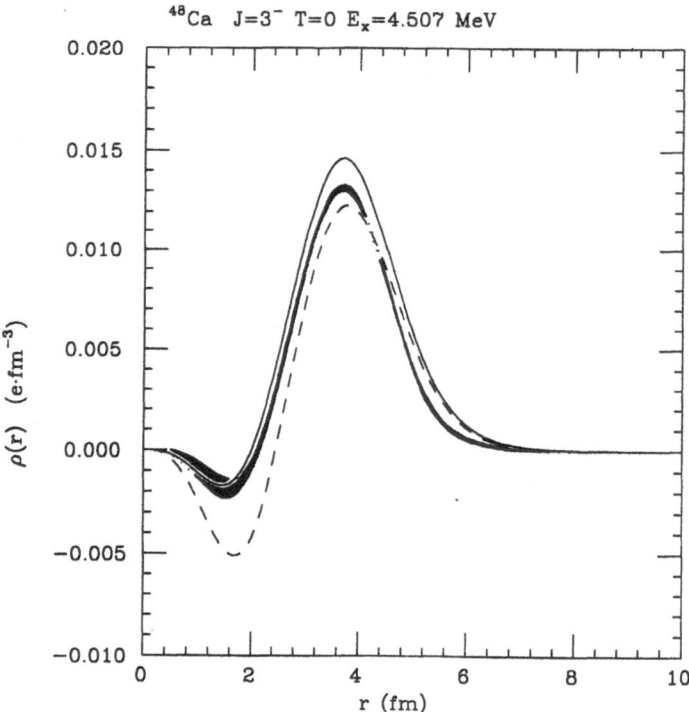

Fig. 4. The Dirac[17] (solid line) and nonrelativistic[20] RPA transition charge densities for the 4.507 MeV 3^- level in ^{48}Ca are compared with the experimental results of Ref. 20.

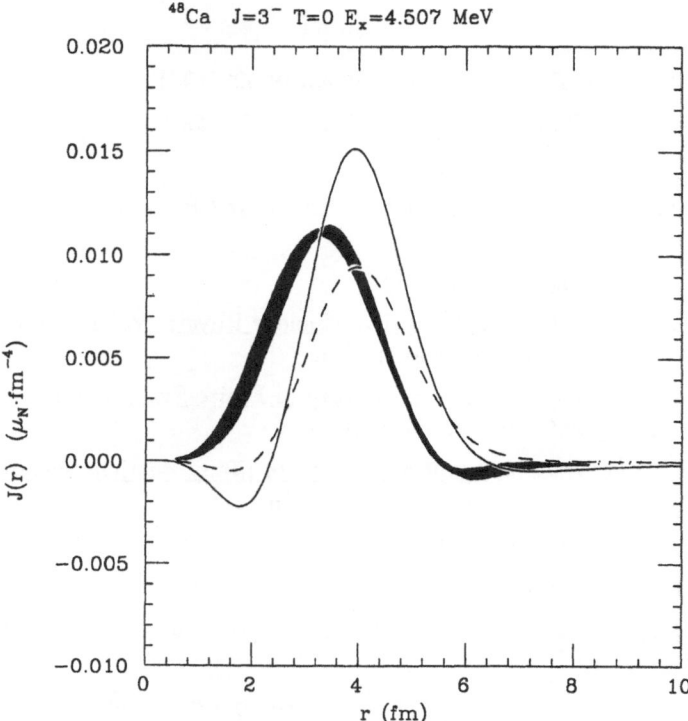

^{48}Ca J=3$^-$ T=0 E$_x$=4.507 MeV

Fig. 5. The Dirac[17] (solid line) and nonrelativistic[20] RPA transition current densities for the 4.507 MeV 3$^-$ level in ^{48}Ca are compared with the experimental results of Ref. 20.

at present but is presumably related to m^* enhancement and frontflow[4,12,14]. In any case, extended comparisons of this kind will be pursued in the near future.

Also among future objectives are extensions and refinements of the Dirac RPA method itself. This includes treatment of π and ρ interactions (and, assuming derivative coupling, their associated contact terms) as well as isobars. Also needed is an open-shell[22] Dirac RPA to treat excitations of non-magic nuclei. Such an RPA would be based on deformed Dirac Hartree ground states such as those discussed by C.Price at this conference. Finally and perhaps most importantly, the dynamics of the Hartree vacuum must be incorporated in the RPA.

We should also be mindful that the RPA **is** an approximation and that it leaves out potentially important aspects of nuclear structure. For example, 2-particle 2-hole contributions (*e.g.*, the last term in the top row of Fig. 1) are known to important in the spin-isospin response of nuclei[23]. Also, meson exchange currents necessarily enter along with charged meson fields. All of these considerations as well as others not mentioned are relevant in the development a realistic picture of nuclear structure in the Dirac framework.

I acknowledge with pleasure the contributions of my collaborators, E.Rost and J.McNeil, to the work presented here. I also acknowledge useful discussions with R.Furnstahl and J.Dawson. This work supported in part by the U.S.D.O.E.

REFERENCES

1. C.J.Horowitz and B.D.Serot, Nucl.Phys.**A368**,503(1981).

2. B.D.Serot and J.D.Walecka, Adv.Nucl.Phys.**16**,1(1986).

3. R.J.Perry,Phys.Lett.**182B**,269 (1986)

4. J.R.Shepard, E.Rost, C.-Y.Cheung and J.A.McNeil,Phys.Rev. **C37**, 1130(1988)

5. D.J.Thouless,Nucl.Phys.**22**,78(1961)

6. A.deShalit and H.Feshbach,Theoretical Nuclear Physics,Vol.1, John Wiley and Sons,N.Y.,1978 and references therein.

7. J.-P.Blaizot and G.Ripka,Quantum Theory of Finite Systems, MIT Press, Cambridge,Mass.,1986 and references therein.

8. A.Fetter and J.D.Walecka,Quantum Theory of Many Particle Systems, McGraw-Hill,N.Y.,1971 and references therein.

9. S.A.Chin,Ann.Phys.**108**,301(1977)

10. R.J.Furnstahl,Phys.Lett.**152B**,313(1985)

11. P.G.Blunden and P.McCorquodale,TRIUMF Preprint TRI-PP-88-04 and to be published.

12. R.J.Furnstahl, this conference; also R.J.Furnstahl and J.Dawson, private communication and to be published.

13. S.Shlomo and G.Bertsch, Nucl.Phys.**A243**,507(1975).

14. R.J.Furnstahl,Indiana Univ.Nucl.Thy.Center,Preprint 87-1

15. S.Iichi,W.Bentz,A.Arima and T.Suzuki,Phys.Lett. **192B**,11(1987).

16. K.Wehrberger and F.Beck,Phys.Rev.**C37**,1148(1988)

17. J.Shepard,E.Rost and J.McNeil, to be published.

18. T.N.Buti *et.al.*,Phys.Rev.**C33**,755 (1986)

19. C.Price and G.E.Walker,Phys.Lett.**155B**,17(1985)

20. J.E.Wise *et.al.*,Phys.Rev.**C31**,1699 (1985)

21. B.Frois and C.N.Papanicolas,Ann.Rev.Nucl.Part.Sci., Vol.37,p.133 (1987)

22. D.J.Rowe and S.S.M.Wong,Nucl.Phys.**A153**,561 (1970)

23. See, *e.g.*, J.Wambach *et.al.*,Phys.Rev.**C37**, 1322 (1988) and J.Wambach, contribution to this conference.

RELATIVISTIC RPA CALCULATIONS IN FINITE NUCLEI: RESULTS

R. J. Furnstahl

Department of Physics and Astronomy
University of Maryland
College Park, MD 20742

INTRODUCTION

There has recently been an increasing interest in the nuclear response predicted in relativistic models. New calculations in both nuclear matter and finite nuclei have ranged from studies of nuclear currents and magnetic moments (elastic response), to inelastic scattering to discrete states, to quasielastic scattering and the Coulomb sum rule. This work is motivated by a desire to extend and test relativistic approaches in new domains and also to search for distinct signatures of relativistic dynamics, as might arise from the reduced effective mass of the nucleon or from vacuum polarization. A variety of nuclear matter calculations, applied to finite nuclei using local density approximations, are discussed by C. Horowitz in these proceedings[1].

In this contribution, I review some recent relativistic RPA calculations that go beyond the local density approximation. Details of the formalism are discussed elsewhere by J. Shepard;[2] I will concentrate on recent results and on what can be expected in the future. The starting point for my discussion is the relativistic Hartree (mean-field) approximation for spherical nuclei. Hartree calculations have successfully described ground state rms radii, charge densities, neutron densities, and spin–orbit splittings[3] and have also provided densities for calculations of elastic proton-nucleus scattering. The predictions of spin observables, using a relativistic impulse approximation in conjunction with Hartree densities to describe the nuclear ground state, have been remarkably successful[4]. A prime motivation for the present RPA calculations is to provide consistent relativistic nuclear structure input for new studies of the elastic and inelastic scattering of polarized protons and other probes.

The particular focus here is on extensions of the Hartree approximation to the ground-states of odd-A nuclei and to the inelastic response of spherical nuclei. Both of these extensions involve the linear response of a spherical nucleus, as calculated in a relativistic RPA. I will survey new predictions for magnetic moments and elastic magnetic scattering, and the energy spectra and electron scattering form factors for discrete particle–hole transitions. I'll conclude with a look beyond Hartree to the Hartree-Fock (HF) and HF-RPA calculations of Blunden and his collaborators[5,6].

All of the results described here were obtained by calculating within a consistent relativistic framework (quantum hadrodynamics), which is described in detail in Ref. 7.

Corrections to the present approximations, including those coming from vacuum polarization effects,[1] can be investigated systematically. In addition, there are no nonrelativistic reductions of nucleon wave functions or interactions in this approach. Finally, I want to emphasize the importance of self-consistency in the relativistic many-body problem; the self-consistent treatment of the large relativistic potentials has been a key ingredient in the past success of relativistic models. In the present context, self-consistent approximations are used to describe nuclear ground states, and excited states are treated using the corresponding linear response. The consistent treatment of the nuclear response increases the predictive power of the models by limiting the phenomenological input and ensures that basic physical principles, such as current conservation, are preserved.

RELATIVISTIC HARTREE-RPA

The most complete finite nucleus RPA calculations to date describe the linear response of a relativistic Hartree ground state. I will call this the Hartree-RPA to distinguish it from the linear response of a Hartree-Fock ground state (HF-RPA). The Hartree-RPA has been applied recently to the calculation of magnetic moments and elastic magnetic scattering,[8,9,10] inelastic scattering to discrete particle-hole excitations,[11,12] and quasielastic electron scattering.[13] Before showing examples of specific calculations, I will briefly review the different techniques used to solve the RPA and some common features of the recent calculations that deserve emphasis.

The single-particle spectrum in the Hartree approximation, namely the eigenvalues of the Dirac equation that describes nucleon motion, includes both positive- and negative-energy states. There are a relatively small number of weakly bound positive-energy states and then a positive-energy continuum starting at the nucleon mass. In contrast, the strong scalar and vector potentials lead to a large number (several hundred) of bound negative-energy states. For example, there are eight bound $s_{1/2}$ negative-energy levels in the Hartree model of Ref. 3. The negative-energy basis states play an important role in the relativistic Hartree-RPA.

The linear response of a Hartree ground state is given diagrammatically by the sum of ring contributions to the polarization propagator (linear response function).[14] It is important to clarify the spectral content of the rings. The ground-state Hartree approximation (or MFT) neglects contributions to the nucleon self-energy from the filled Dirac sea, while including these contributions defines the RHA.[7] Subsequently, the linear response of the RHA ground state includes contributions both from conventional particle–hole configurations and from particle–negative-energy configurations. The latter part can be rewritten as a vacuum polarization piece minus the part of the vacuum response that is Pauli blocked in the nuclear medium (see Fig. 1), and which features hole–negative-energy configurations. The consistent MFT linear response includes particle–hole contributions *and* the Pauli-blocked response.

The physics that is included in the RHA response but missing from the MFT is the overall polarization of $N\overline{N}$ pairs in the vacuum, an effect that depends on the nucleon effective mass M^*. Recent calculations in nuclear matter indicate that these vacuum contributions can have significant effects on the nuclear response, even at low momentum transfers.[15,1] Their inclusion in finite nucleus calculations will be an important extension of the present calculations.

The Hartree-RPA ring diagrams can be summed to all orders with an integral equation. The basic problem is to solve this integral equation efficiently in the desired context. The calculations of the linear response function can be divided into two types: spectral and nonspectral. In the spectral approach, the lowest-order ring

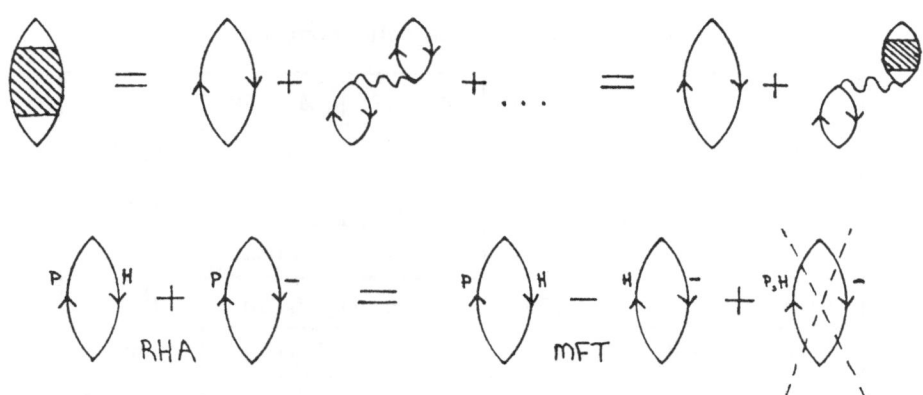

Fig. 1. Integral equation for the linear response function and the spectral content of the RHA and MFT approximations.

is built from Hartree propagators written in a spectral representation. That is, each Green's function is written as a sum over eigenfunctions of the nucleon Dirac equation. The nonspectral approach rewrites the ring in terms of a sum over occupied (positive-energy) states and a nonspectral Green's function. This Green's function is constructed at a given energy from linearly independent solutions of the Dirac equation with appropriate boundary conditions. These two methods are discussed in detail by J. Shepard in these proceedings.[2] In principle, these two approaches should give the same answers, but there are practical advantages and disadvantages to each.

The principal disadvantage of the nonspectral approach is that it is unfamiliar technology and requires care to implement correctly. Alternative calculations using spectral methods are particularly useful in developing nonspectral RPA computer codes. Once implemented, the nonspectral approach has many advantages. In particular, the negative-energy contributions to the response are included automatically and both positive- and negative-energy continuums are treated exactly. As a result, this method is well suited to describing a wide range of inelastic nuclear response, from discrete states through the giant resonance region to the quasielastic region. In contrast, spectral calculations must rely on a discretization and truncation of the continuum. Care must be taken that the response is not distorted by these approximations; as we shall see, large configuration spaces are needed to obtain stable, converged results that agree with nonspectral predictions. However, useful physical insights can be obtained from the spectral expansions because different pieces of the response can be isolated.

ELASTIC CURRENTS

In relativistic models, the reduced nucleon effective mass M^* in the nuclear medium (because of the large scalar field) results in a single-particle convection current enhanced by approximately M/M^* compared to the nonrelativistic value. One might expect strong relativistic signatures because of this difference. For example, the simple shell-model picture of a valence nucleon outside a closed-shell core predicts strong deviations from the Schmidt predictions for isoscalar magnetic moments when applied to relativistic Hartree wave functions.[16] On the other hand, the experimental isoscalar moments of LS closed shell ± 1 nuclei are in reasonable agreement with the Schmidt values (see Table I.).

This "magnetic moment problem" of relativistic models has been the subject of much study during the past few years. The *real* problem is that the simple shell model picture provides a poor approximation to the current of the self-consistent Hartree

Table I. Isoscalar Magnetic Moments

	$A = 15$	$A = 17$	$A = 39$	$A = 41$
Orbital	$1p1/2$	$1d5/2$	$1d3/2$	$1f7/2$
Schmidt	0.187	1.440	0.636	1.940
Valence only	0.350	1.571	1.033	2.275
LDA	0.206	1.435	0.670	1.953
RPA: $+/-$ only	0.204	1.436	0.663	1.950
RPA	0.204	1.436	0.661	1.950
Self-consistent	0.195	1.448	0.655	1.960
Experiment	0.218	1.414	0.706	1.918

ground state for the odd-A system. The valence nucleon is a source of new meson fields, and the response of the core nucleons to these new fields cannot be neglected when considering the current. If the core response is included in nuclear matter, the core wave functions are mixed with negative-energy states (this is the Pauli-blocked vacuum response described above) and the enhancement of the convection current is essentially cancelled.[17-21] When this core response is applied to finite nuclei using a local density approximation, the result is a return to the isoscalar Schmidt moments.[20,21]

There are two paths to self-consistency in a full finite nucleus calculation. The first approach is to solve the Hartree equations for the (deformed) odd-A system directly. Much progress toward such solutions has been made recently[22] and preliminary results are discussed in these proceedings by C. Price. An alternative approach starts with the basis of the spherical core nucleus (which is easily calculated) and includes the core response to the valence nucleon in a linear response approximation (Hartree-RPA). Note that this is only an approximation to the fully self-consistent solution; the valence nucleon is treated as an *external* source of meson fields so that the core response does not act back on the valence wave function.

Several different groups have taken this second approach to isoscalar magnetic moments in finite nuclei,[8-10] working in the $\sigma-\omega$ model with parameters from Ref. 3 but with no isovector mesons. Results from the spectral calculations of Ref. 10 are shown in Table I. (The nonspectral calculations of Refs. 8 and 9 yield similar results.) Magnetic moments generated using only the valence orbital ("Valence only") show large enhancements with respect to the Schmidt values, in disagreement with the experimental results. The calculation including the full core response ("RPA") shows that the enhancement is essentially cancelled by the core contribution so that the isoscalar moments are insensitive to the value of M^*. These results are in reasonable agreement with (unprojected) moments from a self-consistent Hartree calculation of the deformed intrinsic ground state of the odd-A system ("Self-consistent").[22]

The other entries in the table are from a local density approximation ("LDA") to the moments and from an RPA calculation that includes *only* the mixing of negative-energy solutions into the core wave functions ("RPA: $+/-$ only"). It is evident that this part of the response, which can be isolated in a spectral calculation, is responsible

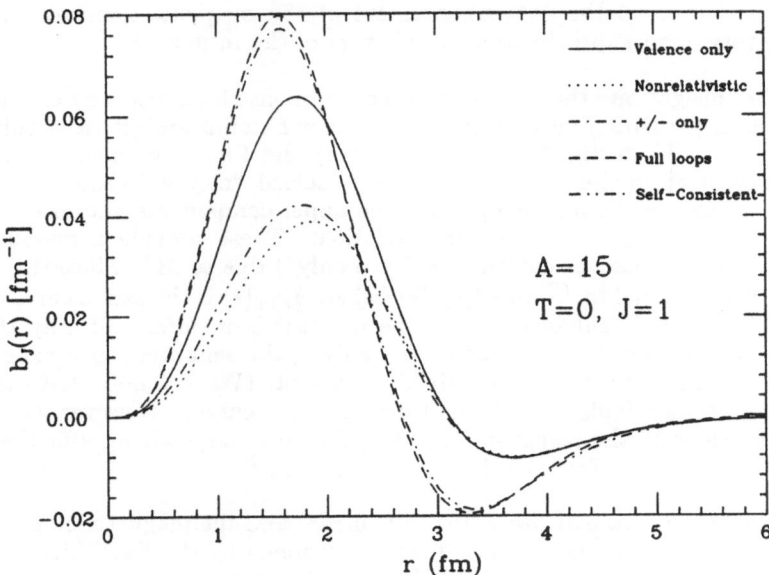

Fig. 2. The isoscalar, $J = 1$ convection current for $A = 15$, in coordinate space.[10]

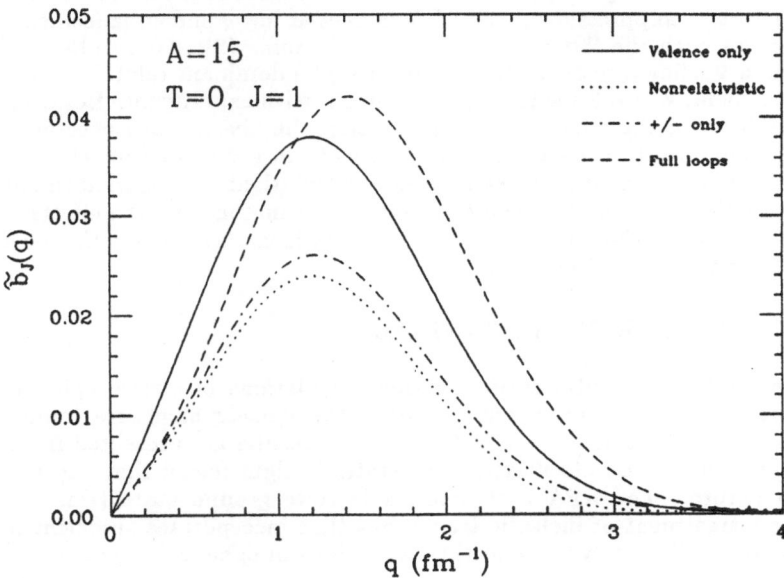

Fig. 3. The isoscalar, $J = 1$ convection current for $A = 15$, in momentum space.[10]

for the cancellation of the M^* enhancement of the magnetic moment. The LDA provides a very good approximation to this part of the response.[10]

Further insight into the nature of the core response is obtained by examining the full convection (or Dirac) current $\psi^\dagger \vec{\alpha} \psi$. The $J = 1$ radial projection of this current is shown for $A = 15$ in Fig. 2 (coordinate space) and Fig. 3 (momentum space), for several different approximations.[10] The curves labeled "nonrelativistic" are obtained from the valence wave functions by fixing the upper components and generating new lower components using the free space relation. These provide a measure of the enhancement of the valence current ("valence only") due to M^*. Since the magnetic moment is proportional to $\int_0^\infty dr \, r \, b_{J=1}(r)$ (where $b_{J=1}(r)$ is the radial current plotted in Fig. 2), the M/M^* enhancement of the moment is manifest. If only the Pauli-blocked vacuum response is included ("$+/-$ only"), the enhancement is cancelled and the current is similar to the "nonrelativistic" current. (We also note that the LDA of Ref. 10 is essentially indistinguishable from the $+/-$ curve.) However, the particle-hole contributions to the response ("full loops") cause large changes in the current; the same result is found for the self-consistent current.[22]

In momentum space at low q, the full current and the magnetic moment (which is proportional to the slope near $q = 0$) are determined by the Pauli-blocked vacuum response to the valence current. The full current is suppressed relative to the valence current only at low q; the particle-hole response leads to *additional* enhancement of the current at higher q. This result is consistent with calculations of the linear response in nuclear matter.[9,10] As discussed by C. Horowitz in these proceedings,[1] the enhancement is a precursor of a particle-hole condensate that appears in nuclear matter at greater than nuclear density.[10,23] This instability disappears when vacuum polarization is included and we may expect qualitative differences in the current in this case.

The Hartree-RPA results are applied to calculations of elastic magnetic electron scattering in Figs. 4 and 5 (see Ref. 10 for more details). In each figure, the solid curve is calculated from the valence wave function only, the dashed curve is a nonrelativistic Schrödinger-equivalent result,[7] and the dotdash curve is the RPA result. The characteristic suppression of the RPA current at low q and enhancement at higher q is evident in both the ^{15}N and the ^{39}K form factors. (Similar results for ^{15}N were found in Ref. 9 using a nonspectral approach.) The dominant relativistic effect is the M^* enhancement, which leads to worse agreement with experiment. However, because of the simplicity of the model, and in particular, the absence of isovector mesons, it would be premature to draw strong conclusions from the results for elastic magnetic scattering. It is important to extend these calculations of magnetic moments and elastic magnetic scattering to include both isovector mesons and the effects of vacuum polarization. Direct self-consistent Hartree calculations may provide the most efficient framework for these investigations.

INELASTIC RPA FOR DISCRETE STATES

To extend the impulse approximation calculations of elastic spin observables to inelastic (p, p'), it is important to derive the nuclear structure input within a consistent relativistic framework. In this section, results are presented from inelastic RPA calculations of discrete particle-hole states in light nuclei that can provide this nuclear structure. The RPA built upon a Hartree ground state (Hartree-RPA) is the simplest treatment of inelastic transitions that incorporates sufficient physics to describe realistically low-lying collective excitations in spherical nuclei.

The mean-field theory (MFT) has been most successful in describing bulk isoscalar properties of nuclei, so we can hope for success in describing isoscalar, low-lying collective excitations (natural parity) using a time-dependent mean-field picture.

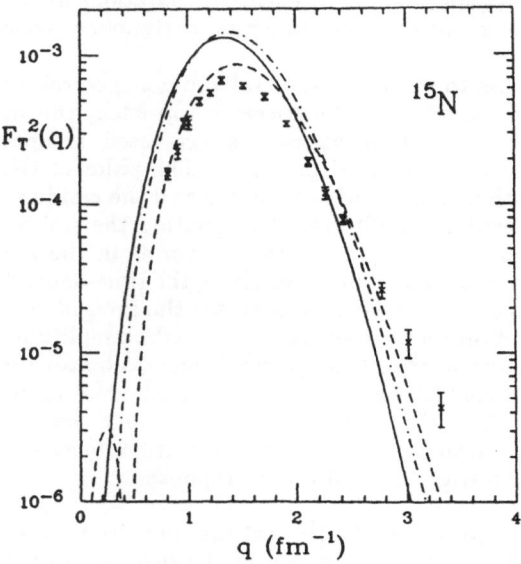

Fig. 4. Transverse elastic form factor for ^{15}N. Valence only (solid), Schrödinger-equivalent (dashed), and RPA (dotdash) calculations are compared with data.[10]

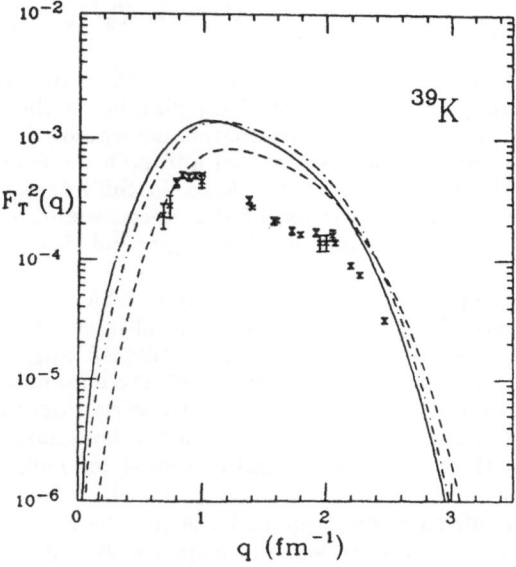

Fig. 5. Transverse elastic form factor for ^{39}K. Valence only (solid), Schrödinger-equivalent (dashed), and RPA (dotdash) calculations are compared with data.[10]

In such a picture, all of the nucleons can contribute coherently to the excitation through the mediation of the mean meson fields. This coherence can only be fully realized in the Hartree-RPA by working self-consistently and with a complete basis. In addition, these requirements ensure that spurious center-of-mass excitations are completely separated from physical excitations and that the transition currents are conserved.

The two approaches to the RPA, spectral and nonspectral, require very different technologies for excited states. In the spectral approach, the integral equation for the particle-hole (polarization) propagator is expressed in spectral form using a discretized and truncated single-particle basis.[14] The poles of this propagator occur at the collective excitation energies of the nucleus and the residues are proportional to the corresponding transition amplitudes. By equating the residues of the poles, one can derive RPA matrix equations of the same form as in the nonrelativistic RPA.[24] (These equations can also be derived by linearizing the time-dependent Euler-Lagrange equations about the static mean-field solution for the ground state.) These matrices are diagonalized to obtain discrete energies and RPA amplitudes in the discretized basis. In contrast, in the nonspectral approach one evaluates the response function at a given (complex) frequency by solving the integral equation directly (by iteration or matrix inversion). One then searches for poles near the real axis (for discrete states) and extracts the residues to determine transition densities. Escape widths are naturally included along with the continuum response.

Both spectral[11] and nonspectral[12] calculations are represented in the present review. All calculations in this section are in the σ–ω model with parameters from Ref. 3 (except as noted for ^{40}Ca; see below). The ground state parameters completely determine both the single-particle basis and the unperturbed energies as well as the particle-hole interaction. This interaction is given simply by (neglecting retardation):[24]

$$V(1,2) = -\beta_1\beta_2 \frac{g_s^2}{4\pi}\frac{e^{-m_s r_{12}}}{r_{12}} + \left(1 - \underset{\sim}{\alpha}_1 \cdot \underset{\sim}{\alpha}_2\right)\frac{g_v^2}{4\pi}\frac{e^{-m_v r_{12}}}{r_{12}} ,$$

where $r_{12} \equiv |\mathbf{r}_1 - \mathbf{r}_2|$. Because this interaction is used in matrix elements of four-component Dirac spinors, its form is deceptively simple. If the interaction were to be reduced for use with conventional two-component spinors, it would become an infinite series of terms with complicated radial and spin dependencies (*e.g.*, tensor, spin–orbit, *etc.*). Note that it is important to keep the full interaction for consistency. In particular, the three-vector interaction of the omega meson cannot be excluded, even though it does not contribute in the spherical ground state.

Let us first consider the spectral calculations, which are described in detail in Ref. 11. The discretized single-particle basis is obtained by applying boundary conditions at a radius several times the nuclear radius. A large configuration space is necessary to ensure that traces of the spurious 1^- are eliminated from physical 1^- states. With a large enough space, energies and transition densities are stable with respect to the size of the basis and independent of the boundary conditions. This is illustrated in Fig. 6 for the spurious state and two low-lying collective states in ^{16}O.

It is important to note that the complete basis includes negative-energy solutions, which contribute to the Pauli-blocked vacuum response. If only conventional particle-hole configurations are used, the spurious state is found at $\pm 5i$ MeV and the other 1^- states have spurious admixtures that distort the transition densities. The configuration spaces for the following spectral calculations (corresponding to the largest spaces in Fig. 6) include particle–hole configurations up to several hundred MeV in the continuum and $+/-$ configurations including the entire bound negative-energy spectrum. However, recent work by J. McNeil using "optimized" single-particle basis functions based on b-splines indicates that consistent and accurate spectral calculations

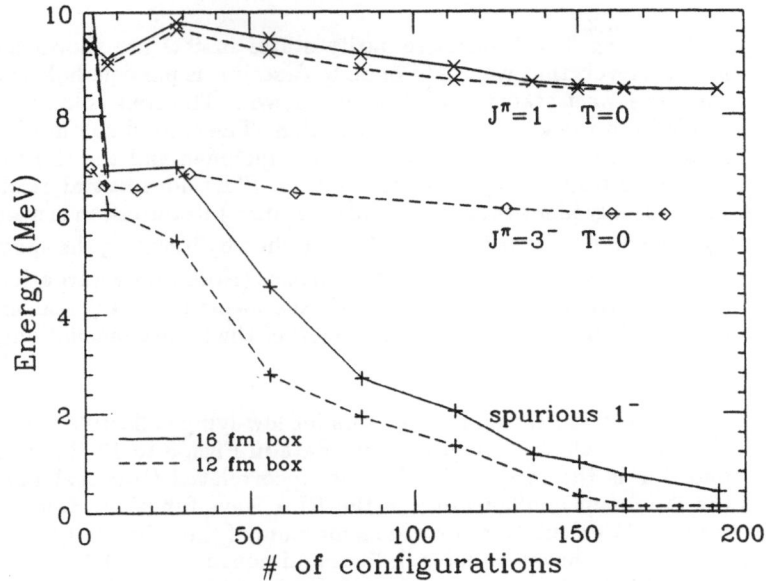

Fig. 6. Energy of the spurious state and other low-lying states in ^{16}O in a spectral RPA calculation.[11]

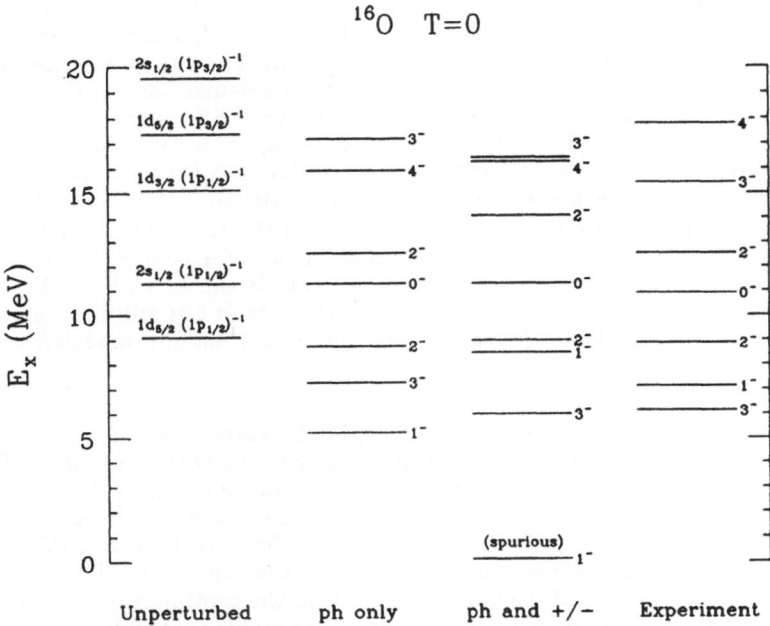

Fig. 7. Selected low-lying states in ^{16}O in a spectral RPA calculation.[11]

with much smaller configuration spaces are practical. He has been able to reproduce the results reported here with much less computational effort.[25]

In Fig. 7, low-lying $T = 0$ negative-parity states in ^{16}O are shown along with certain experimental levels that we might hope to describe as particle-hole excitations. For clarity, only the lowest states of each J are shown. The first column shows the unperturbed levels from the ground-state calculation. The second column is the RPA spectrum when only particle-hole configurations are included and the third column is the full RPA spectrum from the spectral calculation. (The nonspectral calculation of the spectrum is essentially identical to the third column.[12]) Retardation is negligible for the states in Fig. 7 since $E_{ex}^2 \ll m_{s,v}^2$.[24] This was verified by iterating the spectral RPA equations with state-dependent masses for the mesons. (Note that retardation is easily included in the nonspectral calculations.) Overall, the spectrum is very reasonable for a self-consistent RPA calculation, particularly in view of the large cancellations inherent in relativistic models.

Calculated electron scattering form factors for low-lying collective natural parity states in ^{16}O, ^{12}C, and ^{40}Ca are compared with data in Figs. 8 to 12. In each of these figures, the solid line is the form factor for the uncorrelated (Hartree) particle-hole state that is the dominant configuration in the RPA wave function. Comparisons of these curves to the RPA calculations provide a measure of the collectivity of the states. Spectral RPA results are shown as dotdash lines and nonspectral RPA as dashed lines (they are indistinguishable in many cases). The form factors include a center-of-mass correction factor and single-nucleon form factors, as described in Ref. 11. Current conservation has been verified numerically for these states.[12]

The 1^- state shown in Fig. 8 is particularly interesting because it provides a sensitive test of self-consistency. Only a fully self-consistent calculation will completely eliminate spurious contributions to the wave function for this state. A large amount of configuration mixing is required to remove the $J = 1$ character of the form factor at low q; the RPA calculations succeed in this respect! The momentum dependence of the experimental data is well reproduced by the RPA calculations but the strength is a factor of two too low. The 3^- state in ^{16}O shown in Fig. 9 reflects the strong collectivity found experimentally. Again, the momentum dependence is quite reasonable but the strength is about 2/3 the experimental form factor (about the same ratio is obtained in the best nonrelativistic self-consistent RPA calculation). The 2^+ state at 4.44 MeV in ^{12}C (Fig. 10) also shows significant collectivity in the RPA calculations, leading to good agreement with the experimental strength (the RPA energy is about 4 MeV). The transverse electric form factor for the 2^+ state is shown in Fig. 11. The RPA result is quenched with respect to the single-particle result but still lies well above the data. (In these calculations, the carbon ground state is simply modeled as a closed $1p_{3/2}$ shell.)

If we now turn to ^{40}Ca, we find the low-lying 5^- state at 3.9 MeV in the Hartree-RPA with a longitudinal form factor in good agreement with experiment (Fig. 12). However, the collective 3^- state appears at an *imaginary* energy with the parameter set of Ref. 3 and imaginary frequencies in the RPA indicate instabilities of the Hartree ground state. A spectral calculation in a truncated basis reveals that this instability is driven by the particle–hole response from high in the continuum. If we truncate the particle–hole space at several "shells" (~ 50 MeV in the continuum), the energy of the 3^- is real (about 2 MeV) and the form factor is reasonable (dotdash curve in Fig. 10). We note that the energy of this state is very sensitive to the interaction and can be made real in the full calculation by slightly adjusting the parameters of the model. For example, the dashed curve is the result when ^{40}Ca is calculated with the scalar meson coupling squared reduced by two percent.[12] If the scalar meson mass is then adjusted

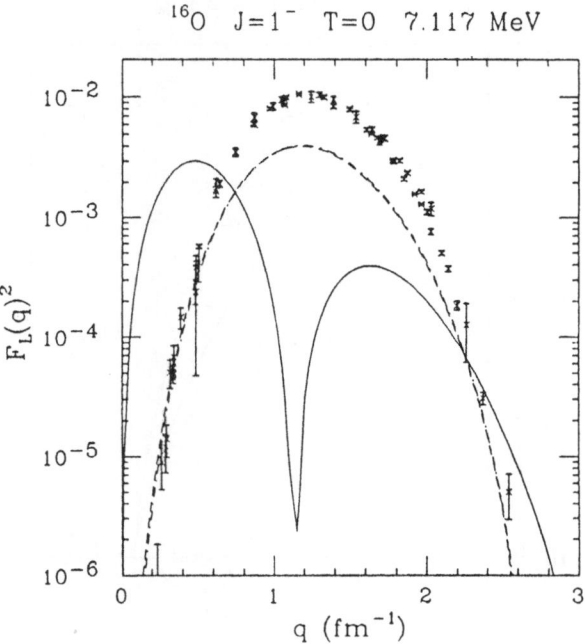

Fig. 8. Longitudinal form factor for the 7.1 MeV isoscalar 1^- state in ^{16}O. Spectral[11] (dot-dashed) and nonspectral[12] (dashed) RPA curves are compared to an unperturbed $2s_{1/2}(1p_{1/2})^{-1}$ pair (solid) and to data.

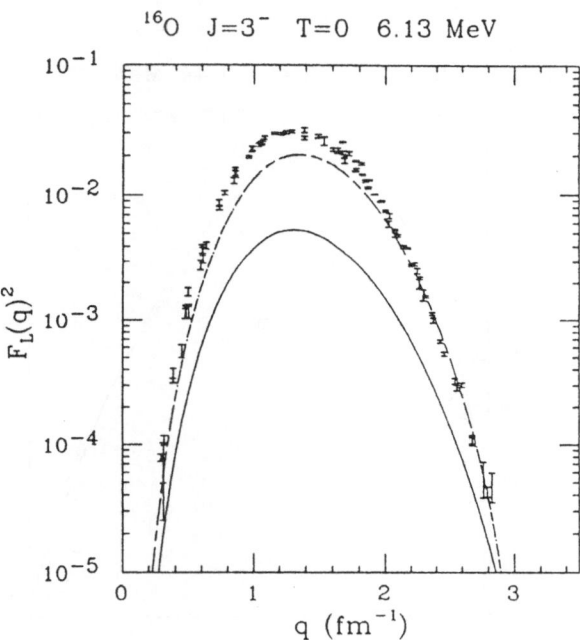

Fig. 9. Longitudinal form factor for the low-lying isoscalar 3^- state at 6.1 MeV in ^{16}O. Spectral[11] (dot-dashed) and nonspectral[12] (dashed) RPA curves are compared to an unperturbed $1d_{5/2}(1p_{1/2})^{-1}$ pair (solid) and to data.

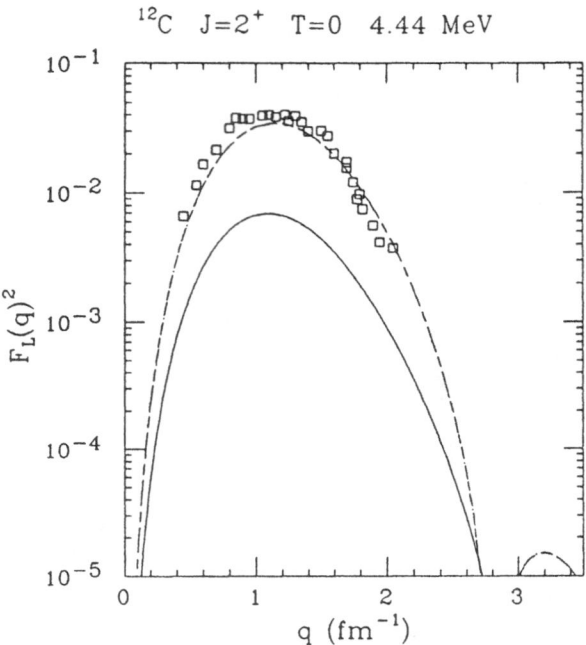

Fig. 10. Longitudinal form factor for the low-lying isoscalar 2^+ state at 4.4 MeV in ^{12}C. Spectral[11] (dot-dashed) and nonspectral[12] (dashed) RPA curves are compared to an unperturbed $1p_{1/2}(1p_{3/2})^{-1}$ pair (solid) and to data.

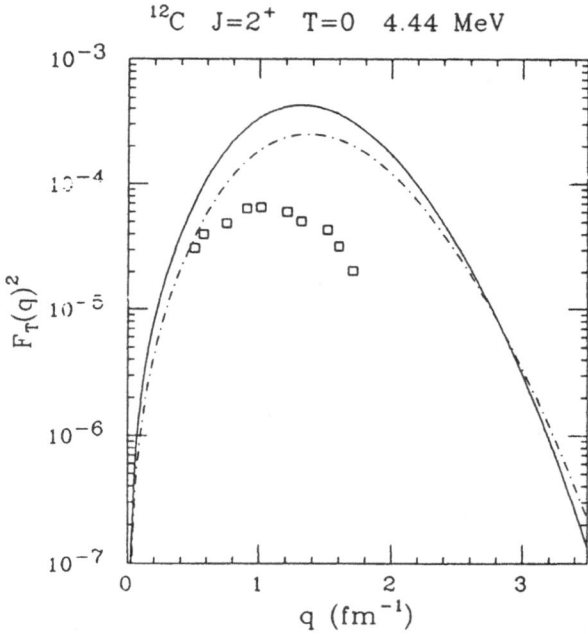

Fig. 11. Transverse form factor for the low-lying isoscalar 2^+ state at 4.4 MeV in ^{12}C. Spectral[11] (dot-dashed) and nonspectral[12] (dashed) RPA curves are compared to an unperturbed $1p_{1/2}(1p_{3/2})^{-1}$ pair (solid) and to data.

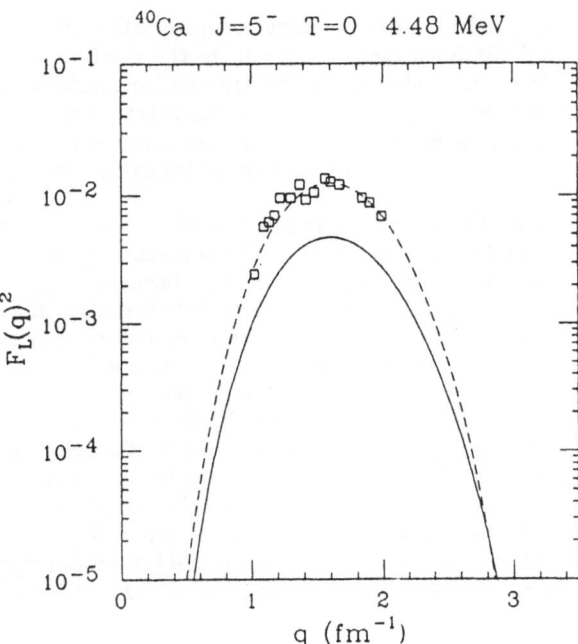

Fig. 12. Longitudinal form factor for the low-lying isoscalar 5^- state at 4.5 MeV in ^{40}Ca. A nonspectral RPA curve (dashed)[12] is compared to an unperturbed $1f_{7/2}(1d_{3/2})^{-1}$ pair (solid) and to data.

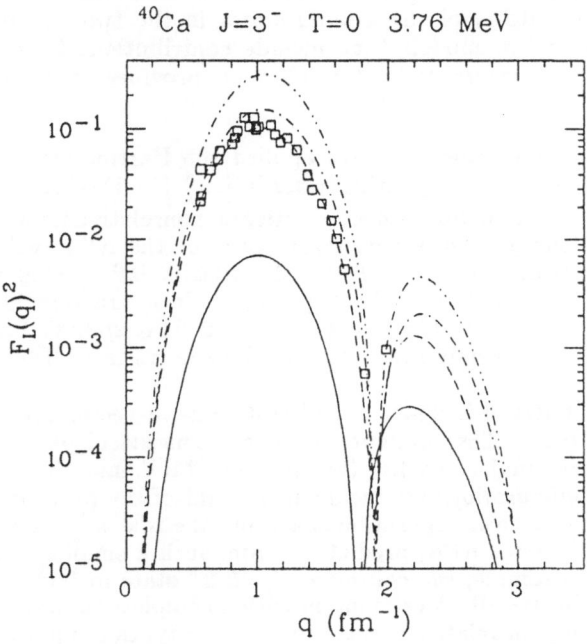

Fig. 13. Longitudinal form factor for the low-lying isoscalar 3^- state at 3.8 MeV in ^{40}Ca (see text).

to reproduce the same nuclear matter saturation properties as in Ref. 3, the state is found near 1 MeV and the form factor is given by the double-dotdashed curve. The sensitivity of this state to small changes in the interaction precludes strong conclusions at this time about the stability of the Hartree ground state in finite nuclei. Additional systematic RPA calculations should be carried out using more realistic interactions (e.g., including Coulomb) and also with vacuum polarization effects added.

Finally, we consider the unnatural parity transitions, where the nuclear response is mediated solely by the three-vector part of the ω meson interaction. The transition currents here are analogous to the elastic current discussed earlier. The transition current (Dirac part only) for the lowest 4^- state in ^{16}O is shown in Fig. 14, with curves for the uncorrelated particle-hole pair (solid), nonrelativistic (dots), RPA with only particle-hole configurations (dashed), and the full spectral RPA (dotdash) calculations. As was seen for elastic currents, the M/M^* enhancement of the valence current is further enhanced at finite q in the RPA. (Note the dramatic enhancement that occurs when only particle-hole configurations are included.) The electron scattering form factors for the unnatural parity states do not describe the experimental data unless significant quenching factors are applied. A typical example is the 4^- state, shown in Fig. 15. The RPA curve reproduces the experimental momentum dependence but requires a quenching factor of .3. Clearly the RPA model does not adequately describe these transitions in detail; however, this conclusion is not unique to relativistic calculations.

HARTREE-FOCK AND RPA WITH EXCHANGE

In this section I will show results for some first steps beyond the Hartree approximation and Hartree-RPA that include the effects of exchange. These new calculations are based on relativistic Hartree-Fock (HF) calculations of spherical nuclei that have recently become available. Blunden and Iqbal have found that for bulk properties Hartree and Hartree-Fock ground states are quite similar, if the model parameters are fitted to the same nuclear matter saturation properties[5] However, the linear response (RPA) of the ground states in the two approximations can be dramatically different in models that include contributions from isovector mesons (π and ρ). The comparison of RPA predictions provides an important test of the interactions.

Blunden and McCorquodale have applied the Hartree-Fock wave functions to spectral RPA calculations, neglecting retardation[6] (In this case, the RPA matrix equations are identical in form to conventional nonrelativistic RPA equations[24]) These wave functions provide a consistent basis for the RPA including both direct and exchange contributions (HF-RPA). (Relativistic RPA using a Hartree single-particle basis but with a Hartree-Fock particle-hole interaction including exchange was considered in Ref. 26.) They have compared RPA spectra based on a Hartree ground state and a direct-only interaction to the spectra from HF-RPA calculations.

The model of Ref. 6 includes σ and ω mesons, a vector-coupled ρ meson, and the pseudovector pion. The configuration space consists of particle–hole configurations including three major "shells" (about 30–50 MeV into the continuum) and no negative-energy configurations are included. In light of the previous discussion about the importance of a large configuration space and the role of the negative-energy configurations in the Hartree-RPA, predictions from such a small space must be viewed with caution. For example, the collective $T = 0$ 2^+ state in ^{12}C is found more than 2 MeV lower in a Hartree-RPA calculation with a complete configuration space. However, if we focus only on relative spectra for the two types of approximation (Hartree and Hartree-Fock), then we can obtain valuable insight into the differences between the interactions.

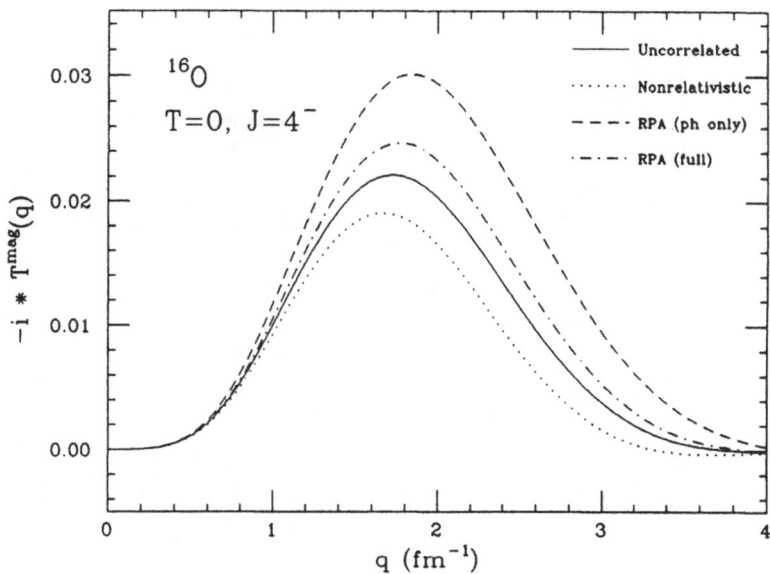

Fig. 14. Dirac current in momentum space for the isoscalar 4^- state in ^{16}O.[11]

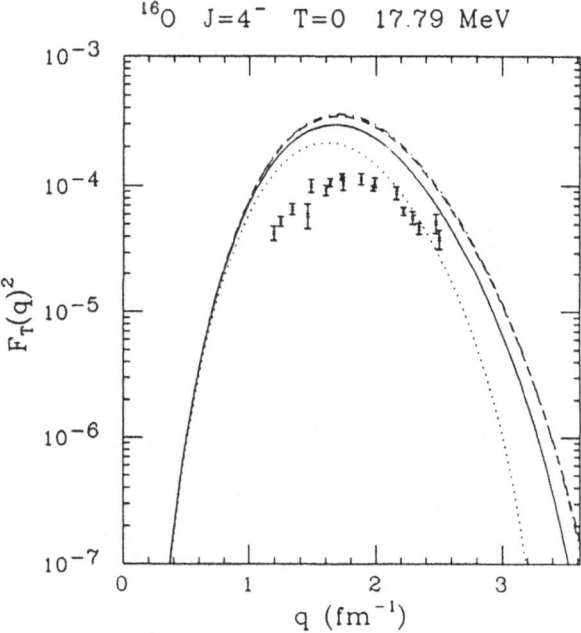

Fig. 15. Transverse form factor for the isoscalar 4^- state in ^{16}O for the same curves as in Fig. 14[11] except that the dashed curve is from a nonspectral RPA calculation.[12]

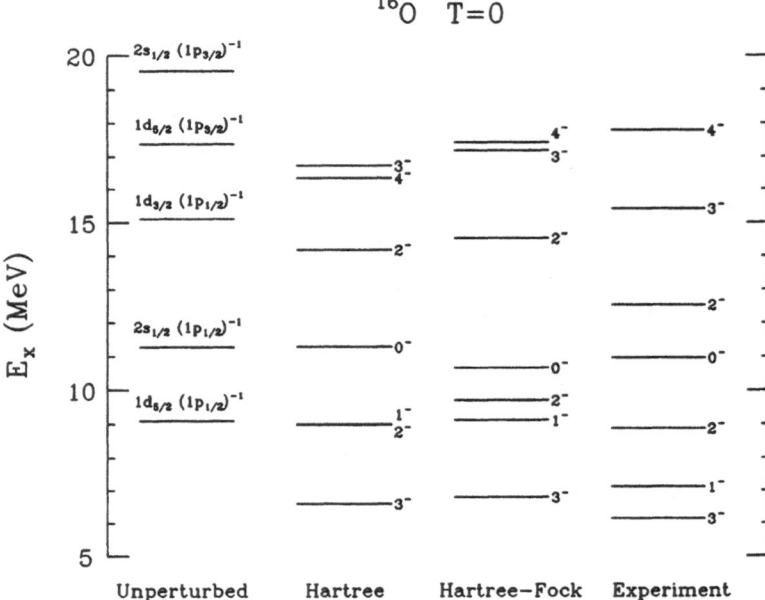

Fig. 16. Selected low-lying, $T = 0$, negative parity states in ^{16}O from calculations in Ref. 6.

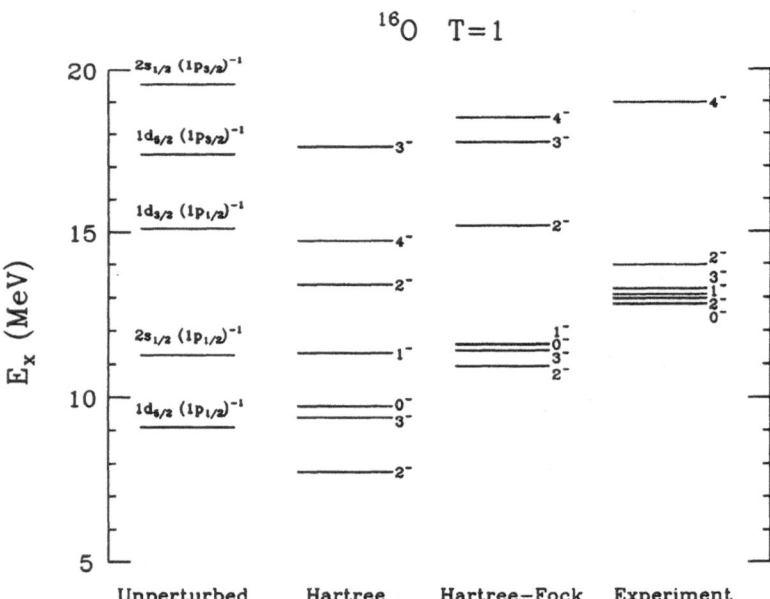

Fig. 17. Selected low-lying, $T = 1$, negative parity states in ^{16}O from calculations in Ref. 6.

Selected states from the low-lying negative-parity spectra in ^{16}O are shown in Figs. 16 ($T = 0$) and 17 ($T = 1$). (More complete spectra can be found in Ref. 6.) Consider first the $T = 0$ states. Note that only the σ and ω mesons contribute to the $T = 0$ particle-hole interaction in the Hartree-RPA while all the mesons contribute in the HF-RPA. Qualitatively the Hartree and Hartree-Fock spectra are quite similar. The principal differences are that the 0^- is shifted by the Hartree-Fock interaction and the HF-RPA 4^- is in better agreement with experiment.

If we now turn to $T = 1$, we see large differences between the Hartree and Hartree-Fock spectra. Here only the pion and the ρ meson contribute to the Hartree particle–hole interaction. The pion provides a very attractive interaction, particularly in unnatural parity states (which have the pion quantum numbers). Note in particular that the isovector unnatural parity states are pushed below their isoscalar counterparts, in strong disagreement with experimental data. In fact, with this interaction and a larger configuration space it would not be surprising to find evidence of pion condensation near normal nuclear densities. In contrast, the exchange contribution to the Hartree-Fock interaction provides repulsion that counteracts the attractiveness of the pion in the direct interaction. The result is a reasonable $T = 1$ spectrum for the HF-RPA. It would seem that nature shows a clear preference for the HF-RPA, at least within this model of the isovector interaction. However, further investigations are necessary to study other models of the isovector interaction and the role of the vacuum in the HF-RPA.

SUMMARY

I have reviewed a variety of recent relativistic RPA calculations in finite nuclei that have been formulated using quantum hadrodynamics. There have been several common threads running through this survey. One such thread is the emphasis on self-consistency. With self-consistency, there is no "magnetic moment problem" (at least for isoscalar moments), currents are conserved, and spurious excitations separate out from physical states. Another theme is the need to extend the current models. QHD provides a framework within which approximations can be tested and the important physics isolated and applied to better models. Finally, I have stressed the role of the vacuum in the relativistic RPA, including both the Pauli-blocked response that is an important ingredient of the present calculations and the full vacuum polarization that is still to be studied in finite nuclei.

A variety of interesting results have been obtained so far. The finite nucleus calculations of elastic currents verify the LDA prediction that isoscalar magnetic moments are insensitive to M^* and fall near the Schmidt values. However, at finite q the M/M^* enhancement persists. In the present models, elastic magnetic scattering is not well described, but much of the important physics is missing. The isovector interaction and the effects of vacuum polarization on the elastic response are the principal areas for future studies.

Inelastic RPA calculations have produced reasonable isoscalar spectra and have demonstrated current conservation and the separation of spurious states from physical excitations. With care taken to ensure complete calculations, spectral and nonspectral approaches show good numerical agreement. Transition charge densities for collective isoscalar excitations are generally well-described by the Hartree-RPA. The cloud on the horizon is the imaginary energy found for the 3^- state in ^{40}Ca; this may be related to instabilities found in nuclear matter in this approximation. The Hartree-RPA also strongly overestimates the magnitude of transition currents for unnatural parity states. As is the case for the elastic RPA calculations, it is important to extend the present calculations to include isovector mesons and the effects of vacuum polarization. The work of Blunden and McCoquodale shows that the isovector interaction including the pion may prove unrealistic in the Hartree-RPA, at least without some phenomenological contact term. The HF-RPA shows promise of resolving these problems, but more study is needed, including the role of the vacuum.

We can also expect new applications of the these models to other domains of the nuclear response. The nonspectral RPA techniques are particularly well-suited to investigations of the giant resonance and quasielastic regions. The authors of Ref. 13 have calculated quasielastic electron scattering in finite nuclei and found significant quenching of the longitudinal response from RPA correlations. Finally, we certainly want to apply current and future RPA transition densities as nuclear structure input to studies of (e, e'), (p, p'), and other reactions. The sampling of RPA calculations presented here is really just a preview of coming attractions. The technology for relativistic RPA technologies is rapidly maturing and many new results can be expected in the near future.

REFERENCES

1. C. J. Horowitz, contribution to these proceedings.
2. J. R. Shepard, contribution to these proceedings.
3. C. J. Horowitz and B. D. Serot, Nucl. Phys. **A368**, 503 (1981).
4. L. G. Arnold, B. C. Clark, R. L. Mercer, and P. Schwandt, Phys. Rev. **C23** (1981) 1949, and references cited.
5. P. Blunden and J. Iqbal, TRIUMF preprint TRI–PP–88–03, 1988.
6. P. Blunden and P. McCoquodale, TRIUMF preprint TRI–PP–88–04, 1988.
7. B. D. Serot and J. D. Walecka, Adv. in Nucl. Phys. **16** (Plenum, New York, 1986).
8. S. Ichii, W. Bentz, A. Arima, and T. Suzuki, Phys. Lett. **192B**, 11 (1987).
9. J. R. Shepard, E. Rost, and C.-Y. Cheung, and J. A. McNeil, Phys. Rev. **C37**, 1130 (1988).
10. R. J. Furnstahl, "Convection Currents in Nuclei in a Relativistic Mean-Field Theory", U. of Md. preprint 88–156, 1988, submitted to Phys. Rev. C.
11. J. F. Dawson and R. J. Furnstahl, Indiana preprint IU/NTC 87-5, 1988 (in preparation).
12. E. Rost, J. R. Shepard, and J. A. McNeil, private communication.
13. K. Wehrberger and F. Beck, to be published in Phys. Rev. C.
14. A. L. Fetter and J. D. Walecka, **Quantum Theory of Many-Particle Systems,** (McGraw-Hill, New York, 1971).
15. R. J. Furnstahl and C. J. Horowitz, Nucl. Phys. A. (in press)
16. L. D. Miller, Ann. of Phys. **91**, 40 (1975).
17. T. Matsui, Nucl. Phys. **A370**, 365 (1981).
18. W. Bentz, A. Arima, H. Hyuga, K. Shimizu and K. Yazaki, Nucl. Phys. **A436**, 593 (1985).
19. H. Kurasawa and T. Suzuki, Phys. Lett. **165B**, 234 (1986).
20. J. A. McNeil, R. D. Amado, C. J. Horowitz, M. Oka, J. R. Shepard and D. A. Sparrow, Phys. Rev. **C34**, 746 (1986).
21. R. J. Furnstahl and B. D. Serot, Nucl. Phys. **A468**, 539 (1987).
22. R. J. Furnstahl and C. E. Price, work in progress.
23. B. L. Friman and P. A. Henning, GSI–88–06, 1988, submitted to Phys. Lett.
24. R. J. Furnstahl, Ph.D. thesis, Stanford University (1985), unpublished; R. J. Furnstahl, Phys. Lett. **152B**, 313 (1985).
25. J. A. McNeil, private communication.

26. R. J. Furnstahl, "Relativistic Shell-Model Calculations," in **Nuclear Structure at High Spin, Excitation, and Momentum Transfer**, Hermann Nann, ed., AIP Conference Proceedings No. 142 (American Institute of Physics, New York, 1986), p. 376.

CHARGE EXCHANGE (P , N) REACTIONS TO THE ISOBARIC ANALOG STATE

S. Hama

Department of Physics
The Ohio State University
Columbus, Ohio 43210

M.J. Iqbal

TRIUMF
4004 Wesbrook Mall
Vancouver, B.C.
Canada V6T 2A3

and

J.I. Johansson, and H.S. Sherif

Department of Physics
University of Alberta
Edmonton, Alberta
Canada T6G 2J1

Relativistic nucleon-nucleus optical models based on the Dirac equation provided successful descriptions for elastic proton nucleus scattering data , especially spin observables, at intermediate energies[1,2]. These elastic scattering studies have so far tested extensively the isoscalar parts of the Lorentz scalar and vector in the pA relativistic optical potentials over a wide range of energies.

For nuclei with $N \neq Z$, the isovector components of the Lorentz scalar and vector potentials are present. These isovector potentials were recently calculated using the original RIA[3-6] by Clark et al[7]. In addition to contributing to elastic scattering, the isovector terms can also induce transitions to the isobaric analog of the target ground states via the charge exchange (p,n) reaction.

In the case of scattering from a spin-zero target nucleus with filled (l, j) subshells but with

$N \neq Z$, the generalized relativistic optical potential[7] becomes,

$$U_{opt}(r) = U_s^{(0)} + \gamma^0 U_v^{(0)} + [U_s^{(1)} + \gamma^0 U_v^{(1)}](\vec{t} \cdot \vec{T})/A \qquad (1)$$

where the subsripts 0 and 1 denote the isoscalar and isovector components, respectively, $\vec{t}\,(\vec{T})$ represents the projectile (target) isospin operator, and A is the mass number of the target nucleus.

In order to calculate the isovector potentials, we used the model of Horowitz and Murdock[8,9]. This model will be denoted in this paper as the "Relativisitic Folding Model (RFM)". The first RIA calculations used the Lorentz invariant nucleon - nucleon scattering amplitudes given by[10]

$$F = F_s I^{(1)} I^{(2)} + F_v \gamma_\mu^{(1)} \gamma^{\mu(2)} + F_t \sigma_{\mu\nu}^{(1)} \sigma^{\mu\nu(2)} + F_p \gamma_5^{(1)} \gamma_5^{(2)} + F_a (\gamma_\mu \gamma_5)^{(1)} (\gamma^\mu \gamma_5)^{(2)}. \qquad (2)$$

Here coefficients of expansion F_i are functions of energy (E) and momentum transfer (q). The pA scalar and vector optical potentials in the original RIA are generated by folding the Lorentz amplitudes F_s and F_v with appropriate nuclear densities. It is well known that those RIA optical optentials fail to reproduce elastic pA observables below 300 MeV [6]. To remedy these deficiencies Horowitz[11] parametrized F as a sum of direct and exchange terms based on the first Born approximation to the meson exchange model of NN interaction. The original RIA calculations are improved by using pseudovector πN coupling in the relativistic NN amplitude, treating exchange terms in the optical potentials explicitly and including medium modification from Pauli blocking.

In this paper we concern ourselves with the isovector optical potential and its application to the (p,n) reaction to isobaric analog states (IAS). The relativistic folding model is described in Ref.9 and 12. In the RFM, the direct part of isoscalar and isovector optical potential can be written as

$$U_{D,i}^{(0)}(r) = -\frac{4\pi i k_{cm}}{m} \int dr\,'[\frac{3}{4}F_{D,i}^{(T=1)}(|\,\vec{r} - \vec{r}\,'\,|) + \frac{1}{4}F_{D,i}^{(T=0)}(|\,\vec{r} - \vec{r}\,'\,|)]$$
$$\times \; [\rho_i^{(n)}(|\,\vec{r}\,'\,|) + \rho_i^{(p)}(|\,\vec{r}\,'\,|)], \qquad (3)$$

and

$$U_{D,i}^{(1)}(r) = -\frac{4\pi i k_{cm}}{m}[\frac{4A}{N-Z}] \int dr\,'[\frac{1}{4}F_{D,i}^{(T=1)}(|\,\vec{r} - \vec{r}\,'\,|) - \frac{1}{4}F_{D,i}^{(T=0)}(|\,\vec{r} - \vec{r}\,'\,|)]$$
$$\times \; [\rho_i^{(n)}(|\,\vec{r}\,'\,|) - \rho_i^{(p)}(|\,\vec{r}\,'\,|)]. \qquad (4)$$

The construction of a local exchange optical potential from the exchange amplitudes is done using the local density approximation of Brieva and Rook[13]. We write the exchange parts of the isoscalar and isovector potentials as

$$U_{E,i}^{(0)}(r) = -\frac{4\pi i k_{cm}}{m} \int dr\,'[\frac{3}{4}F_{E,i}^{(T=1)}(|\,\vec{r} - \vec{r}\,'\,|) + \frac{1}{4}F_{E,i}^{(T=0)}(|\,\vec{r} - \vec{r}\,'\,|)]$$
$$\times \; [\rho_i^{(n)}(\vec{r}, \vec{r}\,') + \rho_i^{(p)}(\vec{r}, \vec{r}\,')]j_0(k_{lab}|\,\vec{r} - \vec{r}\,'\,|), \qquad (5)$$

and

$$U_{E,i}^{(1)}(r) = -\frac{4\pi i k_{cm}}{m}[\frac{4A}{N-Z}] \int dr\,'[\frac{1}{4}F_{E,i}^{(T=1)}(|\,\vec{r} - \vec{r}\,'\,|) - \frac{1}{4}F_{E,i}^{(T=0)}(|\,\vec{r} - \vec{r}\,'\,|)]$$
$$\times \; [\rho_i^{(n)}(\vec{r}, \vec{r}\,') - \rho_i^{(p)}(\vec{r}, \vec{r}\,')]j_0(k_{lab}|\,\vec{r} - \vec{r}\,'\,|), \qquad (6)$$

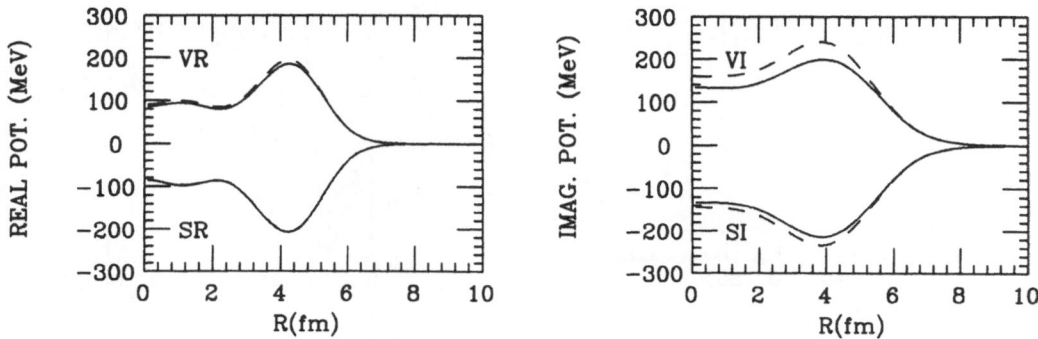

Figure 1. The relativistic folding model isovector potentials for ^{90}Zr at 135 MeV. The left (right) curves show the real (imaginary) components of the potential. The curves denoted S (V) are the scalar (vector) terms. The solid (dashed) curves refer to the calculation with (without) Pauli blocking effects.

where

$$\rho_i(\vec{r}, \vec{r}\,') = \rho_i[\frac{|\,\vec{r} + \vec{r}\,'\,|}{2}][\frac{3}{k_F\,|\,\vec{r} - \vec{r}\,'\,|}]j_1(k_F\,|\,\vec{r} - \vec{r}\,'\,|). \tag{7}$$

Here $j_0(kr)$ and $j_1(kr)$ are the spherical Bessel functions and k_F is the local Fermi momentum, related to the baryon density by

$$\rho_B[\frac{|\,\vec{r} + \vec{r}\,'\,|}{2}] = \frac{2}{3\pi^2}k_F^3. \tag{8}$$

In these calculations, the NN amplitude from SAID program[14] and the relativistic Hartree nuclear densities of Horowitz and Serot [15] were used.

The effect of Pauli blocking on the calculation of the isovector potentials has been included. This is done in the manner prescribed by Murdock and Horowitz[9]. In the Fig.1 we show the real and imaginary parts of the scalar and vector isovector optical potentials for ^{90}Zr at 135 MeV. The dashed curves correspond to calculations without Pauli blocking while the solid curves include Pauli blocking. The radial dependence of the isovector potentials is very

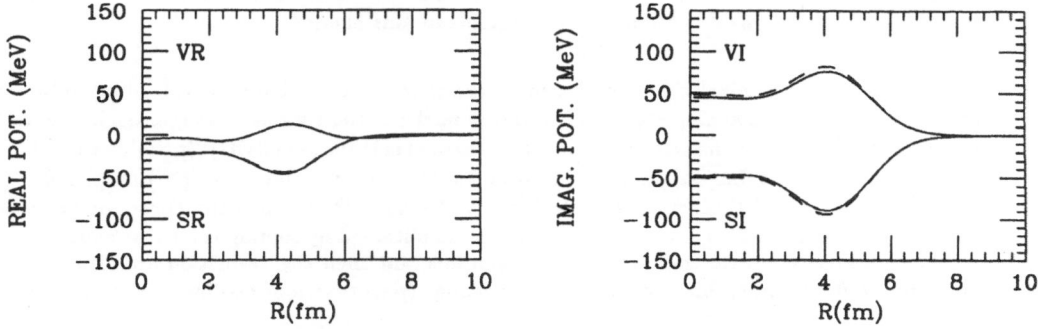

Figure 2. Same as Fig.1, but for ^{90}Zr at 400 MeV.

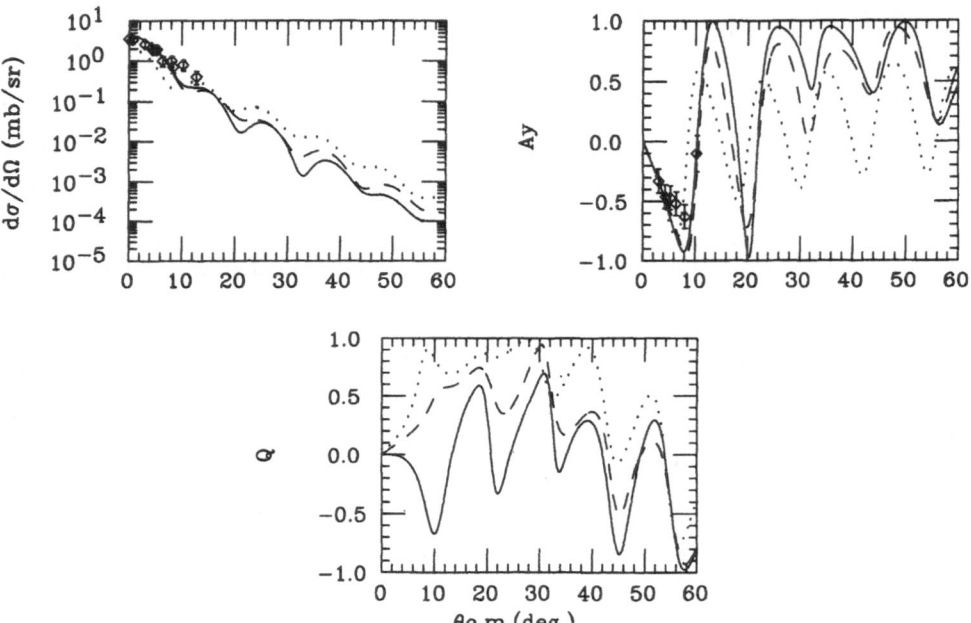

Figure 3. Cross sections, analyzing powers and spin rotation parameters for the reaction
^{90}Zr(p,n)^{90}Nb (IAS) at 160 MeV. The dashed and dotted curves are the results from
the RFM calculation with Pauli blocking and from the RIA calculation, respectively.
The solid curves correspond to the RFM isovector potentials with phenomenological
diagonal potentials. The data are from Ref.17.

different from that of the isoscalar ones. The surface peaking occurs because the isovector
optical potential is proportional to the difference of neutron and proton densities, which peaks
at the nuclear surface. Also, unlike their isoscalar counterparts, the imaginary parts of both
isovector potentials are stronger than the corresponding real component. At this energy, Pauli
blocking leads to a suppression of the imaginary strength in the interior region, but leaves the
tail region unchanged. The real components of the isovector potential are much less sensitive
to Pauli blocking. In Fig.2, we show the isovector optical potential for ^{90}Zr at 400 MeV. The
solid curves includes Pauli blocking effects. We see that both scalar and vector potentials are
much smaller than the 135 MeV case. The strength of the imaginary parts, however, remain
dominant over that of the real parts. The effects of Pauli blocking, which is important for the
imaginary parts at 135 MeV is found to be negligible at this energy.

The isovector potentials derived above can induce charge exchange reactions leading to the
excitation of the isobaric analog states of the target nucleus. Here we use two approaches and
compare them. The one is the relativistic coupled channel calculation (RCC)[7] in which the full
potentials are used in a Lane-type coupled Dirac equation to calculate the (p,n) amplitude.
The other is the relativistic DWBA calculation (RDWBA)[12,16] in which the Dirac distorted
waves for the incident proton and outgoing neutron are obtained by solving the Dirac equation
with the appropriate elastic channel optical potentials and then the transition amplitude is
calculated to obtain observables for the charge exchange (p,n) reactions leading to the isobaric
analog states.

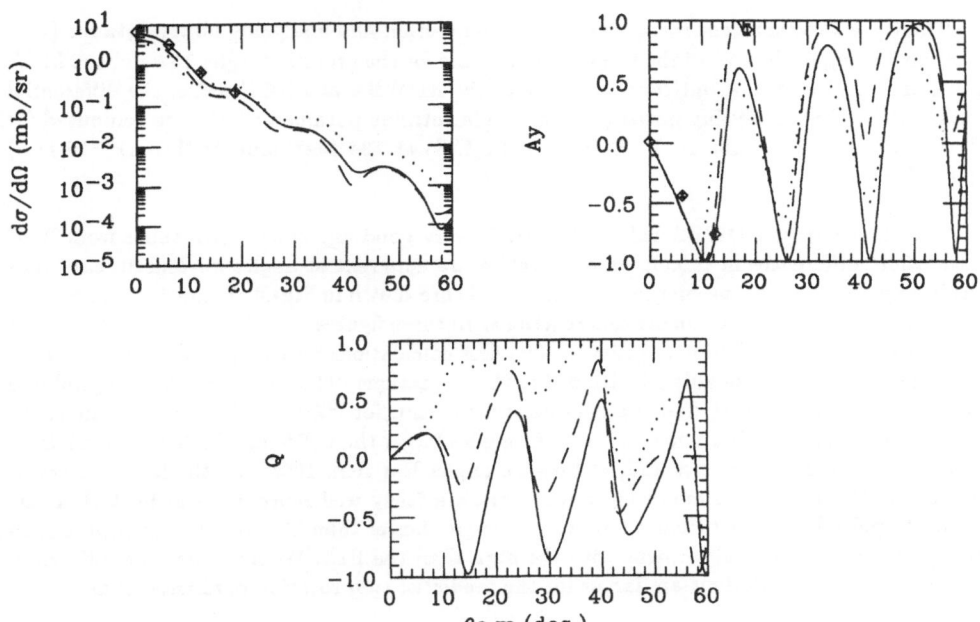

Figure 4. Same as Fig.3, but for ^{48}Ca(p,n)^{48}Sc (IAS) at 134 MeV. The dashed and dotted curves have the same meaning as in Fig.3. The solid curves correspond to the predicted diagonal potentials from a global fit[21]. The data are from Ref.18 and 19.

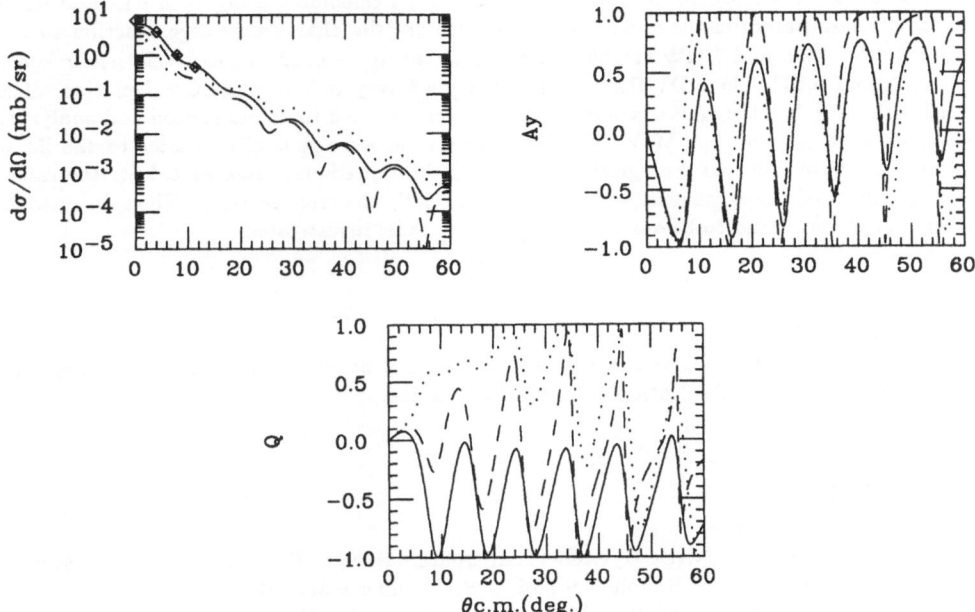

Figure 5. Same as Fig.4, but for ^{208}Pb(p,n)^{208}Bi (IAS) at 134 MeV. The data are from Ref.20.

There are several recent measurement of cross section and analyzing power data for (p,n) reaction leading to the IAS of the target ground state for the proton energies 134 and 160 MeV. Using the RFM and RIA optical potential and the RDWBA and RCC codes, the differential cross section (dσ), analyzing power (A_y), and spin rotation parameter (Q) were computed for ^{90}Zr(p,n)^{90}Nb(IAS) at 160 MeV, ^{48}Ca(p,n)^{48}Sc(IAS) at 134 MeV and ^{208}Pb(p,n)^{208}Bi(IAS) at 134 MeV.

The results from RDWBA calculation are in very good agreement with those from RCC calculation; differences in calculated observables are apparent at angles beyond 30°c.m. The resulting parameter-free prediction from RCC code are shown in Figs.4, 5, and 6 in comparison with the data for the above mentioned reactions. In those figures, the RFM (RIA) calculations are shown by the dashed (dotted) curves. The RIA calculations for the (p,n) cross section data are underestimated by roughly a factor of 2 in those reactions. The RFM potentials reproduced the (p,n) cross section very well at angles less than 8°c.m. for ^{90}Zr(p,n), but underestimate the cross sections for the other two reactions. At angles larger than 10°c.m. the RFM calculations lie well below the RIA results and, at forward angles less than 10°c.m. , the RFM curves lie above the RIA ones. The analyzing power data are fairly well reproduced in both RIA and RFM at angles less than 10°c.m., however, at angles larger than 10°c.m. the RFM potentials did in somewhat better agreement with the data than the RIA. We note that the differences between various calculations are larger for the predicted spin rotation parameter than for Ay.

As noted in Ref.7, the underestimate of the (p,n) cross section data is largely a result of too much absorption in the diagonal optical potentials. Therefore we have used phenomenological optical optentials for ^{90}Zr and the predicted optical potentials from global fits[21] for ^{48}Ca and ^{208}Pb as the diagonal optical potentials. The isovector potentials remain fixed at the RFM values. The computed (p,n) cross sections show much better agreement with data for ^{48}Ca(p,n) and ^{208}Pb(p,n), as indicated by the solid curves.

In summary, the isovector optical potential has been calculated using both RIA and RFM models. These potentials have been used to calculate the charge exchange reaction to the IAS for ^{90}Zr, ^{48}Ca and ^{208}Pb at 160 and 134 MeV using two independent relativistic codes RDWBA and RCC. The RDWBA calculation agreed very well with RCC calculation. The RFM with Pauli blocking gives generally good agreement with the cross section and analyzing power data at 134 and 160 MeV. The RIA calculation were generally inferior to the RFM calculation. Using phenomenological potentials or those predicted from global fits the direct channel improved the agreement with the data especially the cross sections. The spin rotation parameter Q was found to be very sensitive to the model investigated.

Acknowledgement

This research was supported in part by the Natural Sciences and Engineering Research Council of Canada and the National Science Foundation.

REFERENCES

1. B.C. Clark in Relativistic Dynamics and Quark - Nuclear Physics, Wiley - Interscience Publication, Ed by M.B. Johnson and A. Picklesimer p.302 (1986)
2. B.C. Clark, S. Hama and R.L. Mercer in the Interaction Between Medium Energy Nucleons in Nuclei - 1982, AIP Conf. Proc. No.97 Ed. by H.O. Meyer, p.260 (1983)
3. J. Shepard, J.A. McNeil and S.J. Wallace, Phys. Rev. Lett. **50** (1983) 1443
4. J.A. McNeil, J. Shepard and S.J. Wallace, Phys. Rev. Lett. **50** (1983) 1439

5. B.C. Clark, S. Hama, R.L. Mercer, L. Ray and B.D. Serot, Phys. Rev. Lett. **50** (1983) 1644

6. B.C. Clark, S. Hama, R.L. Mercer, L. Ray, G.W. Hoffmann and B.D. Serot, Phys. Rev. **C28** (1983) 1421

7. B.C. Clark, S. Hama, E. Sugarbaker, M.A. Franey, R.L. Mercer, L. Ray, G.W. Hoffmann and B.D. Serot, Phys. Rev. **C30** (1984) 314

8. C.J. Horowitz and D.P. Murdock, Phys. Lett. **168B** (1986) 31

9. D.P. Murdock and C.J. Horowitz, Phys. Rev. **C35** (1987) 1442

10. J.A. McNeil, L. Ray and S.J. Wallace, Phys. Rev. **C27** (1983) 2123

11. C.J. Horowitz, Phys. Rev. **C31** (1985) 1340

12. M.J. Iqbal, J.I. Johansson, S. Hama and H.S. Sherif, U. of Alberta preprint; Submitted to Nucl. Phys. A.

13. F.A. Brieva and J.R. Rook, Nucl. Phys. **A291** (1977) 317

14. R.A. Arndt et al., Phys. Rev. **D28** (1983) 97; Scattering Analysis Interactive Dial-in (SAID) program.

15. C.J. Horowitz and B.D. Serot, Nucl. Phys. **A368** (1981) 503

16. J.I. Johansson, E.D. Cooper and H.S. Sherif, Nucl. Phys. A (in press)

17. E. Sugarbaker, et al. in Proc. of Int. Conf. on Nuclear Structure, Ed. by A. van der Woude and B.J. Verhaar (North Holland, Amsterdam, 1982) p.77

18. B.D. Anderson et al., Phys. Rev. **C31** (1984) 1147

19. B.D. Anderson et al., Phys. Rev. **C34** (1986) 422

20. B.D. Anderson, private communication

21. E.D. Cooper, B.C. Clark, R.E. Kozack, S. Shim, S. Hama, H.S. Sherif and R.L. Mercer (in preparation)

VALIDITY OF THE HARTREE APPROXIMATION IN THE WALECKA MODEL *

Robert J. Perry[†]

Department of Physics, The Ohio State University

174 W. 18th Ave., Columbus, Ohio, 43210

Abstract

This work addresses the issue of when the Hartree approximation is valid in the Walecka model by computing the correlation energy (i.e., two loop corrections or lowest order Fock terms) in nuclear matter. It is shown that these corrections are quite large when one includes vacuum polarization. Two types of convergence and validity are discussed, strong and weak, with the large correlation energy implying that the Hartree approximation is not strongly valid in the Walecka model until one reaches densities near thirty times nuclear matter density. This raises questions about what validity of the Hartree approximation might mean in this model, and about the interpretation of the Walecka model as a field theory as opposed to an effective field theory.

I. INTRODUCTION

As we have seen in many talks at this workshop quantum hadrodynamics (QHD)[1-3] has been quite successful in describing a wide variety of nuclear phenomena. To date most work in this field has employed either the mean field approximation or the relativistic Hartree approximation (i.e., mean field plus the Dirac sea). There are notable exceptions in which the relativistic RPA has been used to understand magnetic moments of closed shell plus one nuclei[4] and the properties of deformed nuclei[5], although here again many of the effects of Dirac sea polarization are omitted. If one wishes to interpret QHD as a legitimate field theory, one must test these approximations.

There are several reasons that one should be suspicious of the Hartree approximation. The coupling constants in QHD are large, and no one has yet determined how to rigorously treat strong-coupling field theories. In addition, the Hartree approximation is inadequate in nonrelativistic descriptions and one must ask why relativity should alter this fact. Why should many-body effects known to be important in nonrelativistic nuclear theory suddenly become negligible when one employs a relativistic field theory? There exists a fairly well defined mapping between parts of the relativistic theory and a nonrelativistic treatment in which potentials arise from meson exchange. Rephrasing the above question, given such a mapping how can one preserve the successes of nonrelativistic many-body descriptions of the nucleus without implementing the approximations which were necessary to gain these successes?

The primary justification for the Hartree approximation has been the claim that it becomes

---
* Supported in part by NSF grant PHY-8719526.
† Presidential Young Investigator

valid at 'high density'[1,2]. Although I will examine this claim in detail below, let me first give a few naive arguments to support it. As one increases nuclear density the expectation values of meson fields grow. The Hartree approximation includes these large expectation values to all orders and neglects only the quantum fluctuations from these values. If the expectation values are large enough these fluctuations should become of decreasing importance. For example the lowest order quantum correction to the Hartree approximation, which includes only one virtual meson, is the exchange energy correction. In the Hartree approximation each nucleon interacts with the entire mean field, including that part of the mean field set up by the nucleon itself. The exchange energy corrects this approximation by subtracting this self-interaction. As the density of the system increases, however, one expects that the contribution of any given nucleon to the mean field will become negligible, allowing one to neglect such a correction.

The interpretation of the results presented in this work requires a detailed consideration of what one means when one says that an approximation is valid. I will consider only approximations which are in theory part of a systematic expansion. The relativistic Hartree approximation can be considered the first term in a variety of expansions, with the loop expansion being the most relevant for this calculation. To discuss the validity of the Hartree approximation I will consider two types of convergence. An expansion is strongly convergent when corrections which are calculated perturbatively are small, and I will say that an approximation to such an expansion is strongly valid. This means that one can determine renormalized parameters and mean field expectation values at a given order, and use these values to determine the magnitude of the leading correction, with this correction being small in a strongly convergent expansion. This is the type of convergence normally considered when testing a perturbation series. For example, when someone says that the Born approximation is valid, it is normally assumed that the second Born terms for all observables of interest are small.

When dealing with a strongly interacting field theory it may be necessary to consider a second, less stringent, type of convergence. It is easiest to define what I mean by weak convergence operationally. When testing for weak convergence in QHD, one should recalculate mean fields self-consistently, and one should adjust the parameters of the theory to refit a given set of observables. If the corrections to all relevant observables are small after such a readjustment of parameters and fields, the expansion is weakly convergent. It should be noted that strong convergence implies weak convergence, but the converse obviously does not hold.

This division between strong and weak convergence is actually not so clear cut. The readjustment of parameters at each order of an expansion is part of renormalization. One should renormalize bare parameters in a field theory not only to cancel infinities, but also to reproduce a fixed set of fundamental observables. In quantum electrodynamics (QED) the charge of the electron is an observable, and one can always express predicted observables in terms of the physical charge. The coupling constants in QHD are not directly observable, so one must determine the 'physical' couplings indirectly. The standard method of fixing parameters in QHD has been to fit saturation properties of nuclear matter and a limited number of properties of finite nuclei (e.g., the charge radius of ^{40}Ca). This means that the 'physical' couplings and masses used in QHD can change from one order of an expansion to the next. My definition of strong convergence does not allow such changes, so that one may question its utility in an evaluation of QHD. Instead, as described below, I fix all coupling constants and masses as if they were directly physical observables.

If one has a constant in which a true perturbative expansion is being made (e.g., $\hbar$), such parameter changes will be of higher order than the order being considered. I have used $\hbar$ as a perturbative parameter to cancel infinities without inducing higher order infinites in the process; however, $\hbar$ is not truly a perturbative parameter. Fermions first enter at $\mathcal{O}(\hbar)$, through the fermion loop, so that to lowest order in an $\hbar$ expansion there is no nuclear matter. When one determines ϕ self-consistently in the relativistic Hartree approximation, one includes terms in the energy of all orders in $\hbar$. When one renormalizes, terms of higher order than the number of loops considered are dropped. This situation is not without precedent, because when one calculates quantum corrections to the spectrum of hydrogen in QED, one includes terms of all orders in the electromagnetic coupling constant when solving the Dirac equation, and then drops terms of a given order in this coupling when renormalizing the corrections*. In QED this process is well-defined because one calculates the effective action perturbatively, and introduces non-perturbative effects (e.g., bound states) by extremizing this action. However, one does not need these non-perturbative effects to renormalize QED because the observables used to fix this theory's parameters are themselves perturbative. In QHD, on the other hand, one wants to employ non-perturbative observables (e.g., the binding energy of nuclear matter) in the process of renormalization, and this makes the perturbation sequence somewhat ill-defined. It also makes it difficult to clearly differentiate between strong and weak convergence of any expansion used in QHD. If this discussion is confusing, the distinction between strong and weak convergence provided above is probably clear, even though it shouldn't be.

The calculations presented in this paper are most relevant to a test of the strong convergence of expansions which begin with the Hartree approximation. Most arguments designed to demonstrate that the Hartree approximation is valid in the high density limit actually address strong convergence. It will be shown that the Hartree approximation is not valid in this sense until one reaches densities which are extremely high. No direct test of the weak validity of the Hartree approximation has been completed, although I will have several comments to make about this issue.

In the next section I shall reproduce some of Chin's work on the exchange energy[2], and examine his justification of the Hartree approximation. Chin calculated only two of six terms which occur in the exchange energy for the Walecka model, having neglected the terms which involve vacuum polarization. In the third section I will present results that include these remaining terms. Finally, in the conclusion I will consider some of the implications of the large corrections presented in the third section.

II. VALENCE CONTRIBUTION TO THE EXCHANGE ENERGY[2]

The expression for the exchange energy corrections to the energy density of nuclear matter in the Walecka model was originally derived by Chin. The Feynman diagrams for these terms are shown in Figure 1, where a virtual scalar meson is exchanged in the first diagram, and a virtual vector meson is exchanged in the second. One can readily see that these are two loop diagrams by counting the number of independent momenta. These two loop corrections are of $\mathcal{O}(\hbar^2)$, whereas the one loop term s which enter the Hartree approximation are of $\mathcal{O}(\hbar^2)$. Thus, $\hbar$ can be used as an expansion parameter when renormalizing these new terms. Throughout this paper I will consider the $\mathcal{O}(\hbar)$ terms as perturbative corrections. This means that the mean meson fields which are employed will extremize the energy density to $\mathcal{O}(\hbar)$, and will not be allowed to change in the presence of the two loop corrections. Note however that, as stated above, a strict perturbative expansion in powers of $\hbar$ is not being made, because in such an expansion fermions do not contribute to lowest order, preventing one from treating the $\mathcal{O}(\hbar)$ terms perturbatively.

The lagrangian for the Walecka model is

$$\mathcal{L} = \bar{\psi}(i\not{\partial} - g_v\not{V} - M^*)\psi + \frac{1}{2}(\partial_\mu\phi)(\partial^\mu\phi) - \frac{1}{2}m_s^2\phi^2 - \frac{1}{4}F_{\mu\nu}F^{\mu\nu} + \frac{1}{2}m_v^2 V_\mu V^\mu \tag{1}$$

where M^*, the effective nucleon mass, is $M_N - g_s\phi$. The vector field in this lagrangian is usually identified with the omega meson, while the scalar field is intended to represent multi-meson resonances (e.g., two pion resonances) which give rise to the intermediate range attraction in the nucleon-nucleon interaction.

Using finite density Hartree propagators for the fermions, and thereby including the large

---

* This is a slight oversimplification, because the binding energy happens to have a simple power law dependence on the electromagnetic coupling constant.

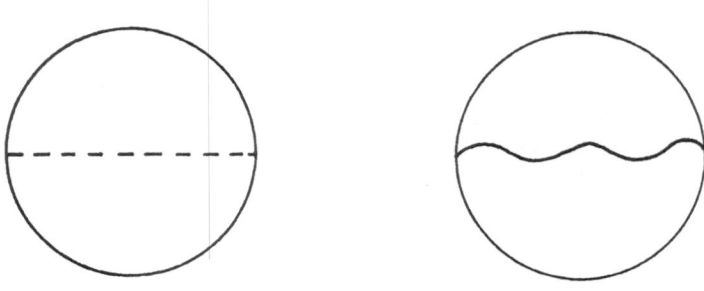

Figure 1. Scalar and vector exchange diagrams.

scalar mean field to all orders, one finds that the exchange energy density is

$$\epsilon_F = \frac{1}{2}\hbar^2 g_s^2 tr \int \frac{d^4k}{(2\pi)^4} \int \frac{d^4q}{(2\pi)^4} \ G(q)G(k)\Delta(k-q)$$
$$+ \frac{1}{2}\hbar^2 g_v^2 tr \int \frac{d^4k}{(2\pi)^4} \int \frac{d^4q}{(2\pi)^4} \ G(q)\gamma_\mu G(k)\gamma_\nu D^{\mu\nu}(k-q) \tag{2}$$

where,

$$G(k) = G_F(k) + G_D(k)$$
$$G_F(k) = \frac{\not{k}+M^*}{k^2 - M^{*2} + i\epsilon}, \ G_D(k) = \frac{i\pi(\not{k}+M^*)}{\sqrt{\mathbf{k}^2+M^{*2}}}\Theta\left(k_F - |\mathbf{k}|\right)\delta\left(k_0 - \sqrt{\mathbf{k}^2+M^{*2}}\right)$$
$$\Delta(k) = \frac{1}{k^2 - M_s^2 + i\epsilon} \tag{3}$$
$$D_{\mu\nu}(k) = \left(-g_{\mu\nu} + \frac{k_\mu k_\nu}{m_v^2}\right)\frac{1}{k^2 - m_v^2 + i\epsilon}$$

$G_F(k)$ is the Feynman propagator familiar from zero density calculations, while $G_D(k)$ is a density dependent part of the fermion propagator. The correlation energy naturally divides into six terms:

$$\epsilon_{FF}^S = \frac{1}{2}\hbar^2 g_s^2 tr \int \frac{d^4k}{(2\pi)^4} \int \frac{d^4q}{(2\pi)^4} \ G_F(q)G_F(k)\Delta(k-q) \tag{4}$$

$$\epsilon_{FD}^S = \hbar^2 g_s^2 tr \int \frac{d^4k}{(2\pi)^4} \int \frac{d^4q}{(2\pi)^4} \ G_F(q)G_D(k)\Delta(k-q) \tag{5}$$

$$\epsilon_{DD}^S = \frac{1}{2}\hbar^2 g_s^2 tr \int \frac{d^4k}{(2\pi)^4} \int \frac{d^4q}{(2\pi)^4} \ G_D(q)G_D(k)\Delta(k-q) \tag{6}$$

$$\epsilon_{FF}^V = \frac{1}{2}\hbar^2 g_v^2 tr \int \frac{d^4k}{(2\pi)^4} \int \frac{d^4q}{(2\pi)^4} \ G_F(q)\gamma_\mu G_F(k)\gamma_\nu D^{\mu\nu}(k-q) \tag{7}$$

$$\epsilon_{FD}^V = \frac{1}{2}\hbar^2 g_v^2 tr \int \frac{d^4k}{(2\pi)^4} \int \frac{d^4q}{(2\pi)^4} \ [G_F(q)\gamma_\mu G_D(k)\gamma_\nu$$
$$+ G_D(q)\gamma_\mu G_F(k)\gamma_\nu] D^{\mu\nu}(k-q) \tag{8}$$

$$\epsilon_{DD}^V = \frac{1}{2}\hbar^2 g_v^2 tr \int \frac{d^4k}{(2\pi)^4} \int \frac{d^4q}{(2\pi)^4} \ G_D(q)\gamma_\mu G_D(k)\gamma_\nu D^{\mu\nu}(k-q) \tag{9}$$

ϵ_{FF}^S and ϵ_{FF}^V are vacuum polarization corrections, which subtract the interaction of fermions in the Dirac sea with that portion of the mean field they generate. These terms correspond to excitations of negative energy nucleons to all positive energy states; however, some of these positive energy states are occupied in nuclear matter. ϵ_{FD}^S and ϵ_{FD}^V provide the Pauli blocking effects needed to correct these last two terms. Finally, ϵ_{DD}^S and ϵ_{DD}^V correspond to particle-hole excitations in the positive energy Fermi sea. These are the only two pieces of the relativistic exchange energy which have a direct classical analog, and they are the only terms computed by Chin. I will examine only these last two valence exchange energy corrections in the remainder of this section, returning to the remaining four terms in the next section.

The evaluation of ε^S_{DD} and ε^V_{DD} is straightforward, and many details which simplify the calculation, along with the results can be found in Chin's paper. The results are:

$$
\varepsilon^S_{DD} = \frac{\hbar^2 g^2_s}{32\pi^4} \left\{ \left[k_F \sqrt{k^2_F + M^{*2}} - M^{*2} \operatorname{asinh}\left(\frac{k_F}{M^*}\right) \right]^2 \right.
$$
$$
\left. - M^{*4} \left(1 - \frac{1}{\alpha_s}\right) \int_1^\eta dx \left(1 - \frac{1}{x^2}\right) \int_1^\eta dy \left(1 - \frac{1}{y^2}\right) \ln\left[\frac{(x-y)^2 + \frac{4}{\alpha_s} xy}{(xy-1)^2 + \frac{4}{\alpha_s} xy}\right] \right\}
\tag{10}
$$

$$
\varepsilon^V_{DD} = \frac{\hbar^2 g^2_v}{16\pi^4} \left\{ \left[k_F \sqrt{k^2_F + M^{*2}} - M^{*2} \operatorname{asinh}\left(\frac{k_F}{M^*}\right) \right]^2 \right.
$$
$$
\left. + \frac{1}{2} M^{*4} \left(1 + \frac{2}{\alpha_v}\right) \int_1^\eta dx \left(1 - \frac{1}{x^2}\right) \int_1^\eta dy \left(1 - \frac{1}{y^2}\right) \ln\left[\frac{(x-y)^2 + \frac{4}{\alpha_v} xy}{(xy-1)^2 + \frac{4}{\alpha_v} xy}\right] \right\}
\tag{11}
$$

Here k_F is the Fermi momentum, $\alpha_s = 4M^{*2}/m^2_s$, $\alpha_v = 4M^{*2}/m^2_v$ and $\eta = (k_F + \sqrt{k^2_F + M^{*2}})/M^*$. M^* is assumed to be positive throughout this paper, but all expressions are valid if M^* is replaced by its absolute value. One can evaluate either one of the two remaining integrals in these expressions, but both cannot be evaluated analytically. Nonetheless, these terms are easily calculated numerically.

To evaluate these corrections one must first fix the coupling constants and masses which enter the problem. I have chosen to use the parameters found in Horowitz and Serot's Hartree calculations of the properties of finite nuclei[6], in which the Dirac sea contribution to the Hartree energy is evaluated using a local density approximation. The parameters are $g^2_s = 54.289$, $g^2_v = 102.770$, $m_s = 458$ MeV, $m_v = 783$ MeV, and the nucleon mass is taken to be 939 MeV. The Hartree approximation for the energy density, after the vector field is chosen to solve its Euler-Lagrange equation of motion, is

$$
\varepsilon_H = \frac{1}{2} m^2_s \phi^2 + \frac{2}{9\pi^4} \frac{g^2_v}{m^2_v} k^6_F + \frac{\hbar}{4\pi^2} \left\{ k_F (k^2_F + M^{*2}) \sqrt{k^2_F + M^{*2}} \right.
$$
$$
\left. - M^{*4} \operatorname{asinh}\left(\frac{k_F}{M^*}\right) \right\}
$$
$$
- \frac{\hbar}{8\pi^2} \left\{ M^{*4} \ln\left(M^{*2}/M^2_N\right) - 2M^3_N (M^* - M_N) - 7M^2_N (M^* - M_N)^2 \right.
$$
$$
\left. - \frac{26}{3} M_N (M^* - M_N)^3 - \frac{25}{6} (M^* - M_N)^4 \right\}
\tag{12}
$$

By minimizing this expression with respect to ϕ, one can determine the scalar field expectation value at finite density, and then use this Hartree value in the perturbative calculation of the exchange energy. The results of such a calculation are shown in Figure 2, in which the solid line is the binding energy per nucleon in the Hartree approximation and the dashed line results from the inclusion of the exchange energy corrections.

The exchange energy corrections are large in comparison to the total binding energy. At saturation ($k_F = 1.3$fm^{-1}) the Hartree binding energy is 15.74 MeV per nucleon, while the

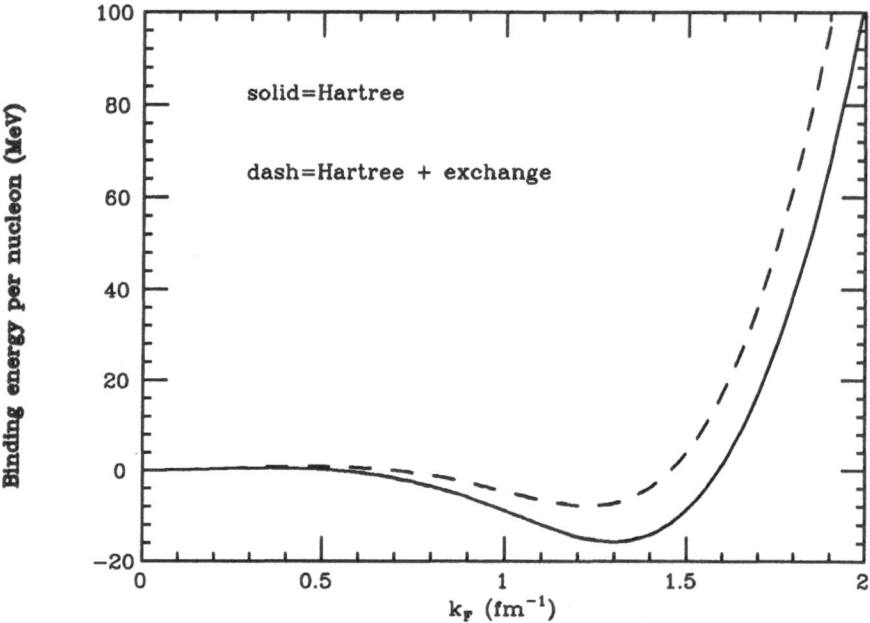

Figure 2. Effect of valence nucleon exchange energy.

exchange corrections decrease this by 8.400 MeV per nucleon. In other words, half of the binding energy is lost when one includes the exchange terms. The contribution to the binding energy from direct meson exchange is the result of large cancellations between vector and scalar exchange. Each of these direct contributions is much larger than the corresponding exchange contributions, and in this sense the exchange contributions are small. It should also be noted that the exchange contributions from scalar and vector mesons are of opposite sign, and therefore cancel one another to some extent. Unfortunately, for many observables such as the energy it is not the individual scalar and vector contributions that are relevant, but the sum; and we are interested in determining whether the exchange contributions to relevant observables are small. Only in this case can we be confident of a result determined using the Hartree approximation.

Why then was Chin able to conclude that these corrections are small and the Hartree approximation is valid? In Chin's direct calculations of the exchange energy he tested only for weak validity of the Hartree approximation. He minimized the full energy with the exchange terms included (i.e., self-consistent exchange corrections), and then readjusted the parameters of the theory to reproduce saturation of nuclear matter at the accepted density and binding energy. When this is done he reports that the curves corresponding to those shown in Figure 2 lie on top of one another. However, even when one does this the relative magnitude of the direct (Hartree) and exchange (Fock) contributions to the energy are not changed. The exchange contribution remains as large as half the direct contribution until one reaches three times nuclear matter density, and remains larger than 10% at thirty times nuclear matter density. More importantly, even after this reparameterization, predictions for the binding energy of neutron matter are significantly altered by the exchange terms well beyond nuclear matter density[2]. One may regard the energy density of neutron matter at the 'low' density of nuclear matter as being of little interest; but the large magnitude of the exchange contributions here raises the question of whether the Hartree approximation is even weakly valid at these densities for observables other than the energy density of symmetric nuclear matter, which is used to refit the parameters of the model.

To summarize, when one calculates the valence nucleon contribution to the exchange energy in nuclear matter perturbatively, one finds that these corrections are of the same magnitude as the total binding energy found in the Hartree approximation. Barring fortuitous cancellation from the remaining vacuum contributions, this allows one to conclude that the loop expansion is not strongly convergent, and that the Hartree approximation is probably not strongly valid. It still remains possible that one can find another expansion in which perturbative corrections to the Hartree approximation are small, but until such an expansion is found no evidence exists for the strong validity of the Hartree approximation at densities relevant to nuclear physics. On the other hand, when the exchange energy is calculated nonperturbatively, and the parameters of the model are refit, one recovers the Hartree results for the binding energy of symmetric nuclear matter. However, even in this case the binding energy of neutron matter is significantly altered for densities up to three times nuclear matter density, leaving the issue of whether the Hartree approximation is weakly valid at such densities unresolved.

III. VACUUM CONTRIBUTIONS TO THE EXCHANGE ENERGY

I want to begin this section with a warning. The calculations presented in this section were completed during the week before the Telluride meeting, and although I have attempted to thoroughly check my calculations throughout the course of their completion, several checks remain unfinished. Two of the remaining terms, ε^S_{FD} and ε^V_{FD}, are relatively easy to check and I am fairly confident of my results. These are the pieces of the exchange energy which represent Pauli blocking of Dirac sea excitations by the valence nucleons. The two terms which correspond to pure vacuum polarization corrections, ε^S_{FF} and ε^V_{FF}, are far more difficult to compute. Indeed, the final renormalized expressions for these terms are sufficiently complicated and long that I have chosen not to list them in this paper.

One of the reasons that Chin did not compute these vacuum terms is that they vanish at all densities when the nucleons have their normal mass, M_N. However, as discussed above, the mean scalar field is large, and when one includes insertions of this field to all orders nucleons acquire an effective mass, $M_N - g_s\phi$, which can deviate from M significantly in nuclear matter. Let me mention that the density dependence of the exchange energy arises from two sources. First there is the explicit dependence on the Fermi momentum in G_D. The largest power law dependence on the Fermi momentum occurs in the terms analyzed in the last section, and at 'high enough' densities one expects this density dependence to dominate. The second source of density dependence is implicit, arising from the effective mass which is a function of density. The terms ε^S_{FF} and ε^V_{FF} contain no explicit dependence on the Fermi momentum; however, they demonstrate large density dependence through their dependence on the effective mass.

Before presenting results I want to briefly discuss the issue of renormalization. Any thorough treatment of this issue would be technical and lengthy, so I will be satisfied with an explicit description of the renormalization scheme used. By counting powers of momentum, one can easily satisfy oneself that ε^S_{FD} and ε^V_{FD} are naively linearly divergent, while ε^S_{FF} and ε^V_{FF} are naively quartically divergent. In addition these last two terms contain nested divergences. To remove these divergences one needs to renormalize the nucleon wave function and mass, the scalar coupling constant, the scalar mass, and introduce linear, cubic and quartic scalar self-interaction counterterms. The nucleon wave function and mass renormalizations are determined by insisting that the fermion propagator which includes the self-energy insertions shown in Figure 3a has a pole at the nucleon mass with unit residue. The scalar coupling constant is renormalized to cancel the corrections shown in Figure 3b. The remaining scalar tadpole, mass and self-interaction renormalizations are adjusted so that in the vacuum there is no linear, cubic or quartic dependence of the energy density on the scalar field; and in addition so that the scalar mass is not altered by the exchange energy. The vector coupling constant is renormalized in exactly the same way as the fermion wave-function because of gauge invariance.

Figure 3a. Scalar and vector contributions to the fermion self-energy.

Figure 3b. Scalar and vector contributions to the scalar coupling renormalization.

With these renormalization prescriptions, one finds that ε_{FD}^{S} and ε_{FD}^{V} reduce to

$$
\varepsilon_{FD}^{S} = \frac{\hbar^2 g_s^2}{8\pi^4} \left\{ \int\limits_0^1 dx \left[(1+x) M^{*2} \ln\left[\frac{x m_s^2 + (1-x)^2 M^{*2}}{x m_s^2 + (1-x)^2 M_N^2} \right] \right. \right.
$$
$$
\left. \left. + \frac{2(1-x) M_N^2 M^*(M_N - M^*)}{x m_s^2 + (1-x)^2 M_N^2} \right] \right\} \times \int\limits_0^{k_F} \frac{k^2\, dk}{\sqrt{k^2 + M^{*2}}}
$$

(13)

$$
\varepsilon_{DD}^{V} = -\frac{\hbar^2 g_v^2}{4\pi^4} \left\{ \int\limits_0^1 dx \left[(2-x) M^{*2} \ln\left[\frac{x m_v^2 + (1-x)^2 M^{*2}}{x m_v^2 + (1-x)^2 M_N^2} \right] \right. \right.
$$
$$
\left. \left. + \frac{4(1-x) M_N^2 M^*(M_N - M^*)}{x m_v^2 + (1-x)^2 M_N^2} \right] \right\} \times \int\limits_0^{k_F} \frac{k^2\, dk}{\sqrt{k^2 + M^{*2}}}
$$

(14)

The remaining integrations are easily completed analytically. As mentioned above the full renormalized expressions for ε_{FF}^{S} and ε_{FF}^{V} are lengthy and unenlightening. These have been calculated by explicitly analyzing the expressions in eqs. (4) and (7), using dimensional regularization to extract the divergences. Renormalization of the fermion wave function and mass, and of the scalar coupling constant, is determined by evaluating the diagrams in Figure 3, and this cancels the nested divergences in ε_{FF}^{S} and ε_{FF}^{V}. Such explicit cancellation of divergences provides one test of these calculations, but the divergent pieces of any diagram are usually the easiest to analyze.

The result of including all four vacuum contributions to the exchange energy are shown in Figure 4. One finds that these corrections can be an order of magnitude larger than the Hartree binding energy. At $k_F = 1.3\text{fm}^{-1}$, the Hartree binding energy is adjusted to be 15.74 MeV. The complete set of exchange energy corrections are: $\varepsilon_{FF}^{S} = 13.36\text{MeV}$, $\varepsilon_{FD}^{S} = 73.63\text{MeV}$, $\varepsilon_{DD}^{S} = 26.51\text{MeV}$, $\varepsilon_{FF}^{V} = -5.81\text{MeV}$, $\varepsilon_{FD}^{V} = -517.3\text{MeV}$, and $\varepsilon_{DD}^{V} = -18.11\text{MeV}$. The largest contributions to the exchange energy come from the Pauli blocking terms, ε_{FD}^{S} and ε_{FD}^{V}. This

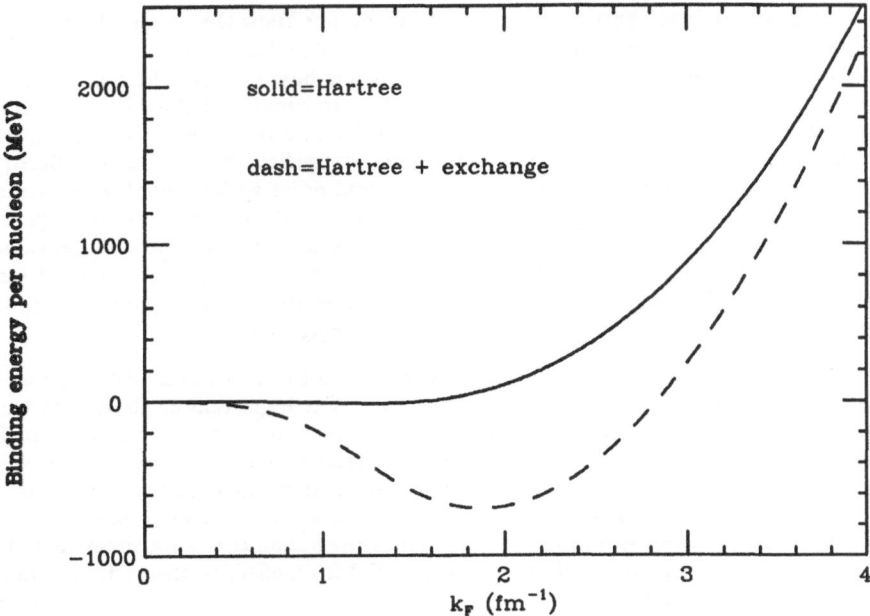

Figure 4. Effect of full exchange energy.

is true for densities up to thirty times nuclear matter density. The exchange energy remains larger than 10% of the Hartree energy up to thirty times nuclear matter density, so that the Hartree approximation does not become strongly valid until one reaches such densities.

In order to test weak validity of the Hartree approximation, one should now self-consistently determine ϕ by reminimizing the full energy, and then adjust parameters to refit the saturation properties of nuclear matter. This calculation has not been completed, and it is probably dangerous to speculate about the results of such work. At this level the interplay between various contributions to the binding energy has become complicated. As an example of this interplay consider how one might correct the fact that nuclear matter saturates at too high a density and is overbound. In the perturbative calculation nuclear matter is overbound primarily due to ε_{FD}^V. This term can be reduced by decreasing the vector coupling constant, but this would also increase the binding energy elsewhere in the calculation by decreasing the repulsion which arises from the direct vector meson exchange, and it might tend to increase the density at which nuclear matter saturates. Whether it is possible to refit the saturation properties of nuclear matter remains to be seen.

IV. CONCLUSIONS

This work addresses an important formal question. Is the Hartree approximation valid at densities of interest to nuclear physics in QHD? To answer this question one must first determine what is required for an approximation to be 'valid'. If a student of first year quantum mechanics were asked to test the validity of the Born approximation, she might do so by calculating the second Born term and comparing its magnitude with the first Born term. This is the approach taken here, and I have chosen to call this a test of strong validity. The Hartree approximation has been shown to fail such a test in the Walecka model at densities less than approximately thirty times nuclear matter density.

The analog of testing for what I have called weak validity is somewhat more complicated. Suppose that we have a simple one body system whose spectrum of eigenvalues has been experimentally determined. Further suppose that we do not know the potential, and have not yet discovered a method of converting our measured eigenvalues into a unique potential. Finally, assume that even when given a potential we don't know how to solve the full Schroedinger equation numerically or otherwise, but we do know how to compute a Born series. One might then proceed by computing the Born series to a given order, and adjust the potential to fit a fixed set of eigenvalues, treating the remaining eigenvalues as predictions. To test the Born series for strong convergence one need simply compute the next order without further adjustment of the potential, and determine whether the new term is small. To test for weak convergence one computes the next order, readjusts the potential to reproduce the fixed set of eigenvalues. If the remaining eigenvalues are shifted by a small amount, one concludes that the Born series is weakly convergent. This work has not determined whether the Hartree approximation to the Walecka model is the first term in a weakly convergent expansion.

It is not surprising that the Hartree approximation to the Walecka model is not strongly valid, because the coupling constants are large. Thus it is important to determine whether this approximation is weakly valid. If it is, one has some hope of systematically calculating quantum corrections, which is one of the primary motivations for employing a renormalizable field theory. If the Hartree approximation is not even weakly convergent, one must question the meaningfulness of the Hartree results obtained to date. The successes achieved in QHD may only be retained in this situation if one abandons the notion that a legitimate field theory is being used, and switches to the philosophy that QHD is an effective theory that is intended to be used only at tree level.

I would like to thank Tom Cohen, Chuck Horowitz, and Brian Serot for useful discussions. I would also like to thank Brian for providing me with his notes on the scalar exchange energy. These were helpful in several instances.

REFERENCES

1. J.D. Walecka, Ann. Phys. **83** (1974) 491.

2. S.A. Chin, Ann. Phys. **108** (1977) 301.

3. B.D. Serot and J.D. Walecka, in: Advances in Nuclear Physics, Vol. 16, eds. J.W. Negele and E. Vogt (Plenum, New York, 1985).

4. T. Matsui, Nucl. Phys. **A370** (1981) 365;

   W. Benta, A. Arima, H. Hyuga, K. Shimizu and K. Yazaki, Nucl. Phys. **A346** (1985) 593;

   J.A. McNeil, R.D. Amado, C.J. Horowitz, M. Oka, J.R. Shepard and D.A. Sparrow, Phys. Rev. C34 (1986) 746;

   R.J. Furnstahl and B.D. Serot, Nucl. Phys. **A468** (1987) 539.

5. R.J. Furnstahl, C.E. Price and G.E. Walker, Phys. Rev. **C36** (1987) 2590.

6. C.J. Horowitz and B.D. Serot, Phys. Lett. **B140** (1984) 181.

7. J.D. Bjorken and S.D. Drell, Relativistic quantum fields, (McGraw-Hill, New York, 1965).

A RELATIVISTIC DESCRIPTION OF DEFORMED NUCLEI[*]

Charles E. Price

Physics Division
Argonne National Laboratory
Argonne, IL 60439

INTRODUCTION

In recent years relativistic mean field calculations have successfully reproduced the properties of spherical nuclei using parameters that are fit to the empirical binding energy and saturation density of nuclear matter.[1] Specifically, the large scalar and vector mean fields that are a characteristic of this model lead to a natural inclusion of the large spin-orbit splittings that are present in real nuclei. Unfortunately, the restriction to spherical symmetry effectively limits the model to calculations in five doubly-magic nuclei. In order to extend the model to non-closed shell nuclei it is necessary to include deformations. This makes it possible to determine if the model can provide a good description of deformed nuclei using the same parameters that reproduce the properties of nuclear matter and spherical nuclei.

In contrast to the successful description of spherical nuclei, early calculations for the magnetic moments of closed shell ± 1 nuclei[2] were in qualitative disagreement with the experimental measurements. In these calculations, the additional particle or hole was treated as a valence particle and was not allowed to effect the closed shell core. This effectively sacrificed the self-consistency of the calculation in favor of simplicity. Subsequent calculations[3-6] have attempted to restore the self consistency by including the effects of the valence particle on the core through a variety of approximations. In general, the results of these calculations are in reasonable agreement with the data and are very similar to the non-relativistic Schmidt moments. In order to carry out a fully self-consistent calculation for these odd-mass nuclei it is again necessary to include deformations.

Another interesting feature of odd-mass nuclei is that there are additional mean fields that are not present for nuclear matter or for spherical nuclei. These fields are the three vector components of all of the vector fields (ω, ρ and photon fields) and the pion field. So in a very real sense, deformed nuclei provide an area for studying new effects that are not present in calculations for spherical nuclei.

In the following section, we will outline the basic model and describe the calculations for deformed nuclei. Then, in sections 3 and 4, we will present the results of our calculations for axially symmetric nuclei in the s-d shell[7,8] and for closed shell ± 1 nuclei[9] respectively. For more detail on the relativistic model see ref. 10 and for the detailed calculations for deformed nuclei refer to refs. 7 – 9.

---

[*] Work supported by the U. S. Department of Energy, Nuclear Physics Division, under contract W-31-109-ENG-38.

FORMALISM

The basic starting point for these calculations is a relativistic quantum field theory Lagrangian,

$$
\begin{aligned}
\mathcal{L} =& \mathcal{L}_{i}(\psi, \phi, V, b, \pi, A) \\
& - g_v \bar{\psi}\gamma_\mu \psi V^\mu + g_s \bar{\psi}\psi\phi - \frac{1}{3!}\kappa\phi^3 - \frac{1}{4!}\lambda\phi^4 \\
& - ig_\pi \bar{\psi}\gamma_5 \vec{\tau} \cdot \vec{\pi}\psi - \frac{1}{2}g_\rho \bar{\psi}\gamma_\mu \vec{\tau} \cdot \vec{b}^\mu \psi - e\bar{\psi}\gamma_\mu \frac{1}{2}(1 + \tau_3)\psi A^\mu
\end{aligned}
\tag{1}
$$

which includes the couplings of the nucleons (ψ) to sigma (ϕ), omega (V), rho (b) and pi (π) mesons and the photon (A). It also allows for explicit nonlinear self-couplings of the scalar meson in the form of terms proportional to ϕ^3 and ϕ^4.

In the Mean-Field or Hartree Approximation, the quantum meson fields are replaced by their expectation values which are classical fields. In this limit, we can write down a simple set of equations of motion for the various fields:

$$
[-i\vec{\alpha} \cdot \vec{\nabla} + \gamma_0 \Sigma_H(\vec{x})]U_\alpha(\vec{x}) = \epsilon_\alpha U_\alpha(\vec{x})
\tag{2}
$$

where

$$
\begin{aligned}
\Sigma_H(\vec{x}) =& - g_s\phi(\vec{x}) + g_v\gamma_\mu V^\mu(\vec{x}) + g_\pi\gamma_5\tau_3\pi_0(\vec{x}) \\
& + \frac{1}{2}g_\rho\gamma_\mu\tau_3 b_0^\mu(\vec{x}) + e\gamma_\mu\frac{1}{2}(1 + \tau_3)A^\mu(\vec{x})
\end{aligned}
\tag{3}
$$

and

$$
\begin{aligned}
(-\vec{\nabla}^2 + m_s^2)\phi(\vec{x}) &= -g_s\, \mathrm{Tr}\,[iG_H(\vec{x}, \vec{x})] \\
(-\vec{\nabla}^2 + m_v^2)V^\mu(\vec{x}) &= -g_v\, \mathrm{Tr}\,[i\gamma^\mu G_H(\vec{x}, \vec{x})] \\
(-\vec{\nabla}^2 + m_\pi^2)\pi_0(\vec{x}) &= -g_\pi\, \mathrm{Tr}\,[i\gamma_5\tau_3 G_H(\vec{x}, \vec{x})] \\
(-\vec{\nabla}^2 + m_\rho^2)b_0^\mu(\vec{x}) &= -g_\rho\, \mathrm{Tr}\,[i\gamma^\mu\tau_3 G_H(\vec{x}, \vec{x})] \\
(-\vec{\nabla}^2)A^\mu(\vec{x}) &= -e\, \mathrm{Tr}\,[i\gamma^\mu\frac{1}{2}(1 + \tau_3)G_H(\vec{x}, \vec{x})]
\end{aligned}
\tag{4}
$$

where

$$
iG_H(\vec{x}, \vec{x}) = \sum_{\alpha}^{occ} U_\alpha(\vec{x})\bar{U}_\alpha(\vec{x}).
\tag{5}
$$

In eqs. (2) – (5), Σ_H is the Hartree self-energy, G_H is the Hartree Green's function, U_α and ϵ_α are the nucleon orbitals and eigenvalues respectively and α represents a complete set of quantum numbers for the individual orbitals. In the case of deformed nuclei, the total angular momentum is not a good quantum number so α is limited to parity and the third component of the angular momentum, m.

For spherical nuclei, the meson fields are angle independent and the angle dependence of the nucleon orbitals is trivial. In this case, these equations can be rewritten in terms of a set of coupled ordinary differential equations that only depend on the radial coordinate. On the other hand, for deformed nuclei the mean fields have an explicit angle dependence and eqs. (2) – (5) form a set of coupled partial differential equations. In principle, these equations could be solved directly; however, in practice it is much more convenient to use a few simple expansions which make it possible to separate out all of the angular dependence.

Table 1. Deformed solutions in ^{20}Ne.

Prolate		Oblate	
Level	Energy	Level	Energy
$\frac{1}{2}^{+}$	-45.6	$\frac{1}{2}^{+}$	-46.4
$\frac{1}{2}^{-}$	-28.6	$\frac{3}{2}^{-}$	-27.0
$\frac{3}{2}^{-}$	-22.6	$\frac{1}{2}^{-}$	-25.4
$\frac{1}{2}^{-}$	-15.6	$\frac{1}{2}^{-}$	-17.1
$\frac{1}{2}^{+}$	-12.1	$\frac{1}{2}^{+}$	- 9.4
$E_0 = -103.3 MeV$		$E_0 = -98.5 MeV$	
$Q = 403 mb$		$Q = -199 mb$	

For the nucleon orbitals we expand in terms of the usual relativistic spin-angle functions:[11]

$$U_\alpha(\vec{x}) = \sum_\kappa \left[\begin{array}{c} \frac{iG_\kappa^\alpha(r)}{r}\Phi_{\kappa m} \\ \frac{-F_\kappa^\alpha(r)}{r}\Phi_{-\kappa m} \end{array} \right] \qquad (6)$$

where the sum over κ effectively mixes components with different total angular momenta but is restricted such that the orbitals have good parity. For the scalar field and the time components of the vector fields we expand in Legendre polynomials:

$$\phi(\vec{x}) = \sum_l \phi_l(r) P_l(\cos\theta)$$

$$V^0(\vec{x}) = \sum_l V_l^0(r) P_l(\cos\theta) \qquad (7)$$

where the sum is restricted to even values of l. And finally, the three vector components of the various vector fields are expanded in terms of vector spherical harmonics:[12]

$$\vec{V}(\vec{x}) = i \sum_l V_l(r)\vec{Y}_{ll1}^0(\Omega) \qquad (8)$$

The resulting angle-independent equations may be solved using standard techniques to obtain the self consistent Hartree ground state. For more detail concerning the solution of these equations see refs. 7, 8 and 9.

RESULTS: EVEN-EVEN NUCLEI

Using the standard Finite Hartree parameters of Horowitz and Serot[1], which have been fitted to the binding energy, saturation density and symmetry energy of nuclear matter, we find both an oblate and a prolate solution for ^{20}Ne. The level energies of the occupied neutron orbitals for each solution are shown in Table 1. As expected for this nucleus, the prolate solution has the lower energy and is therefore the ground state solution. The predicted quadrupole moment of about $400mb$ is somewhat smaller than the experimental result of roughly $550mb$. This may indicate that there is a problem with the relativistic model. To test this possibility, we repeated the calculations for

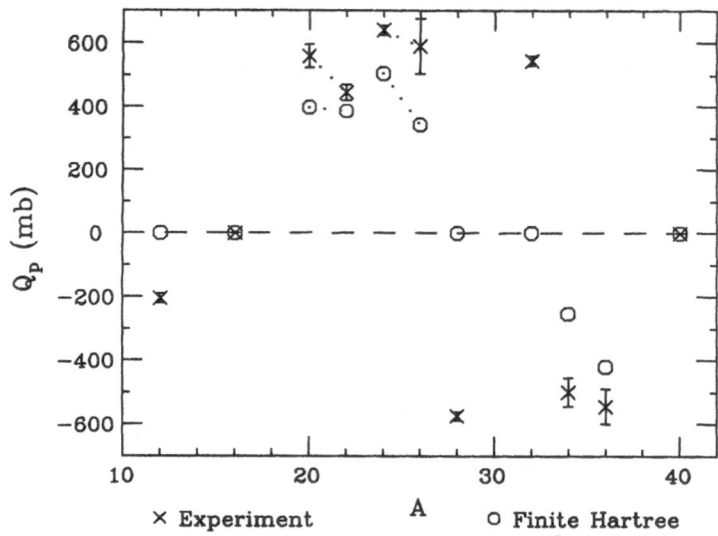

× Experiment   A   ○ Finite Hartree

Fig. 1. Quadrupole moments for nuclei in the s-d shell. The crosses with error bars are the experimental results and the circles are the calculated values.

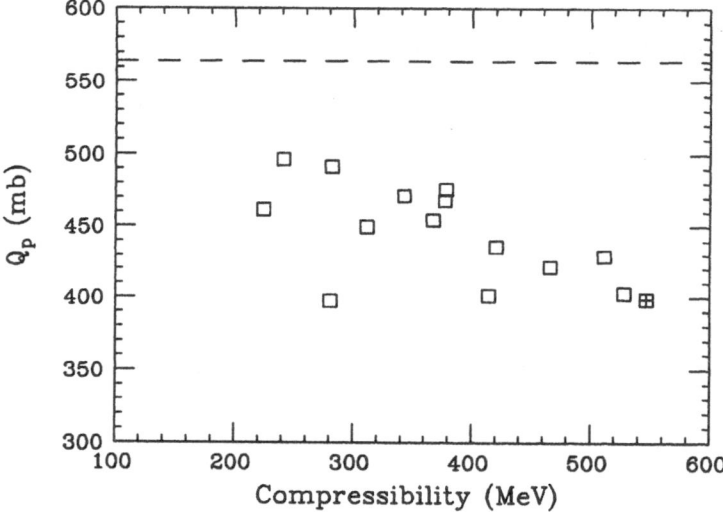

Fig. 2. Compressibility versus quadrupole moment in ^{20}Ne. The dashed line shows the experimental value and the Finite Hartree result is at the extreme right of the figure.

a variety of nuclei throughout the s-d shell. Figure 1 shows the predicted quadrupole moments for these nuclei along with the experimental values.[13]

Clearly, the general trend is that the calculated values are smaller than the experimental results, but more importantly, the relativistic model predicts that all of the closed sub-shell nuclei (e.g. ^{12}C, ^{28}Si, ...) are spherical even though experimentally some of these nuclei are strongly deformed. One possible explanation is that the large value of the compressibility that is a characteristic of this model may be restraining the deformation. A high compressibility indicates that the nucleus is very stiff and has

Table 2. Relativistic Hartree Parameter Sets.

g_s	g_v	g_ρ	$m_s^2(MeV)$	$\kappa(MeV)$	λ	M^*/M	$K(MeV)$	$Q_{ch}(mb)$
109.73	190.59	65.37	520.1	0.	0.	0.54	546.8	399.
108.13	187.15	66.27	519.1	100.	0.	0.55	528.3	403.
101.53	173.73	69.56	514.4	500.	0.	0.58	466.1	421.
84.70	143.00	76.18	497.8	1500.	0.	0.66	367.1	454.
94.01	158.48	73.00	510.0	800.	10.	0.61	420.6	436.
68.32	118.07	80.75	476.7	1500.	50.	0.70	342.9	471.
67.26	118.99	80.59	477.6	500.	100.	0.70	377.3	468.
45.22	80.19	86.65	428.7	2500.	100.	0.78	281.9	491.
27.96	52.44	90.33	368.6	2500.	200.	0.84	240.9	496.
89.95	158.47	73.00	510.0	-1000.	100.	0.61	511.7	429.
52.90	98.65	83.92	455.3	-500.	200.	0.74	378.3	476.
114.93	191.15	65.23	518.9	2000.	-100.	0.54	414.2	401.
95.45	154.93	73.75	505.3	3000.	-100.	0.62	311.4	450.
117.82	188.66	65.88	516.1	4000.	-200.	0.55	280.7	397.
95.11	148.93	74.99	500.8	5000.	-200.	0.63	224.2	461.

a large surface energy, so it is reasonable to expect that this would tend to minimize the degree to which nuclei can deform. Unfortunately, in the standard model, the compressibility cannot be varied; however, if we include the allowed nonlinear scalar coupling terms shown in eq. (1), we can find a variety of parameter sets all of which reproduce the properties of infinite nuclear matter and spherical nuclei but which have different compressibilities. The parameter sets that we have used are shown in Table 2. Notice that some of the parameter sets involve negative values for the quartic coupling constant, λ. These parameterizations are not allowed in the strictest sense because the energy is no longer bounded from below; however, if the parameters are treated phenomenologically, the best fit to all the properties of spherical nuclei is obtained with negative values of λ.[14] For this reason we have chosen to include parameter sets with negative λ so that we can fully investigate the sensitivity of the deformations. The next to last column in Table 2 shows the corresponding values of the compressibility of nuclear matter. Clearly, these parameter sets include a wide range of compressiblitites.

Figure 2 shows the quadrupole moments predicted using the various parameterizations plotted against the compressibilty. Although it is possible to improve the agreement with experiment by included the nonlinear couplings, clearly the compressibility is not strongly correlated with the predicted quadrupole moments. In fact there are four parameter sets with very different compressibilities that all predict quadrupole moments of about $400mb$. Apparently, the naive argument that the surface energy (and hence the compressibility) would be the driving force that determined the deformation is incorrect.

The next most obvious possibility is that the strength of the spin-orbit splitting is responsible for controlling the degree of deformation. In non-relativistic calculations it was seen that the spin-orbit force was very important for correctly predicting

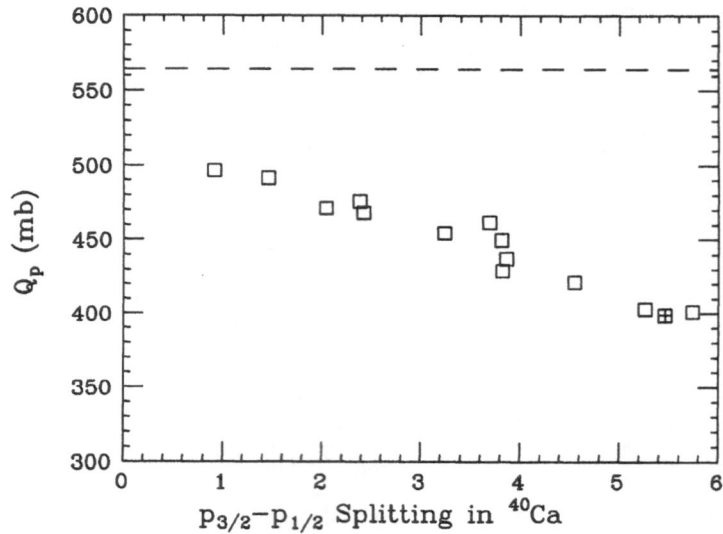

Fig. 3. Spin-orbit splitting in ^{40}Ca versus quadrupole moment in ^{20}Ne. The dashed line shows the experimental value.

Table 3. Relativistic Hartree parameter sets.

Set	g_s	g_v	g_ρ	m_s^2 (MeV)	κ (MeV)	λ	M^*/M	K (MeV)
A	109.73	190.59	65.37	520.1	0.	0.	0.54	546.8
B	94.01	158.48	73.00	510.0	800.	10.	0.61	420.6
C	95.11	148.93	74.99	500.8	5000.	−200.	0.63	224.2

equillibrium deformations, so it is reasonable to expect that the same will be true for the relativistic calculations. In fact, this is an appealing possibility since the spin-orbit force is automatically included in the relativistic model. Figure 3 shows the quadrupole moments in ^{20}Ne, using the various parameter sets, plotted as a function of the predicted spin-orbit splitting in ^{40}Ca. There is a clear correlation between decreasing spin-orbit splittings and increasing quadrupole moments in this nucleus. In fig. 4, we show the quadrupole moments for nuclei in the s-d shell using three selected parameter sets (see Table 3). Set A is the Finite Hartree linear parameterization, Set C is very similar to the non-linear parameterization of Reinhard *et al.*, and Set B is chosen to give similar agreement with the experimental spin-orbit splitting but with positive quartic coupling. The general agreement with experiment (crosses with error bars) is greatly improved for sets B and C, and more importantly, these sets predict large quadrupole moments for the closed sub-shell nuclei. So throughout the s-d shell, the relativistic model predicts the correct sign and magnitude of the equilibrium deformation provided we use a parameter set which provides a good description of the spin-orbit splitting. Furthermore, fig. 5, which shows a comparison between the results of parameter set B and the results from two non-relativistic calculations using Skyrme interactions,[15] indicates that the relativistic model provides a description of deformed nuclei that is in quantitative agreement with non-relativistic calculations.

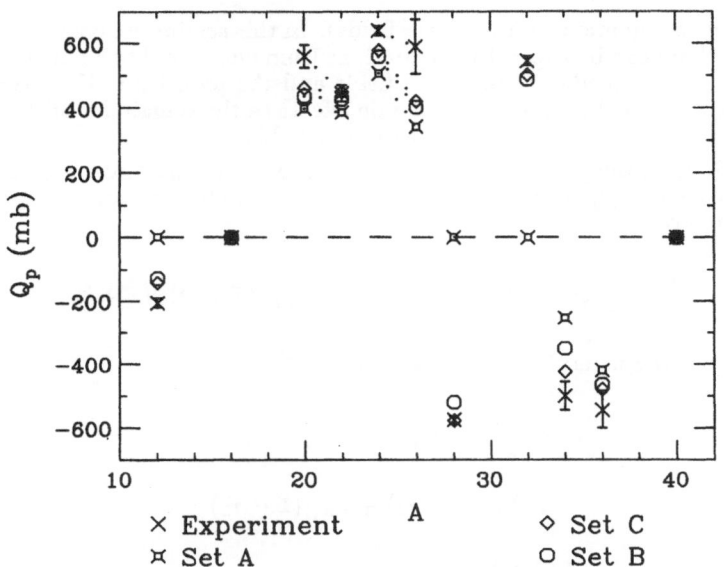

Fig. 4. Quadrupole moments for nuclei in the s-d shell calculated with the parameter sets shown in Table 3.

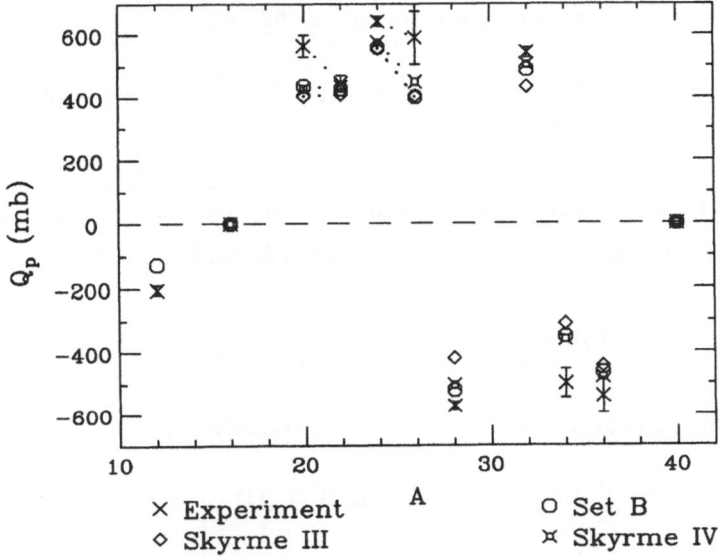

Fig. 5. Comparison of s-d shell quadrupole moments calculated using the relativistic model and non-relativistic Skyrme interactions.

This conclusion is supported by the work of Gambir and Ring.[16] They found that the deformations in the rare earth region can also be correctly predicted in the relativistic model provided that a nonlinear parameterization which gives the correct spin-orbit splitting is used.

RESULTS: CLOSED SHELL ±1 NUCLEI

In the calculations presented in the preceeding section the symmetry of the nuclei limited the mean fields to those that were present for spherical nuclei (i.e. the scalar field and the time components of the vector fields). In this section we consider odd-mass nuclei in which there can be explicit contributions from additional mean fields. Namely, the three vector components of the vector fields and the pion field. Unfortunately, at this time we have no results which include pion effects so the remainder of the discusion will be limited to the inclusion of the three vector fields.

In order to look at the ground state nuclear currents, we must introduce an effective electromagnetic current operator as in refs. 2 and 10. For elastic transitions, the three-vector current operator is

$$\vec{J}(\vec{x}) = \psi^\dagger(\vec{x}) Q \vec{\alpha} \psi(\vec{x}) + \frac{1}{2M} \vec{\nabla} \times (\psi^\dagger(\vec{x}) \lambda \beta \vec{\Sigma} \psi(\vec{x})) \tag{9}$$

where $\vec{\alpha}$ and β are the usual Dirac matrices and

$$Q \equiv \frac{1}{2}(1 + \tau_3)$$

$$\lambda \equiv \lambda_p \frac{1}{2}(1 + \tau_3) + \lambda_n \frac{1}{2}(1 - \tau_3) \tag{10}$$

$$\vec{\Sigma} \equiv \begin{pmatrix} \vec{\sigma} & 0 \\ 0 & \vec{\sigma} \end{pmatrix}.$$

By evaluating this operator in the Hartree ground state, $|J_i\rangle$, the current may be expressed in terms of the nucleon wavefunctions as follows:

$$\langle J_i | \vec{J}(\vec{x}) | J_i \rangle = \sum_\alpha^{occ} U_\alpha^\dagger(\vec{x}) Q \vec{\alpha} U_\alpha(\vec{x}) + \frac{1}{2M} \vec{\nabla} \times \sum_\alpha^{occ} U_\alpha^\dagger(\vec{x}) \lambda \beta \vec{\Sigma} U_\alpha(\vec{x}) \tag{11}$$

The first term in eq. (11) is the convection current and is simply the proton contribution to the three-vector source term in eq. (4), and the second term is the anomalous current.

Using these currents, the transverse elastic form factor is given by[17]

$$F_T^2(q) = \frac{1}{2J_i + 1} \sum_{J=1,3...}^{2J_i} \left| \langle J_i || \hat{T}_J^{mag}(q) || J_i \rangle \right|^2 \tag{12}$$

where q is the magnitude of the three momentum transfer and

$$\hat{T}_J^{mag}(q) = \int d^3x\, j_J(qx) \vec{Y}_{JJ1}^M(\Omega) \cdot \vec{J}(\vec{x}) \tag{13}$$

specifies the transverse magnetic multipoles. In eq. (13), j_J is a spherical Bessel function and $\vec{Y}_{JJ1}^M$ is a vector spherical harmonic. Finally, the ground state magnetic dipole moment may be obtained from the elastic magnetic form factors as follows:

$$\mu = \lim_{q \to 0} \left[-i \frac{2M}{q} \left(\frac{6\pi J_i}{(J_i + 1)(2J_i + 1)} \right)^{\frac{1}{2}} \langle J_i || \hat{T}_{J=1}^{mag}(q) || J_i \rangle \right]. \tag{14}$$

In fig. 6, we show the isoscalar contribution to the convection current of eq. (11). The solid curve is the contribution from the valence particle only, which agrees with the early calculations of ref. 2. The dashed curve is taken from ref. 6, in which the effect of the valence particle on the core orbitals is included in the RPA approximation, and the dash-dotted curve is the result calculated from the self-consistent ground

346

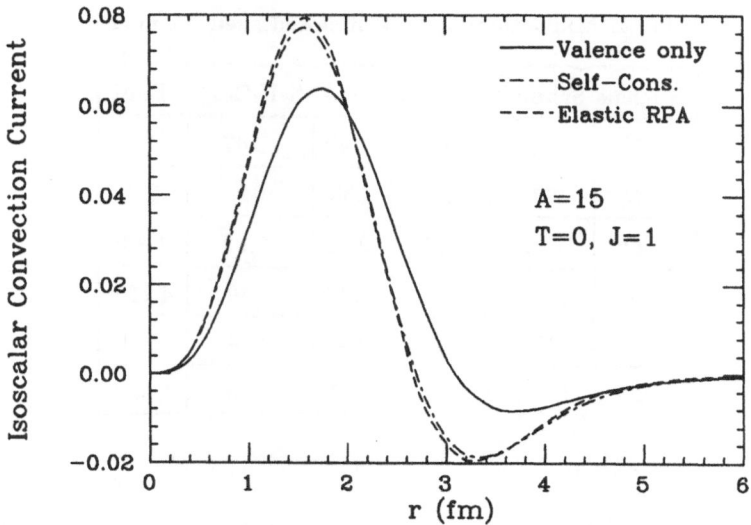

Fig. 6. Isoscalar convection current for the $A = 15$ system. The solid curve is the contribution from the valence orbital, the dashed curve is taken from ref. 6, and the dot-dashed curve is the result obtained from the self-consistent calculations described here.

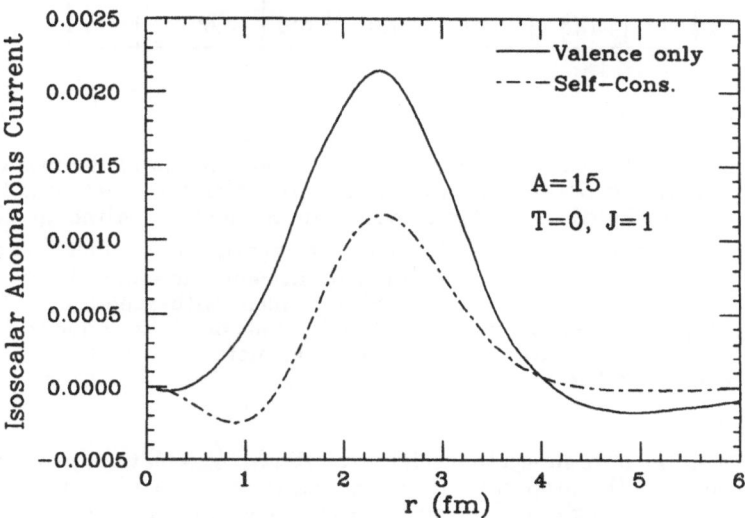

Fig. 7 Isoscalar anomalous current for $A = 15$ as in fig. 6.

state. Clearly the core contribution is significant and cannot be ignored; however, the agreement between the dashed and dash-dotted curves indicates that the RPA is a good approximation for including the core effects in the convection currents. Figure 7 shows the isoscalar contribution to the anomalous current in eq. (11), and again it is clear that the core modifications have a significant effect on the current. Figures 6 and 7 demonstrate the importance of calculating with a self-consistent ground state or at the very least accounting for the self-consistency through an appropriate approximation.

Table 4 shows the magnetic moments for various closed shell ±1 nuclei. The column

Table 4. Magnetic moments of closed shell ±1 nuclei.

Nucleus	Schmidt	Valence	LDA	Self-Cons.	Expt
^{15}N	-.264	.020	-.149	-.247	-.283
^{15}O	.638	.667	.535	.647	.719
^{17}O	-1.913	-1.905	-2.03	-2.03	-1.894
^{17}F	4.793	5.045	4.901	4.89	4.722
^{39}K	.124	.855	.448	.354	.391
^{39}Ca	1.148	1.170	.848	.968	1.022

Table 5. Isoscalar magnetic moments.

Mass	Orbital	Schmidt	Valence	RPA	Self-Cons.	Expt
15	$1p\frac{1}{2}$	.187	.344	.204	.200	.218
17	$1d\frac{5}{2}$	1.44	1.62	1.44	1.43	1.41
39	$1d\frac{3}{2}$	.636	1.01	.661	.691	.707

labeled *Valence* is obtained by including only the contribution of the valence orbital, and the column labeled *LDA* is taken from ref. 5 in which the effects due to the core modifications are included in the local density approximation. Although the valence moments are in clear disagreement with both the experimental values[18] and the non-relativistic Schmidt moments, both the LDA results and the moments obtained from the self-consistent ground states are in good agreement with experiment. This again indicates the importance of including the modifications of the core due to the valence particle. Table 5 shows the corresponding isoscalar moments with the exception that the column labeled *RPA* is obtained from ref. 6.

SUMMARY

In conclusion, we have shown that relativistic Hartree calculations using parameters that have been fit to the properties of nuclear matter can provide a good description of both spherical and axially deformed nuclei. The quantitative agreement with experiment is equivalent to that which was obtained in non-relativistic calculations using Skyrme interactions. We have also shown that the equilibrium deformation is strongly correlated with the size of the spin-orbit splitting, and that parameter sets which give roughly the correct value for this splitting provide the best agreement with the quadrupole moments in the s-d shell. Finally, for closed shell ±1 nuclei, we showed that the self-consistent calculations are able to reproduce the experimental magnetic moments. This was not possible in relativistic calculations which included only the effects of the valence orbital.

ACKNOWLEDGEMENTS

I am pleased to acknowledge the contributions of my collaborators, G. E. Walker and R. J. Furnstahl.

REFERENCES

1. C. J. Horowitz and B. D. Serot, Nucl. Phys. **A368** (1981) 503
2. B. D. Serot, Phys. Lett. **107B** (1983) 334
3. J. R. Shepard, E. Rost and C.-Y. Cheung, University of Colorado preprint
4. J. A. McNeil, R. D. Amado, C. J. Horowitz, M. Oka, J. R. Shepard and D. A. Sparrow, Phys. Rev. **C34** (1986) 746
5. R. J. Furnstahl and B. D. Serot, Nucl. Phys. **A468** (1987) 539
6. R. J. Furnstahl, Indiana University preprint
7. C. E. Price and G. E. Walker, Phys. Rev. **C36** (1987) 354
8. R. J. Furnstahl, C. E. Price and G. E. Walker, Phys. Rev. **C36** (1987) 2590
9. R. J. Furnstahl and C. E. Price, in preparation
10. B. D. Serot and J. D. Walecka, Adv. in Nucl. Phys. **16** (1986) 1
11. J. D. Bjorken and S. D. Drell, **Relativistic Quantum Mechanics**, (McGraw-Hill, New York, 1964)
12. A. R. Edmonds, **Angular Momentum in Quantum Mechanics**, (Princeton University Press, Princeton, 1960)
13. G. Leander and S. E. Larsson, Nucl. Phys. **A239** (1975) 93
14. P.-G. Reinhard, M. Rufa, J. Marunh, W. Greiner and J. Friedrich, Z. Phys. **A323** (1986) 13
15. D. Vautherin, Phys. Rev. **C7** (1973) 296
16. Y. K. Gambir and P. Ring, Phys. Lett. **202B** (1988) 5
17. T. W. Donnelly and J. D. Walecka, Ann. Rev. of Nucl. Sci. **25** (1975) 329
18. C. Lederer, J. M. Hollander and I. Perlman, **Table of Isotopes** (J. Wiley and Sons, New York, 1967)

VIOLENT COLLISIONS OF SPINNING PROTONS

A.D. Krisch

Randall Lab of Physics
The University of Michigan
Ann Arbor, MI 48109-1120

High energy polarized proton beams and polarized proton targets have allowed good measurements of spin effects in high energy elastic proton-proton scattering. Since I have written about this subject several times in recent years, I will only briefly review the earlier work and refer the interested reader to some rather detailed lectures.[1]

From 1979 to 1986 we concentrated heavily on the project of modifying the 30 GeV AGS to allow the acceleration of polarized protons. As shown in Fig. 1 this project required major modifications in almost every part of the AGS.

In early 1986 the Argonne-Brookhaven-Michigan-Rice-Yale team succeeded in accelerating a polarized proton beam to 22 GeV with over 40% polarization. The complex and difficult acceleration process is described in detail in a recent paper.[2]

My group's main goal in initiating this accelerator improvement project was to extend to higher energy the surprising ZGS results[3] shown in Fig. 2.

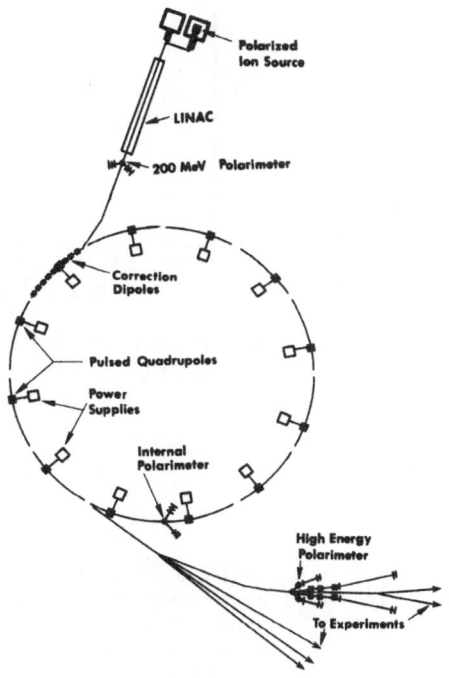

Fig. 1 AGS Layout

Note that in small-$P_\perp^2$ glancing collisions the cross-section seems quite independent of both incident energy and incident spin state. This indicates that for all energies and spin states, whenever two protons have a peripheral collision from the outer surfaces of each other, they see a characteristic size of about 1 fermi which is obtained from the slope of $\exp(-9.3\,P_\perp^2)$.

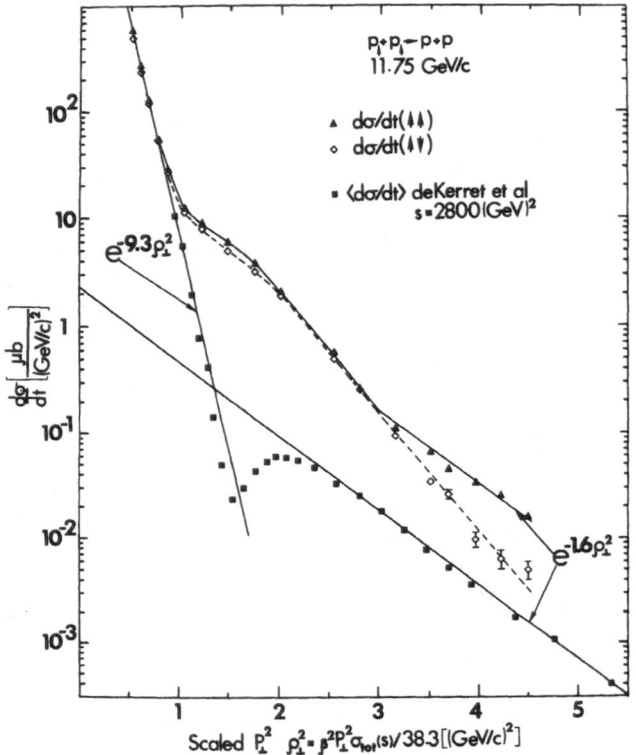

Fig. 2 Spin-Parallel and Spin-Antiparallel proton-proton elastic cross-sections plotted against scaled $P_\perp^2$ variable.

In the $s = 2800$ GeV^2 spin-averaged data from the ISR there is a sharp break followed by a large-$P_\perp^2$ hard-scattering component with the behavior $\exp(-1.6\,P_\perp^2)$, which continues for several decades. This hard-scattering component is probably due to the direct collisions of the protons' constituents. The slope of 1.6 indicates that the effective size of the constituent-constituent interaction is about 1/3 fermi, which includes both the size of the constituents and the range of their interaction.

The 12 GeV ZGS data also has a break which is followed by the same hard-scattering component with the same slope of 1.6. However, this break only occurs when the protons' spins are parallel. When the protons' spins are antiparallel the cross-section keeps dropping. At the maximum $P_\perp^2$ available at the ZGS the ratio dσ/dt (parallel): dσ/dt (antiparallel) has reached a value of 4.

This large ratio causes a serious problem for the popular theory of strong interactions called perturbative QCD. Since this theory assumes that each interacting spin 1/2 proton contains three spin 1/2 quarks, it is most difficult to get a factor of 4 for the spin ratio. Perhaps the easiest way for an experimenter to understand this problem is to view a spin 1/2 proton as a polarized target containing three spin 1/2 quarks; the polarization of this polarized quark target (PQT) is exactly 1/3. Thus, it is impossible to get a proton-proton spin ratio of 4 even if the quark-quark spin-parallel cross-section is infinity and the quark-quark spin-antiparallel cross-section is zero.

While waiting for the AGS polarized beam to operate, we decided to test our apparatus in a low priority experiment which measured A in p + p → p + p with the AGS unpolarized proton beam scattering from our polarized proton target.[4] Perturbative QCD predicts that A should be exactly zero and further states that this prediction should become more reliable as the incident energy and $P_\perp^2$ become larger. Our measurements of A at 28 GeV/c are plotted against $P_\perp^2$ in Fig. 3. Clearly the A = 0 prediction of PQCD is not supported by this data which shows A growing at large $P_\perp^2$.

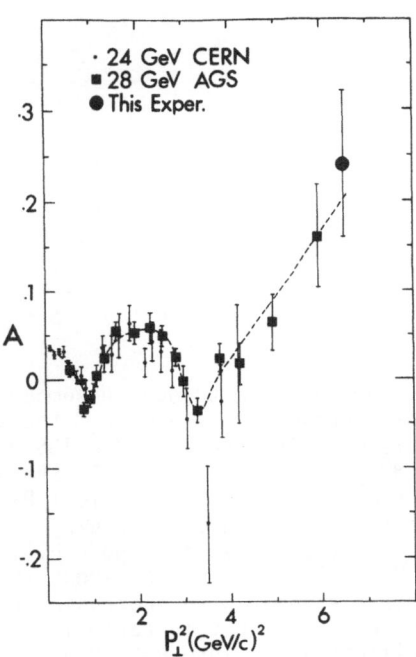

Fig. 3   The Analyzing Power is plotted against $P_\perp^2$ for p + p → p + p at 28 GeV/c.

All spin ratio measurements for high energy p-p elastic scattering are summarized in Fig. 4, which is a 3-dimensional plot of dσ/dt (parallel): dσ/dt (anti-parallel) against incident momentum and against $P_\perp^2$.

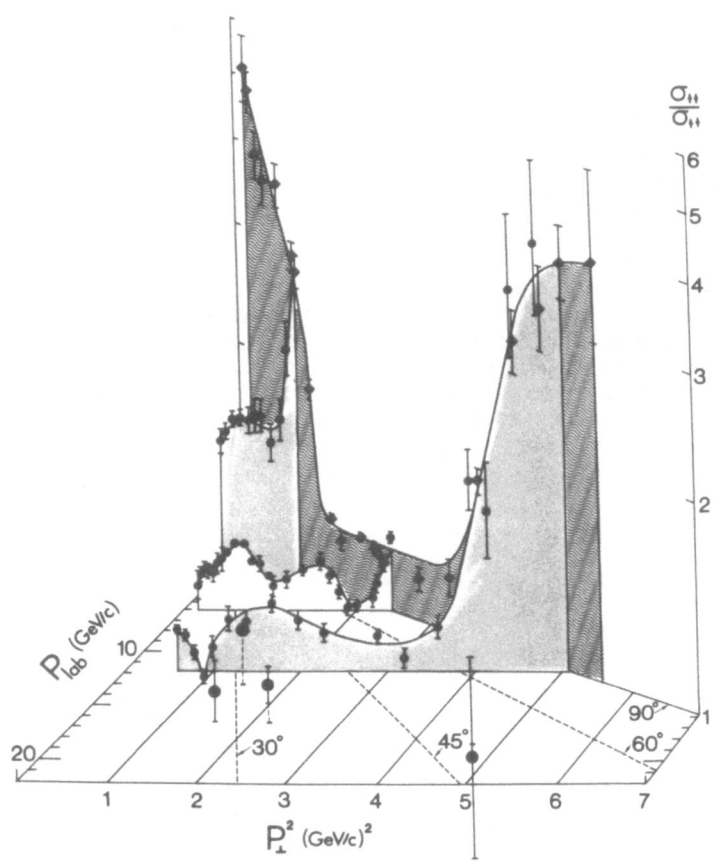

Fig. 4 3-dimensional plot of p + p → p + p spin ratio

This plot includes all our AGS data through November 1987, which is interesting but is a bit sparse. At medium-$P_\perp^2$ the 13.3 and 16.5 GeV/c data is generally consistent with the 12 GeV/c ZGS data. Note that the large-$P_\perp^2$ fixed-angle 90° cm points have exactly the same $P_\perp^2$ behavior as the fixed-energy 12 GeV/c points. This identical $P_\perp^2$ behavior indicates that these large spin effects are indeed large-$P_\perp^2$ hard-scattering effects and not 90° cm particle identity effects, a possibility suggested by Bethe and Weisskopf around 1978. The figure also shows that spin effects certainly do not seem to disappear with increasing P_{lab} and $P_\perp^2$ as suggested by perturbative QCD. The spin ratio in these violent collisions almost seems to be oscillating, an interesting possibility studied by Hendry, Brodsky, Tyurin, and others.

We had a very good AGS run in December 1987/January 1988 with 18.5 GeV/c polarized protons.[5] We obtained rather detailed angular distributions for both A and A_{nn} which are shown in Fig. 5.

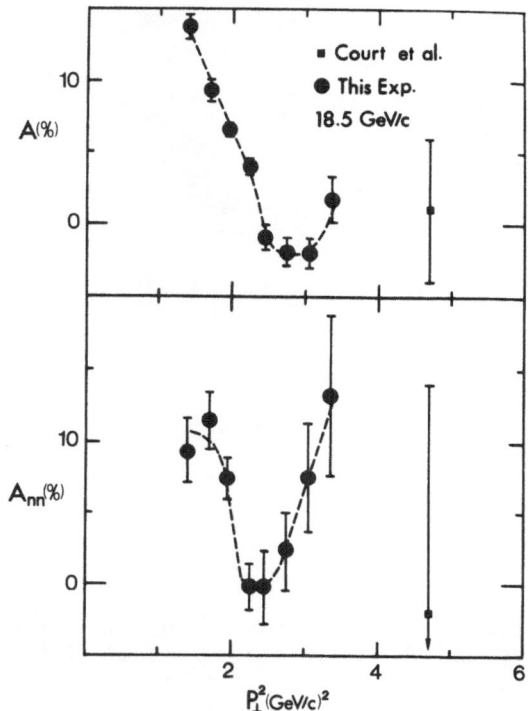

Fig. 5 The Analyzing Power, A, and the Spin-Spin Correlation Parameter, A_{nn}, are plotted against $P_\perp^2$ for p + p → p + p at 18.5 GeV/c

Notice that A has a dip near $P_\perp^2$ = 3 $(GeV/c)^2$ which is quite similar to the dip seen in Fig. 3 at 24 and 28 GeV/c. Note that the errors in the new polarized beam data are generally smaller than in Fig. 3.

The sharp dip in A_{nn} near $P_\perp^2$ = 2.3 $(GeV/c)^2$ at 18.5 GeV/c seems quite interesting. Notice from Fig. 4 that at 12 GeV/c the spin-spin effects have one sharp dip near $P_\perp^2$ = 0.9 $(GeV/c)^2$ where the diffraction peak ends, and a broad dip near $P_\perp^2$ = 3.2 $(GeV/c)^2$ where the hard-scattering component starts. It is not yet clear if the sharp dip near $P_\perp^2$ = 2.3 $(GeV/c)^2$ at 18.5 GeV/c corresponds to the $P_\perp^2$ = 0.9 $(GeV/c)^2$ dip moving up or the large $P_\perp^2$ dip moving down. Our earlier $P_\perp^2$ = 4.7 $(GeV/c)^2$ measurement with A_{nn} near zero had rather large errors and we must await our next run before we are really sure about the large $P_\perp^2$ behavior.

I will end with Fig. 6 which is a plot of the relative pure initial spin state cross-sections:

$$\sigma_{\uparrow\uparrow} \equiv d\sigma/dt(\uparrow\uparrow)/\langle d\sigma/dt\rangle = 1 + A_{nn} + 2A$$

$$\sigma_{\downarrow\downarrow} \equiv d\sigma/dt(\downarrow\downarrow)/\langle d\sigma/dt\rangle = 1 + A_{nn} - 2A$$

$$\sigma_{\downarrow\uparrow} = \sigma_{\uparrow\downarrow} \equiv d\sigma/dt(\uparrow\downarrow)/\langle d\sigma/dt\rangle = 1 - A_{nn}$$

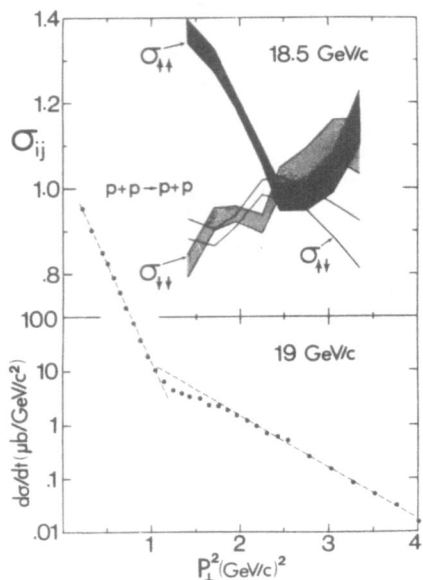

Fig. 6 The relative pure initial spin state cross-sections are plotted against $P_\perp^2$ for $p + p \to p + p$ at 18.5 GeV/c.

Near $P_\perp^2 = 1.4$ (GeV/c)^2, $\sigma_{\uparrow\uparrow}$ is more than 50% larger than both $\sigma_{\downarrow\downarrow}$ and $\sigma_{\uparrow\downarrow}$. All three cross-sections come together near $P_\perp^2 = 2.5$ (GeV/c)^2 and then appear to move apart again at larger $P_\perp^2$. We have also shown for comparison the spin-averaged p-p elastic cross-section $\langle d\sigma/dt\rangle$ at 19 GeV/c[6], which again has a break and dip near $P_\perp^2 = 1$ (GeV/c)^2 followed by the hard-scattering exponential. Note also that $\langle d\sigma/dt\rangle$ seems to drop smoothly near the sharp structure in both A and A_{nn} near $P_\perp^2 = 2.5$ (GeV/c)^2. This suggests that the spin-averaged cross-section is insensitive to large and probably significant forces which appear quite clearly in the pure-spin cross-sections.

REFERENCES

1. A.D. Krisch, Proc. 1985 Osaka Nuclear Polarization Conference J. Phys. Soc. Jpn. 55 Suppl. 31 (1986); Proc. 1986 Serpukhov High Energy Spin Symposium, 41 and 272 (1987).
2. F.Z. Khiari et al., UM HE 88-36, to be submitted to Physical Review.
3. D.G. Crabb et al., Phys. Rev. Lett., 41, 1257 (1978).
4. P.R. Cameron et al., Phys. Rev. Rapid Comm., D32, 3070 (1985).
5. D.G. Crabb et al., UM HE 88-4, submitted to Phys. Rev. Lett. (1988).
6. J.V. Allaby et al., Phys. Rev. Lett. 23B, 67 (1968).

A STATUS REPORT ON THE "t_{20}" EXPERIMENT AT BATES: TOWARD A SEPARATION OF THE CHARGE AND QUADRUPOLE FORM FACTORS OF THE DEUTERON

M. Garçon [a and b], L. Antonuk [c], J. Arvieux [d], D. Beck [e], E. Beise [a], A. Boudard [b], E. B. Cairns [c], J. M. Cameron [c], G. Dodson [a], K. Dow [a], M. Farkhondeh [a], H. W. Fielding [c], R. Goloskie [f], S. Høibraten [a], J. Jourdan [e], S. Kowalski [a], C. Lapointe [c], W. J. McDonald [c], D. Pham [a], R. Redwine [a], N. Rodning [c], G. Roy [c], M. E. Schulze [g], P. A. Souder [g], J. Soukup [c], I. The [a], W. Turchinetz [a], C. F. Williamson [a], K. Wilson [a], S. Wood [a] and W. Ziegler [c].

a) Massachusetts Institute of Technology, Cambridge, MA, USA

b) DPhN/ME, CEN-Saclay, France

c) University of Alberta, Edmonton, Canada

d) Laboratoire National Saturne, France

e) California Institute of Technology, Pasadena, CA, USA

f) Worcester Polytechnic Institute, Worcester, MA, USA

g) Syracuse University, Syracuse, NY, USA

Abstract

We describe the status of an ongoing experiment at the Bates Linear Accelerator Center to measure the tensor polarization (t_{20}, t_{21} and t_{22}) of the recoil deuteron in elastic electron deuteron scattering, for $q^2 = 14.2$, 17.6 and 21.2 fm^{-2}.

I. Introduction

The deuteron being a nucleus of spin 1, its electromagnetic structure is described by three form factors: the electric monopole (or charge) G_C, the electric quadrupole G_Q and the magnetic dipole G_M form factors. All observables of the elastic electron-deuteron scattering are bilinear combinations of these form factors. The measurement of the differential cross-section

$$\frac{d\sigma}{d\Omega} = \left(\frac{d\sigma}{d\Omega}\right)_{NS} (A + B\,\tan^2(\theta/2)) = \left(\frac{d\sigma}{d\Omega}\right)_{NS} I_o$$

at different θ electron angles for a given 4-momentum transfer q^2 allows the separation of

$$A = G_C^2 + \frac{8}{9}\eta^2 G_Q^2 + \frac{2}{3}\eta G_M^2$$

and

$$B = \frac{4}{3}\eta(1 + \eta)\,G_M^2 \ ,$$

where $\eta = \dfrac{q^2}{4\,M_d^2}$ and $\left(\dfrac{d\sigma}{d\Omega}\right)_{NS}$ is the cross-section for structureless particles.

The two electric form factors G_C and G_Q cannot be separated unless at least one other observable is measured; that observable will depend on the polarization of the particles in the initial and/or final state.

With unpolarized electron beams, one may study the dependence of the scattering cross-section upon the spin state of the target deuteron, using a tensor polarized target, or alternatively measure the polarization of the recoil deuteron when the initial deuteron is unpolarized. In the latter, one may in principle extract simultaneously the three tensor moments:

$$t_{20} = -\sqrt{2}\left[\frac{4}{3}\eta\,G_C\,G_Q + \frac{4}{9}\eta^2\,G_Q^2 + \frac{1}{3}\eta(1/2 + (1 + \eta)\,\tan^2(\theta/2))\,G_M^2\right] / I_o,$$

$$t_{21} = \frac{2}{\sqrt{3}} \, \eta \left[\eta + \eta^2 \sin^2 (\theta/2) \right]^{1/2} G_M \, G_Q \, \sec (\theta/2) \, / \, I_o,$$

and $t_{22} = - \dfrac{1}{\sqrt{12}} \, \eta \, G_M^2 \, / \, I_o.$

The vector polarization is identically zero in the one-photon exchange approximation.

t_{20} is the most sizable moment for $q^2 < 25$ fm^{-2} and has thus received most of the attention so far. At sufficiently forward angles, one may neglect the magnetic contribution to t_{20} and write

$$t_{20} (\theta_e) \simeq P = - \sqrt{2} \, \frac{2x + x^2}{1 + 2x^2} , \quad \text{with } x = \frac{2}{3} \, \eta \, \frac{G_Q}{G_C} .$$

P, sometimes called simply the polarization, is not an observable, but is an interesting quantity since it depends only on the ratio x of the quadrupole and monopole form factors. P has a minimum of $-\sqrt{2}$ when x = 1, rises to $-\sqrt{2}/2$ when G_C goes to zero (x $\to \infty$) and to 0 when x = -2.

t_{21}, being proportional to G_Q, also carries new information. This is not the case, however, for t_{22} since G_M^2 is better determined by $B(q^2)$.

There are only two published experimental results for the measurement of t_{20} [1,2]; they have been performed at low momenta ($q^2 < 4$ fm^{-2}) and consequently do not contribute significantly to our knowledge of the deuteron electromagnetic structure. A third experiment at $q^2 = 12.5$ fm^{-2} has been reported in progress [3].

The experiment being reported here is the first attempt to measure the three tensor observables in a region of momentum transfers ($q^2 = 14$ to 21 fm^{-2}) where they are indeed sensitive to the description of the deuteron structure. Above all, irrespective of their interpretation, the measurement of these new observables should allow for the first time a model independent separation of the charge and quadrupole form factors. In particular, one has yet to observe the zero of G_C, predicted by most potential models around $q^2 = 20$ fm^{-2}.

II.   Model Interpretations of the Tensor Polarization

Very schematically, each of the three form factors can be written as a functional F:

$$G_{C,Q,M} = F \{ \text{ nucleon form factors } * \text{ deuteron wave function } \}$$

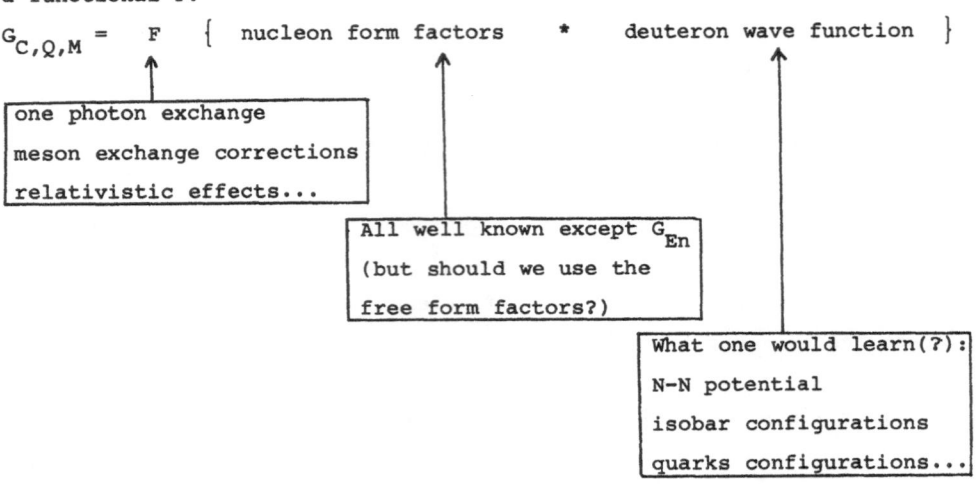

one photon exchange
meson exchange corrections
relativistic effects...

All well known except G_{En} (but should we use the free form factors?)

What one would learn(?): N-N potential isobar configurations quarks configurations...

They are illustrated in Fig.1, as they contribute to $A(q^2)$.   The expected zero crossing of G_C gives rise to rapid variations of the ratio x, and therefore of t_{20} (Fig. 2).   The structure of G_C is not apparent in $A(q^2)$ since this one is

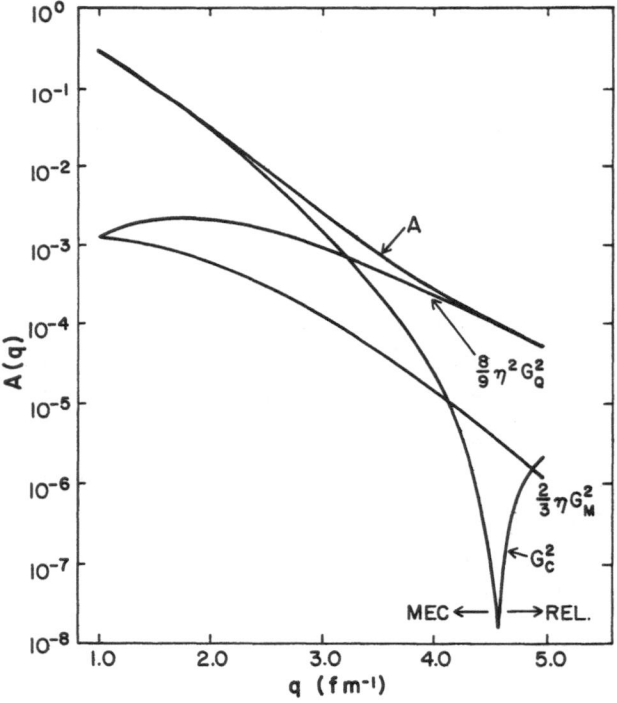

Figure 1.   A(q)

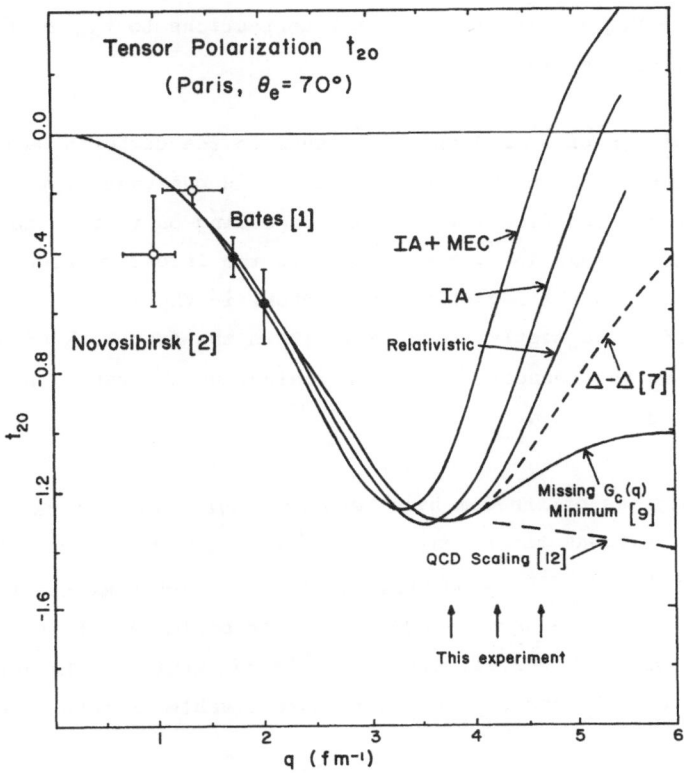

Figure 2. t_{20}

dominated by the G_Q contribution at these momentum transfers. Indeed, different models may have the same prediction for $A(q^2)$ and still differ in G_C and G_Q separately, and therefore in t_{20} and t_{21}.

Note that in the ratio x, in the impulse approximation, the dependence upon the nucleon form factors cancels out.

We will not attempt here to give an exhaustive review of the existing calculations; instead, we will give a flavor of the variety of approaches, especially in the light of the most recent publications.

When a potential model is used to calculate the deuteron wave function, the sensitivity of the polarization observables to differences between some commonly used N-N potentials (Reid vs Paris, for example) is small but measurable at the highest momentum transfers ($q^2 \sim 20$ fm^{-2}), see e.g. [4]. The effect of meson exchange currents and of relativistic corrections are nearly equal and opposite, but the position of the zero of G_C, and therefore the slope of t_{20}, are very sensitive to the details of these corrections.

Models generated with relativistic calculations [5] differ significantly from those using conventional frameworks, even in their prediction
of $A(q^2)$ and $B(q^2)$. The relativistic corrections to t_{20} at $q^2 = 20$ fm^{-2}
have been estimated to be 7% [6].

The addition of $\Delta\Delta$ and NN^* components to the deuteron wave function
has been reexamined in two recent papers: in one case [7], t_{20} is predicted to have significant variations depending on various $\Delta\Delta$ configurations, whereas in [8], the effects of the isobar configurations are
seen primarily in the magnetic form factor; in the latter, however,
t_{20} is found to be mostly sensitive, within the framework of an R-matrix
formalism, to the boundary condition radius, which itself is related to
the size of quark bags.

Quark degrees of freedom have been considered in various ways: in a
hybrid model, it was first conjectured [9] that the effect of a 6q component would be to keep G_C positive up to very high momentum transfers (P
would then never rise above $-\sqrt{2}/2$ after its minimum). More recent
investigations [10] show small effects of 6q admixture in the longitudinal form factors G_C and G_Q, and again more sizable effects in G_M.

In a topological soliton approach [11], G_Q is found significantly
smaller than in other approaches, which translates into a sharper rise of
t_{20} from the negative minimum to positive values, and to a smaller t_{21}.

Finally, assuming helicity conservation, perturbative QCD predicts
$x \rightarrow 1$ and therefore $P \rightarrow -\sqrt{2}$ in the limit of infinite momentum
transfer [12]. Whether our experiment is performed at high enough momentum transfer to test this prediction may be argued, but is unlikely,
since G_M has been found to have a minimum at $q^2 \simeq 50$ fm^{-2} [13].

III. Description of the Experiment

This is a double scattering experiment. After the first scattering
(ed elastic), the deuterons are polarized. The second scattering (dp
elastic and inelastic) analyzes this polarization. Because of the two
scatterings, the counting rates are low; we, therefore, had to achieve a
high luminosity for the first scattering, a high efficiency for the
polarimeter and large solid angles for the detected particles (e and d in
coincidence). Moreover, in order to eliminate the large flux of protons
coming from photodisintegration from entering the polarimeter, a double-

bend transport channel for the deuterons was designed, using a degrader between the two dipoles to remove the protons of the same initial momentum as the deuterons from the entrance of the polarimeter.

Fig. 3 gives a schematic lay-out of the experiment.

In detail, the components of the experiment are:

1) The Liquid Deuterium Target (Fig. 4)

It was designed primarily for this experiment but is meant to be a Bates facility. The deuterium gas is admitted from an external 18.6 m^3 reservoir; it is cooled and condensed by a cryogenically cooled heat exchanger until the system is filled with liquid deuterium (capacity about 10 liters); the liquid is circulated over the heat exchanger by

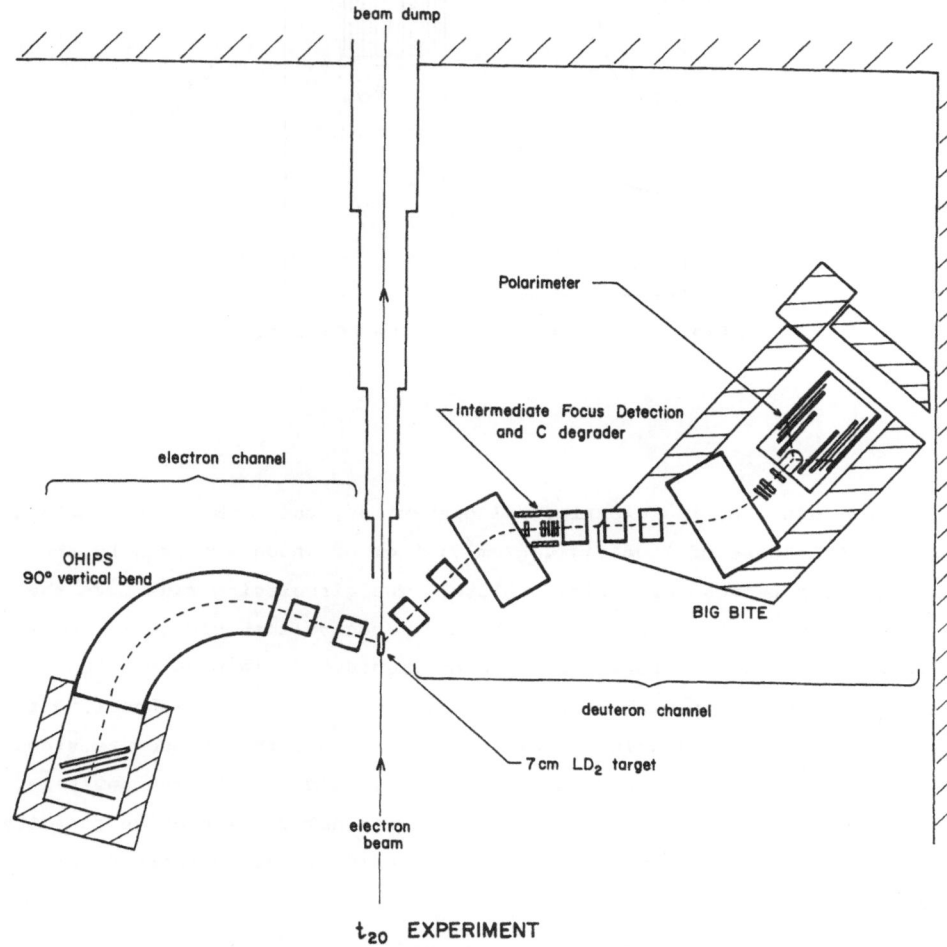

t_{20} EXPERIMENT

Figure 3. Schematic layout of the experiment.

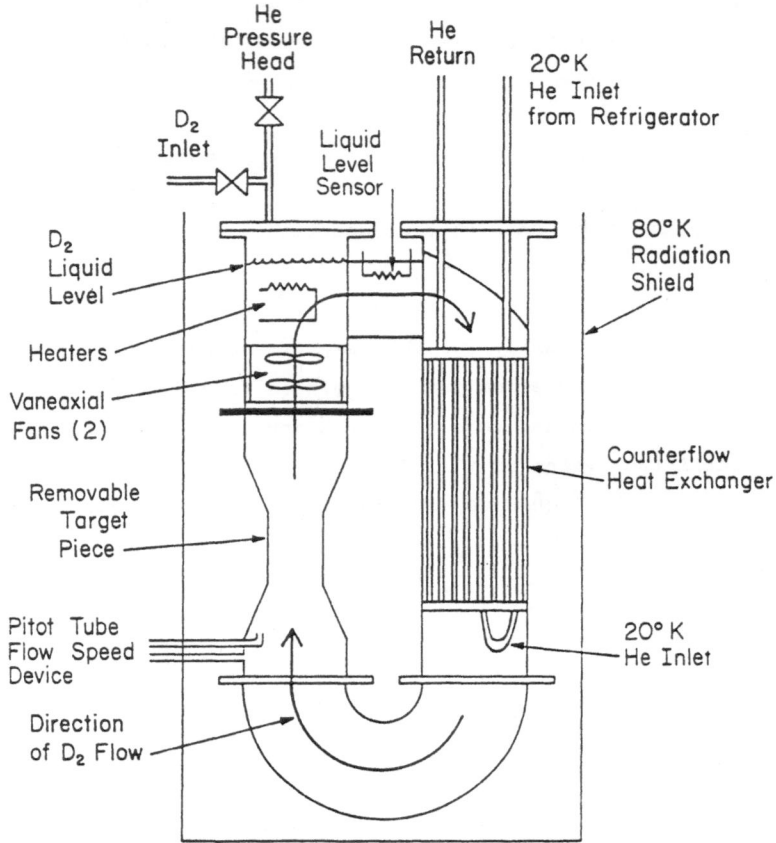

He
Pressure
Head

He
Return

20°K
He Inlet
from Refrigerator

D_2
Inlet

Liquid
Level
Sensor

D_2
Liquid
Level

80°K
Radiation
Shield

Heaters

Vaneaxial
Fans (2)

Counterflow
Heat Exchanger

Removable
Target
Piece

Pitot Tube
Flow Speed
Device

20°K
He Inlet

Direction
of D_2 Flow

Figure 4.   Schematic of the LD$_2$ target.

vaneaxial fans. During experimental operations, the beam passes through
a target thickness of 7 cm (1.14 g/cm^2), 5 cm of which are seen by the
spectrometers because of defining slits. The circulating fans move the
liquid deuterium at a velocity of about 2 m/sec so that successive beam
pulses pass through different parcels of liquid. It is possible to
pressurize the liquid with He gas and operate at about 2 K subcooling to
avoid boiling of the liquid; we did not have to use this procedure since
up to 25 μA (100 W deposited in the target) we did not observe any evidence for boiling. The refrigerator has a nominal refrigeration capacity
of 200 W at 20 K; whether the target can handle higher currents remains
to be tested.

2) The Electron and Deuteron Spectrometers
The spectrometer OHIPS, with its standard detection, is used as an

electron tag for the e-d coincidence. To get the maximum solid angle (18.5 msr calculated), we use the shortest drift distance configuration.

The deuteron channel is designed to best match the electron solid angle and to maximize the deuteron transmission to the polarimeter. The degrader introduces losses due to nuclear scattering and absorption, and to Coulomb multiple scattering and energy loss fluctuations. At the intermediate focus (IF), by the degrader, we installed a hodoscope of ten scintillators (.5" x 5"), and two multi-wire proportional chambers with PCOS-3 read-out. At the entrance of the polarimeter (FE) two scintilla-tors and two MWPC detect the particles incident upon the polarimeter.

The identification of the deuteron is provided by time-of-flight between IF and FE and by dE/dX as well. The e-d coincidence shows as a distinct peak in the time-of-flight between OHIPS and FE (Fig. 5).

Let us define by f the fraction of deuterons (from e-d) entering the polarimeter, to the expected elastically scattered electrons in OHIPS. f is a correction factor taking into account radiation losses, mismatch and transmission losses. It is found very close to our simulations (Fig. 6).

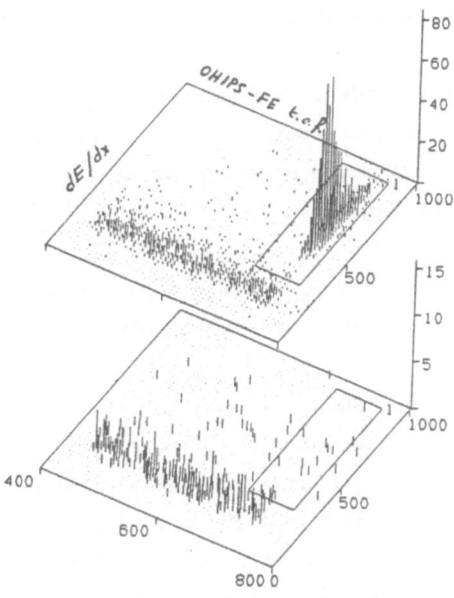

Figure 5. e-d identification (full and empty target).

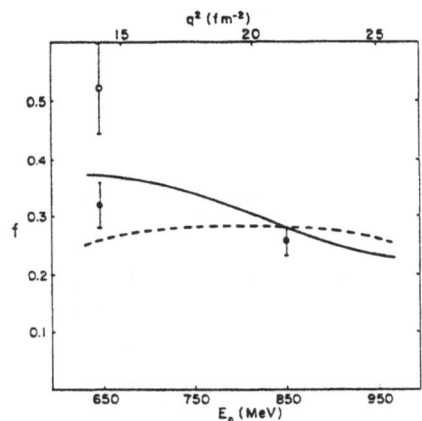

Figure 6. f (see text for defin-ition), the two curves are simulations; full circles: experiment with absorber, open circles: without absorber.

3) The AHEAD Deuteron Polarimeter

The Alberta High Energy/Efficiency Analyzer for Deuterons is designed to analyze deuteron tensor and vector polarization components for an incident energy range 120 MeV $< E_d <$ 250 MeV. The technique is quite different from that used in previous polarimeters. The analyzing reaction is dp elastic scattering. For scattering of tensor polarized deuterons one has:

$$\frac{d\sigma}{d\Omega}(\theta) = \frac{d\sigma_o}{d\Omega}(\theta) \left[1 + t_{20} \, T_{20}(\theta) + 2\, t_{21} \, T_{21}(\theta) \cos\phi + 2\, t_{22} \, T_{22}(\theta) \cos 2\phi\right] \tag{1}$$

where the T_{kq}'s are the dp analyzing powers, $\frac{d\sigma_o}{d\Omega}$ the unpolarized cross-section, θ and ϕ the polar and azimuthal angles of the scattered deuteron.

A complete angular distribution between 50 and 150 degrees c.m. is measured by detecting and identifying either (or both) the proton and deuteron. The lab angle of the the exiting particle is measured relative to the incoming deuteron. This leads to the following advantages:

i) the measurement of the angular dependence allows determination of the t_{2q}'s by a simple comparison of count rates in θ and ϕ and eliminates the need for the absolute measurement of the deuteron flux incident on the polarimeter.

ii) there is little energy dependence in T_{20} over the whole energy range so that one is quite insensitive to energy resolution.

The polarimeter (Fig. 7) consists of a liquid hydrogen target and a cylindrical detection system. The trajectories of the scattered particles are determined by two coaxial cylindrical wire chambers operating in the limited streamer mode. The wires are in 1×1 cm^2 individual cells, the inner and outer walls being formed by thin foam cylinders. Charge division provides a measurement of the position and, therefore, of the polar scattering angle. Particle identification and energy determination are made in DE-E counters (6 DE's and 18 E's).

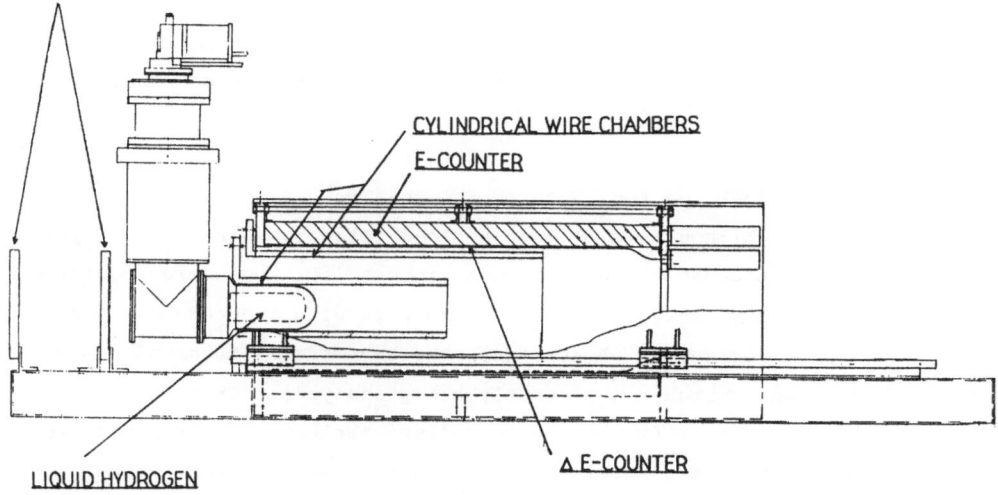

Figure 7. The AHEAD Polarimeter

AHEAD has been fully calibrated with a deuteron beam of known polarization at the Saturne synchrotron (France). The analyzing powers are illustrated in Fig. 8. T_{20} is very much as previously measured at 79 MeV [15] and 191 MeV [16]. T_{21}, for which this is a first measurement at these energies, is as predicted by Tjon [17]. The polarimeter has an efficiency of 2×10^{-3} in the angular domain where the average value of T_{20} is -0.4.

The tensor moments t_{2q} are extracted by a fitting procedure of the observed distribution $N(\theta,\phi)$ to the expected distribution (1) [18]. In this procedure, one gets rid of the absolute normalization and the overall χ^2 is a measure of its validity. The χ^2-test is here a (very constraining!) check that the polarimeter response to unpolarized deuterons has not changed between the calibration and the experiment.

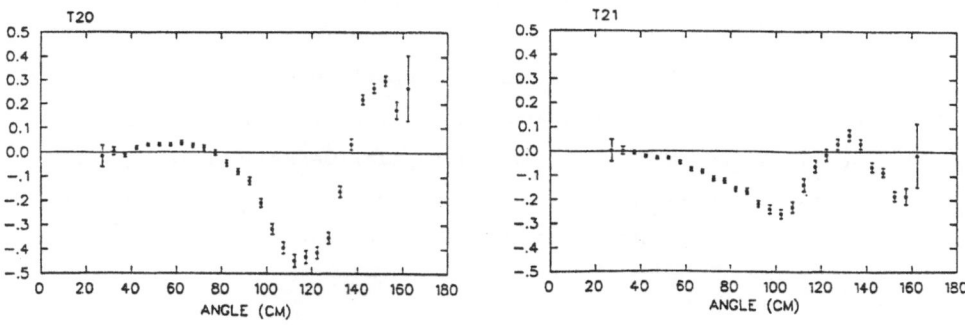

Figure 8. T_{20} and T_{2i} at 145 MeV (dp scattering), preliminary.

IV. Where are we now?

The whole experiment was installed in the Bates South Hall in June 1987. The deuteron channel and the detectors were then successfully tested.

In November 1987, we took data at E_o = 650 MeV, θ_e = 81.0°, q^2 = 14.2 fm^{-2}. Although we do have a clean trigger, we had some pile-up in the chambers because of the numerous protons (for low energy deuterons we cannot use a degrader). We then had to work with reduced beam intensity (1 mA peak current, 5 μA average). We have over a thousand "good" polarimeter events which are presently being analyzed. This point will probably be the most difficult one to analyze because it corresponds to the very low part of the working energy range of AHEAD. For example, the particle ranges in the detector, and many software cuts, may introduce an energy dependence of the polarimeter response which we will have to take into account. Presently, we find tensor moments which are all three in qualitative agreement with most predictions, but with unreasonable χ^2. Since a number of effects have not yet been taken into consideration in the analysis, this is not surprising.

In February 1988, we started tuning for our point at q^2 = 21.2 fm^{-2} (E_o = 850 MeV, θ_e = 76.7°). A new tune of the accelerator was found, with a wider bandpass, allowing a steady operation at 3 mA peak current, 25 μA average; the rates in all detectors were acceptable and the e-d events were easily identified. In the subsequent accelerator studies, the beam intensity was doubled [19]. Note that 850 MeV is a new record for the Bates accelerator, which can, in principle, operate at 1 GeV since a 500 MeV beam was obtained this year in a single pass.

The next run is scheduled for April 1988, when we expect to complete the experiment with two points at 850 and 750 MeV.

We have made, in Ref. 18, a detailed comparison of various past, present or planned experiments to measure t_{20}. Even with a figure of merit (product of luminosity x solid angle x f x polarimeter efficiency x T_{20}^2) three times smaller than our initial objective, this experiment appears at the moment to be the most efficient way to measure t_{20}.

References

1. M. E. Schulze, D. Beck, M. Farkhondeh, S. Gilad, R. Goloskie,
   R. E. Holt, S. Kowalski, R. M. Laszewski, M. J. Leitch, J. D. Moses,
   R. P. Redwine, D. P. Saylor, J. R. Specht, E. J. Stephenson,
   W. Turchinetz and B. Zeidman, Phys. Rev. Lett. $\underline{52}$ (1984) 597.

2. B. B. Voitsekhovskii et al., JETP Lett. $\underline{43}$ (1987) 733.

3. W. Meyer, Nucl. Phys. $\underline{A446}$ (1985) 381c.

4. M. I. Haftel, L. Mathelitsch, and H. F. K. Zingl, Phys. Rev. $\underline{C22}$
   (1980) 1285.

5. R. G. Arnold, C. E. Carlson and F. Gross, Phys. Rev. $\underline{C21}$ (1980) 1426.
   M. S. Zuilhof and J. A. Tjon, Phys. Rev. $\underline{C24}$ (1981) 736.
   M. A. Braun, Sov. J. Nucl. Phys. $\underline{42}$ (1985) 518.

6. A. F. Krutov and V. Troitskii, Sov. J. Nucl. Phys. $\underline{43}$ (1986) 852.

7. R. Dymarz and F. C. Khanna, Phys. Rev. Lett. $\underline{56}$ (1986) 1448.

8. W. P. Sitarski, P. G. Blunden, E. L. Lomon, Phys. Rev. $\underline{C36}$ (1987)
   2479.

9. A. P. Kobushkin and V. P. Shelest, Sov. J. Part. Nucl. $\underline{14}$ (1983) 483.

10. Y. Yamauchi and M. Wakamatsu, Nucl. Phys. $\underline{A457}$ (1986) 621.
    T. S. Cheng and L. S. Kisslinger, Phys. Rev. $\underline{C35}$ (1987) 1432.
    H. Ito and A. Faessler, Nucl. Phys. $\underline{A470}$ (1987) 626.

11. E. M. Nyman and D. O. Riska, Nucl. Phys. $\underline{A468}$ (1987) 473.

12. C. E. Carlson and F. Gross, Phys. Rev. Lett. $\underline{53}$ (1984) 127.

13. R. G. Arnold, D. Benton, P. Bosted, L. Clogher, G. DeChambrier,
    A. T. Katramatou, J. Lambert, A. Lung, G. G. Petratos, A. Rahbar,
    S. E. Rock, Z. M. Szalata, R. A. Gearhart, B. Debebe, M. Frodyma,
    R. S. Hicks, A. Hotta, G. A. Peterson, J. Alster, J. Lichtenstadt,
    F. Dietrich and K. van Bibber, Phys. Rev. Lett. $\underline{58}$ (1987) 1723.

14. J. M. Cameron et al., Third Workshop on Perspectives in Nuclear
    Physics at Intermediate Energies, Trieste, May 1987.

15. E. J. Stephenson, J. D. Brown, M. S. Cantrell, V. R. Cupps,
    D. A. Low, P. Schwandt, J.-Q. Yang, R. J. Holt, E. Ungricht,
    B. Zeidman, D. Beck and M. Schulze, IUCF Scientific and Technical
    Report (1983) 58.

16. M. Garçon, B. Bonin, G. Bruge, J. C. Duchazeaubeneix, M. Rouger,
    J. Saudinos, B. H. Silverman, D. M. Sheppard, J. Arvieux,
    G. Gaillard, N. Van Sen, Y. Yan Lin, J. M. Cameron, W. J. McDonald,
    G. C. Neilson, W. C. Olsen, K. R. Starko, A. Boudard, L. E. Antonuk
    and F. Soga, Nucl. Phys. $\underline{A458}$ (1986) 287.

17. J. A. Tjon, private communication.

18. M. Garçon, Proc. CEBAF Summer Workshop, Newport News, June 1986 (p.583).

19. J. Flanz, private communication.

PHYSICS IN THE GeV REGION WITH POLARIZED TARGETS IN ELECTRON STORAGE RINGS

Roy J. Holt

Physics Division
Argonne National Laboratory
Argonne, IL 60439-4843

INTRODUCTION

During the past five years interest in the internal target method for electron storage rings has intensified. At present there is an active internal target program at the VEPP-3 ring in Novosibirsk and a second ring (NEP) dedicated for nuclear physics is under construction. An internal Ar jet target is in use at the ADONE ring in Frascati in order to produce tagged photons. In addition, plans are being made for an internal target program at the 300-MeV electron ring at Saskatoon. Construction of electron rings at MIT-Bates and NIKHEF is beginning and both laboratories plan internal target facilities. Finally, there are discussions with regard to the PEP and HERA rings for internal target experiments in the Multi-GeV region.

Perhaps the least discussed advantage of the internal target method is the enormous dynamic range available in electron energy. The energy range for experiments varies from 0.22 GeV at the NEP ring to 30 GeV at HERA. Thus, the internal target method is beneficial to a full program of nuclear physics research. For example, a program to study collective modes in the nucleus is planned for the NEP ring, while physics discussions in the GeV region are centered around the emerging fields of nuclear chromodynamics and the effects of color transparency. Since Argonne has embarked upon a joint venture with the Novosibirsk group to measure the tensor analyzing power T_{20} in electron-deuteron elastic scattering at the 2-GeV VEPP-3 ring, I shall draw from examples in this energy region for the discussion today. Finally, there is interest in the study of the structure of the nucleon at the HERA ring in which it is envisaged that polarized electrons and targets will be used to measure the spin structure functions of the proton and neutron.

DEUTERON PHOTODISINTEGRATION IN THE GeV REGION

It is particularly interesting to consider the two-body breakup of the deuteron in the GeV region for three reasons: (i) the electromagnetic interaction with the simplest nucleus, the deuteron, is especially amenable to theoretical interpretation, (ii) at high energy one is probing the short-range part of the deuteron wave function, and (iii) in this energy region there are considerable differences between the predictions of meson-exchange theory[1] and nuclear chromodynamics.

Two hypotheses are inherent within the framework of nuclear chromodynamics: the overall energy dependence of the cross section is known from arguments of dimensional scaling and QCD; and the form factors of the

371

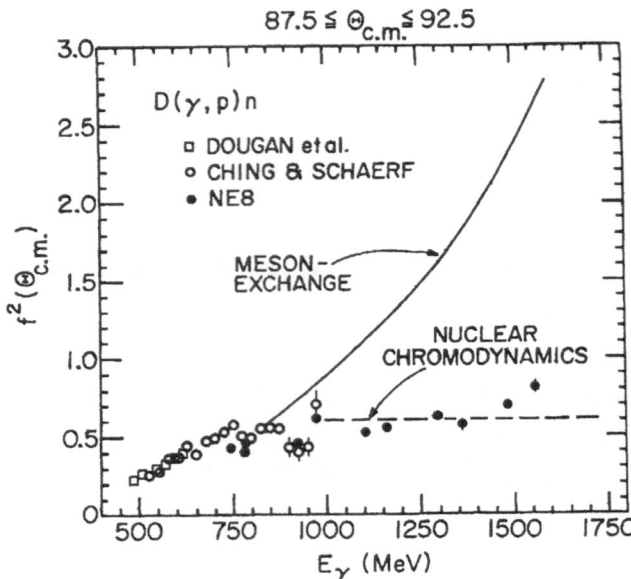

Fig. 1   The solid curve indicates the prediction of the meson-exchange
model while the dashed curve is that of the nuclear chromodynamics
model.   The new data from experiment NE8 at SLAC are given by
the solid circles.   Note that these data are based upon a
preliminary analysis, as described in the text.

two nucleons are factorizable. Thus, the expected[2] energy dependence of the
differential cross section has the form

$$\frac{d\sigma}{d\Omega} \propto \frac{1}{\left[s\left(s-m_d^2\right)\right]^{1/2}}\, F_p^2(t_p)\, F_n^2(t_n)\, \frac{f^2(\theta_{cm})}{p_T^2}$$

where s is the square of the total energy in the center-of-mass frame, m_d is
the deuteron mass, p_T is the transverse momentum and $F_{p(n)}$ are the proton
(or neutron) electromagnetic form factors. Here, $f(\theta_{cm}^{p(n)})$ is the reduced
amplitude which is expected to have no energy dependence where the nuclear
chromodynamic description is valid. This reduced amplitude has been
deduced for existing data at $\theta_{cm} = 90°$ and is illustrated in Fig. 1. Also
shown in Fig. 1 is a recent meson-exchange calculation of T.-S. H. Lee. New
and preliminary data, illustrated as the solid circles in the figure, are from
experiment[3,4] NE8 at SLAC. Clearly, the energy dependence of new data are
better described from the simple arguments of nuclear chromodynamics than
by the meson-exchange theory above 1 GeV. The analysis of these data is
preliminary in that the absolute normalization was taken from an on-line
analysis of electron-proton elastic scattering. However, the energy dependence
of the results is not expected to change significantly in the final analysis.
Polarization experiments should have an important role in the GeV region as
a further test of this finding.

ELECTRON-DEUTERON ELASTIC SCATTERING

M. Garcon in the previous talk[5] gave a thorough introduction to the
motivation for a measurement of T_{20} in electron-deuteron elastic scattering,
and I shall not repeat it here. Briefly, the major issues are: (i) the first
isolation of the charge and quadrupole form factors for the deuteron, (ii)

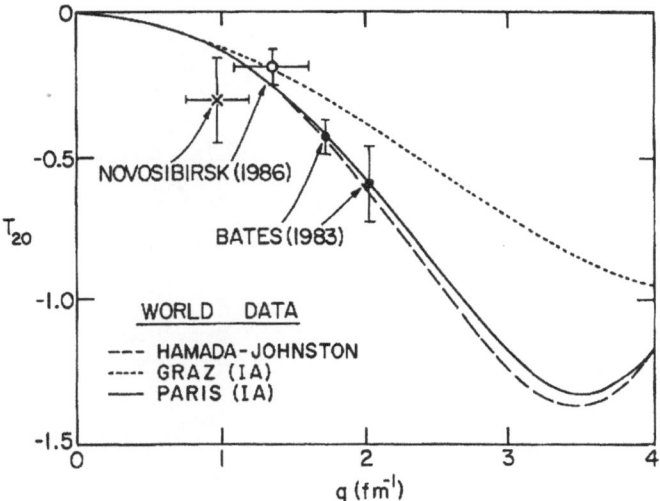

Fig. 2   Summary for all polarization data in electron-deuteron elastic
          scattering.   The two data points at low values of q were determined
          with an internal polarized target method while the two higher q
          points were measured with a deuteron polarimeter.

investigation of QCD effects, and (iii) information on the little understood
isoscalar meson-exchange current.   In fact, it is expected that a measurement
of T_{20} at high momentum transfer would be the most sensitive test of QCD
effects in nuclei, since T_{20} is expected to approach a value of $-\sqrt{2}$ in the
QCD limit.   All existing data for polarization in electron-deuteron scattering
are illustrated in Fig. 2.   Note that this graph represents <u>all</u> existing data for
polarization in electron scattering from nuclei.   The two data points[5] at high
momentum transfer were measured with a deuteron tensor polarimeter at the
MIT-Bates Laboratory, while the lower momentum transfer data[7] were
determined with an internal polarized target at Novosibirsk.   Clearly, these
measurements must be extended to high momentum transfer and efforts are
underway both at MIT and Novosibirsk.

A comparison of the relative scales of these two experiments is indicated
schematically in Fig. 3.   The experiment at MIT required the use of two
large magnetic spectrometers and the attendant shielding to detect the
scattered electrons and deuterons; whereas, the internal target experiment
(shown approximately on the same scale in the inset figure) was performed
with detector arrays.   The Novosibirsk apparatus also is shown with a
magnified scale so that the individual components can be seen.   The detector
hardware is simplified in the internal target geometry because of the purity of
the polarized target, the high duty factor of the ring and the low background
from the thin internal target.

Of course, the primary limitation of the internal target geometry is the
relatively small target thickness, 10^{11} nuclei/cm^2.   Efforts are underway[8] at
Argonne to produce a high flux, laser-driven polarized H or D source and to
develop a storage cell for the polarized atoms in order to increase the
thickness of the polarized target.

ARGONNE-NOVOSIBIRSK COLLABORATION

The news today is that Argonne and Novosibirsk have embarked upon a
collaborative effort in order to measure the tensor analyzing power T_{20} in
electron-deuteron scattering at the highest possible momentum transfer.   This

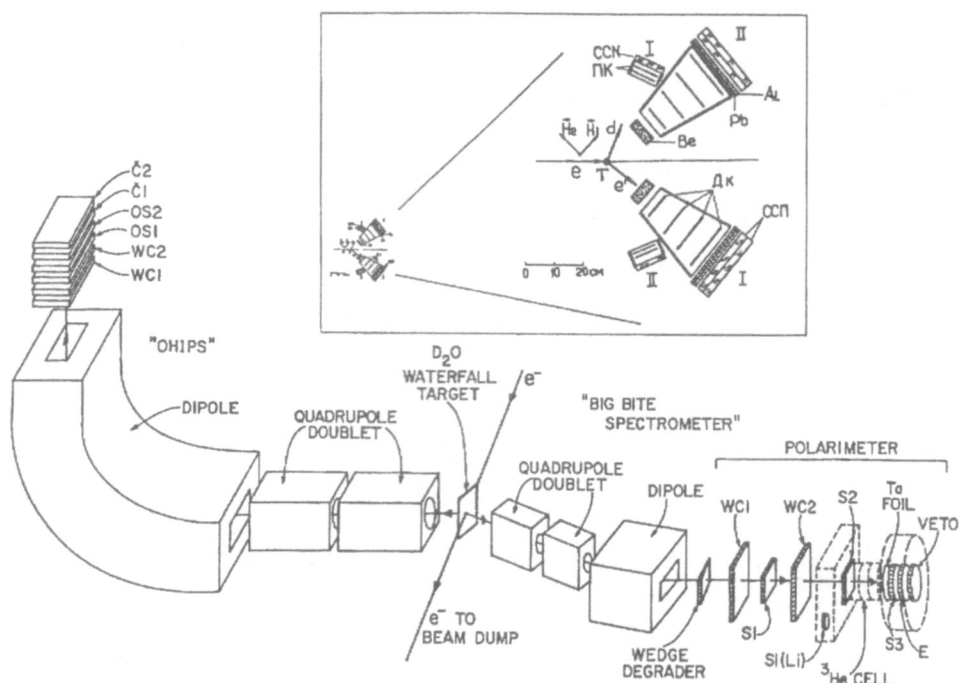

Fig. 3   Schematic diagram of the apparatus required to measure t_{20} in e-d elastic scattering in the external target geometry. Inset figure on left illustrates the apparatus on approximately the same scale for the internal target experiment at Novosibirsk. The figure is also shown with an expanded scale for ease of identification of the components.

collaboration was initiated last month when L. Young and I brought to Novosibirsk a specially fabricated storage cell for polarized atoms in the VEPP-3 storage ring. This collaboration[9] is based upon the experience of the Novosibirsk group in the technology of ultra-thin internal targets in electron storage rings and the experience of Argonne in the development of storage cells for polarized atoms and laser-driven polarized atomic beam sources. Presently, it is perceived that the collaborative effort will proceed in three phases in order to achieve the necessary target thickness and luminosity: (i) use of the Novosibirsk polarized source and a simple ANL storage cell (n = 1×10^{12} cm^{-2}), (ii) use of the Novosibirsk source and a more sophisticated clam shell storage cell (n = 1×10^{13} cm^{-2}), (iii) replace the Novosibirsk source with the ANL laser-driven source (n = 1×10^{14} cm^{-2}).

With regard to storage cells for polarized atoms, the central issue is the surface coating of the cell. A drifilm coating was used for the initial simple ANL storage cell which presently resides in the VEPP-3 ring. Drifilm, the most studied[8,10] surface coating, has the properties that it chemically bonds to the surface and thereby is bakeable at relatively high temperatures. A teflon surface is also expected[11] to inhibit depolarization of H and D atoms, but it is not alkali resistant and cannot be used with the laser-driven source.

INTERNAL TARGET DEVELOPMENT

Now I would like to focus on some recent developments in laser-driven polarized H, D and ^{3}He targets. A high-flux polarized source of H and D atoms based upon an optical-pumping spin-exchange technique is being

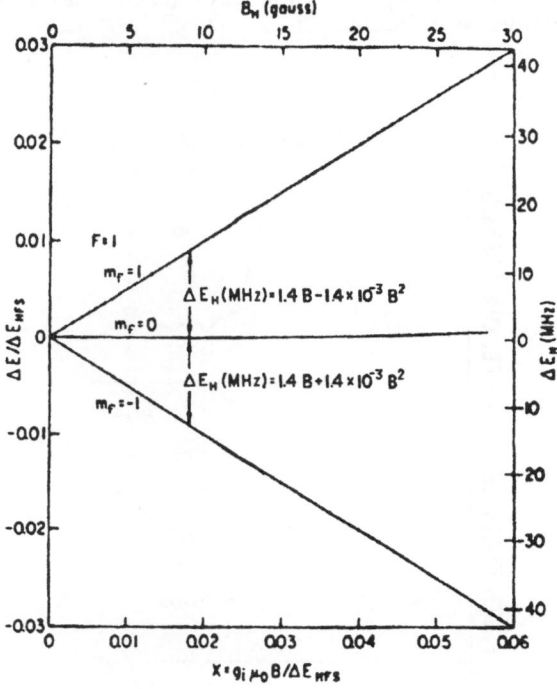

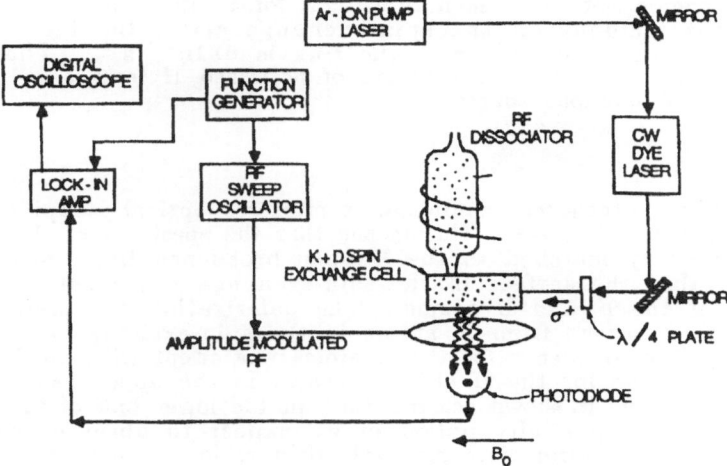

Fig. 4   Breit-Rabi diagram for the H atoms indicating the two available Zeeman transitions. The apparatus to induce and detect these Zeeman transitions is illustrated schematically.

developed at Argonne. The method involves a two-step process in which an ensemble of alkali atoms, K in the present case, are polarized by optical pumping and the polarization is transferred to H or D atoms via spin-exchange scattering. This method is attractive since the flux of H or D atoms is limited, in principle, by available laser power and since the gas load to a storage ring would be minimal.

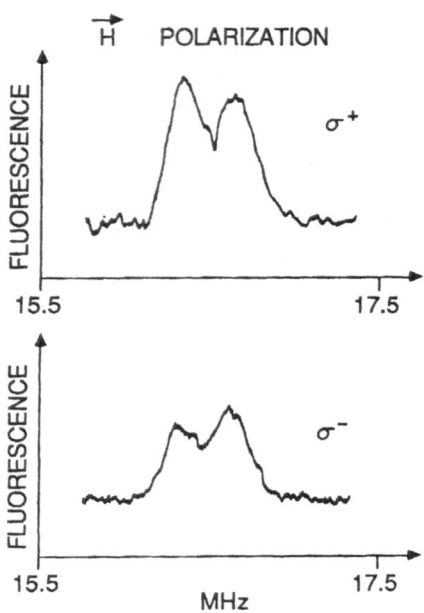

Fig. 5   A comparison of Zeeman transitions for H atoms for σ^+ and σ^- laser light incident on the spin-exchange cell. In this case the H polarization is 10% and the flux is 6×10^{16} s^{-1}. These results represent the first observation of polarized H from a spin-exchange optically-pumped source.

The primary technical challenge is effective optical-pumping of the K atoms without a buffer gas. This means that the spectrum of the laser light must be carefully matched to the Doppler-broadened line width of the K atoms and that the surface in the spin-exchange cell must inhibit spin-relaxation of the K and H atoms. The polarization was determined by observing fluorescence from K atoms in the spin-exchange cell when the Zeeman transition frequency for the H atoms was swept with an RF coil. A Breit-Rabi diagram for the H atom is shown in the upper panel of Fig. 4, while the apparatus is shown schematically in the lower half of the figure. If the H atoms are partially polarized we expect to observe two Zeeman transitions in the intermediate magnetic field region. These transitions are observed and shown in Fig. 5. Clearly, the relative intensity of the two transitions reverses as expected between σ^+ and σ^- light. Here we see the first evidence for H from an optically-pumped, spin-exchange source! Thus far, a vector polarization of 20% with a flux of 1.3×10^{17} s^{-1} for the H atoms has been achieved. With the assumption that the H atoms are in equilibrium with the K atoms, and consequently, that the populations $N(m_F)$

of the atoms in H follow a spin-temperature distribution, i.e. $N(m_F) \propto e^{\beta m_F}$ where β is the spin temperature and m_F is the total magnetic quantum number, the polarization of the protons in the spin-exchange cell is found to be 10% from these data. Further improvement in the polarization is expected as we gain more experience with the source. Since these first Zeeman signals were observed, the laser light was more carefully matched to the Doppler spectrum of the K atoms. The typical Zeeman signals now observed are shown in Fig. 6 for K↑ in the upper panel and H↑ in the lower panel. Clearly, the signals improve as the laser power is increased. In fact, the next step based on these studies is to increase the laser power by adding a second CW dye laser.

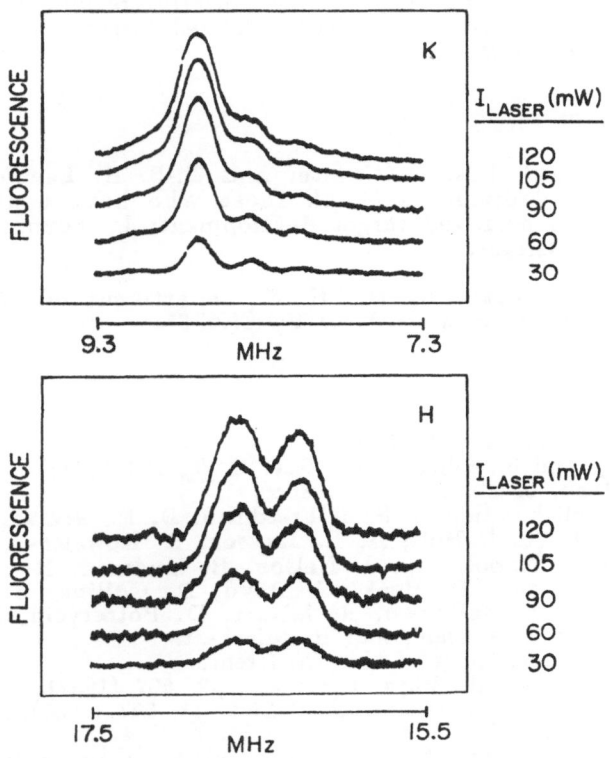

Fig. 6   Zeeman transition signals for K (upper panel) and H (lower panel) as a function of incident laser power.

The goal is to provide a target of D atoms with a thickness of 10^{14} cm^{-2} and a tensor polarization of $t_{20} \gtrsim 0.3$ which could be used in an electron-deuteron scattering experiment. Again, in order to achieve this thickness it is essential to store the atoms in a windowless cell and permit the atoms to leak away before the spins relax. In order to achieve this goal the high flux and polarization expected from the laser-driven source is essential.

A novel internal polarized ^{3}He target is being developed[12] at Caltech. The ^{3}He nuclei are polarized by the well-known method of optically-pumping metastable ^{3}He atoms. This method has benefitted recently from the development of lasers which can operate at 1.083 μm so that $\gtrsim 3 \times 10^{18}$ ^{3}He/s are produced. It is expected that an internal target of ^{3}He with a thickness of $\gtrsim 10^{15}$ cm^{-2} is feasible.

SUMMARY

In summary, it appears that we have evidence from the D(γ,p)n reaction that the meson-exchange model is failing in the GeV region. Surprisingly, it appears that the new D(γ,p)n data favor the energy dependence of the nuclear chromodynamics model rather that that of the meson-exchange model.

Application of the polarization method to electron scattering studies is in its infancy, and it is potentially a very powerful technique. The internal target method coupled with laser-driven polarized targets should represent an important tool for nuclear physics.

ACKNOWLEDGEMENTS

I wish to thank Drs. F. Coester and T.-S. H. Lee for very useful discussions. In addition, I thank those who have contributed most significantly to the polarized target development: L. Young, M. Green, R. Kowalczyk and J. Gregar.

This work supported by the U. S. Department of Energy, Nuclear Physics Division, under contract W-31-109-ENG-38.

REFERENCES

1. T.-S. H. Lee, to be published.
2. S. Brodsky and J. Hiller, Phys. Rev. C **28**, 475 (1983).
3. J. Napolitano, et al., to be published.
4. The NE8 collaboration is: S. J. Freedman, D. F. Geesaman, R. Gilman, M. C. Green, R. J. Holt, H. E. Jackson, R. Kowalczyk, E. Kinney, C. Marchand, J. Napolitano, J. Nelson, B. Zeidman, R. E. Segel, T.-Y. Tung, P. Bosted, D. Beck, G. Boyd, D. Collins, B. Filippone, J. Jourdan, R. D. McKeown, R. Milner, D. Potterveld, R. Walker, C. Woodward, Z.-E. Meziani, and R. Minehart.
5. M. Garcon, Proceedings of this Conference.
6. M. E. Schulze, et al., Phys. Rev. Lett. **52**, 597 (1984).
7. B. B. Woitsekhowski, et al., JETF Lett. **43**, 567 (1986).
8. L. Young, et al., Nucl. Instrum. and Meth. **B24/25**, 963 (1987).
9. The collaborators for Phase I at Novosibirsk are presently: S. G. Popov, B. B. Woitsekhowski, D. M. Nikolenko, D. K. Toponkov, E. P. Tsentalovich, I. R. Rachek, V. F. Dmitriev, V. G. Zelevinsky, and M. Mostovoy; and at ANL: R. J. Holt, L. Young, R. Gilman, and R. Kowalczyk.
10. D. R. Swenson and L. W. Anderson, Nucl. Instrum. and Meth. **B29**, 627 (1988); J. S. Camparo, J. Chem. Phys. **86**, 533 (1987); H. M. Goldenberg, et al., Phys. Rev. Lett. **5**, 361 (1960).
11. W. Haeberli, Bull. Am. Phys. Soc. **31**, 1195 (1986); T. Wise, private communication.
12. R. G. Milner, et al., Nucl. Instrum. and Meth. **A257**, 286 (1987); C. L. Bohler, et al., to be published.

SPIN OBSERVABLES IN P̄P ELASTIC AND PP INELASTIC SCATTERING

R. Bertini

Cern associate, Cern/EP, CH-1211 Geneva 23, Switzerland
Perm. address: DPHN/ME, CEN Saclay, F91191 Gif sur Yvette

1. INTRODUCTION

The measurement of spin observables represents an important tool to investigate specific effects that are not easily observed with the measurement of spin integrated cross sections.Examples of that are: 1) the study of the energy dependence of spin observables in pp scattering that show structures, that have been interpreted in terms of exotic multiquark states or alternatively in the frame of more traditional one boson exchange models. 2) The investigation of the process that produces hyperons strongly polarized in inclusive reactions at large momentum transfers. 3) The search of the appropriate description of the p̄N interaction.This is now given by potential models, that, from contradictory inputs on the annihilation and on the short range cutoff radius, well reproduce the data on the spin integrated cross sections, but predict differing values for spin observables. Experiments related to these three topics performed or in progress at the Saturne and Lear accelerators will be discussed in this talk. Concerning point 1), the discussion is essentially limited to the inelastic channel $pp \rightarrow d\pi^+$, in the energy domain of Saturne, that is below 3 GeV. For this reaction, above .8 GeV, polarisation data are very scarce. New data recently obtained at Saturne will be discussed in section 2.

An other inelastic channel, the $pp \rightarrow p + K^+ + \Lambda$ reaction, has a much smaller contribution to the pp cross section in this energy domain, $\sigma_{tot} = 51 \, \mu$barn at 3.67 GeV/c.Nevertheless the measurement of spin observables in this reaction is very important in view of the understanding of the mecanism that produces polarised Λ hyperons.This point will be discussed in section 3. Measurement of spin observables in p̄p scattering became recently possible possible due to the improvement in the performances of the antiproton beams and of the polarized targets. In this sense the new beams provided by LEAR in the post ACOL era have open new possibilities.The first experiment on spin observables in p̄N elastic scattering, now in progress at CERN and making use of these new beams, will be described in section 4.

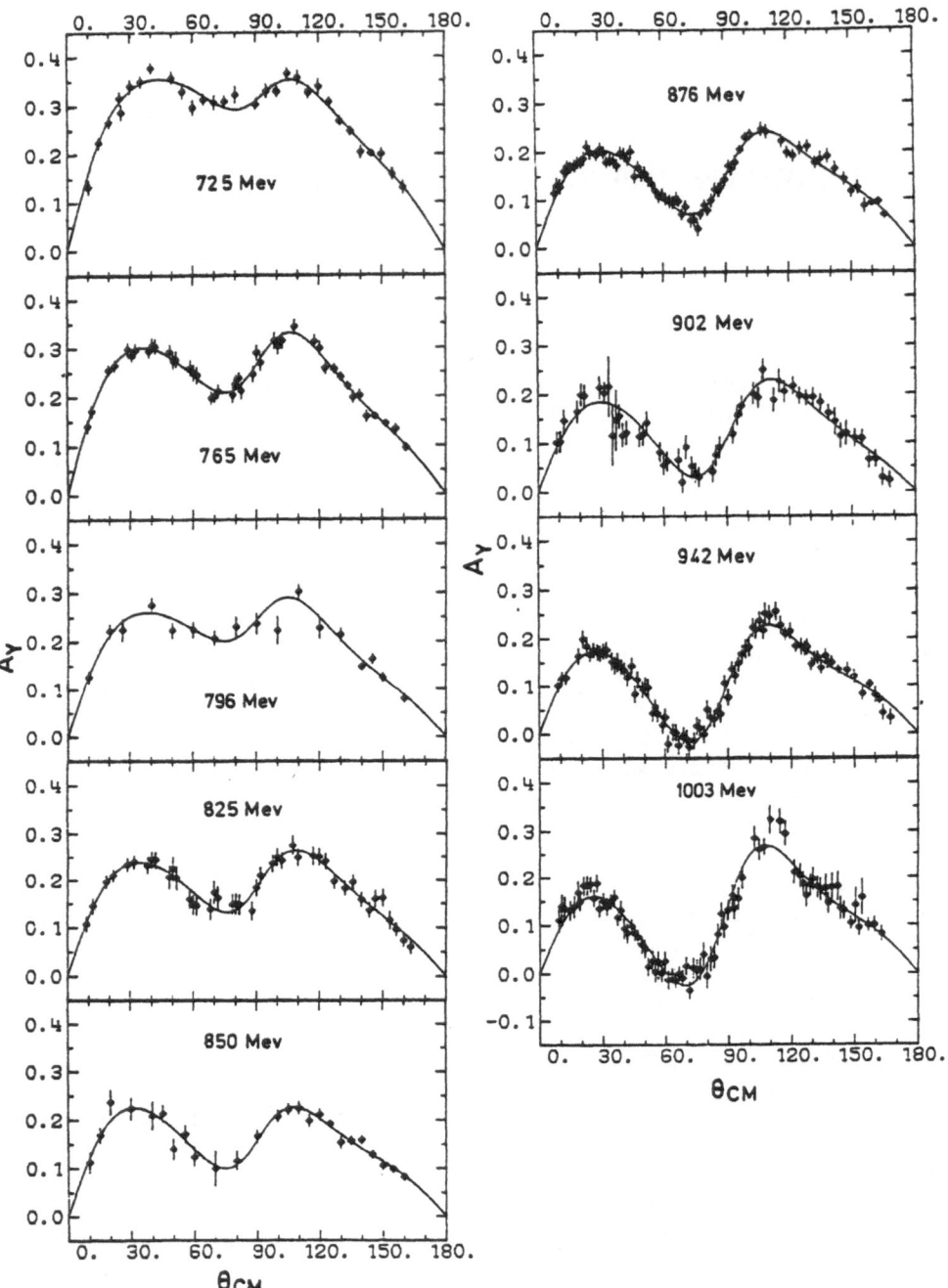

Fig. 1.   Angular distributions of A_{y0}. The data come from ref.[5]. The error bars
include systematic errors. The curves are just a guide to the eye.

380

2. THE PP→Dπ⁺ REACTION

The kinematics of the pp→dπ⁺ reaction is such that it is possible to measure the full angular distribution of a spin observable, in the pion C.M. system $0° < \theta_{CM}{}^\pi < 180°$, detecting the deuteron in a limited forward angular domain. Such an approach has the advantage to provide data free of the large corrections to be applied, due to the decay and absorption, when pions are detected. A large selectivity is needed against competing reactions, because of their large production cross sections. For example, for the pp→dπ⁺π⁰ reaction the total cross section $\sigma_{tot} = 430 \mu$barn. has to be compared to $\sigma_{tot} = 130$ μbarn., for the pp→dπ⁺ reaction. This selectivity can be easily achieved at Saturne, using one of the high resolution spectrometers (SPESI or SPESIV) to measure with good precision the momentum of the deuteron. The experiment, that I am discussing here, was performed with the polarized beam of the Saturne 2 accelerator at a average intensity of 2×10^9 p/burst. The reaction target was liquid H_2 270 mg/cm² thick. Deuterons were detected with the SPES I spectrometer [1] in the beam energy range $.725 \leq T_p \leq 1.003$ GeV and with the SPES IV spectrometer [2] in the energy range $1.2 \leq T_p \leq 2.3$ GeV. Discrimination against other particles was achieved by time of flight over a distance of 5.6 m. with SPES I and over a distance of 16 m. with SPES IV. The deuteron momentum was determined by track reconstruction in order to get, through the reaction kinematics, the pion missing mass. This analysis allowed a very clean background rejection. The remaining contamination was estimated to be lower than 1%. Beam polarisation was reversed each spill and continuosly monitored with a polarimeter placed upstream of the reaction target. The polarimeter which has been previously described [3], measures the left right asymmetry in pp elastic scattering. The average beam polarisation was $P_b = .75$. The ratio of asymmetries on carbon and hydrogen, needed to compute the corrections for the carbon content of the CH_2 polarimeter target, have been measured at each energy. Values of the pp analyzing power A_{pp} have been taken from the compilation of ref. [4]. The angular distributions of the analysing power A_{yo}, measured with SPES I [5] at nine energies: $T_p = .725, .765, .796, .825, .850, .876, .902, .942, 1.003$ GeV, are shown in Fig. 1. The data show a smooth energy dependence of the angular distributions of A_{yo}. The higher energy data [6,7], measured with SPES IV, are shown in Fig. 2. Unlike for energies up to $T_p = 1.2$ GeV the data show a very structurated behaviour. The main trend is that A_{yo} takes negative values in the forward part of the angular distribution ($0° \leq \theta_{CM}{}^\pi \leq 90°$) whereas it assumes positive values at backward angles ($90° \leq \theta_{CM}{}^\pi \leq 180°$). Positive values of A_{yo} at forward angles appear only for the energies $T_p = 2.0$ and $T_p = 2.3$ GeV. In the energy range covered by the present experiment, analyzing power data exist only at $T_p = 1.23$ GeV [8]. They are plotted, together with our $T_p = 1.2$ GeV data, in Fig.2. There are also indicated, for each energy, the angles at which the Mandelstam invariants t and u are equal to zero.

The values of A_{yo}, measured at backward angles ($90° \leq \theta_{CM}{}^\pi \leq 180°$) have been plotted against the Mandelstam invariant u (Fig.3). They cluster in two groups, one corresponding to the energies $T_p = 1.2, 1.4$ and 2.3 GeV, the other one corresponding to the other measured energies. Strong peaks appear centered at u≈ 0.03 GeV².

The forward differential cross section for the reaction pp→dπ⁺ plotted as a function of the C.M. energy shows [9] three bumps at $\sqrt{s_{\pi d}} = 2.17$ GeV, $\sqrt{s_{\pi d}} = 3.0$ GeV and $\sqrt{s_{\pi d}} = 3.7$ GeV, respectively. These structures have been interpreted in the one pion exchange model(OPE) to correspond [10] to excitation of Δ isobars(1232,1950,2420) at the pure $T = 3/2$ $\pi^+ p \to \pi^+ p$ scattering vertex.Excitation of $T = 1/2$ N^* isobars can be obtained by the charge exchange reaction $\pi^0 p \to n\pi^+$ at the same vertex.The contribution of this diagram to the $pp \to d\pi^+$ cross section is 16 times smaller than the $\pi^+ p$ elastic scattering.Consequently it has not been previously

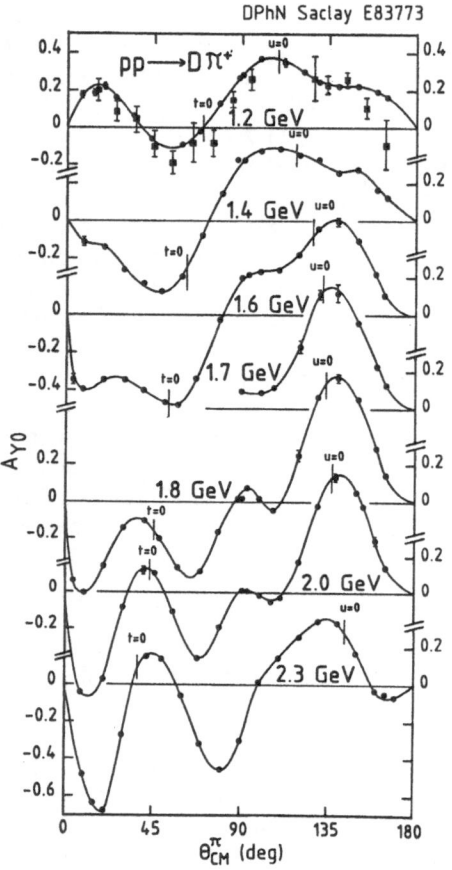

Fig. 2 .

Angular distributions of A_{y0}. The data come from ref.[7] (forward angles) and from ref.[6] (backward angles). The error bars include systematic errors. The curves are just a guide to the eye. Together with the $T_p = 1.2$ GeV data have also been plotted (squares) the values at $T_p = 1.23$ GeV from ref.[8].

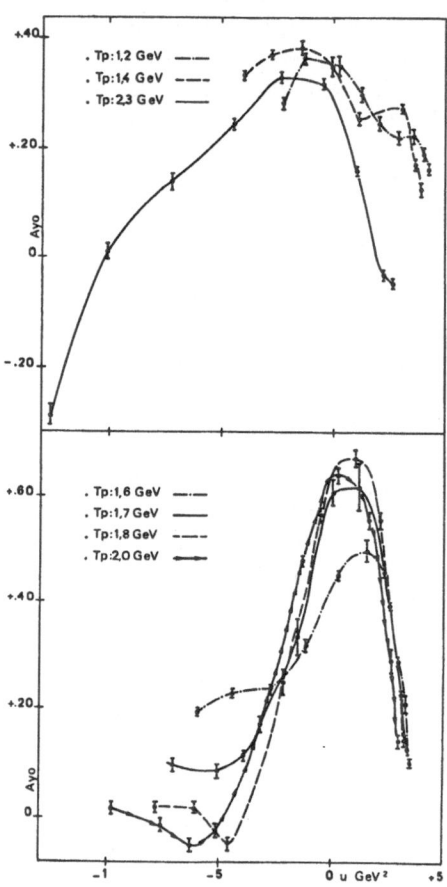

Fig. 3 .

The data of the analyzing power A_{y_0} from ref.[6] are plotted against the Mandelstam invariant $u(\text{GeV}^2)$. The error bars include systematic errors. The curves are just a guide to the eye.

considered in the OPE calculations. However, in our energy region, the forward cross section shows [9], between the first two structures, a dip which is more than ten times less than the adjacent peaks. Therefore we think that the $\pi^0 p \to n\pi^+$ diagram should be taken into account. To see how important can be its contribution, we have plotted in Fig. 4 the values of A_{y0}, taken at $u=0$, as a function of $\sqrt{s_{\pi d}}$.

In the OPE model the CM energies for the $\pi d (\sqrt{s_{\pi d}})$ and $\pi p (\sqrt{s_{\pi p}})$ systems are related through the equation: $2s_{\pi N} = s_{\pi d} + m^2_\pi - 2m^2_N$, where m_π and m_N are the pion and nucleon masses respectively. The related values of $\sqrt{s_{\pi d}}$ and $\sqrt{s_{\pi N}}$ are shown on the horizontal axis of Fig. 4, lower and upper scale respectively. The total cross section for πN scattering $\sigma_{1/2}$ leading to the pure $T=1/2$ state has been computed using the formula : $\sigma_{1/2} = 3/2 \, \sigma_{\pi^- p} - 1/2 \, \sigma_{\pi^+ p}$. The values for the total cross sections for $\pi^- p$ and $\pi^+ p$ scattering come from ref. [11]. The results are also plotted in Fig. 4. Both curves show two bumps, centered at different energies. The bumps in the $\sigma_{1/2}$ curve correspond to the excitation of the N^* resonances 1520 and 1680 MeV respectively. A similarity in shape can be observed for the two curves. However there is a shift of about 60 MeV between the positions of the strucrures in the two curves. This shift can be explained using arguments similar to those developped in ref. [12]. Taking the reasonable assumption that the relative angular momentum in the pN^* is $\mathscr{L}=0$, and that the average Fermi momentum of the nucleon in the deuteron is $q=70$ MeV/c, then a simple kinematical calculation can account for the observed shift. Therefore the structure we observe in Fig. 3 can be interpreted as the excitation of the $N^*(5/2, 1/2)$ resonance at the πN scattering vertex. However the OPE model also predicts equal analyzing powers at the same energy and same scattering angle for the reactions $pd \to dp$ and $pp \to d\pi^+$. Comparison of our data at $T_p=1.6$ GeV with those of ref. [13] at $T_p=1.53$ GeV shows that, whereas we have large positive values, they find negative and smaller values. This discrepancy involves doubts about a straightforward application of the OPE model to the polarisation data.

An alternative interpretation involves the question of dibaryon resonances. A formation reaction mecanism could directly produce such a state that then would decay in the πd channel. Such an hypothesis implies a similar energy dependence of the data taken at the same values of t and u.

In order to study this energy dependence, the values of A_{y0}, at $t=0$, have been plotted as a function of the invariant mass $\sqrt{s}$ (fig. 5a) and compared with the A_{y0} data taken at $u=0$ (fig. 5b). The error bars correspond to the uncertainties in the interpolation procedure.

The forward angle data show a structure (fig. 5a) peaking in between $\sqrt{s}=2.7$ GeV and $\sqrt{s}=2.8$ GeV whereas the backward angle data of A_{y0}, taken at $u=0$, showed [6] a structure peaked at $\sqrt{s}=2.66$ Gev (fig. 5b).

A rapid variation as a function of $\sqrt{s}$ appears also for the A_{y0} data taken at $\theta_{CM}^\pi = 90^0$ (fig. 5c). We observe that they are consistent with zero between $\sqrt{s}=2.6$ GeV and $\sqrt{s}=2.7$ GeV.

All these facts suggest the existence of a resonant state in the pp system at the approximate energy of 2.7 GeV. A structure has indeed been observed at this value of $\sqrt{s}$ in the energy dependence of the spin correlation parameters C_{LL} [14] and C_{NN} [15] in pp elastic scattering around $\theta_{CM}=90^0$. A bump has also been found [16] in the difference between the pp total cross section for antiparallel and parallel longitudinal spin states at 2.7 GeV, with a width of less than 80 MeV and an elasticity of more than 0.1. More recently an even narrower structure (less than 50 MeV) and centered at the same energy has been observed in the energy dependence of A_{y0} in pp elastic scattering, at Saturne [17].

These structures have been tentatively interpreted as the $^1S_0$ six quark state predicted in the cloudy bag model by Lomon [18].However the present data do not rule out an interpretation in terms of a N^* resonance at the πN scattering vertex in the framework of the OPE model. Therefore we plan further measurements involving other spin dependent observables to try to clarify the question [19]. The measurement of A_{yy},for example,can be usefully combined with data on the spin integrated differential cross section σ_{00}.The combination $\sigma_{00}(1+A_{yy})$ depends only on triplet amplitudes (i.e.,initial protons are in a triplet state). The complementary quantity $\sigma_{00}(1-A_{yy})$ depends mainly on singlet amplitudes — the triplet contribution coming only from the $J^\pi=2^-,4^-,...$channels [20].Thus the outcome can,for example,be that the hypothesis of a $^1S^0$ six quark state,that is a pure singlet state,were confirmed or ruled out.

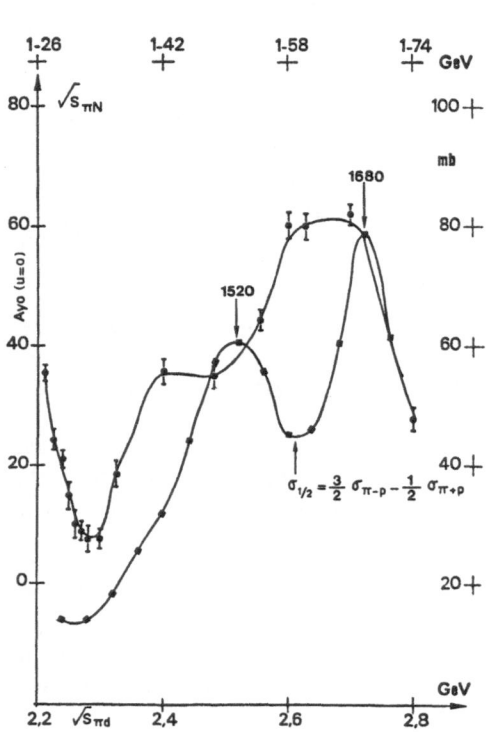

Fig. 4 .

The data of the analyzing power A_{y_0} from ref.[5,6,7] are plotted as a function of the CM energy $\sqrt{s}_{\pi d}$(lower scale). The total cross section for the pure $T=1/2$ state is also plotted as a function of the CM energy $\sqrt{s}_{\pi}N$(upper scale).

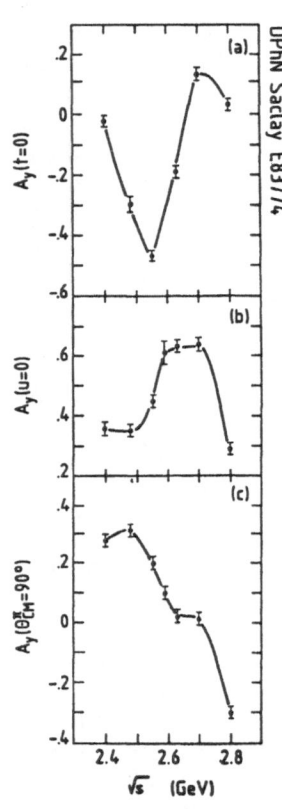

Fig. 5 .

Energy dependence of A_{y0} $(t=0)$, A_{y0} $(u=0)$ and $A_{y0}(\theta_{CM}^\pi=90^\circ)$. The points come from the interpolations of the data shown in fig. 2 when necessary. The curves are just a guide to the eye.

3. SPIN OBSERVABLES IN THE PP→PΛK⁺ REACTION

Since the discovery that the Λ particles,produced at large transverse momenta p_T,were polarised a large amount of data have been collected on inclusive hyperon production.These data show regularities, that have,so far,eluded satisfactory explanation [21]. In order to illustrate these regularities we show in Fig. 6 the hyperon polarisation near $p_T \approx 1$ GeV/c and the Feynman variable $x_F \approx .5$ between $\sqrt{s} = 5$ and 60 GeV/c². We can observe that:

1)   The polarizations P do not depend on $\sqrt{s}$ over the full range.

2)   In the reaction $p \to \vec{\Lambda}$ P is negative (direction $\vec{p}_\Lambda \times \vec{p}_p$) and depends on p_T ($\approx p_\Lambda \theta_\Lambda$) and to some extent on x_F.Under the assumption that the lambda spin orientation is that of the strange quark,quarks produced to the left have negative polarization.

3)   For the reaction $p \to \vec{\Xi}^-$,the polarization of the Ξ is similar to that of the Λ,while for the reaction $p \to \vec{\Sigma}$ the polarization of the Σ has opposite sign to that of the Λ.

4)   The polarization of the $\bar{\Lambda}$ is zero in the reaction $p \to \bar{\Lambda}$ .However the polarization of the $\bar{\Lambda}$ in the $\bar{p} \to \bar{\Lambda}$ is equal,within statistical errors, to that of the Λ in the $p \to \Lambda$ production [22]

5)   No significative difference has been observed so far, in the hyperon polarization,for inclusive production on different nuclear targets.

These surprising observations cannot be explained by perturbative QCD [23].However a number of "ad hoc" models have been constructed to interpret the data [24 − 26]. In these models,the produced and/or scattered quarks become polarized by breaking of strings [24] or Thomas precession [25] or multiple scattering of the strange quarks [26].In this context the prediction is that the Λ polarization in the $K^- \to \vec{\Lambda}$ also should be of the same sign as in the $p \to \vec{\Lambda}$ production. Recent data [22] do not confirm this prediction and show that the polarization in the former case is of the opposite sign and several times larger than in the latter.Thus these models have both successes and failures in predicting the relative hyperon polarization for differing hyperons and differing production reactions. Concerning the regularity 3),one of the models [24] assumes that one quark in the incident proton is lost, through a hard collision,leaving a spectator diquark (uu or ud) which then picks up an s quark to form the forward outgoing hyperon.Thus the uud→uus process produces a Σ^+,while a uud→uds process produces either a Λ or a Σ^0.Assume that the quark is polarized by some unspecified mechanism.Then $P_\Lambda = P_s$,because the (ud) spectator diquark is in a singlet state in the Λ hyperon. For the Σ^+ and the Σ^0 hyperons the non strange quarks must be in a triplet state,so that the polarization of the finite baryon is opposite to that of the strange quark: $P_{\Sigma^+} = P_{\Sigma^0} = -.33$.The observed sign reversal is thus expected from the model,although the expected magnitude $|P_{\Sigma^+}| = .33|P_\Lambda|$ is smaller than the measured one.Regardless of the model,a measurement of additional spin parameters afforded in a study of hyperon production with a polarised proton beam is a crucial step in attempting to understand these phenomena.The only existing experiments of this kind of observables,in the kinematical region ($p_T \approx 1$ GeV/c),where the hyperons are polarized,are D_{NN} (spin transfer) and A_N (analyzing power) for Λ production at 13.3 and 18.5 GeV/c [27].The measured values of A_N and D_{NN} are close to zero as predicted by the quark fragmentation recombination model (QFR)[25].This model predicts that, for Λ production by a proton beam,the proton spin has no effect on the Λ polarization.This is an encouraging success for this model, that must be further checked by measuring spin observables in the production of other hyperons.As a matter of fact,in sharp contrast with the Λ case,this model predicts large spin effects for other hyperons [28].

On the experimental side it can be observed that all the data, that can be found in the literature,have been obtained detecting the anisotropy in the angular distribution of the two particles, produced in the weak decay of the hyperon.In order to discuss the implications of the experimental choices,let look to the $\Lambda \rightarrow p\pi^-$ decay, where two charged particles are detected.The track reconstruction leads to the vertex corresponding to the decay point of the Λ.Through kinematics one checks that the invariant mass of the $p\pi^-$ system is just the Λ mass.However such an analysis does not insure that the Λ has been directly produced and does not originate from the $\Sigma^0 \rightarrow \Lambda\gamma$ decay or possibly from the $\Xi^0 \rightarrow \Lambda\gamma$ decay.Now,in a recent experiment studying inclusive Λ production,it was found [29] that approximately 30 % of the Λ arise from Σ^0 decay.As already mentionned, the QFR model predicts that the incident beam polarisation has a large effect on the Σ^0 polarization,because the transferred ud diquark is in an isotriplet-spin triplet state.The calculation yields [28] that for $\vec{p} \rightarrow \Sigma^0$ the spin transfer D_{NN} should be $+2/3$ and the analyzing power A_N should be of the order of the Σ polarization,opposite in sign to the Λ polarization.It is clear,therefore,that a second generation experiment should select between the different channels leading to Λ production.Moreover in the example discussed here,a Be target was used because a point like source was needed for the hyperon beam.This involves considerable complication for the understanding of the reaction mecanism compared with the case of a H_2 target. A first step to clarify the situation can be done selecting a single reaction channel and measuring different spin observables.In this perspective we are planning to study with the polarised beam of the Saturne accelerator the reaction $pp \rightarrow pK^+\Lambda$.Fluxes larger than 10^{11} polarized protons/burst can currently be obtained.The beam,momentum analyzed, will hit the liquid H_2 target with a spot of about 1 cm^2. We will identify and momentum analyze the scattered proton and the proton and the π^-,decay products of the Λ (64% $\Lambda \rightarrow p\pi^-$).By retracking a complete kinematical reconstruction of the reaction can be obtained.This reconstruction does not need to be very accurate as one has to select between masses,Λ and Σ^0 for example,differing by no more than 80 MeV.The experiment will be performed at 3.7 GeV/c,the highest beam momentum, that can be obtained at Saturne.Under the assumption that the Λ polarization does not depend on $\sqrt{s}$ but only on p_T,the fact that the beam energy is lower than in the quoted experiments is not a limitation for the experiment. Values of $p_T \geq .7$ GeV/c can be obtained at $\theta_\Lambda \geq 25^0$,whereas all the higher energy experiments were performed at θ_Λ not larger than 100 mrad.Large counting rates are expected so that one π_Λ can be measured in about two hors ($\approx 10{,}000$ events for each spin state. A complete measurement of spin observables in hyperon production will require also a spin rotation device for the beam,before the target,and a polarized target.All those devices are available at Saturne.

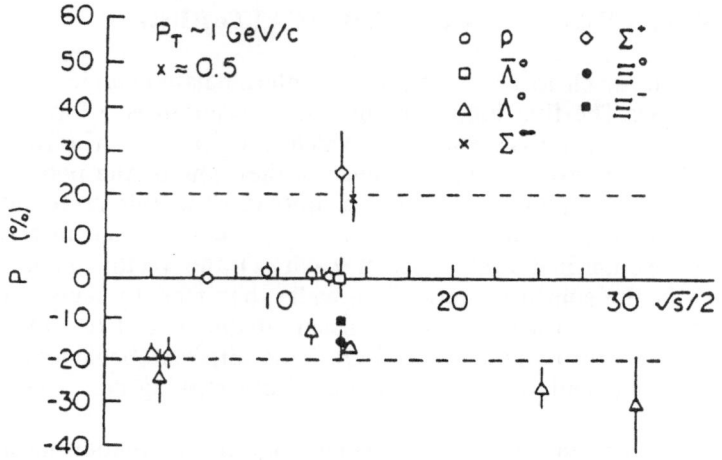

Fig. 6.   Summary of hyperon polarisations near $p_T \approx 1$ GeV/c and $x_F \approx .5$

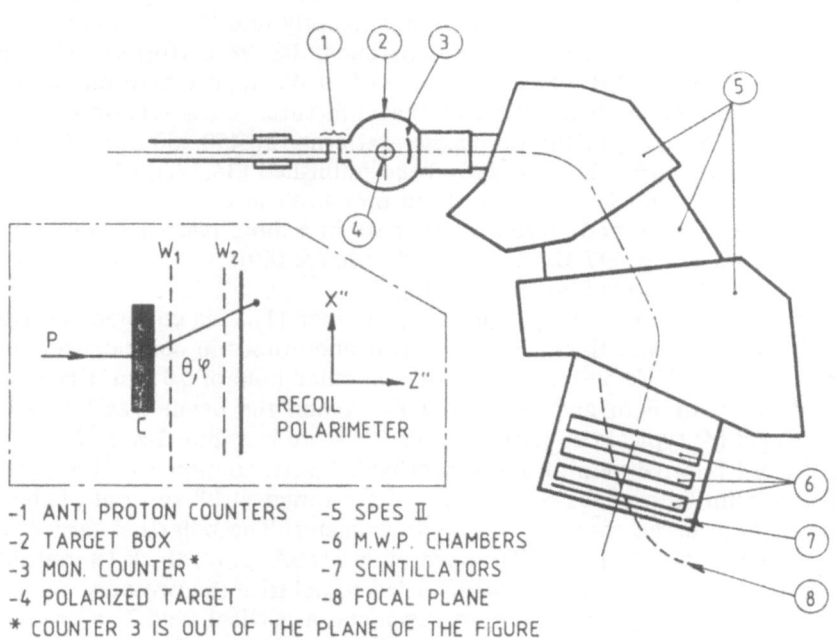

-1 ANTI PROTON COUNTERS   -5 SPES II
-2 TARGET BOX   -6 M.W.P. CHAMBERS
-3 MON. COUNTER*   -7 SCINTILLATORS
-4 POLARIZED TARGET   -8 FOCAL PLANE
* COUNTER 3 IS OUT OF THE PLANE OF THE FIGURE

Fig. 7.   Set up used the $\bar{p}N$ elastic scattering experiment (PS198)

4. SPIN OBSERVABLES IN P̄N ELASTIC SCATTERING

The theoretical approach to N̄N scattering at intermediate energies is mainly based on potential models.The first ingredient of this description is a form of theoretical NN potential, based on meson exchange, which is G-parity transformed to an N̄N potential.This G-parity transformation reverses the signs of the potential contributions of the odd G-parity meson exchanges. Important in this representation is the tendency for all mesons to yield a contribution of the same sign to certain components of the interaction.In the NN system,scalar (ε) and vector (ω)meson exchange add coherently in the spin orbit term V_{LS},while they tend to cancel in the central part V_0.In the N̄N potentials coherences occur in the central ($\omega + \varepsilon$),tensor ($\pi + \rho$) and quadratic spin orbit contributions [30]. This different behaviour leads to the strong N̄N attraction and is expected to have very striking consequences for spin observables.

The second ingredient in N̄N models is some kind of annihilation mechanism. The annihilation cross section is large ($\sigma_{an}/\sigma_{el} \geq 2$) and is responsible for the large imaginary part of the potential.Several different approaches exist for describing the annihilation: one may apply a suitable boundary condition [31],use an optical potential [32,33],do an actual coupled-channel calculation [34],or assume that one $q\bar{q}$ pair annihilating into gluons is the dominant doorway channel for the N̄N annihilation [35].All these approaches fit reasonably well the existing data on the spin-integrated cross sections.For the spin-dependent observables the situation is completly different: the predictions depend consistently on the theoretical inputs.Therefore the measurement of spin observables in the p̄N elastic scattering will provide usefull constraints to define the proper set of parameters of the N̄N potentials.This,provided that the data are obtained over the full angular range,because the different theoretical predictions differ significantly in different angular domains.

This was the main motivation of the experiment PS198 performed with the antiproton beam of the LEAR accelerator at CERN.We report here on the measurement of the full angular distribution of the differential cross section $d\sigma/d\omega$ and of the analyzing power A_{yo} in the p̄p elastic scattering at 450,550 and 702 MeV/c.At this energy data on $d\sigma/d\omega$ have already been published [36,37,38].In regard to A_{yo}, few points,with large error bars, have been measured in a double scattering experiment at 700 MeV/c [39] and,more recently,data in a more restricted angular domain , have been produced at 497,523,679 and 783 MeV/c [40].

The experimental setup is illustrated in Fig.7.

Monitoring of the beam was provided by counter (1).This counter consisted of a thin scintillator F,.3 mm. thick,viewed by two phototubes in coincidence, and of an antihalo scintillator FH, 5mm. thick,with a circular hole of 20 mm.FH was put in anticoincidence with F to give the signal F0. When the beam was focused on the polarized target (4),typical counting rates of FH were less than 2% of F.

An additional relative monitoring was provided with counter (3) [Fig.7],that consisted of two scintillator slabs put in view of the target at 0° and out of the scattering plane and of the acceptance of the spectrometer. The polarized target consisted of a slab of pentanol $\{C_5H_{11}(OH)\}$ doped with EHBA, 5mm thick,18 mm. high and 18 mm. wide. The 2.5 T magnetic field,needed to polarize the protons, was produced by a superconducting split coil magnet supplying a vertical field.The proton polarization was determined comparing the dynamic polarization signal with the natural polarization signal at thermal equilibrium.Typical polarization values $P_T \approx .80$ were obtained. To decrease the influence of the magnetic field on the trajectories of the incoming and outgoing particles we lowered the field of the target to .7 T,operating it in frozen spin mode.The proton relaxation time was about 150 hours.

The scattered particles were detected and momentum analyzed with the magnetic spectrometer SPESII [41].In order to cover the full angular domain in the C.M. system,we detected $\bar{p}$ for the forward C.M. angles and the recoil p for backward C.M. angles, rotating the spectometer and/or reversing its magnetic field to get the suitable C.M. angular set. The detection system of the spectrometer consisted [Fig.7] of four MWPC (CH0,CH1,CH2,CH3),all of them with X and Y planes, and a scintillator counter S[(7) in Fig. 7].With CH0,put at the entrance of the spectrometer,the acceptance of the spectrometer was defined,with CH1,CH2 and CH3 [(6) in Fig. 7] the angle and the momentum of the scattered particles were measured.The counters CH1,CH2,CH3 and S were put on a movable charriot,whose position was changed at each scattering angle so that the focal plane was always acrossing CH2,independently of its kinematical recoil.The trigger was F*S.A precise measurement of time of flight,made with these two counters(F and S),provided the appropriate selection of the required particle ($\bar{p}$ or p) against·the other products, mainly pions,of the interaction of the $\bar{p}$ beam with the target.With the X,Y,θ and ϕ coordinates at the focal plane and the transfer matrix of SPESII a complete retracking was performed at two levels:1) at the level of CH0,entrance of SPESII and 40 cm. downstream of the polarized target,the computed X and Y coordinates of CH0 were compared with the measured ones.From their differences the origin point of the track was determined.In that way spourious tracks coming from the antiproton counter F or from the windows or the thermal screens of the polarized target could be rejected.2) at the level of the target the complete kinematics of reaction was reconstructed,outputting the missing mass.In this way we could select the $\bar{p}p$ elastic channel from the contributions of the nuclear content of the target. Around $\theta_L \approx 45^\circ$,where the energy of the detected particle was minimal,angular and energy straggling considerably deteriorated the resolution on the missing mass.In order to investigate the possible contaminations from quasielastic scattering or other reactions on the nuclear content of the target we added a recoil counter.This one was put at $\approx 90^\circ$ with SPESII and rotated with it.It consisted of a scintillator slab and a MWPC, with X and Y planes.It covered the full acceptance of SPESII and was efficient for the detection of the recoil particle,p or $\bar{p}$, in the angular domain $70^\circ \leq \theta_{CM} \leq 110^\circ$.Through track reconstruction and coplanarity selection it was shown that,even in that angular domain,background contamination was negligible. In a preliminary run with a proton beam we checked our apparatus, measuring $d\sigma/d\omega$ and A_{yo} in pp elastic scattering.Our results were consistent with the phase shift compilations [42]. In the insert in Fig. 7 is scketched the recoil polarimeter, that was put downstream on the focal plane of the spectrometer. It consisted of a carbon scatterer,two X,Y MWPC and a large scintillator.It has been shown [43] that the carbon analyzing power for $\bar{p}$C scattering is about zero,whereas for pC scattering it is large and positive.When protons were detected we were able to measure few points of the Wolfenstein parameter D.The analysis of the data is still in progress.Preliminary results, of $d\sigma/d\omega$ and A_{yo},are available at 700 MeV/c. The $d\sigma/d\omega$ values are in good agreement with the previously published data [36,37].When the A_{yo} are compared with the theoretical predictions,over the whole angular domain, large discrepancies with almost all models.The general trend of the data is reproduced only by the Paris potential [33] and DRI model [44] predictions. A complete discussion could be made only when the final results will be available,because normalisations,like on the target polarisation, could change the data of few percent.

REFERENCES

[1]   Thirion,J.,Birien,P. and Saudinos,J. Note CEA-N-1248 (1970)

[2]   Grorud,E. et al., Nucl. Instrum. Methods **188,** 549 (1981).

[3]   Arvieux,J., AIP Conf. Proc.(ed. A.D.Krisch and A.Y.M.Lin),vol.80,p.185. American Institute of Physics, New York, 1982

[4]   Bystricky,J. et al., Lett. Nuovo Cimento **40,** 466 (1984).

[5]   Mayer,B.,et al., Nucl. Phys. **A337,** 630(1985),and ref. therein.

[6]   Bertini,R.,et al.,Phys.Lett. **162B,** 77 (1985), and ref. therein

[7]   Bertini,R.,et al., Phys. Lett. **203B,** 18 (1988).

[8]   Corcoran,M.D.,et al., Phys. Lett. **120B,** 309 (1983).

[9]   Anderson,H.L.,et al., Phys. Rev. **D3,** 1536 (1971).

[10]   Barry,G.W., Phys. Rev. **D7,** 1441 (1973).

[11]   Flaminio,V. et al.,CERN − HERA 83 − 01

[12]   Bugg,D.V.,J.Phys.G:Nucl.Phys. **10,** 717 (1984).

[13]   Biegert,E.,et al., Phys.Rev.Lett. **41,** 1098 (1978).

[14]   Auer,I.P.,et al., Phys. Rev. Lett. **48,** 1150 (1982).

[15]   Yokosawa,A., Phys. Rep.**64,** 49 (1980); A.Lin et al., Phys. Lett. **74B,** (1978) 273

[16]   Auer,I.P.,et al.,Phys. Rev. **D34,** 2581 (1986).

[17]   Lehar,F. and Perrot,F.,private communication.

[18]   Lomon,E.L.,Nucl.Phys. **A434,** 139c (1985); Gonzalez,P.,LaFrance,P and Lomon,E.L., Phys. Rev. **D35,** 2142 (1987).

[19]   Bertini,R. et al. Saturne exp. N° 130

[20]   Blankleider,B. and Afnan,I.R., Phys. Rev. **C31,** 1380 (1985).

[21]   Bunce,G. et al. ,Phys.Rev.Lett. **36,** 1113 (1976);
         Heller,K. et al.,Phys.Rev.Lett. **41,** 607 (1978) and **45,** 1043(E) (1980);
         Lomanno,F. et al.,Phys.Rev.Lett. **43,** 1905 (1979);
         Abe,F.,et al., Phys. Rev. Lett. **50,** 1102 (1983);
         Pondrom,L.G.,Phys.Rep. **122,** 58 (1985)

[22]   Gourlay,S.A.,et al.,Phys. Rev. Lett. **56,** 2244 (1986)

[23]   Kane,G.L. et al.,Phys.Rev.Lett. **41,** 1689 (1978)

[24]   Andersson,B. et al.,Phys.Lett. **85B,** 417 (1979)

[25]   De Grand,T.A. and Miettinen,H.I.,Phys.Rev. **D24,** 2419 (1981)

[26]   Szwed,J.,Phys.Lett. **105B,** 403 (1981)

[27]   Bonner,B.E.,et al., Phys.Rev.Lett. **58,** 447 (1987)

[28]   De Grand,T.A.,Markannen,J. and Miettinen,H.I, Phys.Rev. **D32,** 2445
         (1985)

[29]   Sullivan,M.W. et al.,Phys.Lett. **142B,** 451 (1984)

[30]   Buck,W.W.,Dover,C.B. and Richard,J.M.,Ann. Physics**121,** 47 (1979);
         Dover,C.B. and Richard,J.M. ibid.,**121,** 70 (1979).

[31]   Delville,A. et al.,Am.J.Phys. **46,** 907 (1978);
         Dalkarov,O.D. and Myhrer,F.,Nuovo Cimento **40A,** 152 (1977).

[32]   Bryan,R.A. and Phillips,R.J.N.,Nucl.Phys. **B5,** 201 (1968);
         Dover,C.B. and Richard,J.M.,Phys.Rev. **C21,** 1466 (1980);
         Ueda,T.,Prog.Theor.Phys. **62,** 1670 (1979).

[33]   Cote,J. et al.,Phys.Rev.Lett. **48,** 1319 (1982).

[34]   Timmers,P.H., et al.,Phys.Rev. **D29** (1984) 1928

[35]   Tegen,R. et al.,Phys.Lett. **182B,** 6 (1986).

[36]   Eisenhandler,E. et al.,Nucl.Phys. **B113,** 1 (1976).

[37]   Kageyama,T. et al.,Phys.Rev. **D35,** 2655 (1987).

[38]   Bruckner,W. et al.,Phys.Lett. **166B,** 113 (1986).

[39]   Kimura,M. et al.,Nuovo Cimento **A71,** 438 (1982).

[40]   Kunne,R.A. et al.,Phys.Lett. **B206,** 557 (1988).

[41]   Bertini,R. et al.,Rep.DPhN. **2070,** 253 (1978).

[42]   Bistricky,J.,Lechoinoine Leluc,C. and Lehar,F. J.Phys. **48,** 199 (1987).

[43]   Beard,C.I.,Physics with antiprotons at Lear in the Acol era,U.Gastaldi,
         R.Klapisch,J.M.Richard and J.Tran Thanh Van eds.,Frontieres (1985) p.
         239

[44]   Dover,C.B. and Richard,J.M., Phys. Rev. **C25,** 1952 (1982).

MEASUREMENT OF 2-nd AND 3-rd ORDER SPIN OBSERVABLES IN ELASTIC SCATTERING OF 1.6 GeV TENSOR AND VECTOR POLARIZED DEUTRON BEAM FROM A POLARIZED HYDROGEN TARGET

V. Ghazikhanian[1,5], B. Aas[1,5], D. Adams, E. Bleszynski[1,6]
M. Bleszynski, J. Bystricky[1,2], G.J. Igo[1], F. Sperisen[1,7]
C.A. Whitten[1], P. Chaumette[2], J. Deregel[2], J. Fabre[2]
F. Lehar[2], A. de Lesquen[2], L. van Rossum[2], J. Arvieux[3]
J. Ball[3], A. Boudard[3,4], and F. Perrot[4]

ABSTRACT

Seven previously unmeasured spin asymmetries and ten previously unmeasured spin transfer observable have been determined in the momentum transfer range, $0.12 < -t < 0.85 \ (\text{GeV}/c)^2$. We used a 1600 MeV polarized deuteron beam incident on a polarized hydrogen target in these measurements. This is equivalent to using an 800 MeV proton beam on a deuteron target. These observables are poorly predicted by the relativistic impulse approximation, which has been used successfully to reproduce spin observables for protons on spin-zero nuclei.

I. Introduction

Due to the phenomenal success of the Relativistic Impulse Approximation (RIA) in reproducing the spin observables in elastic proton scattering from medium and heavy weight nuclei, it has become of interest to test the RIA in other situations. It is well known that elastic proton-nucleus (spin zero) scattering, being a spin 0-$\frac{1}{2}$ system, is only sensitive to two of the five relativistically invariant amplitudes of the N-N interaction, i.e., the spin- independent and pseudoscalar (alternatively, pseudovector) amplitudes. The remaining amplitudes, of tensor character, will enter when the RIA is used to predict proton- deuteron elastic scattering, a spin 1-$\frac{1}{2}$ system. Thus it is possible, in proton-deuteron elastic scattering, to test the RIA in such a way that all five invariant NN amplitudes enter sensitively.

The measurements reported here were made in two 10-day running periods in 1985 and 1986 at the Saturne II accelerator at Saclay.

---

(1) Physics Department, UCLA, Los Angeles, CA 90024
(2) DPhPE, CEN-Saclay, France
(3) LNS, CEN-Saclay, France
(4) DPhN/ME, CEN-Saclay, France
(5) Present Address: Rogland Research Institute, Ullandhung, N4001 Stavanger, Norway
(6) Present Address: Rockwell International Corporation, Los Angeles, CA
(7) Present Address: IUCF, Indiana University, Bloomington, IN 47405

II. Formalism

Spins are expressed in two right-handed coordinate systems $(\hat{S}, \hat{N}, \hat{L})$, one for the incident deuteron and target proton and another for the recoil proton. The polarization of the scattered deuteron was not measured. The longitudinal vector $\hat{L}$ is parallel to $\vec{k}_i$ and the normal vector $\hat{N}$ is perpendicular to the scattering plane along $\vec{k}_i \times \vec{k}_f$. The sideways direction is defined by $\hat{S} = \hat{N} \times \hat{L}$. Note that $\hat{S}$ and $\hat{L}$ change sign under a parity transformation while $\hat{N}$ is unaffected.

The observables are defined in terms of the scattering matrix F:

$$C(a, \alpha; b, \beta) = \frac{\text{Tr}(F\sigma_a J_\alpha F^\dagger \sigma_b J_\beta)}{\text{Tr}(FF^\dagger)} \quad \text{(Vector type)} \tag{1}$$

and

$$C(a, \alpha\eta; b, \beta) = \frac{\text{Tr}(F\sigma_a J_{\alpha\eta} F^\dagger \sigma_b J_\beta)}{\text{Tr}(FF^\dagger)} \quad \text{(Tensor type)} \tag{2}$$

In Eqs. (1) and (2), $\sigma_i(i = L, N, S)$, $J_k(k = L, N, S)$, and $J_{kj}(k, j = L, N, S)$ are the Pauli spin operators and vector and tensor spin operators for the deuteron respectively. The quantities we measure are the relative yield $Y \propto \text{Tr}(\rho_{out})$ and components of polarization of the recoil proton,

$$p_i' = \frac{\text{Tr}(\rho_{out}\sigma_i')}{\text{Tr}(\rho_{out})} \tag{3}$$

where the prime refers to the fact that the polarization is measured in the coordinate frame of the recoil proton. Here ρ_{out} can, as usual, be expressed in terms of the initial density matrix ρ_{in} for the p-d system as $\rho_{out} = F\rho_{in}F^\dagger$. Because of parity restrictions mentioned above, only those coefficients $C(a, \alpha; b, 0)$ and $C(a, \alpha\eta; b, 0)$ involving *even* numbers of spin components, which are either L or S, i.e., a, α, b, in the first and $a, \alpha\eta, b$ in the second, will be non-zero. This is illustrated in Fig. 1 where $C(0, S, 0, 0)$ which can be extracted from the data, is plotted. In this case the sum is equal to unity and the data is consistent with a null value for this observable.

Figure 2 is a schematic design of the N-N area at Saturne where the experiment was done. In the beam line, which is not shown, the asymmetry for quasi-elastic p-p scattering in the collision between the deuterons and a CH_2 target was measured in a polarimeter to determine the vector polarization of the beam. The tensor (and vector) polarization of the polarized beam was measured with another polarimeter before acceleration in Saturne II. In Fig. 2, the position of the frozen spin target is indicated. The polarized material was Pentanol and the typical polarization was 80-95%.

On the exit side of the cryostat a CH_2 target (unpolarized) was attached. Figure 3 shows vertex reconstruction of elastic deuteron events. Events from the polarized target and the CH_2 target are readily distinguishable. Elastic events originating in the CH_2 target were used in the analysis to monitor the relative tensor polarization of the beam.

The four momentum transfer interval of interest $0.12 < -t < 0.85 \ (GeV/c)^2$ could be covered in one angular setting. Elastically scattered deuterons, in the corresponding angular interval, $8.5°$ to $19.5°$, could be readily distinguished from the large breakup cross section in the off-line analysis by the imposition of a cut on the bend angle in the magnet (Gouppillon). Breakup events result in forward- going protons with velocities roughly equal to those of the deuteron and thus bend approximately twice as much. This is illustrated in Fig. 4 where the sharp group (deuterons) is readily distinguishable from the broad group (protons). The N and S components of the recoil proton's spin are determined by measuring the asymmetry produced from scattering in the carbon target shown in Fig. 2. Note that the carbon target was actually wedge-shaped to account for the broad range of energy of the recoil protons corresponding to a range of $-t$ extending from 0.2 to 0.85 $(\text{GeV}/\text{c})^2$.

Since the p-d cross section is more than one order of magnitude smaller than the p-p cross section at 800 MeV, it was necessary to provide a tighter trigger than was necessary in the pp measurements where the OR of the hodoscopes WH and H3 is used to initialize a trigger to fire the MWPC detectors. This was simply done by correlating each of the WH hodoscope elements with two or three of the H3 hodoscope elements which will be struck in a p-d elastic event when account is taken of the kinematics and of the geometry involved. The vector asymmetries C(0,N,0,0), C(N,N,0,0) and C(L,S,0,0) have been measured. The tensor asymmetries C(0,NN,0,0), C(N,NN,0,0), C(0,SS,0,0), C(N,SS,0,0), C(0,SL,0,0), C(N,SL,0,0) and C(L,NS,0,0) have also been measured. Of these, C(L,S,0,0), C(N,N,0,0), C(N,NN,0,0), C(N,SS,0,0), C(0,SL,0,0), C(N,SL,0,0) and C(L,NS,0,0) have not been measured previously in the momentum transfer range of this experiment. As for the triple-spin or spin transfer observables C(0,N,N,0), C(N,N,N,0), C(0,S,S,0), C(N,S,S,0), C(0,NN,N,0), C(0,SS,N,0), C(0,NS,S,0), C(N,NN,N,0), C(N,SS,N,0) and C(N,NS,S,0) have been measured. None of the above spin-transfer observables have been measured in this momentum transfer range before. A publication is being prepared of this work.

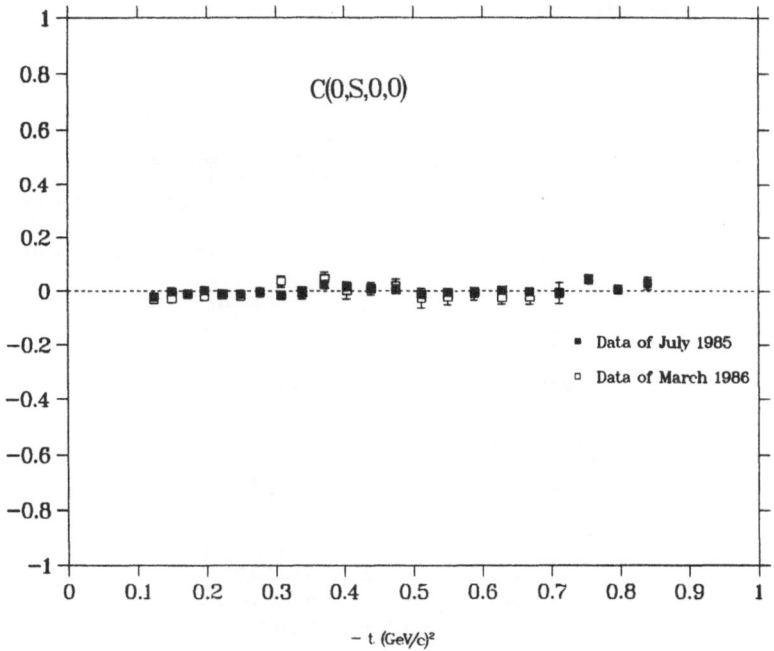

Fig. 1 The null observable C(0,S,0,0).

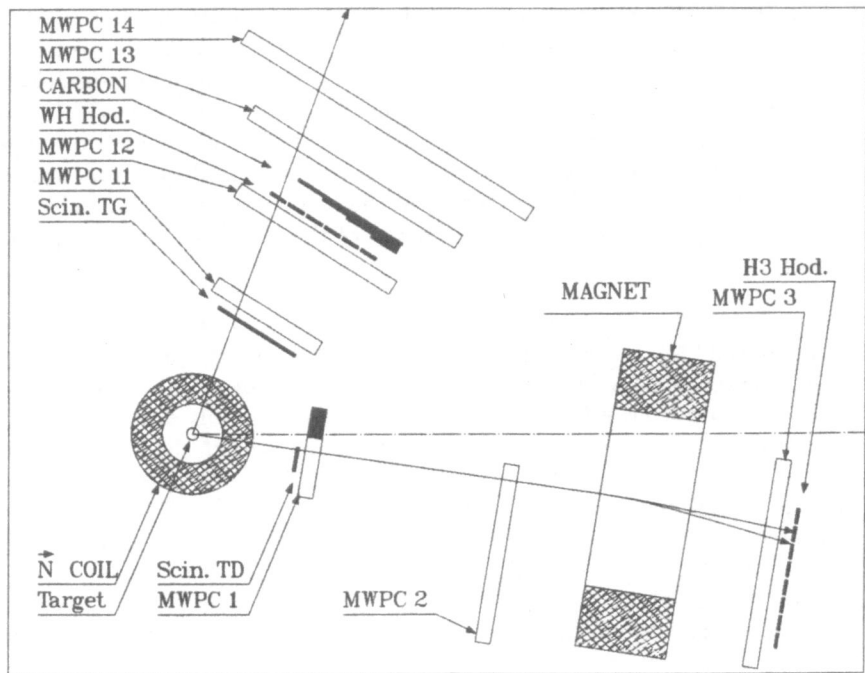

Fig. 2 Schematic diagram of the N-N experimental area at Saturne II. The locations of the spin frozen target, the bending magnet Goupillon, the plastic scintillator hodoscopes and the wedge shape polarimeter target are shown.

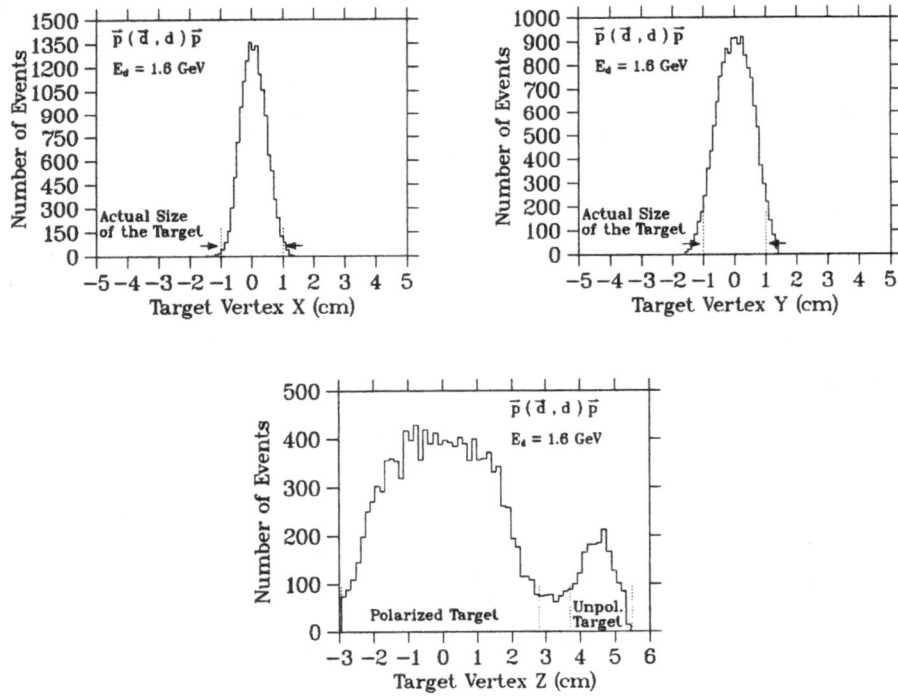

Fig. 3 Vertex reconstruction of the target region.

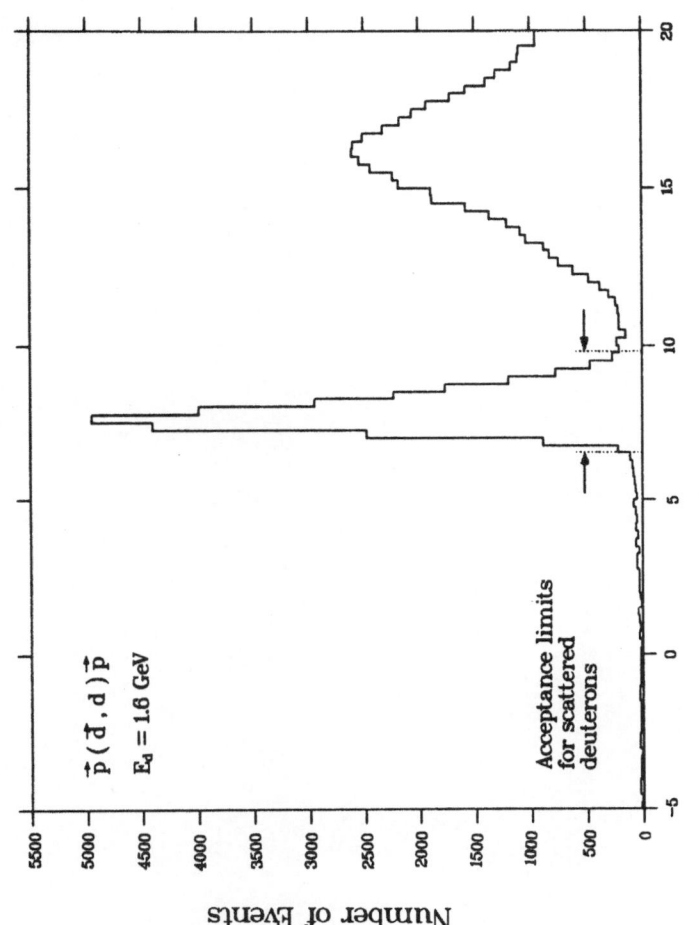

Fig. 4 Horizontal deflection of the forward scattered particles in the magnet Goupillon.

397

NUCLEAR ISOVECTOR ELECTROWEAK INTERACTIONS
IN A RELATIVISTIC MODEL[†]

George E. Walker

Nuclear Theory Center and Physics Department
Indiana University
Bloomington, Indiana   47405

Michael W. Price

Nuclear Theory Center and Physics Department
Indiana University
Bloomington, Indiana   47405
        and
Center for Naval Analyses[‡]
4401 Ford Avenue
Alexandria, Virginia   22302

ABSTRACT

The results of a study of isovector electroweak interactions in a relativistic model are discussed. Adopting the random-phase approximation (RPA) in a relativistic linear response theory, Dirac-Hartree orbitals, residual interactions taken from Quantum Hadrodynamics and the full relativistic form for electroweak transition operators , we have studied electron scattering, muon capture, and beta-decay on the mass 16 system. The results are compared with previous nonrelativistic results. The most dramatic differences are found for $0^+ \leftrightarrow 0^-$ low-lying partial transitions associated with muon-capture and beta-decay. These results are enhanced or reduced compared to a non-relativistic calculation depending on whether pseudo-vector or pseudo-scalar coupling for the πNN vertex is utilized in the residual interaction and in the determination of the weak axial current.

1. INTRODUCTION

Several of the talks at this conference have concentrated on theoretical studies of the application of relativistic field theories to nuclear physics.[1] Recently the study of relativistic quantum field theories and, relatedly, the application of Dirac phenomenology has been an area of intense investigation in nuclear physics. The purpose of the research discussed in this presentation was to study the *isovector* nuclear linear response in a relativistic model.[2] In the study we were particularly interested in the sensitivity of the results to the size of the basis adopted, single-particle (-hole)

[†]Work supported in part by the National Science Foundation.
[‡]Present address.

energies utilized, the relativistic residual interaction used, and the assumed form for the weak axial vector current.

Historically the nucleus has been most often treated nonrelativistically because of the small size of a) the nuclear potential and b) the Fermi-energy compared to the nucleon mass. More recently there has been considerable interest in treating the nucleon as a Dirac particle in the nuclear many-body environment. In such studies the (relatively) shallow nuclear Dirac single-particle potential results mainly from a combination of a large attractive potential ($\simeq 400$ MeV deep) associated with scalar-isoscalar meson exchange and a large repulsive potential ($\simeq 350$ MeV) resulting from exchange of a vector meson. [3,4]

The relativistic description of nuclei yields a natural explanation of the large single-particle nuclear spin-orbit potential and apparently provides superior fits to spin-observables in nucleon-nucleus scattering near 500 MeV. It is encouraging that the input parameters required for Dirac phenomenological fits to nucleon-nucleus elastic scattering are similar to those obtained from using mean field theory in the Hartree approximation and fitting the bulk properties of nuclei. [3-5]

The part of the Dirac potential $U_s(r)$ arising from scalar-isoscalar meson exchange enters in the Dirac equation in such a way that it is natural to combine it with the nuclear mass and define an effective mass term, $M^*(r)$ where

$$M^*(r) = M + U_s(r) \ . \tag{1}$$

Since $U_s(r)$ is deep and attractive the effective mass in the interior of nuclei can be ~60% of the free nucleon mass. The effective mass decrease caused by the deep attractive scalar potential has the effect of enhancing Dirac spinor lower components. One way, perhaps, to observe this enhancement effect and thus, indirectly, obtain evidence for the fertility of the relativistic approach, is to study nuclear transitions induced by operators that couple upper and lower Dirac spinor components. Electroweak nuclear reactions are well-motivated for studying nuclear models because such reactions can usually be treated in lowest-order and the basic electroweak theory is relatively well understood. For electroweak interactions the operators that connect upper and lower components are associated with the spatial components of the vector current and (in its original form) with the time component of the axial current. The trivial observation that one might expect transitions associated with these current components to be enhanced must be tempered by other considerations. For example, some nuclear transition densities are surface-peaked where $M^* \approx M$ so that the enhancement effect is significantly reduced. Moreover, excited states of nuclei are most often not pure particle-hole (p-h) states but at least must be described as linear combinations of such states. The admixture coefficients associated with the configuration mixed p-h states may be different than those obtained nonrelativistically and one expects such considerations may change the predicted transition rates considerably from those obtained from more naive considerations. We stress the importance of using a *consistent* model for obtaining single-particle orbitals, p-h admixture coefficients, and the form of the transition operator in the many-body environment.

In the next section we outline the techniques used to obtain the results presented in Section 3. In Section 4 we summarize the results and make a few concluding remarks regarding future research in this area.

2. THE ELECTROWEAK NUCLEAR LINEAR RESPONSE IN A RELATIVISTIC MODEL-CALCULATIONAL PROCEDURE

We wish to study the nuclear linear response to an electroweak perturbation using the random-phase approximation (RPA). One can obtain the appropriate RPA equations covariantly by using Green function methods (or relatedly, the methods of quantum field theory). If meson propagation effects are included, the density correlation function, which is of central importance in the theory, has a nontrivial frequency dependence.[6] In what follows we assume that the energies, relative to the ground state energy, of the excited states under investigation are much smaller than the masses of the exchanged mesons and that it is therefore a reasonable approximation to suppress retardation corrections (i.e. time delay associated with meson-propagation is ignored).

We also drop terms associated with nucleon-antinucleon pair excitations. Thus we have not included some potentially important backflow[7] (vacuum polarization) corrections to these *isovector* excitations. Inclusion of these terms is a potentially important and interesting area of future investigation. Adopting this set of approximations results in the usual RPA equations for p-h admixture amplitudes but with the change that one keeps the full Dirac four-component nature of the single-particle orbitals and p-h interaction.

We use standard angular momentum and isospin coupling and label the nth excited state of the nucleus as Ψ^n_{JT}, where $J(T)$ denotes the total angular momentum (isospin) of the nuclear excited state. In the RPA approximation, $|\Psi^n_{JT}>$ is obtained from the exact ground state $|\Psi^0>$ by creating or destroying a p-h pair. Thus the excited state can be expanded

$$|\Psi^n_{JT}> = \sum_{ab} \psi^{n^*} \diamond_{JT}(ab)\hat{\varsigma}^+(abJT)|\Psi^0> + \sum_{ab} \phi^n_{JT}(ab)\hat{\varsigma}(abJT)|\Psi^0> \qquad (2)$$

where $\hat{\varsigma}^+(\hat{\varsigma})$ is a particle[a]-hole [b] pair creation (destruction) operator appropriate for a pair having total angular momentum J and isospin T. The RPA admixture amplitudes, ψ^n_{JT} and ϕ^n_{JT}, can be determined by solving the set of coupled algebraic equations

$$\{[E_0 + (\epsilon_a - \epsilon_b)] - E_n\}\psi^n_{JT}(ab) + \sum_{\ell m}[v^{JT}_{ab,\ell m}\psi^n_{JT}(\ell m) + u^{JT}_{ab,\ell m}\phi^n_{JT}(\ell m)] = 0 \qquad (3a)$$

and

$$\{[E_0 - (\epsilon_a - \epsilon_b)] - E_n\}\phi^n_{JT}(ab) - \sum_{\ell m}[v^{JT^*}_{ab,\ell m}\phi^n_{JT}(\ell m) + u^{JT^*}_{ab,\ell m}\psi^n_{JT}(\ell m)] = 0. \qquad (3b)$$

In Eqs. (3a, b) above the (ϵ_a, ϵ_b) refer to single-particle energies and

$$v^{JT}_{ab,\ell m} = -\sum_{J'T'}(2J'+1)(2T'+1)\begin{Bmatrix} j_m & j_a & J' \\ j_b & j_\ell & J \end{Bmatrix}\begin{Bmatrix} \frac{1}{2} & \frac{1}{2} & T' \\ \frac{1}{2} & \frac{1}{2} & T \end{Bmatrix}$$
$$\times [<\ell b J'T'|V|amJ'T'> -(-1)^{1+T'+j_a+j_m+J'} <\ell b J'T'|V|maJ'T'>] \qquad (4a)$$

where

$$u^{JT}_{ab,\ell m} = (-1)^{\frac{1}{2}-\frac{1}{2}-T}(-1)^{j_m-j_\ell-J}v^{JT}_{ab,m\ell}. \qquad (4b)$$

The single-particle wave functions used in evaluating the matrix elements of the effective N-N interaction, V, are Dirac Hartree orbitals associated with Hartree single-particle energy eigenvalues, ϵ. The effective N-N interaction V is written in the general Dirac form

$$V(1,2) = A_S + A_V \gamma^\mu(1)\gamma_\mu(2) + A_{PS}\gamma^5(1)\gamma_5(2)$$
$$+ A_{AV}\gamma^5(1)\gamma^\mu(1)\gamma_5(2)\gamma_\mu(2) + A_T\sigma^{\mu\nu}(1)\sigma_{\mu\nu}(2) \qquad (5)$$

401

where the A's are functions of Lorentz invariants and isospin. The Hartree orbitals, single-particle energies, and the effective interaction V(1,2) comprise the required input for solving Eqs. (3a, b) and obtaining the RPA p-h admixture amplitudes (ψ and ϕ).

The transition amplitude for a one-body operator, $\hat{T}$, defined in the nuclear many-body space, can be determined, given the RPA admixture amplitudes, via[6]

$$< \Psi^n_{JT}\|\hat{T}_{JT}\|\Psi^0 > = \sum_{ab}\{< a\|T_{JT}\|b > \psi^n_{JT}(ab)$$
$$+ (-1)^{\frac{1}{2}-\frac{1}{2}-T}(-1)^{j_b-j_a-J} < b\|T_{JT}\|a > \phi^n_{JT}(ab)\} \tag{6}$$

where the matrix elements are doubly reduced in angular momentum and isospin space.

If one adopts the Tamm-Dancoff approximation (TDA) then the set of equations (3a, b) above is reduced to Eq. (3a) and only the first term in Eq. (6) contributes. (The ground state is assumed to be a closed shell in the TDA hence the ϕ coefficients vanish.)

Note that the single-nucleon orbitals (a,b) in Eq. (6) are of the Dirac type so that the transition operator in Eq. (6) is kept in its full relativistic form. The transition operators appropriate for electron scattering, beta-decay, and muon-capture applications are discussed below.

The RPA Eqs. (3a,b) are an infinite set of linear algebraic equations and must be reduced to a finite set of coupled equations. We adopt the standard approximation for the low-lying excited states studied here by restricting the sum over p-h pairs, ℓm, in Eqs. (3a,b) to particle states, ℓ, in the low-lying unfilled valence shell(s) and hole states m, in the highest-filled orbitals for the closed-shell ^{16}O nucleus.

In this investigation the single-nucleon Dirac-Hartree wave functions of Horowitz and Serot[8] are adopted for the pure p-h basis states. These single-nucleon orbitals are obtained using a Hartree approximation to obtain, self-consistently, the nucleon orbitals and the σ and ω meson mean fields. The procedure used is to assume a Lagrangian containing coupled meson and nucleon fields. The meson fields are treated in the mean field approximation. Some of the Hartree orbitals are associated with unbound nucleons. A continuum orbital such as the d3/2 wave function in ^{16}O can be handled by discretizing the continuum. We have adopted the method followed by Furnstahl who has also systematically studied collective excitations arising from various p-h bases incorporating continuum Hartree wave functions.[9] Our bases A and C (see below) will correspond to Bases A and C used by Furnstahl.

In what follows, bases A and B will consist of hole states taken from the 1p3/2 and 1p1/2 orbitals and particle states from the (bound) 1d5/2 and 2s1/2 valence orbitals as well as the lowest 1d3/2 orbital in the discretized Hartree continuum. This corresponds closely to the usual "$1 - \hbar\omega$" truncated basis often adopted in previous nonrelativistic calculations. For basis A we use the calculated Hartree single-particle binding energies and, for comparison, in basis B we adopt binding energies obtained from neighboring nuclei with one more or one less neutron.

In basis C, we include the particle and hole states in basis A plus additional particle states from the d3/2, d5/2 and s1/2 orbitals in the discretized continuum that result in unperturbed p-h configuration energies below 50 MeV. For this basis the calculated Hartree single-particle energies are used to obtain the unperturbed p-h configuration energies. A summary of the orbitals included and the associated single-particle energies for each basis is given in Table 1.

Table 1

Table taken from Ref. 2

BASIS A		BASIS B		BASIS C	
Orbital[a] S.P.	Energy[a]	Orbital[a] S.P.	Energy[b]	Orbital[a] S.P.	Energy[a]
PARTICLES:					
				4d3/2	28.56
				5s1/2	26.95
				4d5/2	25.93
				3d3/2	15.93
				4s1/2	14.01
				3d5/2	13.89
				2d3/2	6.71
				2d5/2	5.29
				3s1/2	4.62
1d3/2	2.63	1d3/2	0.98	1d3/2	2.63
2s1/2	-1.21	2s1/2	-3.28	2s1/2	-1.21
1d5/2	-3.38	1d5/2	-4.14	1d5/2	-3.38
HOLES:					
1p1/2	-12.49	1p1/2	-15.67	1p1/2	-12.49
1p3/2	-20.77	1p3/2	-21.83	1p3/2	-20.77

[a]See ref. 8.
[b]See ref. 11.

Yukawa-Dirac Particle-Hole Interactions

Previous investigators have denoted a specialized form of the Lagrangian that contains only σ and ω meson couplings as QHD-I.[3] An N-N interaction naturally resulting from the QHD-I Lagrangian contains terms arising only from the exchange of σ and ω mesons. Thus we have utilized an isospin-independent interaction containing Lorentz scalar and vector terms of the form

$$V(1,2) = -\frac{g_\sigma^2}{4\pi}\frac{e^{-m_\sigma r_{12}}}{r_{12}} + \frac{g_\omega^2}{4\pi}\frac{e^{-m_\omega r_{12}}}{r_{12}}\gamma^\mu(1)\gamma_\mu(2) \tag{7}$$

The couplings and masses adopted are listed in refs. 2 and 3.

If one includes πNN and ρNN couplings in the meson-nucleon Lagrangian, the model previously denoted QHD-II results.[3] Since one can couple the πNN vertices with either pseudo-scalar (γ_5) or pseudo-vector ($\not{q}\,\gamma_5$) coupling, which are inequivalent in the many-body environment, we shall consider each coupling separately.

i) Pseudo-scalar coupling of pions to nucleons—QHD-II (PS). For the case of pseudo-scalar coupling the form of the N-N interaction is given by

$$\begin{aligned}V(1,2) &= -\frac{g_\sigma^2}{4\pi}\frac{e^{-m_\sigma r_{12}}}{r_{12}} + \frac{g_\omega^2}{4\pi}\frac{e^{-m_\omega r_{12}}}{r_{12}}\gamma^\mu(1)\gamma_m u(2) \\ &+ \frac{g_\pi^2}{4\pi}\frac{e^{-m_\pi r_{12}}}{r_{12}}\gamma^5(1)\gamma_5(2)\tau_1\cdot\tau_2 + \frac{g_\rho^2}{4\pi}\frac{e^{-m_\rho r_{12}}}{r_{12}}\gamma^\mu(1)\gamma_\mu(2)\frac{\tau_1\cdot\tau_2}{4}\end{aligned} \tag{8}$$

where τ is the usual nucleon Pauli isospin operator.

ii) Pseudo-vector coupling of pions to nucleons—QHD-II (PV).

Starting from a $\not q \, \gamma_5$ coupling one can use the generalized Dirac equation (now containing the effective mass, $M^*(r)$, see Eq. (1), to eliminate derivatives in favor of the γ_5 coupling renormalized by the factor $M^*(r_1)M^*(r_2)/M^2$.) Thus we obtain, assuming Hartree orbitals,

$$
\begin{aligned}
V(1,2) = &-\frac{g_\sigma^2}{4\pi}\frac{e^{-m_\sigma r_{12}}}{r_{12}} + \frac{g_\omega}{4\pi}\frac{e^{-m_\omega r_{12}}}{r_{12}}\gamma^\mu(1)\gamma_\mu(2) \\
&+\frac{g_\pi^2}{4\pi}\frac{M^*(r_1)}{M}\frac{M^*(r_2)}{M}\frac{e^{-m_\pi r_{12}}}{r_{12}}\gamma^5(1)\gamma_5(2)\tau_1\cdot\tau_2 \\
&+\frac{g_\rho^2}{4\pi}\frac{e^{-m_\rho r_{12}}}{r_{12}}\gamma^\mu(1)\gamma_\mu(2)\frac{\tau_1\cdot\tau_2}{4} \quad .
\end{aligned}
\tag{9}
$$

The couplings and masses adopted for the QHD-II interactions are given in refs. 2 and 3.

Formulae for Evaluating Electroweak Nuclear Reactions

We shall present results for the nuclear linear response to electron scattering, muon-capture, and beta-decay using the wave functions obtained using the procedures discussed above. Briefly summarized below are formulae used for evaluating these reactions.

Inelastic Electron Scattering

For the process of inelastic electron scattering the nuclear ground state is linked to the nuclear excited states considered in the p-h model by the one-body electromagnetic operator, $\hat{J}^\mu(x)$,

$$
\hat{J}^\mu(x) = \bar{\hat{\psi}}(x)\gamma^\mu Q\hat{\psi}(x) + \partial_\nu[\bar{\hat{\psi}}(x)\frac{\lambda}{2M}\sigma^{\mu\nu}\hat{\psi}(x)] \, ,
\tag{10}
$$

where the $\hat{\psi}(x)$ are baryon Heisenberg field operators. The nuclear current density includes contributions from the anomalous magnetic moment, λ,

$$
\lambda = \lambda_p\frac{1}{2}(1+\tau_3) + \lambda_n\frac{1}{2}(1-\tau_3) \, ,
\tag{11}
$$

where $\lambda_p(\lambda_n)$ is the proton (neutron) magnetic moment and M is the nucleon mass. The charge operator, Q, is given by

$$
Q = \frac{1}{2}(1+\tau_3) \, .
\tag{12}
$$

If we denote the four-momentum [three-momentum] transferred from the nucleus as q_μ [q], the differential cross-section for inelastic electron scattering in the plane-wave Born approximation may be written[10]

$$
\frac{d\sigma}{d\Omega} = 4\pi\sigma_M[1 + (2\epsilon\sin^2(\theta/2)/M_{\text{target}})]^{-1}F^2 \, ,
$$

where σ_M is the point nucleon Mott cross section and

$$F^2 = (\frac{q_\mu^2}{q^2})^2 F_L^2 + \left[\frac{1}{2}\left(-\frac{q_\mu^2}{q^2}\right) + \tan^2(\theta/2)\right] F_T^2$$

$$F_L^2 = (2J_i + 1)^{-1} \sum_{J=0}^{\infty} |<J_f| |\hat{M}_J(q)| |J_i>|^2$$

$$F_T^2 = (2J_i + 1)^{-1} \sum_{J=1}^{\infty} [|<J_f| |\hat{T}_J^{el}(q)| |J_i>|^2 + |<J_f| |\hat{T}_J^{mag}(q)| |J_i>|^2] .$$

$$(13)$$

In Eq. (13), ϵ is the incident energy of the electron, θ is the scattering angle and M_{target} is the mass of the target. Single-nucleon form-factors and center-of-mass correction factors, defined in refs. 10 and 11, are included in the calculations.

Using the transition operators defined above, squared form-factors have been calculated for inelastic electron scattering leading to $T = 1$ excited states in ^{16}O. In the results section we actually plot

$$\tilde{F}^2(q) = \frac{F^2(q)}{[1 + \tan^2(\theta/2)]} ,$$

$$(14)$$

where $F^2(q)$ is defined in Eq. (13).

Muon-Capture

We will exhibit in the next section muon-capture rates resulting from the capture of a 1s Bohr orbital muon by an ^{16}O nucleus. For this process the leptonic weak current is coupled to the weak nuclear current, $\hat{J}_\mu^{(-)}(x)$:

$$\hat{J}_\mu^{(-)}(x) = \hat{J}_\mu^{v(-)}(x) + \hat{J}_{\mu 5}^{(-)}(x) .$$

$$(15)$$

In Eq. (15), the first term, $\hat{J}_\mu^{v(-)}(x)$, is the charge-changing weak vector current,

$$\hat{J}_\mu^{v(-)}(x) = \bar{\hat{\psi}}(x)\gamma_\mu\tau_-\hat{\psi}(x) + \partial^\nu[\bar{\hat{\psi}}(x)\frac{\lambda}{2M}\sigma_{\mu\nu}\hat{\psi}(x)] ,$$

$$(16)$$

where $\lambda \equiv (\lambda_p - \lambda_n)\tau_-$, and the second term is the weak axial-vector current, $J_{\mu 5}^{(-)}(x)$.

There is an induced pionic contribution to the axial-vector current. Since the nucleon can be coupled to the pion with either pseudo-scalar (γ_5) or pseudo-vector $\not{q} \gamma_5$ coupling, we employ currents based on both methods of coupling. For the case of pseudo-scalar coupling, we employ the following form for the weak axial current

$$\hat{J}_{\mu 5}^{(\pm)}(x) = \bar{\hat{\psi}}(x)\gamma_\mu\gamma_5 w^{(\pm)}\hat{\psi}(x)$$
$$+ \frac{2M^*(x)}{m_\pi^2 - q_\mu^2}q_\mu\bar{\hat{\psi}}(x)\gamma_5 w^{(\pm)}\hat{\psi}(x)$$

$$(17)$$

where

$$w^{(\pm)} = -1.23\tau_\pm .$$

For the case of pseudo-vector coupling we adopt

$$\hat{J}_{\mu 5}^{(\pm)}(x) = \bar{\hat{\psi}}(x)\gamma_\mu\gamma_5 w^{(\pm)}\hat{\psi}(x)$$
$$+ \frac{q_\mu}{m_\pi^2 - q_\mu^2}\bar{\hat{\psi}}(x)\not{q}\gamma_5 w^{(\pm)}\hat{\psi}(x) .$$

$$(18)$$

405

The muon-capture rate is given by the formula[12]

$$
\omega_{fi} = \frac{G^2\nu^2}{2\pi}|\phi_{1s}|^2_{av}\frac{4\pi}{2J_i+1}\Big\{ \sum_{J=0}^{\infty}|<J_f|\,|\hat{\mathcal{M}}_J(\nu)-\hat{\mathcal{L}}_J(\nu)|\,|J_i>|^2
$$
$$
+ \sum_{J\geq 1}|<J_f|\,|\hat{\mathcal{T}}_J^{mag}(\nu)-\hat{\mathcal{T}}_J^{el}(\nu)|\,|J_i>|^2\,\Big\}\,.
$$

(19)

In Eq. (19) G is the Fermi coupling constant, ν is the momentum of the outgoing neutrino and $|\phi_{1s}|^2_{av}$ is the square-modulus of the muon wave function averaged over the nucleus. The multipole operators in Eq. (19) are defined in terms of the current in Eq. (15) as follows:

$$
\hat{\mathcal{M}}_{JM}(q) \equiv \hat{M}_{JM} + \hat{M}_{JM}^5 = \int d\vec{x}[j_J(qx)Y_{JM}(\Omega_x)]\hat{J}^0(x)
$$

(20)

$$
\hat{\mathcal{L}}_{JM}(q) \equiv \hat{L}_{JM} + \hat{L}_{JM}^5 = \frac{i}{q}\int d\vec{x}[\vec{\nabla}(j_J(qx)Y_{JM}(\Omega_x))]\cdot\hat{\vec{J}}(x)
$$

(21)

$$
\hat{\mathcal{T}}_{JM}^{el}(q) \equiv \hat{T}_{JM}^{el}(q) + \hat{T}_{JM}^{el5}(q) = \frac{1}{q}\int d\vec{x}[\vec{\nabla}\times j_J(qx)\vec{Y}_{JJ1}^M(\Omega_x)]\cdot\hat{\vec{J}}(x)
$$

(22)

$$
\hat{\mathcal{T}}_{JM}^{mag}(q) \equiv \hat{T}_{JM}^{mag}(q) + +\hat{T}_{JM}^{mag5}(q) = \int d\vec{x}j_J(qx)\vec{Y}_{JJ1}^M\cdot\hat{\vec{J}}(x)\,.
$$

(23)

where Y_{JM} and $\vec{Y}_{JJ1}$ are the usual spherical and vector spherical harmonics, respectively. The parity of the multipole operators defined above is given by

$$
(-1)^J\ ;\ \hat{M}_{JM},\hat{L}_{JM},\hat{T}_{JM}^{el},\hat{T}_{JM}^{mag5}
$$
$$
(-1)^{J+1}\ ;\ \hat{T}_{JM}^{mag},\hat{M}_{JM}^5,\hat{L}_{JM}^5,\hat{T}_{JM}^{el5}\,.
$$

(24)

Using (24) we note that for $0^+ \leftrightarrow 0^-$ transitions only M_0^5 and L_0^5 contribute.

Beta-Decay

In the next section we present predictions for the beta-decay of the first excited 0^- state in ^{16}N to the 0^+ g.s. of ^{16}O. This state has an excitation energy of $E_x = .120$ MeV relative to the ^{16}N g.s. Defining k_μ as the electron four-momentum and $\vec{\nu}$ as the three-momentum of the neutrino, the three-momentum transfer $\vec{q}$, from the nucleus may be written as

$$
\vec{q} = \vec{k} + \vec{\nu}\,.
$$

(25)

We denote the electron energy by ϵ and define β as

$$
\vec{\beta} \equiv \frac{\vec{k}}{\epsilon}\,.
$$

(26)

The rate for $0^- \to 0^+$ beta-decay transition discussed in the next section may be written[12]

$$
d\omega = \frac{G^2}{2\pi^3}k\epsilon(W_0-\epsilon)^2d\epsilon\frac{d\Omega_k}{4\pi}\frac{d\Omega_\nu}{4\pi}
$$
$$
\times 4\pi\{(1+\hat{\nu}\cdot\vec{\beta})|<0^+|\,|\hat{M}_0^5(q)|\,|0^->|^2
$$
$$
+ [1-\hat{\nu}\cdot\vec{\beta}+2(\hat{\vec{\nu}}\cdot\hat{q})(\hat{q}\cdot\beta)]|<0^+|\,|\hat{L}_0^5(q)|\,|0^->|^2
$$

(27)

$$
- \hat{q}\cdot(\hat{\nu}+\vec{\beta})2Re<0^+|\,|\hat{L}_0^5(q)|\,|0^-><0^+|\,|\hat{M}_0^5(q)|\,|0^->^*\}
$$
$$
\times 2\pi\eta/[\exp(2\pi\eta)-1]\,.
$$

This expression is integrated over the electron energy, ϵ, from the electron rest-mass to W_0, the maximum electron energy. Neglecting the recoil of the nucleus, W_0 is taken as the excitation energy of the excited state relative to the 0^+ g.s. We use the long-wave length reductions of the transition operators appearing in Eq. (27) in the actual evaluation of the beta-decay rate.

3. RESULTS

We show in Figs. 1 and 2 representative results for the energy level spectrum of $T = 1$ states of ^{16}O obtained using the RPA approximation, Dirac-Hartree orbitals and the QHD-I and QHD-II residual interactions given in section 2. Although individual contributions to the residual interaction may be large there is considerable cancellation so that the magnitude of the matrix elements of the total residual interaction are similar to those found in nonrelativistic calculations. Furnstahl[9] first noted that the QHD-II interaction (which includes π and ρ exchange as well as the σ and ω exchange contained in QHD-I) tends to depress unnatural parity (pion-like) states excessively. The use of PV coupling results in a weaker effective pion exchange interaction (compared to PS coupling), leading to less depression of unnatural parity states. For low-lying states the spectrum predicted adopting basis A or C is similar. The use of QHD-I and phenomenological single-particle energies (basis B) yields the low lying predicted $T = 1$ quartet relatively closer to experimental energy levels (note the ordering is still incorrect).

Electron scattering form-factors for $T = 1$ states in ^{16}O have been calculated using the formulae and procedures outlined in the previous section. We have grouped states nearby in excitation energy into complexes. The states included in each complex are given in Table 2. We show typical results in Figs. 3 and 4 for basis C using the QHD-I interaction. In each figure the dashed curves are nonrelativistic TDA results of Donnelly and Walker[11] calculated adopting unperturbed configuration energies obtained from the masses of neighboring nuclei and assuming harmonic oscillator orbitals and a Serber-Yukawa residual interaction.

We find that when states are grouped into complexes the overall isovector electron scattering response calculated in the present relativistic model is not qualitatively different than the nonrelativistic predictions. Although there can be differences for some individual states we found no existing theoretical-experimental discrepancies to be reconciled by the relativistic model. For example the stretched $4^-, T = 1$ state is still predicted to be significantly more strongly excited than is observed.[2,11]

The results for muon-capture rates on ^{16}O are shown in Tables 3 and 4. Previous nonrelativistic calculations for the total rate have given results 20-30% higher than the experimental data. We note from Tables 3 and 4 that total rates predicted in the relativistic model vary on the order of 10% depending on which basis, residual interaction or type of coupling (pseudo-scalar or pseudo-vector) is adopted. (We have also carried out relativistic TDA calculations and compared them with the relativistic RPA results. We find that the RPA results for total rates are $\lesssim$ 10% smaller than the TDA predictions.) The most significant differences in the results shown in Table 3 occur between PV and PS predictions for 0^- and 2^- states. (The PS and PV predictions for the 1^- state are identical since the $\hat{T}_{JM}^{mag\ 5}$ operation has the same form for both couplings.) The PV form of the weak current satisfies PCAC in the many-body environment while the PS form does not (because of terms involving the derivative of the effective mass, $M(x)$). Therefore, at this time, we place more significance on the PV results. An important point is that we find no large enhancements of the relativistic total rate predictions compared to nonrelativistic predictions for the *total* rate.

The $^{16}O_{g.s.}(O^+ \leftrightarrow O^-)^{16}N$ transition is of special interest in muon-capture and beta-decay. As pointed out earlier only M_0^5 and L_0^5 contribute to these tran-

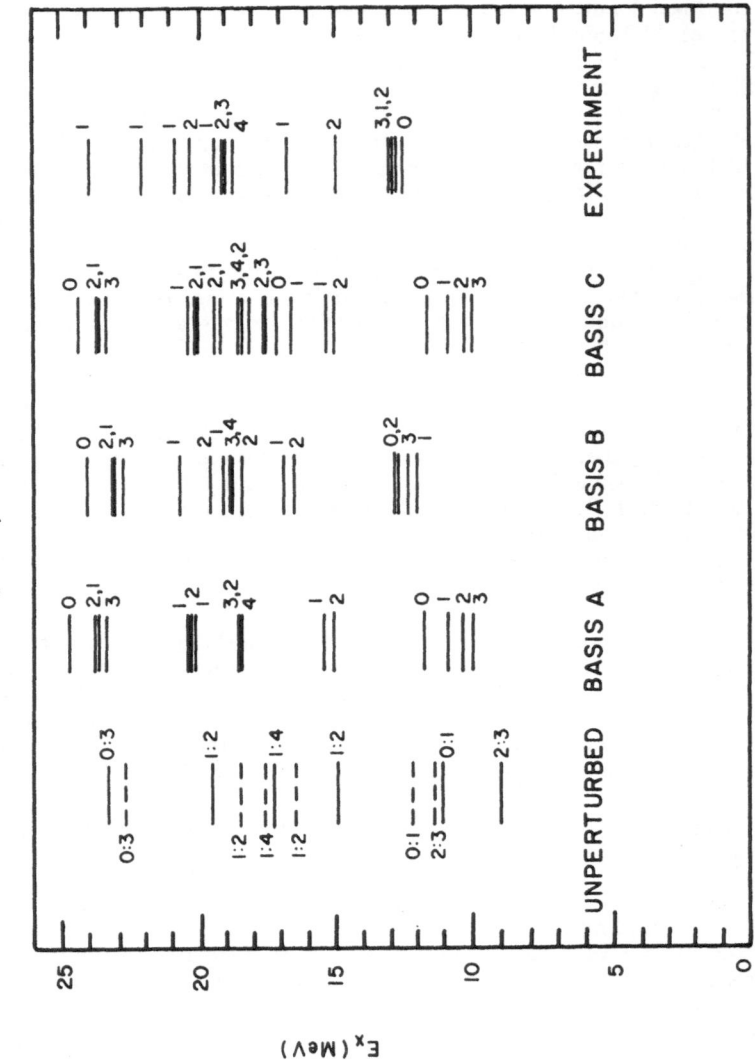

Fig. 1. Negative parity $T = 1$ RPA excited-state energy spectrum calculated using bases A, B, and C and a QHD-I particle-hole interaction. The solid line unperturbed levels shown at the far left are obtained from the Hartree single-particle energies (basis A). The dashed line unperturbed levels shown at the far left are obtained using the masses of neighboring nuclei (basis B). Figure redrawn from ref. 2.

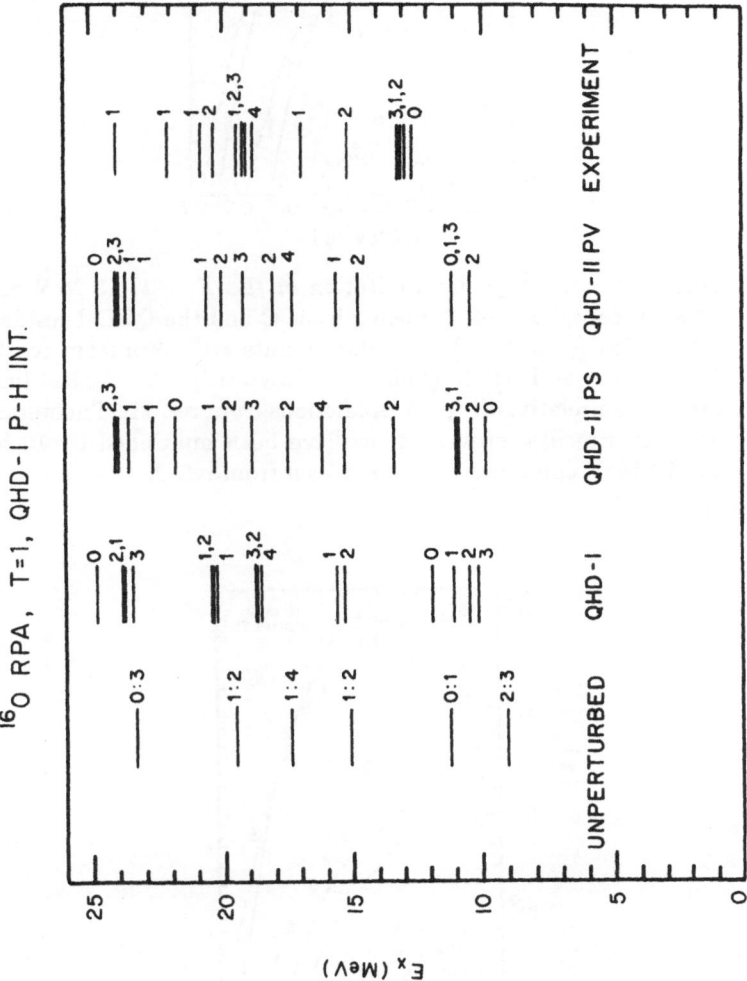

Fig. 2. Basis A negative parity $T = 1$ RPA excited-state energy spectrum calculated using QHD-I and QHD-II particle-hole interactions. Figure redrawn from ref. 2.

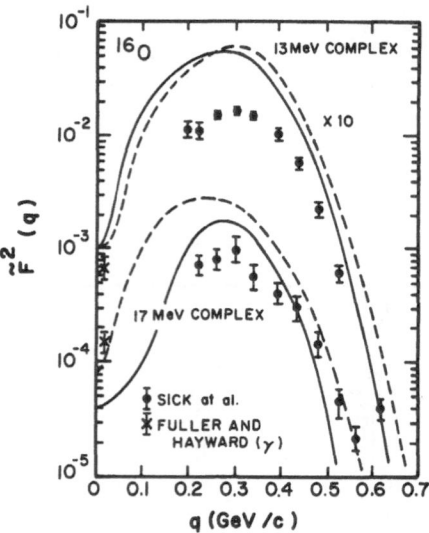

Fig. 3. Form factors, $\tilde{F}^2(q)$, for excitation of the $T = 1$, 13 VeV and 17 MeV complexes of ^{16}O using basis C and the QHD-I residual interaction (solid lines). The data points are taken from ref. 21 (Sick et al.) and ref. 22 (Fuller and Hayward). The dashed lines are the nonrelativistic TDA predictions from ref. 11. Theoretical and experimental cross-sections have been multiplied by 10 for the 13 MeV complex. Figure redrawn from ref. 2.

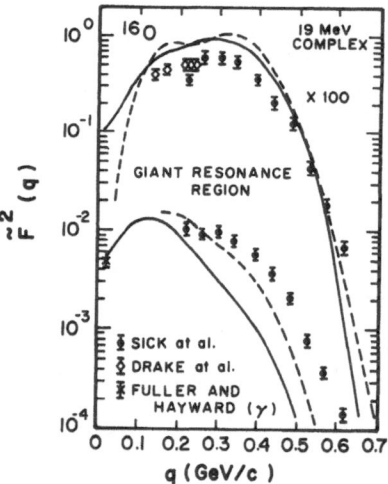

Fig. 4. Same as figure 3 except $T = 1$, 19 MeV and giant resonance complexes are shown. The data points denoted Drake et al. are taken from ref. 23. Theoretical and experimental cross-sections have been multiplied by 100 for the 19 MeV complex. Figure redrawn from ref. 2.

Table 2

The $T = 1$ constituent states of the 13, 17 and 19 MeV and giant resonance complexes excited in (e,e'). Table take from Ref. 2.

	RPA Energy Eigenvalue (MeV)			NR TDA[a]
	BASIS A	BASIS B	BASIS C	
13 MeV:				
1^-	11.14	12.24	11.11	14.38
2^-	10.56	12.95	10.51	13.59
3^-	10.22	12.62	10.20	13.57
17 MeV				
1^-	15.65	17.10	16.88	18.46
2^-	15.32	16.75	15.31	18.45
19 MeV				
1^-	20.21	19.20	20.16	20.73
2^-	18.61	18.58	18.26	19.77
2^-	20.29	19.68	20.24	20.96
3^-	18.67	18.98	18.64	19.17
4^-	18.59	18.89	18.58	19.86
GR:				
1^-	20.47	20.79	20.45	23.26
1^-	23.82	23.19	23.76	26.13
2^-	23.86	23.23	23.81	24.28
3^-	23.56	22.92	23.56	25.30

[a]See ref. 11.

sitions. Thus the time-component of the axial current (which connects upper- and lower-components) contributes strongly for both transitions and therefore one might anticipate enhancements in the relativistic predictions compared to nonrelativistic calculations. Apparently, using the RPA or TDA approximations, nonrelativistic calculations yield predictions for the $0^+ \rightarrow 0^-$ muon-capture rate that are slightly larger than experiment while the predictions for beta-decay are lower than experiment. Hence, the ratio of the muon-capture rate to the beta-decay is approximately a factor of two greater than the experimental ratio. Other authors have suggested there may be significant enhancements in both the muon-capture and beta-decay $0^+ \leftrightarrow 0^-$ rates due to the presence of $M^\star$ in the lower-components for relativistic calculations.[13,14] In this case while both the muon-capture and beta-decay rates may be at variance with experiment, the ratio of the rates may be consistent[13] with the experimental data. The relativistic predictions[13,14] were only schematic and did not use Hartree orbitals or consistent realistic relativistic residual interactions. Thus it is of interest to see what a more complete relativistic calculation predicts. Results obtained within the framework discussed in Section 2 are shown in Table 5. We find the results are not strongly dependent on whether basis A, B, or C is adopted but are strongly dependent on which residual interaction (QHD-I or QHD-II) is adopted and whether PS or PV coupling is assumed. As extremes, adoption of QHD-I and PS coupling results in a muon-capture rate prediction somewhat *lower* than the nonrelativistic TDA prediction and roughly consistent with experiment while use of QHD-II and PV coupling results in a prediction higher than the nonrelativistic prediction and about a factor of two larger than experiment. For the case of beta-decay there are also significant differences depending on whether the QHD-I or QHD-II residual interaction is adopted. The results using QHD-I (PS) and QHD-II (PS and PV) are enhanced, relative to the nonrelativistic

Table 3

Partial muon-capture rates (in s^{-1}) to the 0^-, 1^- and 2^- states of ^{16}O calculated in the RPA using basis A and the QHD-I residual interaction.

J^π	E_x (MeV)	PS	PV	NR	TDA[a]	Experimental Range
0^-	11.97	1,492	2,260		1,835	1,100−1,560[b]
	11.97	5,302	6,376		9,629	
1^-	11.14	2,878	2,878		2,722	1,400−1,850[b]
	15.65	9,521	9,521		5,808	
	20.21	3,285	3,285		3,060	
	20.47	18,974	18,974		20,144	
	23.82	10,473	10,473		27,879	
2^-	10.56	23,802	21,274		13,261	
	15.32	299	185		8	
	18.61	16,140	11,250		1,765	
	20.29	12,600	11,736		26,843	
	23.86	1,942	2,066		8,594	
Total		106,708	100,278		121,516	97,000 ± 5,000[c]
						98,000±3,000[d]

[a]See ref. 15, [b]See ref. 16, [c]See ref. 17, [d]See ref. 18.

Table 4

Total RPA capture rates in $10 \times s^{-1}$ from combining partial rates to the 0^-, 1^-, and 2^- states using the basis and residual interaction indicated.

Basis	Residual Int.	PS	PV
B	QHD-I	105	98
B	QHD-II	111	98
C	QHD-I	117	111
C	QHD-II		113

prediction. In particular, assuming QHD-I (PS) and using basis A and B, or assuming QHD-II (PV) and bases A-C yields results that are close to experiment. We note that QHD-I and PS coupling is required for consistency with *both* the muon-capture and beta-decay $O^+ \leftrightarrow 0^-$ experimental rates. An important conclusion is that whether one obtains significant enhancements over nonrelativistic predictions depends in detail on whether the QHD-I or QHD-II residual interaction is adopted and whether PS or PV coupling is assumed. We also note that for *all* cases we do *not* find significant enhancements (for this $0^+ \leftrightarrow 0^-$ transition) if the TDA, or RPA and QHD-I are used to make the relativistic predictions for β decay or muon-capture.[2]

Table 5

Partial RPA muon-capture rate Λ_μ (in s^{-1}) for the lowest $0^+ \to 0^-$ transition in ^{16}O and the beta-decay rate Λ_β (in s^{-1}) for ^{16}N$(0^-) \to {}^{16}$O g.s. Results taken from Ref. 2.

	BASIS A		BASIS B		BASIS C	
Λ_μ	PS	PV	PS	PV	PS	PV
QHD-I	1492	2260	1424	2172	1262	1924
QHD-II	2712	2863	2511	2748		3557
Λ_β	PS	PV	PS	PV	PS	PV
QHD-I	.376	.318	.363	.306	.308	.259
QHD-II	.691	.422	.648	.407		.555
$(\Lambda_\mu/\Lambda_\beta) \times 10^3$	PS	PV	PS	PV	PS	PV
QHD-I	4.0	7.1	3.9	7.1	4.1	7.4
QHD-II	3.9	6.8	3.9	6.7		6.4

	Λ_μ	Λ_β
Nonrel. TDA	1835[a]	.3[b]
Exp.	1110–1560[c]	.33–.54[d]

[a]See ref. 15   [b]See ref. 19   [c]See ref. 16   [d]See ref. 20.

IV. CONCLUSIONS AND DISCUSSION

We have studied the isovector electroweak nuclear response of the mass 16 system using a relativistic RPA model. The single-particle orbitals and residual interaction are obtained using a consistent Lagrangian-based mean field model. The results for inelastic electron scattering indicate that while individual states may have different energies and excitation strengths (compared to nonrelativistic predictions), the overall isovector response of complexes is not qualitatively different than that obtained nonrelativistically. The stretched $4^-, T = 1$ state is still too strongly excited compared to experiment.

There is considerable variation in the RPA weak linear response for the lowest $0^+ \leftrightarrow 0^-$ transition in muon capture and beta-decay depending on whether QHD-I or II residual interactions and PS or PV coupling is adopted. Although comparisons of calculated[2,9] and experimental level spectra indicate a preference for a QHD-II p-h interaction based on PV coupling, we find that such an interaction with PV coupling also in the weak axial current (PCAC satisfied) gives rise to Λ_μ and $\Lambda_\mu/\Lambda_\beta$ values that are at variance with experiment. Moreover, the same situation holds when PV coupling in the weak axial current is assumed with a QHD-I interaction. Overall, the RPA QHD-I and II PV rates are larger (smaller) for Λ_μ (Λ_β) than the corresponding PS rates. Similarly QHD-II with PS coupling (PCAC not satisfied) leads to values of Λ_μ and Λ_β that are larger than the experimental values. The use of QHD-I on the other hand, with PS coupling in the weak axial current (PCAC not satisfied) gives values of Λ_μ, Λ_β and $\Lambda_m u/\Lambda_\beta$ that are within the range of experimental values when bases A and B are employed. (Λ_β is just below the lowest experimental value for basis C.) The success one encounters with a simple isospin-independent QHD-I interaction in determining level spectra and electroweak reaction rates should not be overlooked. Of equal interest is the degradation of agreement with experiment (level spectra and reaction rates) when one attempts to include the pion (QHD-II) in a theory containing the σ meson. Thus comparison with experiment, while interesting, cannot lead to definitive conclusions until a proper chiral theory incorporating the pion satisfactorily into the relativistic theory is available. Of course even assuming the existence of a satisfactory relativistic theory incorporating pions there are still the usual complications that exist for both relativistic and nonrelativistic theories involving the basis truncation (the RPA may simply be inadequate for studying the linear response quantitatively) and the inclusion of meson exchange currents in a consistent framework. Thus three interesting and important problems for future relativistic investigations include a) incorporation of pions in a satisfactory manner, b) utilization of a larger basis including vacuum polarization effects and multi-particle multi-hole states, and c) calculation of exchange currents in a consistent relativistic model.

REFERENCES

1. See, for example, talks at this conference by P. Blunden, R. Furnstahl, C. Horowitz, R. Perry, C. Price, and J. Shepard.

2. M. W. Price, Ph.D. Thesis (1986), Indiana University (unpublished) and M. W. Price and G. E. Walker to be submitted for publication.

3. B. D. Serot and J. D. Walecka, in Advances in Nuclear Physics 16 (eds. J. W. Negele and E. Vogt) (1986) and references contained therein.

4. M. R. Anastasio et al., Phys. Reports C100 327 (1983) and references contained therein.

5. L. G. Arnold et al., Phys. Rev. C19, 917 (1979); L. C. Arnold et al., Phys. Rev. C23, 1949 (1981); J. R. Shepard et al., Phys. Rev. Lett. 50, 1443 (1983).

6. A. L. Fetter and J. D. Walecka, *Quantum Theory of Many-Particle Systems* (McGraw-Hill, New York, 1971) pp. 558-566.

7. R. J. Furnstahl and B. D. Serot, Nucl. Phys. A468, 539 (1987).

8. C. J. Horowitz and B. D. Serot, Nucl. Phys. A368, 503 (1981).

9. R. J. Furnstahl, Ph. D. Thesis (1985), Stanford University (unpublished), Phys. Lett. 152B, 313 (1985).

10. T. de Forest Jr. and J. D. Walecka in: *Advances in Physics 15*, 1 (1966).

11. T. W. Donnelly and G. E. Walker, Ann. Phys. 60, 209 (1970).

12. J. D. Walecka in: *Muon Physics* (eds. Vernon W. Hughes and C. S. Wu) Academic Press, New York, 1975.

13. G. Do Dang et al., Phys. Lett. 153B, 17 (1985).

14. H. P. C. Rood, Phys. Rev. C33, 1104 (1986).

15. G. E. Walker, Phys. Rev. I74, 1290 (1968), muon-capture results are for a Serber-Yukawa interaction and have been scaled to be consistent with the Fermi coupling constant used in relativistic calculations.

16. P. A. M. Guichon et al., Phys. Rev. **C19**, 987 (1979).

17. J. Barlow et al., Phys. Lett. **9**, 84 (1964).

18. M. Eckhouse, Ph. D. Thesis, Carnegie Institute of Technology (1962).

19. Calculated using wave functions from ref. 11.

20. L. Palffy et al., Phys. Rev. Lett. **34**, 212 (1975).

21. I. Sick et al., Phys. Rev. Lett. **23**, 1117 (1969).

22. E. G. Fuller and E. Hayward, in *Nuclear Reactions*, edited by P. M. Endt and P. B. Smith (North Holland, Amsterdam) Vol. II, p. 113 (1962).

23. T. E. Drake, E. L. Tomusiak, and H. S. Caplan, Nucl. Phys. **A118**, 138 (1968).

ENERGY DEPENDENT CORRECTIONS TO SPIN OBSERVABLES

IN NUCLEON-NUCLEUS SCATTERING

J. Piekarewicz

Nuclear Theory Center
Indiana University
Bloomington, IN 47405

Several speakers at this conference have shown that spin observables can uncover subtle effects that are otherwise hidden in the study of unpolarized phenomena. In this contribution results will be presented that show important modifications to spin observables once the explicit energy dependence of the effective nucleon-nucleon (NN) interaction is taken into account. The main ideas used will be illustrated by concentrating on a very simple, nevertheless non trivial, example, the $0^+(\vec{p}, \vec{p}')0^-$ reaction.

First, let us consider some model independent results. The most general form of the $J = 0 \rightarrow J = 0$ amplitude, consistent with rotational invariance, can be written in terms of the four scalar operators,

$$1, \quad \vec{\sigma} \cdot \hat{n}, \quad \vec{\sigma} \cdot \hat{q}, \quad \vec{\sigma} \cdot \hat{K}$$

where

$$\vec{K} = \frac{1}{2}(\vec{k} + \vec{k}'), \quad \vec{q} = \vec{k} - \vec{k}', \quad \vec{n} = \vec{q} \times \vec{K}$$

are the average momentum, momentum transfer, and vector perpendicular to the scattering plane respectively, written in terms of the initial(final) momentum $\vec{k}(\vec{k}')$ of the probe. We can further restrict the form of the amplitude by imposing parity invariance. The scattering amplitude, for the two possible transitions, will then take the following form

$$A(0^+ \rightarrow 0^+) = A_0 + A_n(\vec{\sigma} \cdot \hat{n})$$
$$A(0^+ \rightarrow 0^-) = A_q(\vec{\sigma} \cdot \hat{q}) + A_K(\vec{\sigma} \cdot \hat{K})$$

where the $A's$ are invariant (complex) amplitudes of the energy and momentum transfer. Since the amplitude is defined up to an overall phase, three independent measurements, for each transition, are enough to completely determine the scattering amplitude. In particular, for the $0^+ \rightarrow 0^-$ reaction, the full set of observables is given by,[1]

$$\frac{d\sigma}{d\Omega} = |A_q|^2 + |A_K|^2 \qquad \frac{d\sigma}{d\Omega} D_{nn} = -|A_q|^2 - |A_K|^2$$

$$\frac{d\sigma}{d\Omega} D_{qq} = |A_q|^2 - |A_K|^2 \qquad \frac{d\sigma}{d\Omega} D_{KK} = -|A_q|^2 + |A_K|^2$$

$$\frac{d\sigma}{d\Omega} D_{0n} = 2I_m(A_q A_K^*) \qquad \frac{d\sigma}{d\Omega} D_{n0} = -2I_m(A_q A_K^*)$$

$$\frac{d\sigma}{d\Omega} D_{qK} = 2R_e(A_q A_K^*) \qquad \frac{d\sigma}{d\Omega} D_{Kq} = 2R_e(A_q A_K^*) ,$$

where we have used the standard definition of the transfer coefficients, namely,

$$\left(\frac{d\sigma}{d\Omega}\right) D_{\alpha\beta} = \frac{1}{2}T_r\left[\sigma_\alpha A \sigma_\beta A^\dagger\right] \qquad \alpha, \beta = (0, n, q, K)$$

In particular we obtain the following relations between $0^+ \rightarrow 0^-$ spin observables,

$$D_{n0} \equiv P = -A_y \equiv -D_{0n}; \qquad D_{qK} \equiv Q = -B \equiv D_{Kq}; \qquad D_{nn} = -1 ,$$

where notice that deviations from the elastic scattering relations, i.e., $P = A_y$, $Q = B$, and $D_{nn} = +1$, are maximal. There has been considerable interest in understanding possible sources of P-A_y. It is interesting to point out that whatever these sources might be, they will already be present in the study of the simplest to measure spin observable, the analyzing power.

Let us now move away from model independent results and mention the approximations that will be used throughout this talk. It will be assumed, as in the case of elastic scattering, that $\hat{q} \cdot \hat{K} = 0$. At the energies and momentum transfers of interest the corrections due to a finite Q value in the reaction are very small. For example, at $T_{lab} \simeq 500$ MeV, $\hat{q} \cdot \hat{K} < \frac{1}{10}$ for $q > \frac{1}{2}$fm^{-1}. Distortions will also be neglected. Even though distortions are very important for elastic spin observables, observables driven by $\vec{\sigma} \cdot \hat{q}$ and $\vec{\sigma} \cdot \hat{K}$ amplitudes are relatively insensitive to distortion effects.[2] We can gain some analytic understanding of this result by using an eikonal approximation in the evaluation of the distortions. While spin-orbit distortions, χ_{so}, "rotate" elastic-like amplitudes among themselves, i.e.,

$$\begin{pmatrix} A_0 \\ A_n \end{pmatrix}_{DWIA} = \begin{pmatrix} \cos\chi_{so} & i\sin\chi_{so} \\ i\sin\chi_{so} & \cos\chi_{so} \end{pmatrix} \begin{pmatrix} A_0 \\ A_n \end{pmatrix}_{PWIA}$$

this is not the case for $0^+ \rightarrow 0^-$-like amplitudes, where instead

$$\begin{pmatrix} A_q \\ A_K \end{pmatrix}_{DWIA} = \begin{pmatrix} 1 & 0 \\ 0 & 1 \end{pmatrix} \begin{pmatrix} A_q \\ A_K \end{pmatrix}_{PWIA}$$

We therefore expect small distortions effects in all observables driven by $\vec{\sigma} \cdot \hat{q}$ and $\vec{\sigma} \cdot \hat{K}$ amplitudes. These are precisely the observables we are most interested in, like P-A_y or the asymmetric part of the correlation function measured in $(p, p'\gamma)$ experiments as was shown previously by Ken Hicks.[3]

The Franey-Love effective NN interaction[4] is used in the calculations discussed below. The parameters in the interaction have been chosen to reproduce, in Born approximation, NN observables in the intermediate energy region. By using this prescription, however, one suppresses all the energy dependence that is usually generated in the iteration of the potential in a Lippmann-Schwinger equation. This deficiency in the model is partially corrected by endowing the coupling strengths with explicit energy dependence. This energy dependence is one of the main focus of this talk.

The interaction between a projectile and a bound nucleon is written in terms of explicit direct and exchange contributions,

$$t_{12} = < \vec{k}'; \phi' | V (1 - P_{12}) | \vec{k}; \phi >$$

$$= \int \frac{d\vec{p}}{(2\pi)^3} \phi'^*(\vec{p} + \vec{q}) \left[\frac{\alpha}{q^2 + \mu^2} - \frac{\alpha}{Q^2 + \mu^2} \right] \phi(\vec{p})$$

where α is the energy dependent strength of the interaction, $\vec{Q} = \vec{Q}(\vec{p}) = \vec{k} - \vec{p}' = \vec{k}' - \vec{p} = \frac{1}{2}(\vec{k} + \vec{k}') - \frac{1}{2}(\vec{p} + \vec{p}') \equiv \vec{K} - \vec{P}$ is the exchange momentum, and $\phi(\phi')$ is the initial(final) bound state wavefunction. We might first attempt to calculate the amplitude using an optimally factorized form for the amplitude, i.e., one that neglects off-shell effects and evaluates the NN t matrix with Breit frame kinematics,[5]

$$\vec{P} = 0; \quad \vec{p} = -\frac{\vec{q}}{2}; \quad \vec{p}' = +\frac{\vec{q}}{2}.$$

The $0^+ \rightarrow 0^-$ transition amplitude can now be written in terms of an on-shell parameterization of the NN t matrix

$$t_n = A + B\vec{\sigma} \cdot \vec{\sigma}_n + \imath q C (\vec{\sigma} + \vec{\sigma}_n) \cdot \hat{n}$$

$$= q^2 D (\vec{\sigma} \cdot \hat{q})(\vec{\sigma}_n \cdot \hat{q}) + E (\vec{\sigma} \cdot \hat{K})(\vec{\sigma}_n \cdot \hat{K})$$

as,

$$A(0^+ \rightarrow 0^-) = < 0^- | \sum_{n=1}^{A} e^{\imath \vec{q} \cdot \vec{x}_n} t_n | 0^+ >$$

Some remarks are in order. First, since the $0^+ \rightarrow 0^-$ transition is an unnatural parity excitation, only spin flip amplitudes, i.e., amplitudes containing $\vec{\sigma}_n$, can contribute to the reaction. Second, as mentioned before, the amplitude must be proportional to the spin operator of the projectile $\vec{\sigma}$. Hence, we immediately see that there will be no contribution to this reaction from the spin-orbit amplitude, usually the main driving term behind a nonzero analyzing power.

After some straightforward manipulations one obtains the following form for the scattering amplitude,

$$A(0^+ \rightarrow 0^-) = (B + q^2 D) < 0^- | \sum_{n} e^{\imath \vec{q} \cdot \vec{x}_n} (\vec{\sigma}_n \cdot \hat{q}) | 0^+ > (\vec{\sigma} \cdot \hat{q})$$

We observe that this reaction samples the longitudinal part of the nuclear spin density. More importantly however, is the absence of the $\vec{\sigma} \cdot \hat{K}$ amplitude. In particular, this model will predict zero for the analyzing power and in general for all observables consisting of the interference between the A_q and A_K amplitudes, in sharp contrast with experimental results.[6]

It is interesting to compare the predictions of a relativistic plane wave impulse approximation (RPWIA) calculation that also uses on-shell amplitudes. The scattering amplitude in a RPWIA treatment can be written as,

$$A = \bar{U}(\vec{k}') < 0^- | \sum_{n=1}^{A} \sum_{\nu=1}^{4} e^{i\vec{q}\cdot\vec{x}_n} [f_\nu + g_\nu \vec{\sigma} \cdot \vec{\sigma}_n] \Gamma_\nu \Gamma_\nu(n) |0^+ > U(\vec{k})$$

where $U(\vec{k})$ and $U(\vec{k}')$ are free Dirac spinors, f_ν and g_ν are Lorentz invariant amplitudes, and the Γ's are four 4×4 coupling matrices. In the present relativistic calculation there is an important contribution from the central-like term f_ν. This might seem inconsistent with our previous statement that only amplitudes proportional to both $\vec{\sigma}$ and $\vec{\sigma}_n$ should contribute to the reaction. One has to remember, however, that we are now dealing with relativistic four component wavefunctions that intrinsically carry the spin operator $\vec{\sigma}$. Schematically, we can write relativistic states in terms of nonrelativistic ones as,

$$|0^+> = \begin{pmatrix} 1 \\ \frac{\vec{\sigma}\cdot\vec{p}}{2M} \end{pmatrix} |0^+_{NR}>$$

Off-diagonal matrices Γ_ν, that generate coupling between upper and lower components will then be able to generate an amplitude proportional to both $\vec{\sigma}$ and $\vec{\sigma}_n$. In particular, Γ_4 generates a nonzero A_K term,

$$\bar{U}(\vec{k}')\Gamma_4 U(\vec{k}) = \frac{\vec{\sigma}\cdot\vec{K}}{2M}, \qquad \Gamma_4 = \begin{pmatrix} 0 & 1 \\ -1 & 0 \end{pmatrix}$$

that directly couples to the spin-convection current density,

$$< 0^- | \sum_n e^{i\vec{q}\cdot\vec{x}_n} \Gamma_4 |0^+ > = < 0^-_{NR} | \sum_n e^{i\vec{q}\cdot\vec{x}_n} (\vec{\sigma}_n \cdot \vec{J}_n) |0^+_{NR} >$$

In fig. 1. results are shown from relativistic, on-shell, calculations for the analyzing power and the spin rotation function $B = -Q$ for the first 0^-, T=0 state in ^{16}O. The experimental data were obtained from reference 6. One may be tempted to conclude that these results represent a clear manifestation of relativistic effects. Let us remember, however, that so far we have totally neglected off-shell effects. It is conceivable, and in fact true, that even though a relativistic on-shell calculation shows a richer structure that an equivalent nonrelativistic calculation, the differences will not be as dramatic once off-shell effects are incorporated into both calculations. In order to understand the implications of off-shell effects let us start by performing a Taylor series expansion of the exchange part of the interaction around the optimal momentum $\vec{Q} = \vec{K}$,

$$\frac{1}{Q^2 + \mu^2} \simeq \frac{1}{K^2 + \mu^2} \left[1 + \frac{2MK}{K^2 + \mu^2} (\hat{K} \cdot \vec{J}_n) \right] .$$

We observe that the first order correction brings into the calculation the two essential pieces that generated a nonzero value for A_K in the relativistic treatment, namely, the average momentum $\hat{K}$ and the convection current $\vec{J}_n$. Off-shell corrections will then be able to generate, as in the relativistic case, $A_K \neq 0$. Spin observables will no longer vanish, and the hope of using spin observables in the $0^+ \rightarrow 0^-$ reaction as a clear signature of obvious relativistic effects has disappeared.

ANALYZING POWER

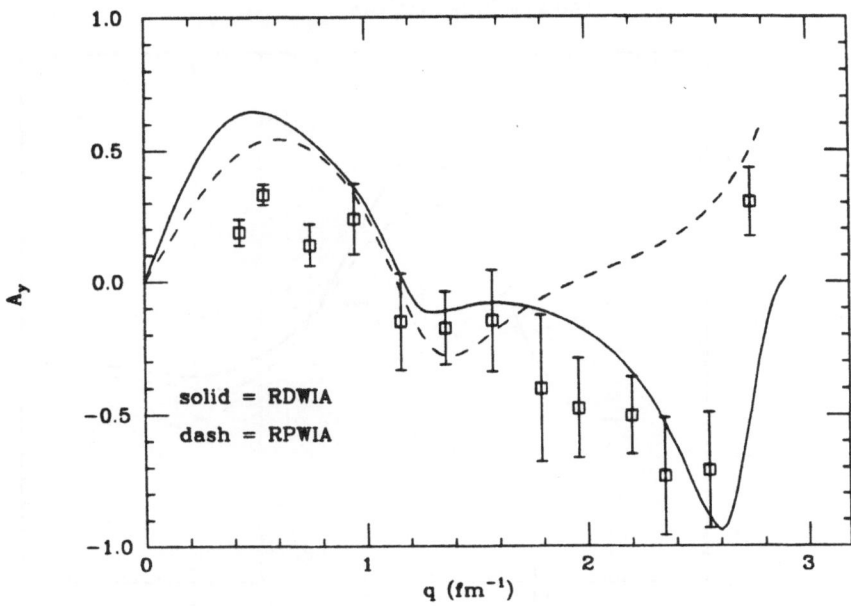

B OBSERVABLE

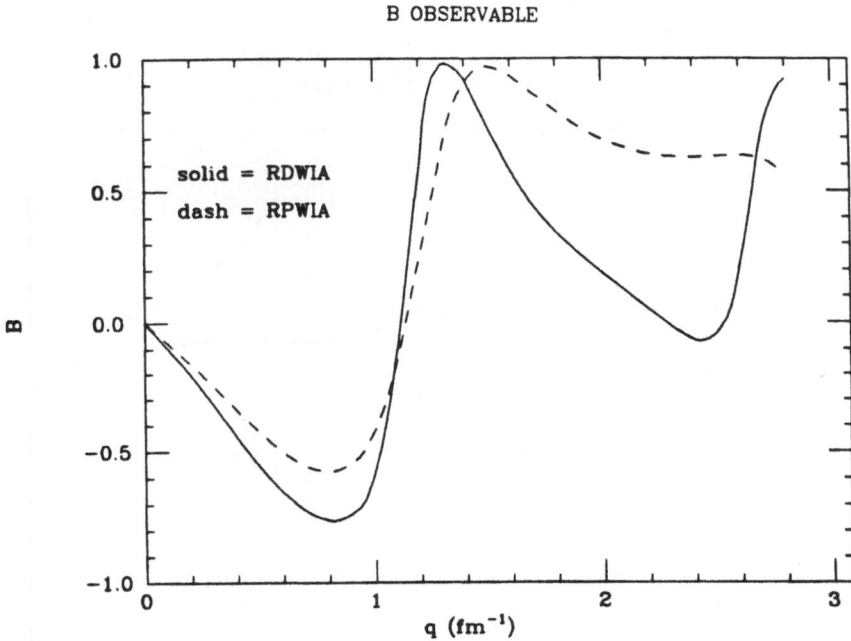

Fig. 1. Analyzing Power and spin rotation function $B \equiv -D_{Kq}$ for the excitation of the 0^-,T=0, state in ^{16}O by T_{lab}=180 Mev protons. Solid(dash) curves are result of relativistic DWIA(PWIA) calculations. Experimental data are from Ref. 6.

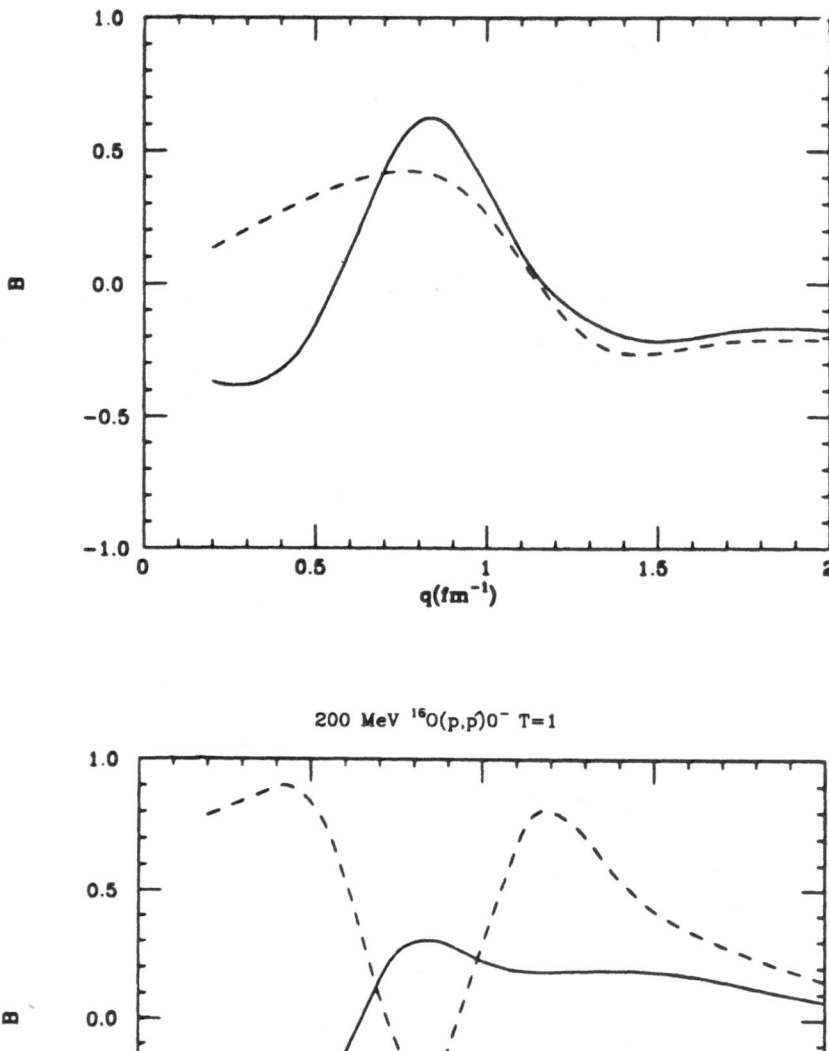

Fig. 2. Spin rotation function $B \equiv -D_{Kq}$ for the excitation of the 0^-,T=0 and T=1, states in ^{16}O by T_{lab}=200 Mev protons. The solid(dash) curves show nonrelativistic PWIA results with(without) energy dependent corrections.

We have observed that off-shell effects are crucial in generating nonzero spin observables in a nonrelativistic framework. As we mentioned before, off-shell effects are present in both the exchange part, as well as in the energy dependence, of the effective NN interaction. While off-shell exchange is now present in almost all theoretical calculations, the energy dependence of the effective NN interaction has, to date, been totally ignored. The energy dependence of the πN interaction, however, has been previously addressed by Siciliano and Walker in πA scattering.[7] Unfortunately, a spinless probe cannot uncover all those subtle effects that one observes in the study of spin observables.

The first off-shell correction due to the energy dependence of the interaction can be written in the following way,

$$\alpha \simeq \alpha_0 + \alpha_0'(\hat{\mathbf{K}} \cdot \vec{\mathbf{J}}_n), \quad \alpha_0' \equiv -K\frac{\partial \alpha}{\partial T_{\text{lab}}}$$

and leads to precisely the same modification to the operator structure of the amplitude as was obtained from exchange. It is therefore inconsistent to keep one effect while apriori neglecting the other.

In fig. 2. results are shown for the spin rotation parameter $B = -Q$ for the excitation of, both, the T=0 and the T=1 states in ^{16}O by 200 MeV protons. We observe that *energy dependent corrections change dramatically the predictions*. We have also investigated energy dependent corrections to other processes. In particular, we have found negligible effects in elastic scattering as well as in transition to stretched states. As was illustrated in the case of the $0^+ \rightarrow 0^-$ reaction, energy dependent corrections become important in low momentum transfer inelastic reactions. In conclusion, the elucidation of details of either the reaction mechanism, or nuclear structure, without incorporating energy dependent corrections is potentially fatally flawed.

REFERENCES

1. J. Piekarewicz, Phys. Rev. C **35**, 675 (1987).
2. J. Piekarewicz, and G.E. Walker, to be submitted for publication.
3. K. Hicks, contribution to these proceedings.
4. M.A. Franey, and W.G. Love, Phys. Rev. C **31**, 488 (1985).
5. S.A. Gurvitz, J.P. Dedonder, and R.D. Amado, Phys. Rev. C **19**, 142 (1979).
6. J.J. Kelly, Ph.D. thesis, **MIT** (1981).
7. E.R. Siciliano, and G.E. Walker, Phys. Rev. C **23**, 2661 (1981).

ACKNOWLEDGMENTS

This work was done in collaboration with G.E. Walker and supported in part by the National Science Foundation.

THE EMPIRICAL CONNECTION BETWEEN (p,n) CROSS SECTIONS AND

BETA DECAY TRANSITION STRENGTHS

T.N. Taddeucci

Los Alamos National Laboratory
Los Alamos, NM 87545

ABSTRACT

A proportionality is assumed to exist between 0° (p,n) cross sections and the corresponding beta decay transition strengths. The validity of this assumption is tested by comparison of measured (p,n) cross sections and analogous beta decay strengths. Distorted waves impulse approximation calculations also provide useful estimates of the accuracy of the proportionality relationship.

INTRODUCTION

Standard reaction theory and experimental observations support the idea that there should be a measure of proportionality between (p,n) cross sections and beta-decay transition strengths.[1-3] This correspondance derives from the similarity of the operators involved in each type of reaction. The central isovector terms in the effective nucleon-nucleon interaction that mediate low momentum transfer spin-flip (S = 1) transitions,

$$\sum_i V_{\sigma\tau}(r_{ip})\vec{\sigma}_i \cdot \vec{\sigma}_p \vec{\tau}_i \cdot \vec{\tau}_p,$$

and non-spin-flip (S = 0) transitions,

$$\sum_i V_{\tau}(r_{ip})\vec{\tau}_i \cdot \vec{\tau}_p,$$

425

are similar to the corresponding operators

$$G_A \sum_i \vec{\sigma}_i \vec{t}_i^{\pm}$$

and

$$G_V \sum_i \vec{t}_i^{\pm}$$

for Gamow-Teller (GT) and Fermi (F) beta decay, respectively. There are numerous beta transitions with known decay rates for which the analogous (p,n) cross section can also be measured. The existence or absence of a useful proportionality can therefore be tested empirically. The initial results of such an investigation have been reported by Goodman et al.[2] for E_p = 120 MeV and were highlighted at the very first Telluride conference. Much additional data has become available since then, and the experimental situation up to about 1986 has been reviewed in another paper.[4] In this article I will summarize previous results and present some new data.

From general considerations, the (p,n) cross section should depend on the bombarding energy E_p, the number and type of nucleons in the target nucleus A(N,Z), the asymptotic three-momentum transfer q, the energy loss ω, where

$$\omega = E_x + (M_A' - M_A + M_n - M_p)c^2 ,$$

$$= E_x - Q_{gs},$$

and the specific nuclear structure relationship between initial and final nuclear states. The dependence on these quantities can be expressed as a product of three factors:

$$\sigma = \hat{\sigma}_\alpha(E_p,A)F_\alpha(q,\omega)B(\alpha), \tag{1}$$

where α = F or GT. The proportionality factor $\hat{\sigma}$, which I shall call the "unit cross section," can be bombarding energy and target dependent and is the factor of primary interest. The factor $F(q,\omega)$ describes the shape of the cross section distribution. At fixed ω it is approximately an exponential function of q^2 (for $q < 0.5$ fm^{-1}),

$$F(q,\omega) \simeq C(\omega)\exp\left(-\frac{1}{3}\langle r\rangle^2 q^2\right)$$

and goes to unity in the limit of zero momentum transfer and energy loss:

$$F(q,\omega) \to 1 \quad \text{as} \quad (q,\omega) \to 0.$$

The beta decay transition strengths $B(\alpha)$ are obtained from beta decay lifetimes according to

$$(G_V)^2 B(F) + (G_A)^2 B(GT) = \frac{K}{ft} \qquad (2)$$

where

$$\frac{K}{(G_V)^2} = 6166 \pm 2 \text{ sec}$$

and

$$\left(\frac{G_A}{G_V}\right)^2 = (1.260 \pm 0.008)^2.$$

The coupling constant values used here are those recommended by Wilkinson.[5]

It is useful to distinguish between comparisons of (p,n) and beta decay for different transitions starting from the same parent state, which I shall call specific proportionality, and comparisons of cross sections for transitions originating from different target nuclides, which I shall call general proportionality. In the former case a knowledge of the A-dependence of $\hat{\sigma}$ is not required. Application of the more general proportionality relationship will require the A-dependence of $\hat{\sigma}$ to be smooth or at least calculable.

The distorted waves impulse approximation (DWIA) calculations described here include "exact" knock-on exchange amplitudes and were performed with the code DW81.[6] The calculations employed relativistic kinematics but are otherwise consistent with the standard non-relativistic Schrodinger equation. The effective interaction used was the nucleon-nucleon t-matrix parametrization of Franey and Love (FL).[7] The 175-MeV version of this interaction was used for the reaction calculations at 160 MeV. Single-particle wave functions were calculated in a harmonic oscillator basis and are labeled by the notation $j_> = \ell + 1/2$ and $j_< = \ell - 1/2$, where ℓ is the orbital angular momentum. The optical potential parameters used in the DWIA calculations were those of Meyer et al.[8] for A = 6-18, Olmer et al.[9] for A = 28, and Schwandt et al.[10] for A = 18-208. The Schwandt parameters were extended to the lower mass range for comparison purposes.

The results of the DWIA calculations for E_p = 160 MeV are shown in Fig. 1. The variations in $\hat{\sigma}$ for different particle-hole configurations are not large for 1^+ transitions with the full single-particle GT strength. The dashed line in this figure represents the average mass dependence of $\hat{\sigma}_{GT}(A)$ and can be used to assess the implicit accuracy of the proportionality of Eq. (1), as predicted in the DWIA. The standard deviation of the DWIA values of $\hat{\sigma}_{GT}$ with respect to this average mass dependence is $\Delta\hat{\sigma}/\hat{\sigma}$ = 7%. To the extent that the DWIA variations model the expected variations in nature, this value of $\Delta\hat{\sigma}/\hat{\sigma}$ thus represents the smallest level of uncertainty that can be achieved in the experimental determination of GT transition strengths through the use of a proportionality relation such as Eq. (1).

In contrast to the GT unit cross sections, large variations are observed in the calculated Fermi unit cross sections. Central-interaction exchange amplitudes alone cause $j_> j_>^{-1}$ 0^+ transitions to have unit cross sections larger than those for $j_< j_<^{-1}$ transitions. This difference increases with target mass from about 5% for A = 12 to about 13% for A = 90 and vanishes when the abnormal parity J[L,S] = 0[1,1] amplitude is set to zero. Love, Nakayama, and Franey[11] pointed out that an interference between the microscopic $V_{LS\tau}$ and V_τ interactions in the presence of an optical-potential spin-orbit force has a large destructive effect upon $j_< j_<^{-1}$ 0^+ transitions and further enhances the separation of the Fermi cross sections into two distinct "bands", as plotted in Fig. 1. The empirical evidence for this latter effect is somewhat inconclusive and will be discussed later in comparisons to data.

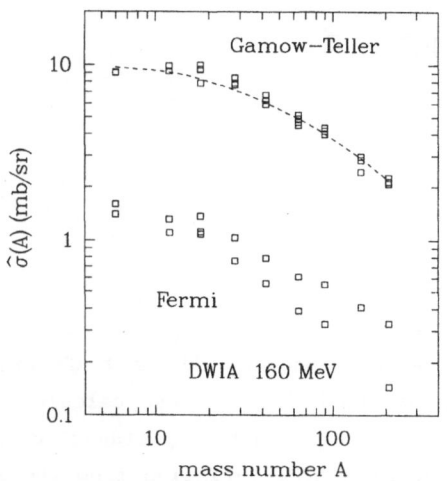

Fig. 1 Distorted-waves impulse approximation unit cross sections (squares). Multiple boxes for a given value of A correspond to different particle-hole configurations; the smallest Fermi cross sections for a given value of A correspond to $j_< j_<^{-1}$ transitions. The dashed line represents the average A dependence of the GT unit cross sections. (See Ref. 4).

The calculations of Fig. 1 employ full single-particle strengths. However, an important concern in the discussion of proportionality is the range of transition strengths over which the relationship is valid. The proportionality must obviously fail when the L=0 central interaction amplitude becomes so weak that competing amplitudes are comparable in magnitude. This issue is best addressed by comparing individual beta decay strengths to relevant single-particle strengths, given by:

$$
B(GT)_{sp} = \left\{
\begin{array}{ll}
\dfrac{(j_> + 1)}{j_>} & j_> j_>^{-1} \\[2ex]
\dfrac{(2j_> + 1)}{j_>} & j_> j_<^{-1} \\[2ex]
\dfrac{(2j_> - 1)}{j_>} & j_< j_>^{-1} \\[2ex]
\dfrac{j_<}{(j_< + 1)} & j_< j_<^{-1} .
\end{array}
\right.
\tag{3}
$$

Figure 2 shows calculations of the unit GT cross section for two different mass values at 160 MeV. In these calculations the GT amplitude was decreased while holding other amplitudes constant. The dotted horizontal lines represent a ±10% variation from the average value of $\hat{\sigma}_{GT}$ for full single-particle strength. With the exception of $j_< j_<^{-1}$ transitions, which are strongly affected by tensor exchange amplitudes, the calculated unit cross sections remain within the 10% limit to quite small values of the GT strength relative to the full single-particle strength.

The ^{27}Al(p,n) reaction provides a good empirical example of weak transitions for which the proportionality appears to be valid. Very good correspondence is observed between beta transition strengths and (p,n) cross sections for the transitions to the 2.16-MeV and 2.65-MeV levels in ^{27}Si (Fig. 3). These transitions carry only 5.1% and 2.6% of the $d_{3/2}d_{5/2}^{-1}$ and $d_{5/2}d_{5/2}^{-1}$ single particle strength, respectively.

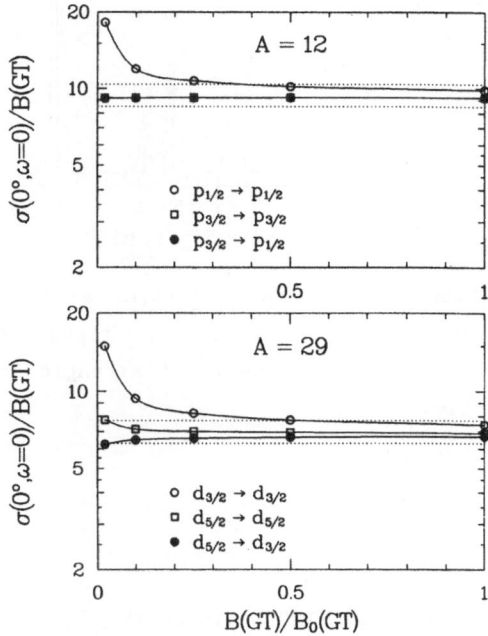

Fig. 2 Values of the GT unit cross section for A=12 and A=29. Starting
with the three pure single-particle transitions indicated
$(B(GT)/B_0(GT) = 1)$, the GT amplitude was decreased while holding
the other amplitudes fixed. $B_0(GT)$ represents the full
single-particle strength. The dotted horizontal lines indicate
a ±10% variation from the average value at $B(GT)/B_0(GT) = 1$.
The FL 175-MeV interaction containing central, spin-orbit, and
tensor terms was used in the calculations.

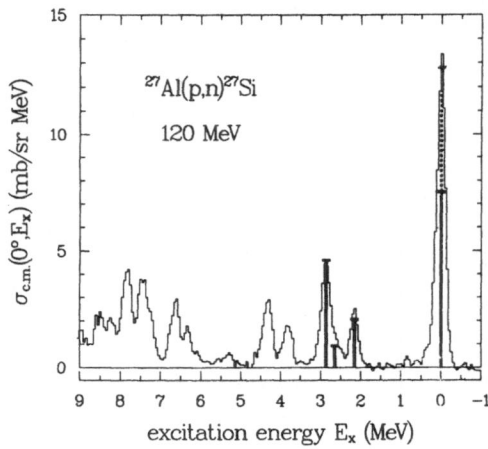

Fig. 3 Cross section spectrum for ^{27}Al(p,n) at 0° and 120 MeV. The vertical bars represent the GT transition strengths for analogous beta decays. The Fermi strength is indicated by the dashed vertical line.

EMPIRICAL EVIDENCE FOR SPECIFIC PROPORTIONALITY

Specific proportionality implies that all GT (p,n) transitions originating from a given target nuclide will have the same beta-decay proportionality factor (unit cross section). Equivalently, there will be a fixed proportionality between these GT (p,n) cross sections and the Fermi component of the cross section for the isobaric analog state transition. In contrast to the large configuration-dependent variations in the ratio $\hat{\sigma}_{GT}/\hat{\sigma}_F$ as displayed in Fig. 1, experimental studies of GT and F transitions have shown a well defined ratio between GT and F cross sections in the energy range 50 - 200 MeV. This ratio can be conveniently parametrized as

$$\hat{\sigma}_{GT}/\hat{\sigma}_F = (E_p/E_0)^2 \qquad\qquad\qquad\qquad (4)$$

where $E_0 = 55.0 \pm 0.4$ MeV. A summary of the data available up to 1986 is shown in Fig. 4. The standard deviation of the data points plotted in this figure is $\Delta E_0 = 1.7$ MeV. Note that this implies an uncertainty in the ratio of unit cross sections of about 6%.

In addition to the ^{27}Al(p,n) transitions illustrated in Fig. 3, some more examples that appear to demonstrate the validity of specific proportionality are ^{13}C(p,n), ^{18}O(p,n), ^{26}Mg(p,n), and ^{34}S(p,n). Spectra for (p,n) reactions on these target nuclides are shown in Figs. 5-8 for a bombarding energy of 120 MeV. The ratio of unit cross sections defined by Eq. (4) is assumed in plotting the relative sizes of the GT and F transition strength bars in these figures. That is, to within an overall

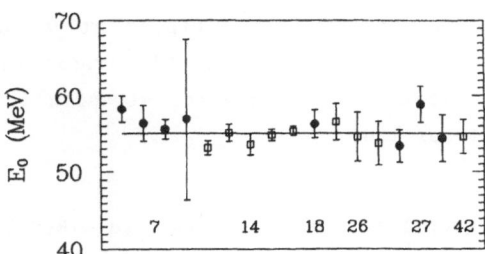

Fig. 4   The parameter $E_0 = E_p/(\hat{\sigma}_{GT}/\hat{\sigma}_F)^{1/2}$. Solid circles are alternated with open squares to indicate data points for different nuclides. For a given nuclide, the bombarding energy increases from left to right. From the left, the data correspond to ^{7}Li, ^{14}C, ^{18}O, ^{26}Mg, ^{27}Al, and ^{42}Ca. The solid horizontal line represents the weighted average $E_0 = 55.0 \pm 0.4$.

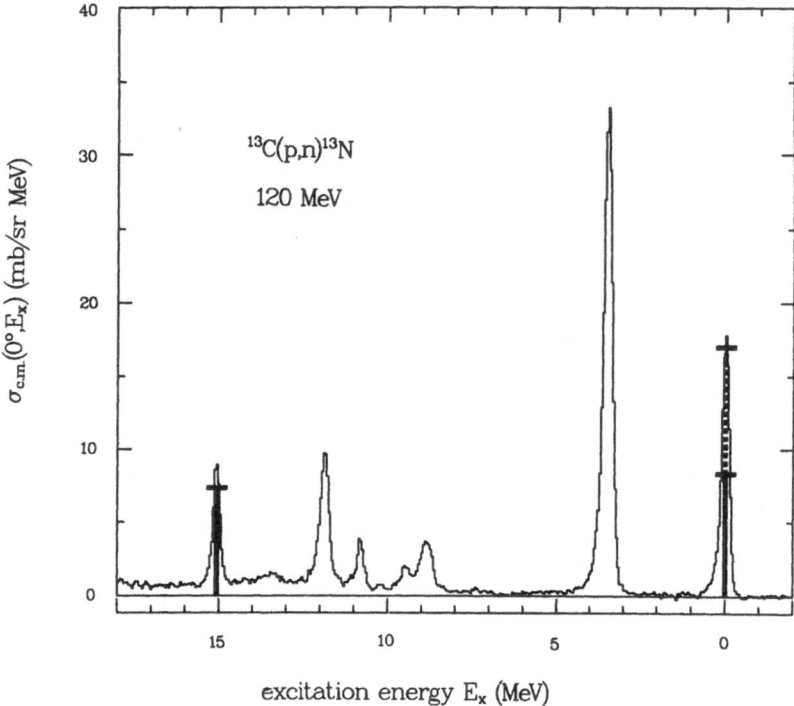

$^{13}C(p,n)^{13}N$

120 MeV

$\sigma_{c.m.}(0°, E_x)$ (mb/sr MeV)

excitation energy E_x (MeV)

Fig. 5 Cross section spectrum for $^{13}C(p,n)$ at 0° and 120 MeV. The vertical bars represent the GT transition strengths for analogous beta decays. The Fermi strength is indicated by the dashed vertical line.

scale factor the quantities plotted are $F(q,\omega)B(GT)$ (solid bars) and $F(q,\omega)B(F)(E_0/E_p)^2$ (dashed bars).

EMPIRICAL EVIDENCE FOR GENERAL PROPORTIONALITY

If the nuclide-specific proportionality factors between (p,n) cross sections and beta-decay transition strengths were to follow a smooth or at least predictable trend, then a more generally useful proportionality would exist. In such a case a limited number of (p,n) transitions could be calibrated against analogous beta decay transitions to determine the proportionality constant $\hat{\sigma}_{GT}$, and this empirically established

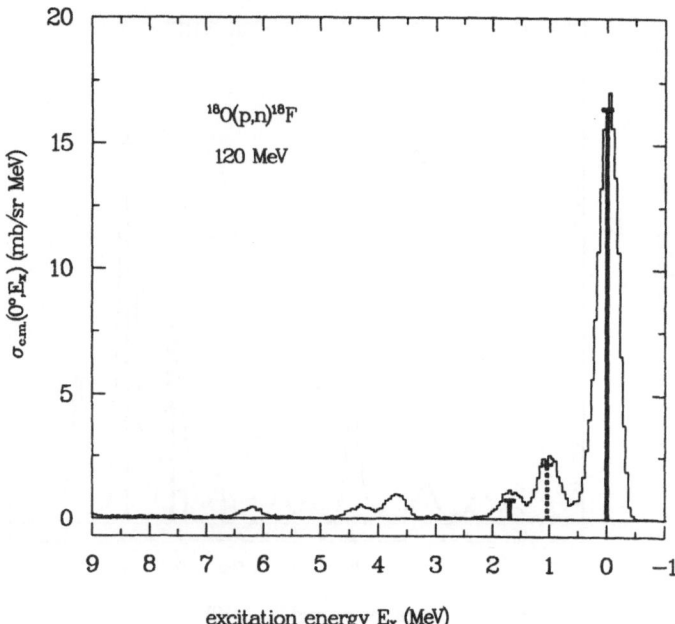

Fig. 6   Cross section spectrum for ^{18}O(p,n) at 0° and 120 MeV. The solid vertical bars represent the GT transition strengths for analogous beta decays and the dashed bar represents the Fermi strength. The overall normalization for this spectrum is poorly determined.

proportionality factor could then be used to measure GT strengths for transitions in other nuclides.

The empirical results for $\hat{\sigma}_{GT}(A)$ and $\hat{\sigma}_F(A)$ are plotted for $E_p = 160$ MeV in Fig. 9. Also plotted is the average value of $\hat{\sigma}_{GT}(A)$ determined from the DWIA calculations of Fig. 1. While the experimental points seem to follow the general trend of the DWIA mass dependence, it is very obvious that the scatter in the experimental values is much larger than the scatter in the calculated values (Fig. 1). The relative-uncertainty-weighted standard deviation of the experimental GT points with respect to the normalized average DWIA mass dependence is $\Delta\hat{\sigma}/\hat{\sigma}$ = 22% for $E_p = 160$ MeV. This spread, if treated as statistical, is too

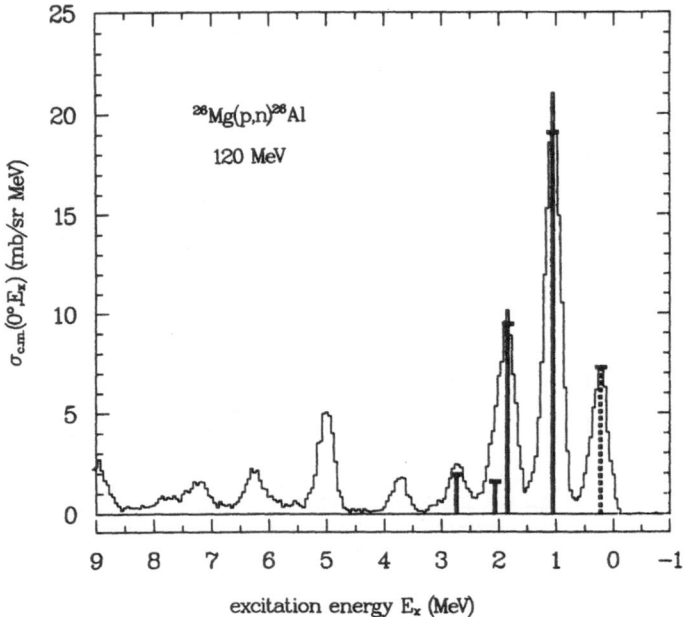

Fig. 7 Cross section spectrum for ^{26}Mg(p,n) at 0° and 120 MeV. The vertical bars represent the GT transition strengths for analogous beta decays. The Fermi strength is indicated by the dashed vertical line.

large to be accounted for by the estimated experimental relative normalization uncertainty of about 8%.

The scatter in the experimental points is central to the investigation of general proportionality; it is therefore important to establish the relative cross sections accurately. Present evidence strongly supports a nonstatistical origin for the observed scatter. A subset of the points displayed in Fig. 9 consists of several independent measurements (i.e., different targets and detector configurations) which yield consistent results. In particular, I shall focus the discussion on the results for ^{12}C, ^{13}C, and ^{14}C, which exhibit variations in $\hat{\sigma}_{GT}$ of as much as 50% from isotope to isotope.

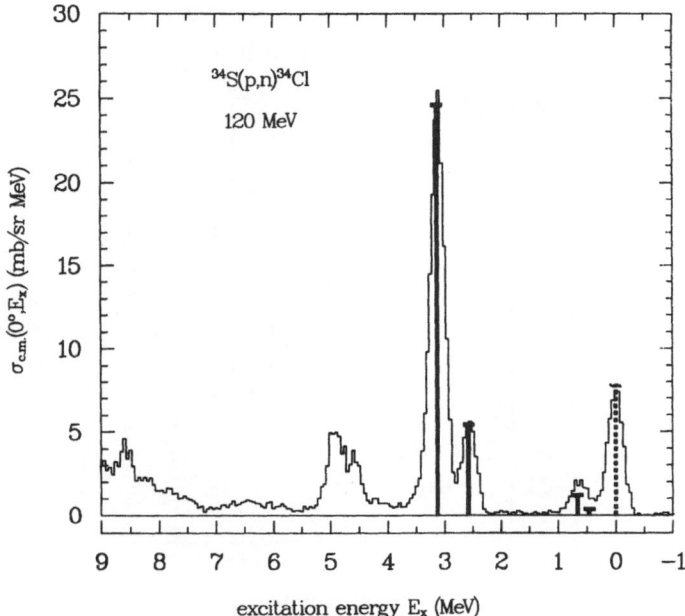

Fig. 8 Cross section spectrum for ^{34}S(p,n) at 0° and 120 MeV. The vertical bars represent the GT transition strengths for analogous beta decays. The Fermi strength is indicated by the dashed vertical line.

The relative cross sections for ^{12}C(p,n) and ^{13}C(p,n) have been verified through measurements with natural carbon targets, which are 1.11% ^{13}C. The large difference in reaction Q values for these two isotopes allows a very clean observation of the low excitation ^{13}C peaks in the natural carbon spectrum. The relative cross sections determined in this way agree well with cross sections measured with isotopically-enriched carbon targets.

The ^{14}C(p,n) cross sections were measured with a target constructed by mixing amorphous carbon (enriched to 88.9% ^{14}C) with a polystyrene (CH$_2$) binder and pressing the resulting mixture into a solid disk. The target thus contained known quantities of both ^{12}C and ^{14}C. The ^{12}C cross

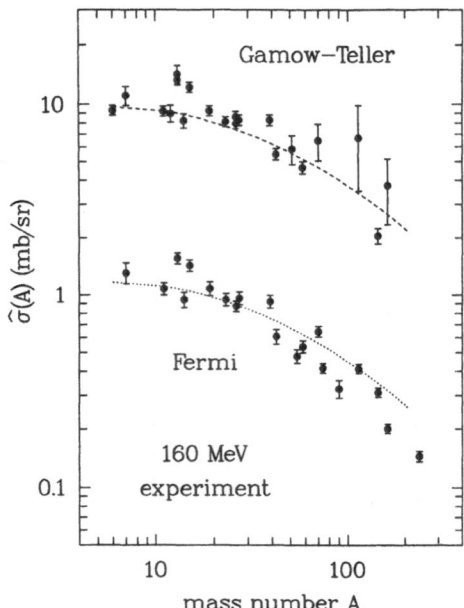

Fig. 9 Experimental unit cross sections for Gamow-Teller and Fermi transitions at E_p = 160 MeV. The dashed line represents the mass dependence of the DWIA GT unit cross sections calculated with the FL 175-MeV interaction and has not been normalized to the data. The dotted line is the dashed line divided by $(160/55)^2$.

sections obtained from measurements with this target agree well with the cross sections obtained with isotopically-pure ^{12}C and natural-carbon targets.

The evidence summarized in the preceding discussion lends strong support to the relative experimental values of $\hat{\sigma}$ determined for ^{12}C, ^{13}C, and ^{14}C. An explanation of the observed differences must therefore be sought in the reaction dynamics.

The surprisingly large value of $\hat{\sigma}$ for ^{13}C compared to that for ^{12}C or ^{14}C is not easily explained in the context of the standard DWIA. Reasonable variations of model parameters, e.g., optical potentials, harmonic oscillator parameter, etc., cannot reproduce the observed difference. Indeed, it appears that even <u>un</u>reasonable parameter variations cannot explain the difference! It would be easy to dismiss the effect as an overlooked subtlety of nuclear structure or reaction mechanism unique to the ground state of ^{13}C were it not for the fact that the 15.1-MeV transition shows the same large value of $\hat{\sigma}$. It is also significant that a similar enhancement is seen in other nuclides such as ^{15}N and possibly ^{39}K. Additionally, cross sections for the analogous ^{13}C(p,p') 15.1-MeV transition also appear to be enhanced relative to ^{12}C(p,p') 15.1-MeV cross sections.[12,13]

SUMMARY AND CONCLUSIONS

Two major problems are yet unresolved in the comparison of (p,n) cross sections to analogous beta decay strengths. First, the proportionality constant that relates (p,n) cross sections to beta decay transition strengths does not have a smooth target nuclide dependence, nor is the dependence presently calculable in some cases to better than about 50% in the context of the standard DWIA reaction model. This observation has several important implications. Until the origin of the fluctuations is understood, extrapolation or interpolation of proportionality constants from one target nuclide to another must be regarded as uncertain at the 20% – 50% level. Quantitative conclusions based upon comparisons of measured cross sections and DWIA calculations should be especially unreliable. This uncertainty must apply as well to related reactions such as (p,p').

A second problem is the predicted sensitivity of 0^+ transitions, particularly those of the $j_< j_<^{-1}$ type, to tensor exchange and spin-orbit amplitudes. Relative cross section systematics for 0^+ and 1^+ transitions show no clear evidence for this effect. Two Fermi transitions which ought

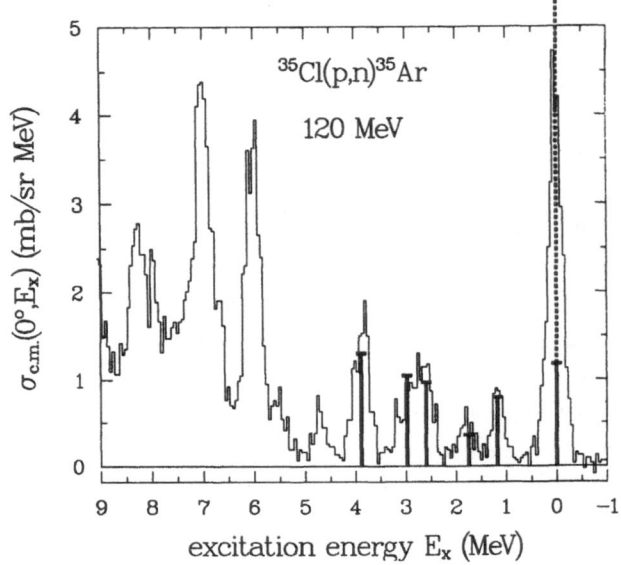

Fig. 10 Cross section spectrum for $^{35}Cl(p,n)$ at $0°$ and 120 MeV. The vertical bars represent the GT transition strengths for analogous beta decays. An estimate of the Fermi cross section based on the relationship of Eq. (3) is indicated by the dashed vertical line and goes off scale.

to show the effect are those in $^{14}C(p,n)$ and $^{34}S(p,n)$. These should be predominantly $p_{1/2}p_{1/2}^{-1}$ and $d_{3/2}d_{3/2}^{-1}$, respectively, yet exhibit $\hat{\sigma}_{GT}/\hat{\sigma}_F$ ratios consistent with other Fermi transitions of $j_>j_>^{-1}$ character. An interesting counterexample to these two cases is provided by $^{35}Cl(p,n)^{35}Ar$. A spectrum for this reaction at 120 MeV is displayed in Fig. 10. The vertical bars in this spectrum represent the corresponding beta decay strengths in the same manner as in Figs. 3,5-8. Also, the dashed vertical line is meant to represent the Fermi strength according to the relative normalization of Eq. (3). Clearly, if the supposedly "universal" ratio of Eq. (3) is applied to this case, the Fermi cross section is considerably overestimated relative to the GT cross sections. In other words, this simple comparison seems to indicate that the Fermi cross section is much smaller than that predicted by Eq. (3). Since this should be a $d_{3/2}d_{3/2}^{-1}$ transition, this reduction is consistent with the calculated effect presented in Fig. 1. However, this comparison is complicated by the fact that the GT transitions for this case are all very weak. In fact, shell model calculations by Brown and Wildenthal[14] indicate that all but one of these transitions can be attributed largely to ℓ-forbidden amplitudes of the type $1s_{1/2}0d_{3/2}^{-1}$. Simple comparisons of the sort just made may therefore be very misleading. More data for Fermi transitions of $j_<j_<^{-1}$ character are clearly desirable.

ACKNOWLEDGEMENT

This work was supported in part by the U.S. Dept. of Energy.

REFERENCES

1. A.K. Kerman, H. McManus, and R.M. Thaler, Ann. Phys. (NY) 8 (1959) 551.
2. C.D. Goodman, C.A. Goulding, M.B. Greenfield, J. Rapaport, D.E. Bainum, C.C. Foster, W.G. Love, and F. Petrovich, Phys. Rev. Lett. 44 (1980) 1755; C.D. Goodman, in The (p,n) reaction and the Nucleon-Nucleon Force, edited by Charles D. Goodman, Sam M. Austin, Stewart D. Bloom, J. Rapaport, and G.R. Satcher (Plenum, New York, 1980), p. 149.
3. F. Petrovich, W.G. Love, and R.J. McCarthy, Phys. Rev. C 21 (1980) 1718.
4. T.N. Taddeucci, C.A. Goulding, T.A. Carey, R.C. Byrd, C.D. Goodman, C. Gaarde, J. Larsen, D. Horen, J. Rapaport, and E. Sugarbaker, Nucl. Phys. A469 (1987) 125.
5. D.H. Wilkinson, Nucl. Phys. A377 (1982) 474; See also P. Bopp et al., Phys. Rev. Lett. 56 (1986) 919.
6. Program DWBA70, R. Schaeffer and J. Raynal (unpublished); extended version DW81 by J.R. Comfort (unpublished).
7. M.A. Franey and W.G. Love, Phys. Rev. C 31 (1985) 488.
8. H.O. Meyer, P. Schwandt, W.W. Jacobs, and J.R. Hall, Phys. Rev. C 27 (1983) 459.
9. C. Olmer, A.D. Bacher, G.T. Emery, W.P. Jones, D.W. Miller, H. Nann, P. Schwandt, S. Yen, T.E. Drake, and R.J. Sobie, Phys. Rev. C 29 (1984) 361.
10. P. Schwandt, H.O. Meyer, W.W. Jacobs, A.D. Bacher, S.E. Vigdor, M.D. Kaitchuck, and T.R. Donoghue, Phys. Rev. C 26 (1982) 55.
11. W.G. Love, K. Nakayama, and M.A. Franey, Phys. Rev. Lett. 59 (1987) 1401.
12. S.F. Collins, G.G. Shute, B.M. Spicer, V.C. Officer, I. Morrison, K.A. Amos, D.W. Devins, D.L. Friesel, and W.P. Jones, Nucl. Phys. A380 (1982) 445.
13. J.R. Comfort, G.L. Moake, C.C. Foster, P. Schwandt, C.D. Goodman, J. Rapaport, and W.G. Love, Phys. Rev. C 24 (1981) 1834.
14. B.A. Brown and B.H. Wildenthal, Atomic Data and Nuclear Data Tables 33 (1985) 347.

SUMMARY TALK—INTERNATIONAL CONFERENCE OF SPIN OBSERVABLES OF NUCLEAR PROBES

Gerald T. Garvey

Los Alamos National Laboratory, MS H836, Los Alamos, New Mexico 87545*

A selected summary of the presentation and discussions at the 4th Telluride Conference is presented. The summary deals mainly with the effects of nuclear spin and isospin on the interaction between nucleons and their consequences in nuclear structure.

1. INTRODUCTION

I have been asked by several people if I had prepared this summary talk before coming to the Conference. Unfortunately, I am no longer in such close contact with this field that I could pretend to have created a presummary. As today was such a nice day, I could not resist skiing and so the more appropriate question is, "Am I prepared at all?"

Before launching into the summary, a few sociological observations are in order. I see new, young faces and hear new voices at this conference; happily, many of them are young women. It has taken far too long for us to obtain extensive participation of women in physics. This new development is most welcome and bodes well for the future. Being joined by Soviet colleagues once again is another excellent development at this meeting, welcomed by all of us. The even better news in this regard is that U.S. and Soviet scientists are not just attending meetings together, but are jointly working together on real physics projects using the effective strengths of our two societies to carry out significant experiments. At Los Alamos we have joined with a very strong Soviet group at the Institute for Nuclear Study (INS) in Moscow to carry out a measurement of the low-energy solar neutrino flux using Ga. I see that the Argonne group is to work on internal polarized targets at the electron accelerator at Novisibirsk. It is very important to continue working to further these collaborations. It will not be simple as the bureaucracy on both sides is still uneasy with the notion. However, with some hard work the obstacles will be overcome and science will remain in the vanguard of international relationships.

2. SUMMARY

Although I did not come to Telluride with the summary talk to hand, I did come expecting to learn about the following questions.

---

* Work performed under the auspices of the U.S. Department of Energy.

2.1 *Where has the $\sigma\tau$ cross section gone?*

2.2 *Why is the longitudinal structure function in (e, e') quasielastic scattering only 60% of the Coulomb "sum rule?"*

2.3 *Why is the ratio of the (p, p') quasielastic cross section for longitudinal to transverse polarization approximately unity independent of A and ω?*

2.4 *How close is the correspondence between the (p, n) and (n, p) cross sections and corresponding weak interaction processes?*

2.5 *Where are the "antinucleons" in the nucleus?*

Not surprisingly, given the organizers of the conference and its history, much of the conference was devoted to these questions. I will try to summarize what has taken place over the past four days by dealing with these questions and recalling some physics presented here that was entirely new to me.

2.1 *Where has the $\sigma\tau$ cross section gone?*

Immediately following the discovery of the so-called giant Gamow-Teller (GT) resonance $(\sigma\tau^{\pm})$ via its large yield at $0°$ in p, n charge exchange it became abundantly clear that the observed yield for the process was well below the sum rule for the operator

$$+\beta_{GT}^{-} - \beta_{GT}^{+} = \frac{1}{3}(N - Z) \ .$$

A variety of explanations sought to explain the observed deficiency such as the strength being transferred up to delta energies because at the quark level both GT transitions and the formation of a delta $(T = 3/2, S = 3/2)$ involve change of the spin and isospin projection of a quark. Other, less exotic explanations argued the depletion of strength as arising from the two-body tensor force which simply spreads the strength to somewhat higher energies, thereby rendering it difficult to observe. The contribution of two-step processes, meson exchange effects, etc., are all believed to be far too small to account for the 40% shortfall in the observed yield. Two experimental tacts have been taken to address the question as to where the missing strength might be. One employs polarized beams and measures the spin of the outgoing particle so that the transferred spin can be inferred. This might be a sensitive way to find small pieces of the spin-flip strength. The other tact is to measure the (n, p) cross section to determine the β_{GT}^{+} term. In earlier discussions it was assumed to be zero and as it must be a positive definite quantity, this quantity can only make the spin flip deficit more severe. Let me first discuss the spin-transfer measurements.

The spin-flip probability S_{NN} is defined as

$$S_{NN} = \frac{\sigma^{+-} + \sigma^{-+}}{\sigma^{++} + \sigma^{+-} + \sigma^{-+} + \sigma^{--}} = \frac{\sigma^{+-} + \sigma^{-+}}{\sigma}$$

where the first superscript indicates the incident particle spin projection and the second the spin projection of the outgoing nucleon. An examination of the matrix elements involved leads us to expect

$S_{NN} \approx 0$ for natural parity giant resonances ($\Delta S \approx 0$);

$S_{NN} \approx 0.5$ for $\Delta S = 1$ giant resonances;

$S_{NN} \approx 0.25$ for quasifree NN due to approximately equal parts of $\Delta S = 1$ and 0.

Kevin Jones presented data on S_{NN} measurements from p, p' forward angle inelastic scattering at 318 MeV on several nuclei. Figures 1 and 2 show the data from ^{40}Ca and ^{90}Zr,

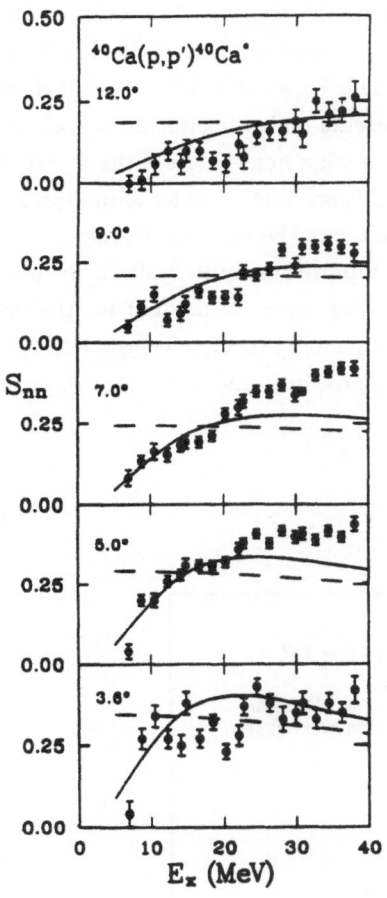

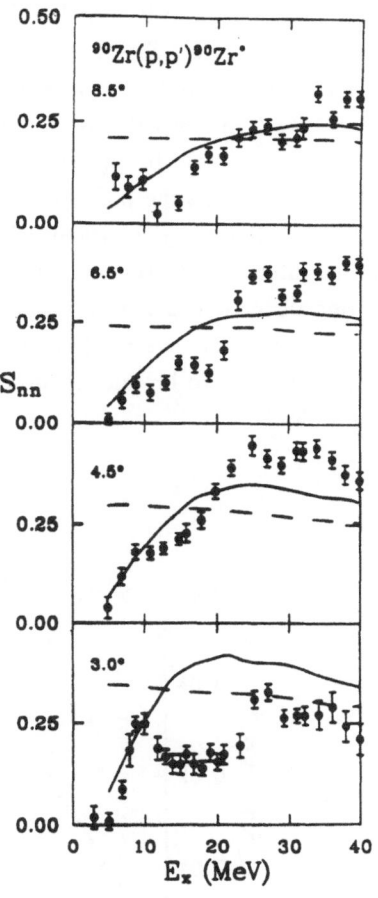

FIGURE 1

The spin-flip amplitude for inclusive quasielastic (p, p') scattering from ^{40}Ca at $E_p = 319$ as a function of residual excitation energy in ^{40}Ca for different scattering angles. The dashed line is the isospin averaged free nucleon-nucleon S_{nn} value, while the solid line represents a slab model response function value for S_{nn}.

FIGURE 2

The spin-flip amplitude for inclusive quasielastic (p, p') scattering from ^{90}Zr at $E_p = 319$ as a function of residual excitation energy in ^{90}Zr for different scattering angles. The dashed line is the isospin averaged free nucleon-nucleon S_{nn} value, while the solid line represents a slab model response function value for S_{nn}.

both showing rather similar behavior. At small energy transfers to the nucleus the value of S_{NN} is near zero and gradually increases up to a value the order of 0.25, the value expected for quasifree scattering. At scattering angles between 4.5 and 7.5° where the momentum transfer is larger, there appears to be a larger value of S_{NN} at excitation energies in excess of 20 MeV. The explanation for this is likely the relatively larger role of the $\Delta S = 1$ giant resonances with $L = 1, 2$. This conjecture is supported by theoretical work reported on in this conference by R. Smith as well as detailed response function calculations carried out by Boucher and Wombach. The $\Delta S = 0$ yield is brought down to lower excitation energy by collective effects relative to the yield for $\Delta S = 1$. This leaves the $\Delta S = 1$ yield dominant at $E_x > 20$ MeV, and $q \approx 0.5$ fm^{-1}. This turn of events makes it difficult for polarization transfer measurements to be especially helpful in uncovering small pieces of the $L = 0$, $S = 1$, $\Delta T = 1$ strength unless the $\vec{p}, \vec{n}$ charge exchange yields are much simpler to interpret. They may well be, as there is no isoscalar collectivity to obscure the spin-flip character.

Another interesting feature of (p, p') calculations reported on by R. Smith is the necessity of requiring distortion in the initial and final scattering states to account for the yield of final-state excitation above 35 MeV. The yield corresponding to these energy losses, shown in Fig. 3, seems not to be due to two or more step processes and would not occur in the plane-wave limit.

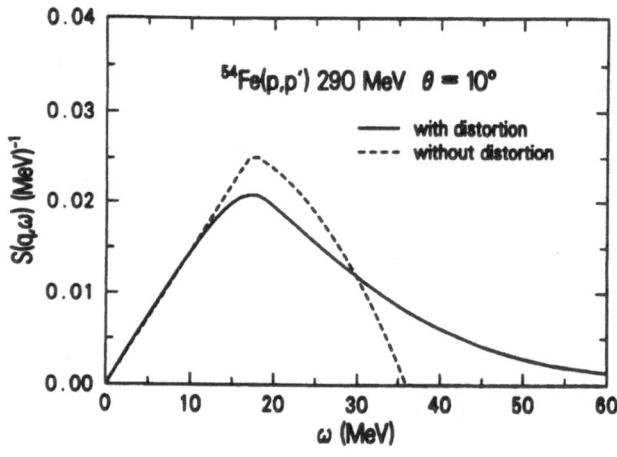

FIGURE 3

Calculation of the energy transferred in a quasielastic (p, p') collision as a function of the energy transfer. The figure shows the important role played by distortion in allowing higher energy transfers.

Peter Jackson presented some beautiful data from TRIUMF on n, p charge exchange. The data were taken at 460-MeV neutron energy on ^{9}Be, ^{13}C, ^{54}Fe, ^{90}Zr, and ^{208}Pb. The ^{54}Fe(n, p) data were very impressively fit with a set of $L = 0$, 1, and 2 multipoles, Fig. 4. An interesting feature of these measurements on ^{13}C$(n, p)^{13}$B ($T = 3/2$ g.s.) is that it yields a value of 10.97 ± 0.56 mb/sr for the Gamow-Teller unit cross section compared to the analogous ^{13}C$(p, n)^{13}$N ($T = 3/2$, 15.1 MeV) reaction of 14.7 ± 1.1 mb/sr. This discrepancy is interesting and should be resolved as it is difficult to see how it can be fundamental; most likely there is a mistake lurking somewhere in this result, or else there is a very nasty isospin violation. Jackson said very little about the impact of the Fe, Zr, and Pb measurements on the sum rule, presumably because it is again difficult to measure the $L = 0$, $\Delta S = 1$ yield at energies above the "Gamow-Teller" resonance.

2.2 Why is the longitudinal structure function in (e, e') quasielastic scattering only 60% of the Coulomb "sum rule?"

One of the puzzling results challenging our simple picture of the nucleus is the failure to find the Coulomb sum rule in the longitudinal

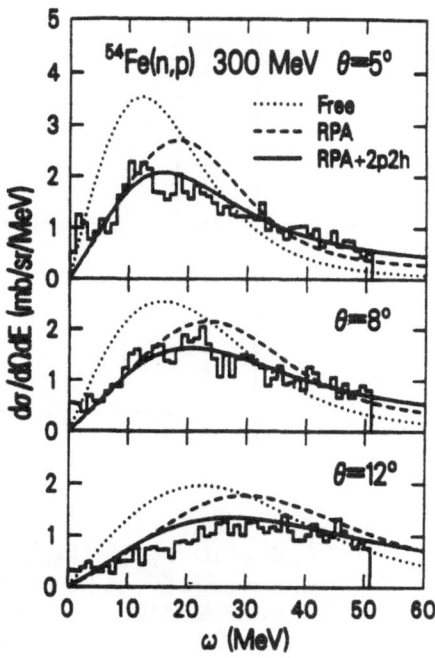

FIGURE 4

The (n, p) spectrum from ^{54}Fe at $E_n = 300$ MeV. The histogram is the actual data, while the dotted line is free n, p scattering. The dashed line is RPA with $\Delta L = 0, 1, 2$, while the solid line includes the $2p$-$2h$ correlation in addition to RPA.

structure function. In a simple and apparently naive picture, the inclusive longitudinal structure function for quasielastic electron scattering when integrated over excitation energy should count the number of protons in the target nucleus. In the lightest nuclei, this sum rule is realized, but in nuclei as light as ^{12}C the integrated structure function falls short of the total charge. In heavier nuclei only 60% of the sum rule (Z) is observed. Strangely enough, the transverse structure function, which is more complicated, appears to be near the predicted value. This situation is depicted in Fig. 5 where the ratio of the transverse to longitudinal structure function is shown. The ratio should be $\mu_{p/1} = 2.79$, but because the longitudinal structure function is less than expected, this ratio appears to be more nearly 3.5. There are many discussions that ascribe this deficiency to two-body correlations, but while this may put some at ease, I find it absolutely unconvincing. It seems that every major difficulty in nuclear structure calculations failing to produce experimentally observed rates is ascribed to

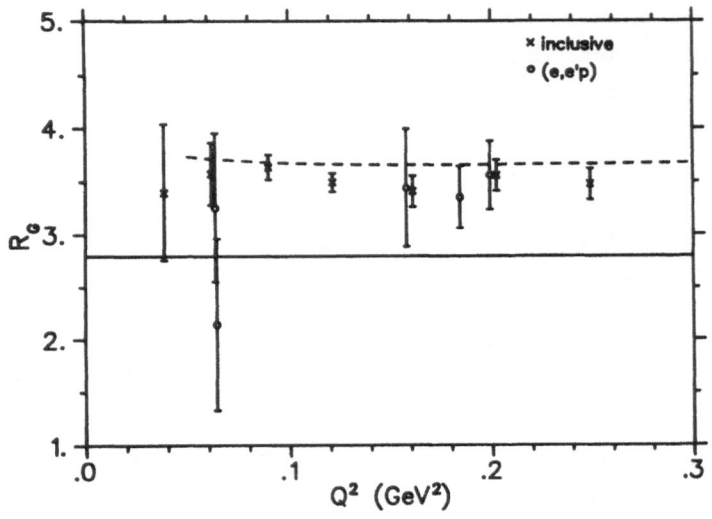

FIGURE 5

Ratio of the $F_T(Q^2)/F_L(Q^2)$ for (e, e') inclusive and (e, e', p) as a function of Q^2. In the latter case one would expect the ratio to be the ratio of the proton magnetic moment (2.79) to the proton charge (1).

two-body correlations, but whenever experiments to measure these correlations are carried out, they escape detection. To my mind, this shortfall of the longitudinal structure function is just another example of missing yield relative to what one expects from the present phenomenological description of nuclei as interacting neutrons and protons. There is likely something very seriously wrong with that picture. If it indeed is the case that short-range correlations shift a larger fraction of the simple wave function hundreds of MeV above the ground state, then the picture of orderly, well-behaved nucleons moving in shell-model orbitals is simply an artifact of convenience having little to do with reality. It must be the case that these strong effects do not materially affect the regularities observed in the observables associated with low-lying states other than the fact that absolute rates are never correctly predicted. However, there may be several major modifications to this simple picture. This issue should be resolutely pursued and squared with a less phenomenological description of the structure of nuclei.

2.3 *Why is the ratio of the (p, p') quasielastic cross section for longitudinal to transverse polarization approximately unity independent of A and ω?*

The longitudinal ($\underset{\sim}{\sigma} \cdot \underset{\sim}{p}$) and transverse ($\underset{\sim}{\sigma} \times \underset{\sim}{p}$) responses of a nucleus in proton quasielastic scattering are expected to be quite different. The strongly attractive p-wave pion-nucleon coupling was believed by some theorists to be sufficiently attractive to bring about pion condensation in the nucleus at an appropriate density. On the other hand, the repulsive nature of the ρ-nucleon coupling appears to thwart the formation of the pion condensate. The pseudoscalar nature of the pion is manifest in the longitudinal coupling $(\underset{\sim}{\sigma} \cdot \underset{\sim}{p})$ while the vector nature of the ρ requires that it show up in the transverse coupling $(\underset{\sim}{\sigma} \times \underset{\sim}{p})$. In finite nuclei, these modes are mixed but examination of the separated longitudinal and transverse modes were expected to reveal the underlying roles of π and ρ exchange on the nucleon-nucleon interaction as modified by the nuclear medium. That is, the longitudinal mode should appear stronger relative to the transverse mode at moderate momentum and energy transfers.

Thus, the expected ratio of the longitudinal $[R_L(q,\omega)]$ to transverse $[R_T(q,\omega)]$ response functions at fixed q as a function of ω is shown by the dotted curve in Fig. 6. In the $(\vec{p}, \vec{p}\,')$ quasielastic scattering the effect is diluted because the scattering is a mixture of $T = 0$ and $T = 1$ interaction. The above discussion pertains only to the $T = 1$ amplitudes; however, the effect was still believed to be readily discernible.

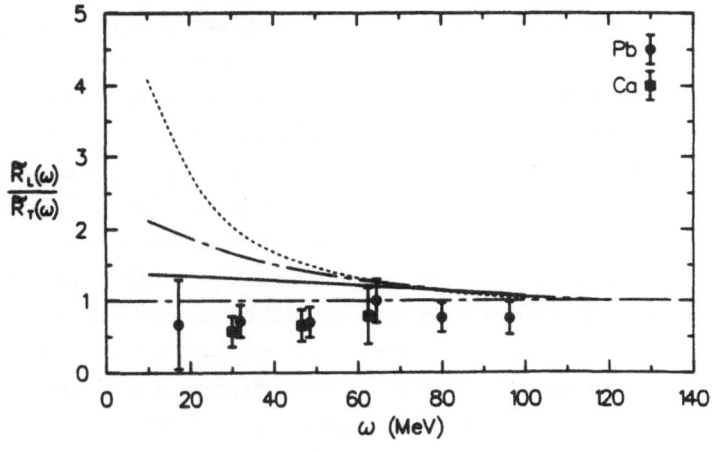

FIGURE 6

Ratio of the ratio of the longitudinal response of quasielastic scattering in Pb and Ca to ^{2}H to the same ratios for the transverse response. The ratio of these ratios is plotted as a function of the residual excitation energy in the final nucleus. The experiments were carried out at $q = 1.8$ fm^{-1}.

Recall that earliest EMC data showed an excess at small x ($x < 0.25$) for the quark distribution function per nucleon in deep inelastic scattering from Fe as compared to deuterium. Many theorists attributed this excess as arising from the pion exchange processes within nuclei. Hence, these same theorists were surprised to see that the ratio of the ratio of the longitudinal to transverse structure functions of Pb to ^{2}H as a function of energy is constant and very near 1. Some have interpreted this as evidence for a lack of pion excess in nuclei; however, there are several reasons why any energy dependence in this ratio of the ratios should be suppressed in (p, p'). First, as mentioned above, the scattering proceeds via both $T = 0$ and 1 amplitudes, thus there is a dilution of the effect being sought. Next, adsorption tends to keep contributions to this channel in the nuclear surface and, lastly, the finite size of the nucleus tends to mix the longitudinal and transverse modes. All these considerations tend to wash out the effect, though most calculations leave residual effects that should be observable.

Experimentally, for the future one can do little except measure the same quantities in $\vec{p}, \vec{n}$ quasielastic-elastic where the scattering will be pure isovector and re-examine the issue when these data are in hand. The newly installed NTOF system at LAMPF, in conjunction with a new high-intensity polarized ion source (OPPIS), will be crucial to this program.

2.4 How close is the correspondence between the (p, n) and (n, p) cross sections and corresponding weak interaction processes?

We now come to the relationship between the (p, n) and (n, p) cross sections and the rates for the corresponding charge-changing weak processes. The correspondence arises from the nuclear initial and final states being the same for the strong and weak processes. There is no *a priori* reason to expect any relationship except that the matrix elements involved appear similar at the level of the nonrelativistic impulse approximation involving nucleons only. As there has been no theoretical formulation of the role of meson-exchange currents in the strong charge exchange reactions, it is difficult to formulate a very penetrating analysis. It is known that the effects of meson exchange are usually small in the case of allowed weak transitions that proceed at near full strength. That is, they are less than 1% of an *unhindered* Fermi or Gamow-Teller transition. Effects arising from finite binding in analogous states are much more serious, as witness the difference between the GT transitions $^{12}\text{B} \rightarrow {}^{12}\text{C} + \beta^- + \tilde{\nu}_e$ and $^{12}\text{N} \rightarrow {}^{12}\text{C} + \beta^+ + \nu_e$. Isospin invariance would lead one to expect that ft values of these transitions would be equal. The observed 10% difference is attributed to the nearly unbound nature of the last proton in ^{12}N. Although this is most likely correct, quantitatively calculating the size of effect is *very* difficult. This difficulty, of course, does not occur in the issue at hand as the initial and final states are common, but the weak process senses the entire nuclear volume while the strong charge exchange process $\{(p, n), (n, p)\}$ is much more sensitive to the nuclear wave function at the nuclear surface because of adsorption effects in the incident and outgoing channels.

As there is no theoretically well-founded description of the strong scattering process, there is no alternative at this time but to compare the observed weak decay rates and corresponding charge exchange cross sections. Terry Taddeucci and his collaborators have done that in a rather extensive manner as we heard in the previous talk this evening. They define a "unit cross section," $\hat{\sigma}_\alpha(E_p, A)$, via

$$\sigma_\alpha \exp(E_p, A, q, \omega) = \hat{\sigma}_\alpha(E_p, A) F_\alpha(q, \omega) B_\alpha$$

where $\sigma_\alpha(\exp)$ is the experimentally measured p, n cross section at $0°$ for a transition α which is either Fermi or Gamow-Teller. B_α is the square of the matrix element measured via weak decay which equals

$$B_{GT} \equiv \frac{1}{2J_i + 1} |\langle J_f, A \| \sigma\tau^- \| J_i, A\rangle|^2$$

$$B_F = \frac{1}{2J_i + 1} |\langle J_f, A \| \tau^- \| J_i, A\rangle|^2 \ .$$

$F(q, \omega)$ is a factor depending on the momentum (q) and energy (ω) transfer and

$$F(q, \omega) \underset{q \to 0, \omega \to 0}{\longrightarrow} 1 \ .$$

These cross sections show a decrease with A for both Fermi and Gamow-Teller transitions. The decrease is ascribable to distortion and absorption.

One of the interesting outcomes of this study is the dependence on incident proton energy of the relative sizes of Fermi and Gamow-Teller cross sections. Fig. 7 shows the results obtained for ^{14}C over a range of energies. ^{14}C is a very good case for study, as the F and GT transitions are pure and well separated. The curious fact is that the square

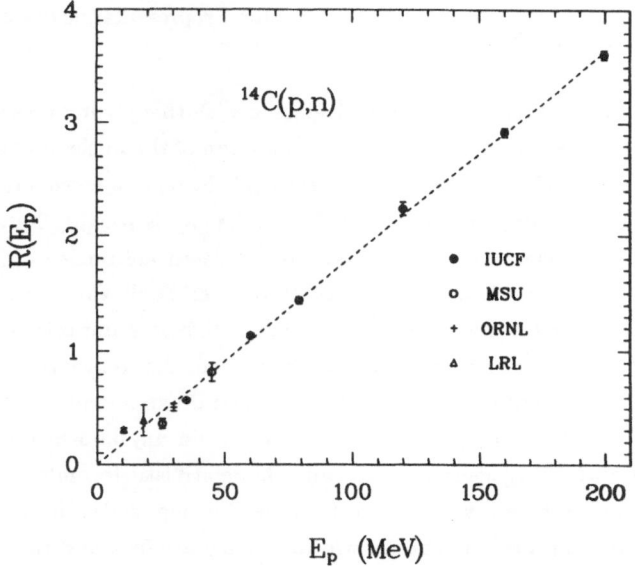

FIGURE 7

Ratio of the square root of the cross section at $0°$ of the "Gamow-Teller transition" to the "Fermi transition" in ^{14}C$(p, n)^{14}$N$^+$ as a function of proton energy. The dotted line is $R(E_p) = E_p/55$.

root of the GT to F yield as a function of energy falls on a straight line $[E_p/55 \text{ (MeV)}]$ for $55 \leq E_p \leq 200$ as shown in Fig. 7. This very interesting fact is partially understood at a more fundamental level and is due largely to a reduction in the two-body V_τ. The density dependence has also been shown to be very important. Love and collaborators have recently worked out a G-matrix approach based on the Bonn potential. The density dependence of the G matrix for the various effective couplings is shown in Fig. 8. A great deal of progress has been made and is being made in this area, but improvement by factors of 2 to 5 is needed in the quantitative understanding and the evaluation of reliability if charge exchange is to provide the matrix elements so badly needed in other areas of nuclear physics investigations.

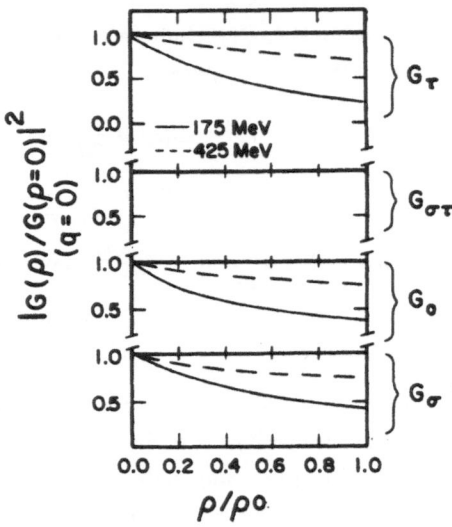

FIGURE 8

Plot of the density dependence of the various effective two-body interactions in nuclei as a function of density. The value 1 represents the density of nuclear matter.

2.5 Where are the "antinucleons" in the nucleus?

Unfortunately, this still remains a theoretical issue with there being no supporting experimental data. The commonly used relativistic formulation of the nucleon-nucleus interaction with its very attractive scalar and repulsive vector fields leads to a considerable reduction in the nucleon-antinucleon energy gap in the nucleus. The gap is roughly halved to $\sim$400-500 MeV. It is by no means clear how to uncover the increased antinucleon presence that the gap reduction would predict. Deep inelastic scattering, or Drell-Yan, experiments will not produce convincing, if any, evidence for $\overline{N}$ in nuclei. It seems important to demonstrate the reduced gap if we are to have real faith in the present day relativistic formulation. At the present moment, the contact is entirely through certain spin-dependent effects that are obtainable more naturally in a Dirac formulation than via any known nonrelativistic prescription. The rest of the baggage that comes with the relativistic formulation is very difficult to deal with. I certainly take my hat off to those few strong souls who are re-establishing all of nuclear structure in a relativistic description. They are few, and the job is enormous.

In a global sense, establishing a nuclear theory in terms of a finite number of hadronic types that will work up to energy and momentum transfer of a few GeV would be a great achievement because one could then employ QCD to push on to higher energy. We would then have a way of proceeding from low energy up to the TeV scale.

3. NEW MATERIAL

Among the new material presented at this conference, the report by Roy Holt on photodisintegration of the deuteron done at NPAS and the two reports on delta production by Ellegard and Dimitriv were most interesting to me.

Let's start with the deuteron photodisintegration. The experimental result is a result of a collaboration between Argonne/Caltech/NPAS referred to as NE8. To predict the behavior of the expected form factor at large momentum transfer, there are several ways to proceed. In the context of the parton model at asymptotically large Q^2, it is necessary for the struck quark to share its momentum with the remaining quarks; each of these quark-quark interactions introduces a factor of $1/Q^2$ due to hard gluon exchange. In the case of the deuteron, this involves five quarks in addition to the one that is struck to take up the momentum so that at very large Q^2 we expect that the deuteron form factor would scale as

$$\underset{Q^2 \to \infty}{F_d(Q^2)} \sim Q^{-10} \ .$$

At lower than asymptotic Q^2 a more detailed model is required. One commonly used procedure is to simply calculate the cross section with a hadronic model that includes all the known hadron dynamics and form factors. Alternatively, Brodsky and collaborators have set up an ansatz based on QCD that produces a scaling behavior well below where one would expect scaling to work. In their approach, the nucleons share equally in the momentum transfer and are correlated via a gluon exchange. While the correctness of these assumptions can be easily called into question, it provides a specific recipe that often agrees with experiment and leads to what is often referred to as precocious scaling because it sets in long before one would expect any scaling behavior based on QCD. In this case, one would have for the matrix element in deuteron photoabsorption

$$M_{\gamma d}^{(Q^2)} \sim Q F_d(Q^2) \sim Q\, F_N^2(Q^2/4)\frac{1}{Q^2}$$

where the last factor of $1/Q^2$ accounts for the gluon exchange between the nucleons. As $\underset{Q^2 \to \infty}{F_N(Q^2)} \to Q^{-4}$, the predicted behavior shows the same power law dependence as the predicted behavior shows the same power law dependence as the asymptotic case when $Q^2 \to \infty$. In a hadronic description, one has nucleon form factors for the photon adsorption and the momentum is shared via pion exchange so that the matrix element has the following form

$$M_{\gamma d} \approx Q\, F_N(Q^2)\, F_{\pi NN}^2(Q^2)\frac{1}{Q^2} \ .$$

As $F_{\pi NN}^2(Q^2)$ goes as $[\frac{\Lambda^2}{\Lambda^2+Q^2}]$ there are clearly differences at finite Q^2.

In a model-independent format the cross section for $D(\gamma, p)n$ at large E_γ is written as

$$\frac{d\sigma}{d\Omega} = \frac{1}{S}\, F_p^2\, F_n^2\, \frac{f(\theta_{cm})}{P_T^2}$$

where $f(\theta_{\rm cm})$ is an energy-independent "reduced" amplitude. The task of any model would then be to compute $f(\theta_{\rm cm})$. Figure 9 shows the observed yield for as a function of photon energy as reported by the NE8 collaboration. Above 0.8 GeV their results appear to agree much better with the predictions of chromodynamics than with the specific meson exchange model shown in the figure. However, caution is the order of the day as meson exchange calculations are often very model dependent. It is, however, another interesting example of apparent precocious scaling.

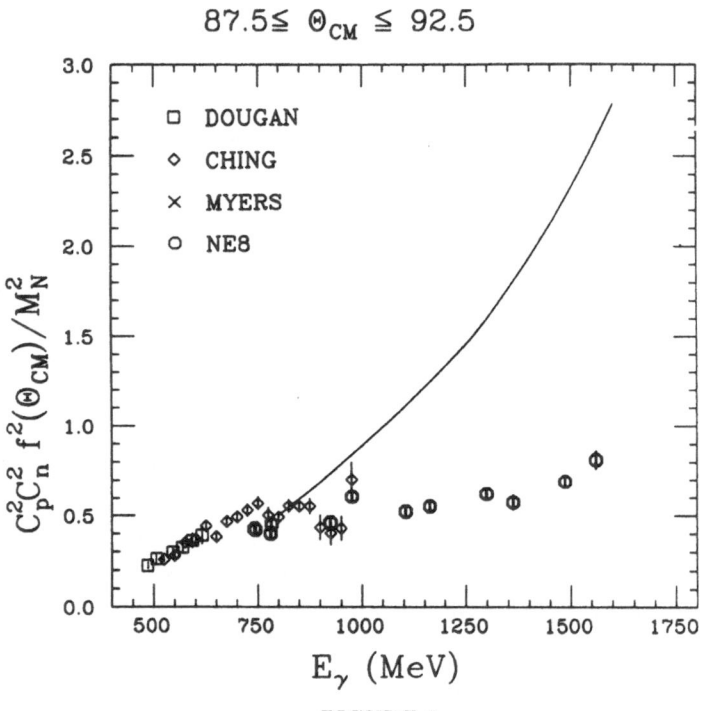

FIGURE 9

Plot of the photodisintegration cross section of the deuteron as a function of incident photon energy.

Delta production is a further example of the $\sigma\tau$ operator and several interesting presentations were made regarding delta production with complex projectiles on nuclei. The reports by Dimitriv and Ellegard on delta production as seen via charge exchange reactions at Dubna and Saterne, respectively. The Dubna experiments involve $P(^3{\rm He},t)\Delta^{++}$ and $^{12}{\rm C}(^3{\rm He},t){\rm X}$. These experiments were carried out at ^{3}He incident momenta of 4.40, 6.81, and 10.79 GeV/c. The observed $0°$ cross sections at 6.81 and 10.79 GeV/c bombarding energy are shown in Figs. 10(a) and (b). The cross section for this process seems large, for example, at 10.79 GeV/c the yield from a ^{12}C target at $0°$ is about 85 mb/sr, a factor of two above the yield on a free nucleon. The measured polarization is roughly one-half of

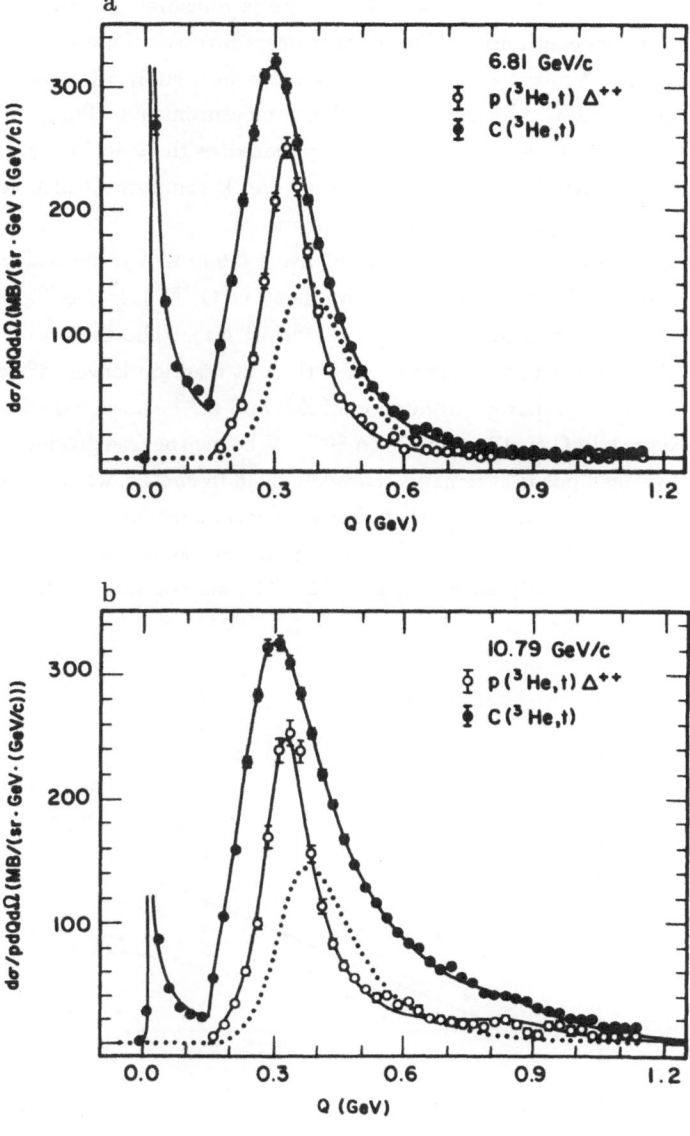

FIGURE 10

(a) Plot of the triple differential cross section for the $(^3\text{He},t)$ reaction at $0°$ for $P_{^3\text{He}} = 6.81$ GeV/c. The cross section is plotted as a function of residual excitation energy in the target system. (b) Plot of the triple differential cross section for the $(^3\text{He},t)$ reaction at $0°$ for $P_{^3\text{He}} = 10.79$ GeV/c. The cross section is plotted as a function of residual excitation energy in the target system.

what would be expected from delta production via pion exchange. Hence, half the cross section must be due to other processes, possibly ρ exchange. Dimitriv asserts that there are two processes involved in Δ production. One is quasifree Δ production with the Δ being produced in the continuum. The other process involves the creation of the Δ in the field of the nucleus. Processes of the first kind can be readily identified via observation of the normal products of Δ decay in coincidence measurements. The peak resulting from coincidence measurement occurs at higher energy than does the Δ inclusive spectrum. Hence, the noncoincident contribution is at lower energy and is reminiscent of a Δ-hole excitation in the nuclear system, supporting Dimitriv's assertion.

The Saterne measurements are carried out using 0.900 GeV/amu heavy-ion beams and involve proton charge changing processes via (He,t), $(^{16}O,^{16}N)$, $(^{20}Ne,^{20}F)$, and also a case of neutron charge exchange in the beam via $(^{20}Ne,^{20}Na)$. The first three reactions leave behind Δ^+ or Δ^{++} in the target nucleus while the last reaction leaves Δ^0 or Δ^-. The peak associated with charge exchange production of Δ^+ and Δ^{++} shows a downward shift of the delta resonance peak in C that is some 70 to 80 MeV below the free production and then as a function of A the peak position remains relatively fixed in energy, while the charge exchange producing Δ^0 and Δ^- yields a peak that shows a continuous downward shift. These effects must represent the combined effect of Coulomb plus nuclear binding effects on the delta in nuclei. This is schematically shown in Fig. 11. The size of these delta production cross

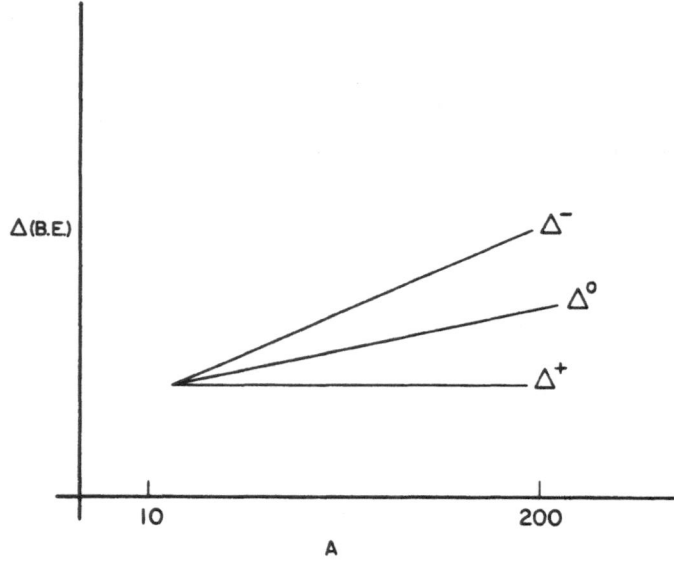

FIGURE 11
Schematic plot of the apparent binding energy of various charged
states of the Δ in nucleon systems as a function of A. The
binding energy increases upward.

sections is again very large and would be larger if the giant GT resonance occurred as a bound state in the outgoing projectile. It is likely that the total Δ production cross section is on the order of hundreds of millibarns.

4. CONCLUSION

There were a host of new experimental undertakings reported on in this meeting that were once thought too difficult to carry out. There is genuine progress in what we are able to both consider and execute. For example, the tensor polarization T_{20} in electron elastic scattering on the deuteron, (n,p) reactions, $(\vec{p}, \vec{n})$ are now being carried out. We will soon be gathering data on $\vec{A}(\vec{p}, \vec{p})$, $A(\vec{p}, \vec{n})$, $(e, e', \vec{p})$, and (e, e', n). Spin-dependent deep inelastic scattering, as well as neutral current elastic scattering should reveal much about the partonic structure of the nucleon. All of these experiments require extensive effort and a high degree of effective collaboration. They represent the kind of effort that the nuclear physics of tomorrow will require.

This has been an excellent meeting. We should thank Chuck Horowitz and Charles Goodman for creating a 4th Telluride Conference that retains all the vitality and importance of the preceding conferences!

CONTRIBUTORS

Ableev, V.G.
 Joint INstitute for Nuclear
 Research, Dubna, USSR

Alberico, W.M.
 Dipartimento di Fisica Teorica
 Università di Torino and
 Istituto Nazionale di Fisica
 Nucleare, Torino, Italy

Bertini, R.
 Cern/EP, CH-1211 Geneva 23
 Switzerland and
 DPHN/ME, CEN Sarclay
 F91191 Gif sur Yvette, France

Blunden, P.G.
 TRIUMF, 4004 Wesbrook Mall
 Vancouver, BC, Canada V6T 2A3

Boucher, P.M.
 Department of Physics
 Queen's University
 Kingston K7L 3N6 Canada

Chanfray, G.
 Institut de Physique Nucléare
 Universite Lyon-1 43
 Bd du 11 Novembre 1918
 69622 Villeurbanne, Cedex, France

Cooper, E.D
 Dept. of Physics, McGill Univ.
 3600 University St.
 Montreal, Quebec, Canada H3A 2T8

Dmitriev, V.F.
 Institute of Nuclear Physics
 Novosibirsk, USSR

Donnelly, T.W.
 Center for Theoretical Physics
 Massachusetts Institute of Tech.
 Cambridge, Mass. 02139

Ellegaard, C.
 Niels Bohr Institut
 University of Copenhagen
 DK-2100 Copenhagen, Denmark

Ershov, S.N.
 Joint Institute for Nuclear
 Research, Dubna, USSR

Furnstahl, R.J.
 Department of Physics and Astronomy
 University of Maryland
 College Park, MD 20742

Garcon, M.
 Massachusetts Institute of Tech.
 Cambridge, Mass. 02139

Garvey, G.T.
 Los Alamos National Lab., MS H836
 Los Alamos, New Mexico 87545

Ghazikhanian, V.
 Physics Department, UCLA
 Los Angeles, Calif. 90024

Hama, S.
 Department of Physics
 The Ohio State University
 Columbus, Ohio 43210

Hicks, K.H.
 TRIUMF, 4004 Wesbrook Mall
 Vancouver, BC, CAnada V6T 2A3

Holt, R.J.
 Physics Division
 Argonne National Laboratory
 Argonne, Ill. 60439-4843

Horowitz, C.J.
 Physics Department
 and Nuclear Theory Center
 Indiana University
 Bloomington, Indiana 47405

Ichimura, M.
  Institute of Physics
  University of Tokyo
  Komaba, tokyo 153, Japan

Iqbal, M.J.
  TRIUMF, Wesbrook Mall
  Vancouver, BC, Canada V6T 2A3

Jackson, K.P.
  TRIUMF, Wesbrook Mall
  Vancouver, B.C., Canada V6T 2A3

Jain, B.K.
  Nuclear Physics Division
  Bhabha Atomic Research Centre
  Bombay 400 085, India

Jeppesen, R.G.
  Simon Franser University
  Burnaby Mountain
  Burnaby, BC, Canada V5A 1S6

Jones, K.W.
  Los Alamos National Laboratory
  Los Alamos, New Mexico 87544

Krisch, A.D.
  Randall Laboratory of Physics
  The Univeristy of Michigan
  Ann Arbor, Michigan 48109-1120

McClelland, J.B.
  Los Alamos National Laboratory
  Los Alamos, New Mexico 87545

Perry, R.J.
  Department of Physics
  The Ohio State University
  174 W. 18th Ave.
  Columbus, Ohio 43210

Piekarewicz, J.
  Nuclear Theory Center
  Indiana University
  Bloomington, Indiana 47405

Price, C.E.
  Physics Division
  Argonne National Laboratory
  Argonne, Illinois 60439

Shepard, J.R.
  Department of Physics
  University of Colorado
  Boulder, Colorado 80309

Smith, R.D.
  Los Alamos National Laboratory
  Los Alamos, New Mexico 87545

Tadeucci, T.N.
  Los Alamos National Laboratory
  Los Alamos, New Mexico 97545

Ullmann, J.L.
  Los Alamos National Laboratory
  Los Alamos, New Mexico 87545

Van Orden, J.W.
  Continuous Electron Beam
  Accelerator Facility
  Newport News, Virginia   23606

Walker, G.E.
  Nuclear Theory Center and Physics
  Department, Indiana University
  Bloomington, Indiana 47405

Ward, T.E.
  Neutral Particle Beam Division
  Department of Nuclear Energy
  Brookhaven National Laboratory
  Upton, New York 11973

Watson, J.W.
  Kent State Univesity
  Kent, Ohio 44242

INDEX